REGULATION OF PROCARYOTIC DEVELOPMENT

Structural and Functional Analysis of
Bacterial Sporulation and Germination

REGULATION OF PROCARYOTIC DEVELOPMENT

Structural and Functional Analysis of Bacterial Sporulation and Germination

Editors

Issar Smith

Department of Microbiology, The Public Health Research Institute
New York, New York

Ralph A. Slepecky

Department of Biology, Syracuse University
Syracuse, New York

Peter Setlow

Department of Biochemistry, University of Connecticut Health Center
Farmington, Connecticut

AMERICAN SOCIETY FOR MICROBIOLOGY
Washington, D.C.

Copyright © 1989 American Society for Microbiology
1325 Massachusetts Ave., N.W.
Washington, DC 20005

Library of Congress Cataloging-in-Publication Data

Regulation of procaryotic development : a structural and functional analysis of bacterial sporulation and germination / editors, Issar Smith, Ralph Slepecky, Peter Setlow.

   p.   cm.
   Based on the Tenth International Spores Conference, held at Woods Hole, Mass., Mar. 1988.
   Includes bibliographies and index.
   ISBN 1-55581-011-X
   1. Bacillus (Bacteria)—Genetics—Congresses. 2. Bacillus subtilis—Genetics—Congresses. 3. Spores (Bacteria)—Congresses. 4. Germination-Congresses. I. Smith, Issar. II. Slepecky, Ralph. III. Setlow, Peter. IV. International Spore Conference (10th : 1988 : Woods Hole, Mass.)
   [DNLM: 1. Bacillus Subtilis—physiology—congresses. 2. Gene Expression Regulation—congresses. 3. Genes, Bacterial—congresses. 4. Spores, Bacterial—physiology—congresses. QW 52 R344 1988]
QR82.B3R44   1989
589.9′087322—dc20

Contents

Preface vii

Introduction ix

1. Revised Genetic Map of *Bacillus subtilis* 168 • Patrick J. Piggot 1

2. Spore Thermoresistance Mechanisms • Philipp Gerhardt and Robert E.
   Marquis 43

3. Methods for Genetic Manipulation, Cloning, and Functional Analysis of
   Sporulation Genes in *Bacillus subtilis* • Philip Youngman, Harald Poth,
   Brian Green, Karen York, Gabriela Olmedo, and Kerry Smith 65

4. The Trigger Mechanism of Bacterial Spore Germination • Simon J. Foster and
   Keith Johnstone 89

5. Metabolic Regulation of Sporulation and Other Stationary-Phase Phenomena
   • Abraham L. Sonenshein 109

6. Subtilisin: a Redundantly Temporally Regulated Gene? • Fernando Valle and
   Eugenio Ferrari 131

7. The Competence Regulon of *Bacillus subtilis* • David Dubnau 147

8. Sigma Factors and the Regulation of Transcription • Charles P. Moran, Jr. 167

9. Initiation of Sporulation • Issar Smith 185

10. Forespore-Specific Genes of *Bacillus subtilis*: Function and Regulation of
    Expression • Peter Setlow 211

11. Dependence Pathways for the Expression of Genes Involved in Endospore
    Formation in *Bacillus subtilis* • Richard Losick and Lee Kroos 223

12. Temporal and Spatial Control of Gene Expression during Sporulation: from
    Facts to Speculations • Patrick Stragier 243

13. Role, Structure, and Molecular Organization of the Genes Coding for the
    Parasporal δ-Endotoxins of *Bacillus thuringiensis* • Didier Lereclus,
    Catherine Bourgouin, Marguerite M. Lecadet, André Klier, and Georges
    Rapoport 255

14. Sporulation in *Streptomyces* • Keith F. Chater 277

Index 301

Preface

Jackson Foster, a pioneer sporologist, wrote a seminal review in 1956, "Morphogenesis in Bacteria: Some Aspects of Spore Formation" (*Q. Rev. Biol.* **31**:102–118, 1956), in which he stated, "if one contemplates the infinite resourcefulness nature has adopted to perpetuate its biological representatives, one can hardly avoid a profound reverence for the artfulness she has employed. Numberless are the mechanisms whose ultimate effect is to enable the species to adapt to its changing environment yet none is more marvelous than that device which confers relative indestructibility upon an organism, namely sporogenesis in bacteria." That wonder, and the appreciation of sporulation as a model system for differentiation, stimulated H. Orin Halvorson to organize a conference in 1956 solely devoted to bacterial spores. Evidence of the attractiveness this system still holds for biologists from many subdisciplines is the continuation of this series of conferences, with the most recent being the Tenth International Spores Conference, held at Woods Hole, Massachusetts, in March of 1988.

The organizers of this conference (Roy Doi, Richard Losick, Patrick Piggot, and Abraham Sonenshein in addition to the Editors of this volume) express their appreciation to the International Spores Conference Advisory Council, to Arnold Demain, Peter Fortnagel, Alessandro Galizzi, Harlyn Halvorson, James Hoch, Alex Keynan, Andre Klier, Yasuo Kobayashi, Anne Moir, Hiuga Saito, Derek Smith, Anatoly Stepanov, and Jekisiel Szulmajster, and to various groups for financial support for the meeting. The latter included the National Institutes of Health, the National Science Foundation, U.S. Army Research Office, Amgen, Inc., Campbell Soup Company, E. I. DuPont de Nemours, Inc., Ethicon, Inc., Fujisawa Pharmaceutical Co., Genecor, Inc., Gist-Brocades, Kodak, Merck Sharp and Dohme, Miles Laboratory, Syntro Corporation, Abbott Laboratories, Bristol-Meyers, Lab-Line Instruments, Monsanto, New England Biolabs, Inc., Pioneer Hi-Bred International, Inc., and Wyeth Laboratories. A highlight of this meeting was the recognition of past contributions to the field by recent retirees Philip Fitz-James, Hiuga Saito, Pierre Schaeffer, and Jekisiel Szulmajster.

This volume, the 10th volume in the series issuing from the International Spores Conferences, does not follow the usual format, i.e., a series of short research papers that would, most likely, appear in scientific journals at some later time. Because of many recent developments in the sporulation field, it was felt that this book would be a good opportunity to review the current status of our knowledge in this area. Thus, a small number of review articles written by people with specific expertise in the areas covered are presented. While these authors stress their own work, they cover also the pertinent literature within a conceptual framework so that mechanisms of the various aspects of sporulation and late growth control, i.e., germination, metabolic control, temporal regulation of enzyme synthesis, competence, sporulation initiation, late spore gene expression, and spore resistance, are presented. Since differentiation in *Streptomyces* spp. has some features related to *Bacillus* differentiation, a review on that topic is included as well. Thus, a near-comprehensive picture of the spore field is given. While there are already a number of recent reviews on spores, the authors were chosen so that emphasis would be placed on aspects not yet covered in detail or on recently reviewed topics where another viewpoint might be useful.

The editors would like to express their gratitude to their scientific colleagues for helpful discussions and insight in the preparation of this volume. They also give special thanks to Annabel Howard for expert secretarial assistance and to Dennis Burke and Eleanor Tupper of the ASM Publications Department for their patience and assistance in the editing of this book.

Issar Smith
Department of Microbiology
The Public Health Research Institute
New York, New York 10016

Ralph A. Slepecky
Department of Biology
Syracuse University
Syracuse, New York 13244

Peter Setlow
Department of Biochemistry
University of Connecticut Health Center
Farmington, Connecticut 06032

Introduction

One of the most important questions in modern biology deals with differentiation. How does a cell of one type give rise to one or more cells with completely different morphology, physiology, and function? In higher organisms, this problem has been discussed since the time of Aristotle, but only in the last few years has eucaryotic development been examined on the molecular level. Studies on *Drosophila* development have recently demonstrated the physical existence of molecules which are involved in the establishment of the long-postulated anteroposterior and dorsoventral gradients (2, 3). There have been striking discoveries in mammalian and even human development through the use of model systems like F9 stem cells. In this undifferentiated tissue, development into mesodermal structures can be induced by the binding of retinoic acid to specific receptors, which then enter the nucleus and become positive transcriptional regulators for genes which are activated as differentiation ensues (5). As exciting and fundamental as these eucaryotic systems are for explorations into mechanisms of development, microorganisms, with their well-studied genetic systems and ease of manipulation, are ideal for this type of study. In these life forms, development is usually, but not exclusively, induced by some form of environmental stress (the alternation of cell types, i.e., stalked and flagellated, in *Caulobacter* and related species is an example of environmentally independent development). Nutrient stress will induce mating and meiosis in fungi and the elaboration of various extracellular degradative enzymes and secondary metabolites in the gram-positive, soil-dwelling genera *Streptomyces* and *Bacillus,* as well as the appearance of spores in these diverse procaryotes and eucaryotes.

Sporulation in the genus *Bacillus* is the best-studied example of procaryotic differentiation, for a number of reasons. Historically, the study of this process is over 100 years old. In 1877, Cohn and Koch discovered the existence of spores in *Bacillus subtilis* and *Bacillus anthracis,* respectively (4). The significance of spores for the food industry led to much work on *Bacillus* spores and mechanisms of spore resistance. The chapter by Philipp Gerhardt and Robert Marquis presents the current state of knowledge in this area. The unique property of the *Bacillus* differentiation system, however, is its genetics. The discovery of DNA-mediated transformation in *B. subtilis* (7), coupled with the development of rapid mapping techniques for chromosomal genes (1), has led to an extensive genetic map with over 700 genes, including at least 100 developmental loci. The chapter by Patrick Piggot gives the latest description of this map, with special emphasis on genes involved in germination and sporulation. The advent of molecular cloning has been especially fruitful for studies of sporulation in *B. subtilis*. Clones of developmental genes can be directly selected for, and reverse genetics can be used to create new mutations in developmental genes. These powerful techniques are being improved at a rapid rate, and the chapter by Philip Youngman and co-workers gives a state-of-the-art discussion of cloning techniques and genetic strategies.

The next chapters are arranged in a quasi-temporal sequence so that the chapter by Simon Foster and Keith Johnstone, which covers spore germination, the actual beginning of the growth cycle, is followed by the chapter on intermediary metabolism. This contribution by Abraham Sonenshein relates biochemical and physiological events occurring during vegetative growth and the transition to the stationary phase. At the commencement of this stage in the *Bacillus* life cycle, several phenomena occur, among which are the secretion of extracellular enzymes (which is of importance for the industrial production of enzymes and the secretion of heterologous proteins) and the development of competence, i.e., the ability of cells to take up DNA and to be transformed. The chapters by Fernando Valle and Eugenio Ferrari and by David Dubnau discuss these temporally regulated processes and their relationships to sporulation. The question of nonsporulation responses to nutrient stress is important, not only for bacilli but for other microorganisms as well. Much recent work in *Escherichia coli,* for example, indicates that there is induction of several global responses after severe nutrient starvation.

One of the unique aspects to the control of development in *B. subtilis* is the involvement of several RNA polymerase holoenzymes. This fact led to a most provocative hypothesis concerning the regulation of sporulation: the σ factor "cascade" (6). The chapter by Charles Moran, Jr., brings this theory of transcriptionally controlled development up to date and illustrates that newly appearing RNA polymerases (which contain different σ factors) do control late growth development and that the process is much more complex than originally envisaged. It is noteworthy that alternative σ factors (the heat shock and the Ntr σ factors), involved in responses to changing environmental conditions, are also found in gram-negative bacteria.

The initiation of sporulation is the subject of the next chapter, by Issar Smith, and the concept of alternative pathways of late growth development regulated by developmental switches is presented. The next major stage in sporulation is the separation of the forespore and the spore mother cell. The de-

velopment of the forespore is discussed in the chapter by Peter Setlow. The chapter by Richard Losick and Lee Kroos and the one by Patrick Stragier present comprehensive analyses of the entire sporulation process and concentrate on differentiation into the two cell types and their independent development. One of the unusual aspects of differentiation in certain *Bacillus* strains is the appearance of parasporal crystals in *Bacillus thuringiensis* and related species. The commercial significance of these natural insecticides has led to great interest in the nature of the toxins included in the crystals and the regulation of their genes. The penultimate chapter by Didier Lereclus and colleagues discusses these areas.

Sporulation in gram-positive bacteria is not confined to the genus *Bacillus* and related bacteria, e.g., *Clostridium. Streptomyces* species also sporulate, and this genus is of vast economic importance. Recent years have been marked by the emergence of elegant cloning techniques to complement the fundamental genetic studies pioneered by David Hopwood, Keith Chater, and their co-workers. It is appropriate that the final chapter of this book on procryotic development ends with a chapter on *Streptomyces* differentiation, written by Keith Chater. This area is developing very quickly, and perhaps the next review volume on procaryotic development will consist of 13 chapters on *Streptomyces* spp. with the final one dealing with *Bacillus.*

LITERATURE CITED

1. **Dedonder, R. A., J. A. Lepesant, J. Lepesant-Kejzlarova, A. Billault, M. Steinmetz, and F. Kunst.** 1977. Construction of a kit of reference strains for rapid genetic mapping in *Bacillus subtilis* 168. *Appl. Environ. Microbiol.* 33:989–993.

2. **Driever, W., and C. Nusslein-Volhard.** 1989. The bicoid protein is a positive regulator of *hunchback* transcription in the early *Drosophila* embryo. *Nature* (London) 337:139–143.

3. **Ingham, P. W.** 1988. The molecular genetics of embryonic pattern formation in *Drosophila. Nature* (London) 335:25–34.

4. **Keynan, A., and N. Sandler.** 1984. Spore research in historical perspective, p. 1–48. *In* A. Hurst and G. W. Gould (ed.), *The Bacterial Spore.* Academic Press, Inc. (London), Ltd., London.

5. **LaRosa, G. J., and L. J. Gudas.** 1988. An early effect of retinoic acid: cloning of an mRNA (Era-1) exhibiting rapid and protein synthesis-independent induc-tion during teratocarcinoma stem cell differentiation. *Proc. Natl. Acad. Sci. USA* 85:329–333.

6. **Losick, R., and J. Pero.** 1981. Cascades of sigma factors. *Cell* 25:582–584.

7. **Spizizen, J.** 1958. Transformation of biochemically deficient strains of *Bacillus subtilis* by deoxyribo-nucleate. *Proc. Natl. Acad. Sci. USA* 44:1072–1078.

Regulation of Procaryotic Development
Edited by Issar Smith, Ralph A. Slepecky, and Peter Setlow
© 1989 American Society for Microbiology, Washington, DC 20006

Chapter 1

Revised Genetic Map of *Bacillus subtilis* 168

Patrick J. Piggot

It is now four years since the last *B. subtilis* genetic maps appeared (371, 534). In the intervening period a number of new loci have been described, and almost 700 have now been identified (Table 1). There has been a dramatic increase in the number subjected to molecular analysis. More than 300 loci have been cloned (i.e., a clone isolated and a restriction map determined), and among these more than 180 have been sequenced. These are indicated in Table 1. This does not, however, bring out the importance of the early clone banks (203, 389), particularly the Charon 4A bank of Ferrari et al. (110); there is no attempt to list all the clones detected in those banks. I must apologize in advance if clone identification is less than comprehensive. It has become increasingly clear that browsing in the library is no longer an efficient way to produce a map. It is nevertheless the method that I have used, greatly aided by information from a large number of generous colleagues.

The map (Fig. 1) was built on the framework of the previous map (371). There was no systematic attempt to sift through 30 years of *B. subtilis* genetics to recheck every linkage. Such a check might have improved accuracy marginally. It would not overcome the basic problem that much of the map depends on two-factor transduction crosses which are inadequate to give an unambiguous linkage map. Cloning and sequencing should largely remove these uncertainties. However, until all involved actually perform the backcrosses, Southern blots, and other such checks to confirm that they have not swapped mutations, rearranged clones . . . The original literature should be consulted to assess the accuracy of the map in areas of interest.

The main method for gross linkage analysis continues to be PBS1-mediated transduction, commonly using as reference the kit of strains constructed by Dedonder et al. (86) to provide selectable markers spanning the chromosome. Vandeyar and Zahler (479) have recently produced a set of strains with Tn*917* inserts at strategic points around much of the chromosome; this should substantially aid transduction mapping. A physical map could give a useful alternative to the rather fraught procedure of transduction mapping. *B. subtilis* DNA yields approximately 26 and 52 fragments, respectively, when digested with the endonucleases *Sf*I and *Not*I. These provide the basis for constructing a physical map (J. Castro, A. Muhammad, H. Sandoval, J.-J. Wu, and P. J. Piggot, unpublished observations). All these methods will slip into planned obsolescence if and when the complete DNA sequence of the chromosome is known.

Patrick J. Piggot • Department of Microbiology and Immunology, Temple University School of Medicine, Philadelphia, Pennsylvania 19140.

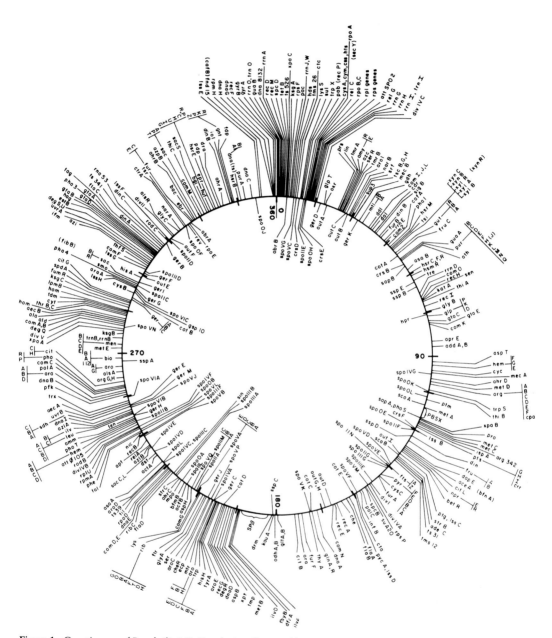

Figure 1. Genetic map of *B. subtilis* 168. For clarity, ribosomal loci in the 10° to 15° region are not shown individually.

Table 1
Genetic Markers of *B. subtilis*

Gene symbol	Map position[a]	Phenotype, enzyme deficiency, or other characteristic	Reference[b]
abrA	326	Partial suppressor of stage 0 phenotypes; may be same as *rev-4*	474
abrB	3	Partial suppressor of stage 0 phenotypes; same locus as *cpsX* and probably *absA*, *absB*, and *tolA*	157, 361*, 474, 475
abrC		Weak intragenic suppressors of *spo0A*	474
absA		Partial suppressor of stage 0 phenotypes; see *abrB*	214
absB		Partial suppressor of stage 0 phenotypes; see *abrB*	214
aceA	126	Pyruvate dehydrogenase defect; defective in E1 (pyruvate decarboxylase) component of pyruvate dehydrogenase	41, 199
acfA	230	Resistant to acriflavin, also to ethidium bromide and distamycin; sensitive to streptomycin	21, 211, 432
acfB	215	Same as *acfA*	432
ada	NM	Adaptive response to alkylating agents	315
addA	86	Subunit of ATP-dependent DNase; *recE5* now considered an allele of this locus	91, 245†
addB	86	Subunit of ATP-dependent DNase	245†
adeC	130	Adenine deaminase	101, 338
adeF	NM	Adenine phosphoribosyltransferase	101
aecA	252	Aminoethylcysteine resistance; close to, or in, structural gene for aspartokinase II	68*, 291; Q
aecB	282	Aminoethylcysteine resistance	291
ahrA	342	Arginine hydroxamate resistance, linked to *cysA*	169, 317
ahrB	328	Arginine hydroxamate resistance	265, 317
ahrC	219	Arginine hydroxamate resistance	317, 443†
ahrD	99	Arginine hydroxamate resistance	25
ala	281	Alanine auxotrophy	291
ald	280	L-Alanine dehydrogenase	124, 265, 476
alsA	263	Acetolactate synthase	531
alsR	317	Constitutive acetolactate synthase	531
amm	245	Glutamate requirement	94
amt		3-Aminotyrosine resistance; part of or very close to *tyrA* locus	374
amyB		Control of amylase synthesis; probably identical to *sacQ* and *pap*; see *sacQ*	424, 451
amyE	25	Amylase structural gene; also called *amyA*	518*, 529
amyR	25	Control of amylase synthesis; also called *amyH*	521, 529
aprE	84	Structural gene for subtilisin E; map order (*hpr glyB*)-*aprE*-*metD* (formerly *sprE*)	448*, 513
apt	236	Adenine phosphoribosyltransferase	417
ara	172	Arabinose utilization	357
araA	256	Arabinose utilization; L-arabinose isomerase	357; D†
araB	256	Arabinose utilization; L-ribulokinase	357; D†
araC	294	Arabinose utilization regulation	357, 412
araD	256	Arabinose utilization; L-ribulose-5-phosphate 4-epimerase	357; D†
argA	102	Arginine requirement; identified by complementation of corresponding locus in *Escherichia coli*; formerly *argO*	318†, 319†
argB	102	Arginine requirement; identified by complementation of corresponding locus in *E. coli*; formerly *argO*	318†, 319†
argC	102	Arginine requirement; identified by complementation of corresponding locus in *E. coli*; formerly *argO*	318†, 319†

(Continued on next page)

Table 1—*Continued*

Gene symbol	Map position[a]	Phenotype, enzyme deficiency, or other characteristic	Reference[b]
argD	102	Arginine requirement; identified by complementation of corresponding locus in *E. coli*; formerly *argO*	318†, 319†
argE	102	Arginine requirement; identified by complementation of corresponding locus in *E. coli*; formerly *argO*	318†, 319†
argF	102	Arginine requirement; identified by complementation of corresponding locus in *E. coli*; formerly *argC*	318†, 319†
argG	260	Arginine requirement; identified by complementation of corresponding locus in *E. coli*; formerly *argA*	318, 319
argH	260	Arginine requirement; identified by complementation of corresponding locus in *E. coli*; formerly *argA*	318, 319
arg-342	117	Arginine-ornithine or citrulline requirement	191
aroA	264	3-Deoxy-D-arabinoheptalosonic-7-phosphate synthase	193, 328
aroB	203	Dehydroquinate synthase	281†, 328
aroC	206	Dehydroquinate synthase	193, 328, 494†
aroD	226	Shikimate dehydrogenase	4, 191, 193; U†
aroE	202	3-*enol*-Pyruvylshikimate-5-phosphate synthase	178*, 328, 331
aroF	203	Chorismate synthase	193, 328
aroG	264	Chorismate mutase; isozyme 3	294, 328
aroH	203	Chorismate mutase; isozymes 1 and 2	274, 328
aroI	26	Shikimate kinase	112†, 265, 352†
aroJ		Tyrosine and phenylalanine; see *hisH*	330
asaA	227	Arsenate resistance	1, 2
asaB	57	Arsenate resistance (derived from *B. subtilis* W23)	3
aspA		Pyruvate carboxylase; see *pycA*	52, 191
aspB	198	Aspartate aminotransferase	191
aspH	215	Constitutive aspartase	205, 500
aspT	92	Aspartate transport	500, 501
ath	55	Adenine-thiamine requirement; probably an allele of *purM*	99*, 530
attSPβ	190	Integration site for phage SPβ	532
attSPO2	13	Integration site for phage SPO2	210, 441
attφ3T		Integration site for phage φ3T; probably maps between *kauA* and SPβ	509
attφ105	243	Integration site for phage φ105	363, 407
azc	25	Resistant to azetidine-2-carboxylic acid; *azc* 90% cotransformed with *aroI*	135
azi	304	Resistance to sodium azide	L
azlA	247	4-Azaleucine resistance; derepressed leucine biosynthetic enzymes	279†, 491†, 496
azlB	234	4-Azaleucine resistance	496, 497
azpA		Resistance to azopyrimidines; alteration of DNA polymerase III, see *polC*	76
azpB	329	Resistance to azopyrimidines	59
bac	327	Bacilysin biosynthesis	187
betR	128	Resistant to betacin produced by SPβ lysogens	174
bfmA	126	Lacks branched-chain-keto-acid dehydrogenase; isolated in *aceA* strain, and not separated genetically from *aceA*	41, 507
bfmB	216	Requires branched-chain fatty acid, valine, or isoleucine; maps between *strC* and *lys*	41, 507
bgl	334	Endo-β1,3-1,4-glucanase (lichenase)	38†, 60†, 322*, 345
bglC	NM	Endo-β1,4-glucanase (carboxymethyl cellulase)	278*, 326*, 398*

(*Continued on next page*)

Table 1—*Continued*

Gene symbol	Map position[a]	Phenotype, enzyme deficiency, or other characteristic	Reference[b]
bioA	268	7-Oxo-8-aminopelargonate:7,8-diaminopelargonate aminotransferase	353
bioB	268	Biotin synthetase	353
bio-112	268	Early defect in biotin synthesis	353
bry		Bryamycin (thiostrepton) resistance; maps in ribosomal protein cluster	168, 441
bsr		Restriction, modification by *B. subtilis* R; see *hsrA*	472
but	NM	5-Bromouracil tolerant	33
cafA	46	Caffeine resistant	58
cafB	355	Caffeine sensitive; also sensitive to nalidixic acid	58
cal	12	Chalcomycin resistant	427
cam-2	45	Resistant to chloramphenicol	15
carA, *carB*		See *cpa*	
catA	75	Hyperproduction of extracellular proteases; can sporulate in presence of glucose; see *hpr*	184, 215, 265, 304
cdd	NM	Deoxycytidine-cytidine deaminase	395
cdr	38	Cadmium resistance	533
che	155	Chemotaxis; includes 20 complementation groups, A through Q and S through U	346†, 347†
cheR	205	Chemotactic methyltransferase	346
citB	173	Aconitate hydratase	89*, 404†, 406, 531
citC	259	Isocitrate dehydrogenase	194, 406, 364†
citF		Succinate dehydrogenase; now split into three loci, *sdhA*, *sdhB*, and *sdhC*	172, 194, 406
citG	289	Fumarate hydratase	303*, 308†, 406
citH	259	Malate dehydrogenase	140
citK	181	α-Ketoglutarate dehydrogenase complex, enzyme E1; see *odhA*	
citL	127	Lipoamide dehydrogenase; E3 component of both pyruvate dehydrogenase and α-ketoglutarate dehydrogenase complexes	190
citM	181	Dihydrolipoyltransuccinylase; E2 component of 2-ketoglutarate dehydrogenase complex; see *odhB*	
cml		Chloramphenicol resistance, caused by mutations in one of at least five 50S ribosomal proteins	349
comA	280	Competence; early block; *com-9* (107) may map in *comA* or *comB*	107; A*
comB	280	Competence; early block	A*
comC	258	Competence	A*
comD	224	Competence	A*
comE	224	Competence	A*
comF	304	Competence	A*
comG	213	Competence; may include *com-71* (107); DNA sequence contains seven open reading frames	A*
comI	37	Competence; codes for 17-kDa[c] nuclease	488*; A
comJ	37	Competence; codes for 18-kDa protein that is isolated with the ComI protein	488*; A
comK	80	Competence; may include *com-30* (107)	107; A
comL	37	Competence	A
comM	330	Competence; previously called *com-104*	107; A

(*Continued on next page*)

Table 1—*Continued*

Gene symbol	Map position[a]	Phenotype, enzyme deficiency, or other characteristic	Reference[b]
comN	159	Competence	A, S
comO	68	Competence	A, S
com-9		See *comA*	
com-30		See *comK*	
com-71		See *comG*	
com-104		See *comM*	
cotA	52	Spore coat protein	92†
cotB	290	Spore coat protein	92†
cotC	168	Spore coat protein	92*
cotD	200	Spore coat protein	92*
cotE	150	Spore coat protein	O*
cotF	NM	Spore coat protein	O*
cpa	102	Carbamoyl-phosphate transferase-arginine; also designated *carA,B*	318†, 319†
cpsX		See *abrB*	
crk	NM	Cytidine kinase	395
crr	118	Phosphocarrier protein for glucose in the phosphotransferase system; also referred to as *ptsX*	154†; V*
crsA	224	Carbon source-resistant sporulation; resistant to novobiocin and acridine orange during sporulation; mutation in *rpoD*; has similar phenotype to *rvt* mutations	428, 460
crsB	56	Requires high glucose for sporulation	460
crsC	217	Carbon source-resistant sporulation; maps close to or in *spo0A*	460
crsD	8	Carbon source-resistant sporulation	460
crsE	12	Carbon source-resistant sporulation; maps in *rpoBC* operon; *rfm-11* suppresses *crsE*	460, 461
crsF	119	Carbon source-resistant sporulation	460
css		Cysteine sensitivity; see *cysA*	227
cta	149	Required for expression of cytochrome *aa₃*	320†; X*
ctc	7	Expressed at the end of exponential growth under conditions in which the enzymes of the tricarboxylic acid cycle are repressed	204*
ctrA	324	CTP synthetase; requirement for cytidine in the absence of ammonium ion	265, 470†, 471*, 531
cyc	95	D-Cycloserine resistant	136
cym		Requirement for cysteine or methionine; see *cysA*	227, 367
cysA	11	Cysteine requirement, serine transacetylase; a complex locus consisting of *css* (cysteine sensitivity), *cym* (cysteine or methionine), *hts* (hydrogen sulfide excretion), and *cysA*	227, 367
cysB	292	Cysteine requirement	94
cysC	140	Cysteine, methionine, sulfite, or sulfide requirement	191, 524
cyt	284	Requires cytidine	200
dac	NM	D-Alanine carboxypeptidase; PBP 5	467*
dal	38	D-Alanine requirement; alanine racemase	88†, 97, 111*, 124
dapE	125	N Acetyl-LL-diaminopimelic acid ligase	54
dcd	NM	Deoxycytidine-5'-monophosphate deaminase	312
dck	NM	Deoxycytidine/deoxyadenosine kinase	311, 395
ddd	NM	Deoxycytidine kinase	395
ddl	36	D-Alanyl-D-alanine ligase	54
degQ	279	Degradation enzyme regulation; previously called *sacQ*	263, 517*

(Continued on next page)

Table 1—*Continued*

Gene symbol	Map position[a]	Phenotype, enzyme deficiency, or other characteristic	Reference[b]
degR	200	Degradation enzyme regulation; previously called *prtR*	325*, 519*
degS	306	Degradation enzyme regulation; part of *sacU*	18†, 182*, 248*, 249, 264, 266, 451, 465*
degU	306	Degradation enzyme regulation; part of *sacU*; also called *iep*	18†, 182*, 248*, 249, 264, 266, 451, 465*
deoA	NM	Thymidine phosphorylase	292
dfrA	195	Dihydrofolate reductase; coordinately regulated with *thyB*	217*, 323†
din	121	DNase inhibitor	379
dinA	310	DNA damage inducible; possibly the same as *uvrA*	148
dinB	43	DNA damage inducible	148
dinC	313	DNA damage inducible	148
divI	144	Temperature-sensitive cell division; formerly *divD*	478
divII	317	Temperature-sensitive cell division; formerly *divC*	478
divIVA	144	Minicell production	391
divIVB	242	Minicell production	391
divIVC	15	Minicell production; formerly *divA*	478
divV	278	Temperature-sensitive cell division; formerly *divB*	478
dnaA	165	DNA synthesis; ribonucleotide reductase	17, 26, 228, 294, 358†, 394; U†
*dnaA*BS		See *dnaH*	
dnaB	256	DNA synthesis; initiation of chromosome replication; probably more than one gene	201*, 208, 228, 294, 298, 341*
dnaC	351	DNA synthesis	14, 228, 294
dnaD	199	DNA synthesis, initiation of chromosome replication	228, 294
dnaE	224	DNA synthesis	228, 294, 492*
dnaF		DNA synthesis; DNA polymerase III; see *polC*	228, 294
dnaG	0	DNA synthesis; homologous to *dnaN* of *E. coli*	228, 294, 316*, 340*, 342*, 358†
dnaH	0	DNA synthesis; homologous to *dnaA* of *E. coli*; also called *dnaA*BS	16, 228, 294, 316*, 340*, 342*, 475
dnaI	248	DNA synthesis	228, 294
dnaA(Ts)	347	DNA synthesis	165, 294
dnaB(Ts)	347	DNA synthesis	165, 294
dnaZY	NM	DNA polymerase III	457*
dna-8132	0	DNA synthesis; initiation of chromosome replication	165, 167, 475
dpa	148	Requires dipicolinic acid for heat-resistant spores; probably the same as *spoVF*	20
dra	339	Deoxyriboaldolase	459
drm	182	Phosphodeoxyribomutase	459
dst	231	Resistant to distamycin and acriflavin	432
D-*tyr*		Resistance to D-tyrosine; maps within the *tyrA* locus; see *tyrA*	65
ebr	326	Ethidium bromide resistance	33
ecp	204	Resistant to 2-amino-5-ethoxycarbonyl pyrimidine-4(3*H*)-one	390
efg	10	Elongation factor G	5, 95, 241
eno	NM	Enolase	E*
epr	335	Minor extracellular protease	436*
ery		Erythromycin resistance; ribosomal protein L22; see *rplV*	464, 466

(*Continued on next page*)

Table 1—*Continued*

Gene symbol	Map position[a]	Phenotype, enzyme deficiency, or other characteristic	Reference[b]
estB	306	Esterase B defect	183
ethA	NM	Ethionine resistance	489
fdpA	344	Fructose-bisphosphatase	127
fibA	307	Macrofiber formation; weak linkage to hisA	416
fibB	295	Macrofiber formation	416
flaA	152	Defect in flagellar synthesis; autolysin deficient	155, 377
flaB	152	Defect in flagellar synthesis	155, 377
flaC	310	Defect in flagellar synthesis	155, 377
flaD	222	Defect in flagellar synthesis; autolysin deficient	377, 423†
fruA	123	Fructose transport	145
fruB	123	Fructose-1-phosphate kinase	145
fruC	51	Fructokinase	143, 145
ftr	208	Fluorotryptophan resistance	27
ftsA	134	Homologous to *E. coli* cell division gene *ftsA*; probable location of *spoIIN279*	29*, N
ftsZ	134	Homologous to *E. coli* cell division gene *ftsZ*	29*
fumR	288	Regulation of fumarate hydratase	L
fun	12	Close to, or in, *rpsL*	176
furA	141	Resistance to 5-fluorouracil	94
furB	41	Fluorouracil resistance	372
furC	326	Resistance to 5-fluorouracil in the presence of uracil	B
furE	326	Resistance to 5-fluorouracil in the presence of uracil	BB
furF	170	Resistance to 5-fluorouracil in the presence of uracil	84
fus		Fusidic acid resistance; see *efg*	241
gap	NM	Glyceraldehyde-3-phosphate dehydrogenase	E*
gca	NM	L-Glutamine-D-fructose-6-phosphate aminotransferase	125
gdh	34	Structural gene for glucose dehydrogenase	67, 259*, 484†
gerA	289	Germination defective; defective in germination response to alanine and related amino acids; consists of three genes	108*, 308†, 309, 310, 409, 437
gerB	314	Germination defective; defective in germination response to the combinations of glucose, fructose, asparagine, and KCl	310, 437
gerC	203	Germination defective; temperature-sensitive germination in alanine	310, 437; P†
gerD	16	Germination defective; defective germination in range of germinants	310, 372, 437
gerE	251	Germination defective; defective germination in range of germinants; may be a spore coat defect	77*, 170†, 219†, 220†, 307, 310, 437
gerF	301	Germination defective; defective germination in a range of germinants	310, 372, 437
gerG	294	Germination defective; mutant lacks phosphoglycerate kinase (*pgk*) activity; germinates poorly in alanine; sporulates poorly	123, 310, 381, 437
gerH	246	Defective germination in a range of germinants	372
gerI	296	Defective germination in a range of germinants	372
gerJ	206	Defective germination in a range of germinants; allele *gerJ51* (also called *tzm*) is present in many laboratory strains	281†, 495
gerK	32	Defective germination response to glucose	212
gerM	251	Defective germination and sporulation	410†
glnA	167	Glutamine synthetase structural gene	39, 84, 117†, 118, 137†, 393, 456*

(Continued on next page)

Table 1—*Continued*

Gene symbol	Map position[a]	Phenotype, enzyme deficiency, or other characteristic	Reference[b]
glnR	167	Negative regulator of glutamine synthetase	117†; U*
glpD	75	Glycerol-3-phosphate dehydrogenase	270
glpK	75	Glycerol kinase	270
glpP	75	Pleiotropic glycerol mutant	270
glpT	18	Fosfomycin resistant, glycerol phosphate transport defect	235
gltA	180	Glutamate or aspartate requirement, glutamate synthase	34†, 84, 191, 313†; U*
gltB	180	Glutamate synthase	34†, 313†; U*
gltC	180	Positive regulator of *gltAB*	U*
glyA	207	Glycine requirement	233–235
glyB	74	Glycine requirement	167
glyC	320	Glycine requirement	59
gntK	344	Gluconate utilization; gluconate kinase	127, 129*, 130
gntP	344	Gluconate utilization; gluconate permease	127, 129*, 130
gntR	344	Gluconate utilization; repressor	127, 128*, 129*, 130
gntZ	344	Gluconate utilization; unknown function	129*
gra	25	Glucose-resistant amylase production	335*
groEL	NM	Phage growth	62
gsp-10	294	Spore outgrowth	H*
gtaA	309	Glucosylation of teichoic acid; lacks UDPglucose:poly(glycerol-phosphate) glucosyltransferase	265, 378, 523
gtaB	308	Glucosylation of teichoic acid; lacks UDPglucose pyrophosphorylase	378, 523
gtaC	77	Glucosylation of teichoic acid; lacks phosphoglucomutase	378, 523
gtaD	308	Glucosylation of teichoic acid	378
gtaE	77	Glucosylation of teichoic acid	378
guaA	54	GMP synthetase; previously *guaB*	418
guaB	0	IMP dehydrogenase; previously *guaA*	418, 475
guaC	NM	GMP reductase	101
guaF	NM	Hypoxanthine-guanine phosphoribosyltransferase	101
guaP	NM	Inosine-guanosine phosphorylase	101
gutA	50	D-Glucitol permease	64, 143†
gutB	50	D-Glucitol hydrogenase	64, 143†
gutR	50	Constitutive synthesis of D-glucitol permease and D-glucitol dehydrogenase; regulatory gene for glucitol catabolism	64, 143†
gyrA	0	DNA gyrase (*nalA*)	59, 256†, 258†, 316*
gyrB	0	DNA gyrase (*novA*)	59, 256†, 258† 316*
hag	307	Flagellar antigen	155, 287
hds	6	Pleiotropic extragenic suppressors of DNA mutations	433
hemA	244	δ-Aminolevulinate synthase	239
hemB	244	Porphobilinogen synthase	30
hemC	244	Porphobilinogen deaminase	30
hemD	244	Uroporphyrinogen-III cosynthase	301
hemE	94	Uroporphyrinogen decarboxylase	300
hemF	94	Coproporphyrinogen oxidase	300
hemG	94	Ferrochelatase	300
hisA	298	Histidine requirement; probable location of all histidine enzymes except *hisH*	37, 102, 491†

(*Continued on next page*)

Table 1—*Continued*

Gene symbol	Map position[a]	Phenotype, enzyme deficiency, or other characteristic	Reference[b]
hisH	202	Histidinol-phosphate aminotransferase, tyrosine and phenylal-anine aminotransferase	1, 178*, 209†, 330
hom	284	Threonine and methionine requirement, deletion lacking ho-moserine dehydrogenase	356*
hos	12	Suppresses the temperature-sensitive phenotype of elongation factor G mutants	290
hpr	75	Overproduction of proteases; shown to be the same as *catA* and *scoC*	184, 360*
hsmR	60	Modification methylase of the *Bsu*R system	238*
hsrB	345	Host-specific restriction and modification of *B. subtilis* IAM 1247; endonuclease *Bsu*B; isoschizomer of *Pst*I	207, 429
hsrC	59	Host-specific restriction and modification of *B. subtilis* 1247 II; endonuclease *Bsu*C	207
hsrE	337	Host-specific restriction and modification of *B. subtilis* IMA 1231; endonuclease *Bsu*E; isoschizomer of *Fnu*DII	207, 224
hsrF	59	Endonuclease *Bsu*F; isoschizomer of *Hpa*II	207, 224
hsrM	47	Host-specific restriction and modification of *B. subtilis* Mar-burg; probably identical to *nonB*; endonuclease *Bsu*M; iso-schizomer of *Xho*I	207, 224
hsrR	59	Endonuclease *Bsu*R; isoschizomer of *Hae*III	47, 224, 238*, 472
hts		Excretion of hydrogen sulfide; see *cysA*	227
hutC	335	Histidine utilization; control locus	66, 237
hutG	335	Histidine utilization; formiminoglutamate hydrolase	66, 237
hutH	335	Histidine utilization; histidase	66, 237, 339*
hutI	335	Histidine utilization; imidazolonepropionate hydrolase	66, 237
hutP	335	Histidine utilization; positive control element	66, 237, 339*
hutR	335	Histidine utilization; control locus	66, 237
hutU	335	Histidine utilization; urocanase	66, 237, 339*
iep		See *degU*	
ifm	304	Hypermotility; suppresses *flaA* and *flaD* mutations	155, 377
igf	343	Deletion covering *iol*, *gnt*, *fdpA*, and *hsrB* (formerly *fdpA1*)	127
ilvA	194	Threonine dehydratase	24, 190
ilvB	247	Condensing enzyme	24, 279†, 491†, 496
ilvC	247	α-Hydroxy-β-keto-acid reductoisomerase	24, 279†, 491†, 496
ilvD	196	Dihydroxyacid dehydratase	151
infA	11	Initiation factor I	T*
infB	149	Initiation factor II	N*
inh		Inhibition by histidine; probably within *tyrA* locus; see *tyrA*	328, 331
iol	343	Inability to grow on *myo*-inositol; possibly *myo*-inositol dehy-drogenase gene	127
ispA	117	Intracellular serine protease	243*
kan	10	Kanamycin resistance; maps in the ribosomal protein cluster	153
katA	70	Catalase	273
kauA	182	Branched-chain α-keto acid transport	151
kir		Probably mutation in the structural gene for elongation factor Tu; see *tuf*	439
ksgA	4	High-level kasugamycin resistance	57, 469, 475, 510
ksgB	277	Low-level kasugamycin resistance	469
ksgC	287	Fumarate hydratase defective; kasugamycin resistance	L

(*Continued on next page*)

Table 1—*Continued*

Gene symbol	Map position[a]	Phenotype, enzyme deficiency, or other characteristic	Reference[b]
leuA	247	α-Isopropylmalate synthase	24, 94, 279†, 324†, 491†, 496, 497
leuB	247	Isopropylmalate isomerase	279†, 324†, 491†, 496
leuC	247	β-Isopropylmalate dehydrogenase	24, 209*, 279†, 324†, 491†, 496, 497
leuD	247	Possibly a subunit of isopropylmalate isomerase	279†, 280, 491†
lin	24	Lincomycin resistance	152, 168
lpm		Lipiarmycin resistance; RNA polymerase; see *rpoC*	445
lpmB	286	Lipiarmycin resistance	444
lssA	35	Thermosensitive lysis; linked to *ddl* by transformation	45
lssB	120	Thermosensitive lysis; may be an allele of *dapE*	45
lssC		Thermosensitive lysis; may be an allele of *ptg*	45
lssD	149	Thermosensitive lysis; closely linked to *pyc* by transformation	45
lssE	354	Thermosensitive lysis	45
lssF	316	Thermosensitive lysis	45
lssG	303	Thermosensitive lysis	45
lssH	293	Thermosensitive lysis	45
lys	210	Lysine requirement, diaminopimelate decarboxylase	205, 223†, 233
lysS	8	Lysyl-tRNA synthetase	388
lyt		Autolytic enzymes; see *flaD*	109, 374
mdh		Malate dehydrogenase; see *citH*	140
mecA	97	Medium-independent expression of competence	A
mecB	30	Medium-independent expression of competence	A
mecC	117	Medium-independent expression of competence	A
menB	273	Menaquinone deficient; multiple aminoglycoside resistant; dihydroxynaphthoate synthase	297, 305†, 463
menC,D	273	Menaquinone deficient; multiple aminoglycoside resistant; blocked in formation of *o*-succinylbenzoic acid from chorismic acid	305†, 463
menE	273	Menaquinone deficient; multiple aminoglycoside resistant; *o*-succinylbenzoyl-coenzyme A synthetase	297, 305†, 463
metA	110	Responds to methionine, cystathionine, or homocysteine	449
metB	197	Responds to methionine or homocysteine	12
metC	115	Responds to methionine	94
metD	100	Responds to methionine	523
metE	272	*S*-Adenosylmethionine synthetase	489†
mic		Resistance to microccocin; see *rplC*	442
mit	UC	Responds to mitomycin C; maps near *rplV*	218
mpo	218	Membrane protein overproduction, temperature-sensitive sporulation	289
mtlA	34	Lacks mannitol transport; maps near *mtlB*	240; I
mtlB	34	Mannitol-1-phosphate dehydrogenase	67†, 265
mtr	204	Resistance to 5-methyl-tryptophan; derepression of the tryptophan biosynthetic pathway	196, 198
nalA		Resistance to nalidixic acid; see *gyrA*	167
narA	320	Inability to use nitrate as a nitrogen source	265, 531
narB	28	Inability to use nitrate as a nitrogen source	94, 531
nea		Neamine resistance; see ribosomal protein cluster	152
neo		Neomycin resistance; see ribosomal protein cluster	152, 168

(Continued on next page)

Table 1—*Continued*

Gene symbol	Map position[a]	Phenotype, enzyme deficiency, or other characteristic	Reference[b]
nic	240	Nicotinic acid requirement	113†, 189, 233
nonA	UC	Permissive for bacteriophage SP10 and φNR2; closely linked to *rfm*	408
nonB		Permissive for bacteriophage SP10 and φNR2; see *hsrM*	408
novA	0	Resistance to novobiocin, *gyrB*	168, 475
nprE	127	Structural gene for neutral (metallo) protease	477, 483, 520*
nprR	127	Regulatory gene for neutral protease	181*, 468*, 477
odhA	181	2-Oxoglutarate dehydrogenase, E1 subenzyme of the 2-oxo-glutarate dehydrogenase complex; originally designated *citK*	61†, 190, 406; K*
odhB	181	Dihydrolipoamide transsuccinylase, E2 subenzyme of the 2-oxoglutarate dehydrogenase complex; originally designated *citM*	61†, 116; K*
ole		Oleandomycin resistance; see ribosomal protein cluster	152, 168
ordA	NM	Ornithine:2-oxo-acid aminotransferase	500
outA	20	Blocked in outgrowth after RNA, protein, and DNA syntheses have started; previous designation, *gspIV*	9, 134, 372
outB	28	Blocked in outgrowth before most macromolecular syntheses have started; previous designation, *gsp-81*	7, 8*, 112†, 132, 372
outC	28	Blocked in outgrowth after RNA and protein syntheses have started, but before DNA synthesis; previous designation, *gsp-25*	9, 372
outD	158	Blocked in outgrowth; protein and DNA syntheses reduced; previous designation, *gsp-1*	133, 372
outE	300	Blocked in outgrowth; RNA synthesis normal; protein synthesis reduced; DNA synthesis prevented; previous designation, *gsp-42*	9, 372
outF	316	Blocked in outgrowth; RNA and protein syntheses reduced; DNA synthesis prevented; previous designation, *gsp-4*	9, 372
outG	160	Identified by outgrowth-specific transcript	147†
outH	160	Identified by outgrowth-specific transcript	147†
outI	129	Identified by outgrowth-specific transcript	147†
oxr	19	Oxolinic acid resistant	485
pab	10	*p*-Aminobenzoic acid requirement; subunit A of *p*-aminobenzoate synthase	226
pac	5	Resistance to pactamycin	168, 475
pap		Hyperproduction of proteases and amylase; see *degQ*	451, 521
pbp	NM	Penicillin-binding protein; mutations affect PBPs 2a, 2b, 3 and 4; probably four loci; see also *dac*	48, 431
pbuG	55	Purine base uptake	417
pdp	337	Pyrimidine nucleoside phosphorylase	292†, 405
pfk	255	Phosphofructokinase	143, 145
pgk	NM	3′-Phosphoglycerolkinase	G
pha-1	47	Resistance to phage SPO1	265
pha-2	280	Resistance to phage SPP1	413
pha-3	308	Resistance to phage of group III	106
pha-4	295	Resistance to phage	D
pheA	240	Phenylalanine requirement; prephenate dehydratase	24, 113†, 526†
phl	12	Phleomycin resistance; mutator strain linked to *rpsL*	206
phoA		*B. subtilis* has several alkaline phosphatases; the likely genes for some of these have been cloned and are listed as *xpaA*, -*B*, and -*C*; the *phoA* designation is not yet used for any of these	M

(Continued on next page)

Table 1—*Continued*

Gene symbol	Map position[d]	Phenotype, enzyme deficiency, or other characteristic	Reference[b]
phoP	258	Regulation of alkaline phosphatase and phosphodiesterase; homologous to *phoB* of *E. coli*	261, 421*, 516
phoR	258	Regulation of alkaline phosphatase and phosphodiesterase; homologous to *phoR* of *E. coli*	260, 261, 302, 422*
phoS	114	Constitutive alkaline phosphatase	373
phoT	245	Constitutive alkaline phosphatase	R
pig		Sporulation-associated pigment; allele of *cotA*	399
polA	257	DNA polymerase I	140, 252, 358†
polC	147	DNA polymerase III; azopyrimidine resistance	17, 76, 276, 344†, 396
pro	115	Proline requirement	53, 135
prs	24	Phosphoribosylpyrophosphate synthetase	336†
prs-3	74	Protein secretion	244
prs-11	298	Protein secretion	244
prs-33	150	Protein secretion	244
prtR		See *degR*	
ptg	129	Peptidoglycan biosynthesis	54
ptm	107	Pyrithymine resistance	H
ptsG	118	An open reading frame with homology to the *E. coli ptsG* locus (glucose-specific enzyme II of the phosphotransferase system) has been identified adjacent to *crr* and *ptsH*	V*
ptsH	118	Phosphoenolpyruvate phosphotransferase, Hpr	144, 154†, 334; V*
ptsI	118	Phosphoenolpyruvate phosphotransferase, enzyme I	144, 154†, 334; V*
ptsX	118	See *crr*	
pupA	NM	Adenosine phosphorylase	164
pupG	NM	Inosine-guanosine phosphorylase	338†
pupI	NM	Inosine phosphorylase	164
purA[d]	348	Adenine requirement	351, 376†
purB[d]	55	Purine requirement; adenylosuccinate lyase; previously designated *purE*	99*, 274, 418†
purC[d]	55	Purine requirement; phosphoribosylaminoimidazolesuccino-carboxamide synthetase	99*, 274
purD[d]	55	Purine requirement; 5′-phosphoribosyl-1-glycinamide synthetase	99*, 418†
purE[d]	55	Purine requirement; 5′-phosphoribosyl-5-aminoimidazole carboxylase I	99*
purF[d]	55	Purine requirement; amidophosphoribosyl transferase; previously designated *purB*	99, 274, 282*, 343
purH(J)[d]	55	Purine requirement; phosphoribosylaminoimidazole-carboxamide formyltransferase/IMP cyclohydrolase	99*, 418†
purK[d]	55	Purine requirement; 5′-phosphoribosyl-5-aminoimidazole carboxylase II; previously designated *purD*	99*, 274
purL[d]	55	Purine requirement; 5′-phosphoribosyl-*N*-formylglycinamide synthetase II	99*, 418†
purM[d]	55	Purine requirement; 5′-phosphoribosyl-5-aminoimidazole synthetase	99*, 418†
purN[d]	55	Purine requirement; 5′-phosphoribosyl-1-glycinamide formyltransferase	99*
purQ[d]	55	Purine requirement; 5′-phosphoribosyl-*N*-formylglycinamide synthetase II	99*

(Continued on next page)

Table 1—*Continued*

Gene symbol	Map position[a]	Phenotype, enzyme deficiency, or other characteristic	Reference[b]
pycA	149	Pyruvate carboxylase	52, 191, 320†
pyrA	139	Carbamoyl-phosphate synthetase	267†, 380
pyrB	139	Aspartate carbamoyltransferase	267†, 268*, 380
pyrC	139	Dihydroorotase	267†, 380
pyrD	139	Dihydroorotate dehydrogenase	267†, 380
pyrE	139	Orotate phosphoribosyltransferase; also called *pyrX*	267†, 380
pyrF	139	Orotidine-5′-phosphate decarboxylase	267†, 380
pyrG	NM	CTP synthetase	395
recA	157	Genetic recombination and radiation resistance	189, 294
recB	240	Genetic recombination and radiation resistance	189, 294
recC		Genetic recombination; indirect effect of bacteriophage SPO2 lysogeny; see *attSPO2*	93, 139, 294
recD	0	Genetic recombination and radiation resistance	166, 294
recE	157	Genetic recombination and radiation resistance; ATP-dependent nuclease; complemented by *E. coli recA*	87, 93, 284†, 293, 294
recF	0	Genetic recombination and radiation resistance	10, 166, 257†, 294, 316*, 358†, 475
recG	200	Genetic recombination and radiation resistance	293
recH	69	Genetic recombination and radiation resistance	294, 329
recI	72	Genetic recombination and radiation resistance	294, 329
recL	UC	Genetic recombination and radiation resistance; linked to *cysA*	93, 293, 294
recM	0	Genetic recombination and radiation resistance	93, 293, 294
recP	11	Reduced recombination and high sensitivity to mytomycin C; prevents φ105 restriction in *BsuR*⁺ hosts; formerly *rec-149*	10, 375
relA	235	ATP:GTP 3′-phosphotransferase	440, 462
relC		See *tsp*, *rplK*	440
relG	13	Defect in glucose uptake	384, 385
rev-4	324	Suppressor of some pleiotropic effects (but not asporogeny) of *spo0* mutations; suppresses effect on sporulation of various drug-resistant mutations; may be same as *abrA*	425, 426, 470*
rfm		Rifampin resistance; RNA polymerase; see *rpoB*	162, 362
rgn	UC	Improved protoplast regeneration; maps near *cysA*	6
ribA	209	Riboflavin requirement	234, 386†
ribB	209	Riboflavin requirement	46, 386†
ribC	222	Riboflavin requirement	46
ribD	209	Riboflavin requirement	46, 386†
ribF	209	Riboflavin requirement	387†
ribG	209	Riboflavin requirement	387†
ribH	209	Riboflavin requirement	387†
ribO	209	Riboflavin requirement	46, 386†
ribT	209	Riboflavin requirement	387†
rna-53	312	Temperature-sensitive RNA synthesis	397
rnpB	NM	RNase P, RNA component	392*
rodB	243	Cell wall defective	229
rodC	314	Cell wall defective; shown to be the same as *tag*	229, 378
rodD	198	Cell wall defective	R
rplA	12	Ribosomal protein BL1, chloramphenicol resistance II	82, 349, 350
rplB	12	Ribosomal protein L2 (BL2)	438
rplC	12	Ribosomal protein L3 (BL3); probably micrococcin resistance	80, 350, 442
rplE	12	Ribosomal protein L5 (BL6)	80

(Continued on next page)

Table 1—*Continued*

Gene symbol	Map position[a]	Phenotype, enzyme deficiency, or other characteristic	Reference[b]
rplF	12	Ribosomal protein L6 (BL8)	80
rplJ	12	Ribosomal protein L10 (BL5)	82
rplK	12	Ribosomal protein L11 (BL11); thiostrepton resistance; *relC*	82, 350, 362, 504
rplL	12	Ribosomal protein L12 (BL9); chloramphenicol resistance VI	82, 349, 350
rplO	11	Ribosomal protein L15; chloramphenicol resistance III	349, 350; T*
rplQ	11	Ribosomal protein L17 (BL15)	438; T*
rplU	242	Ribosomal protein L21 (BL20)	78
rplV	12	Ribosomal protein L22 (BL17); erythromycin resistance	350, 438, 464
rplX	12	Ribosomal protein L24 (BL23)	438
rpmA	241	Ribosomal protein L27 (BL24)	78, 114*
rpmH	0	Ribosomal protein L34	316*
rpmD	11	Ribosomal protein L30 (BL27)	79
rpmJ	11	Ribosomal protein B	T*
rpoA	11	RNA polymerase α subunit	458*
rpoB	12	β Subunit of RNA polymerase; rifampin resistance	162, 382†, 446
rpoC	12	β′ Subunit of RNA polymerase; streptolydigin resistance	162, 382†, 446
rpoD	224	RNA polymerase σ⁴³ subunit; see *sigA*	383
rpoE	325	RNA polymerase δ subunit	255*
rpsC	12	Ribosomal protein S3 (BS3)	438
rpsD	12	Ribosomal protein S4 (BS4)	177, 185
rpsE	11	Ribosomal protein S5; spectinomycin resistance	79, 168, 263, 348; T*
rpsF	4	Ribosomal protein S6 (BS9)	81
rpsG	12	Ribosomal protein S7 (BS7)	82
rpsH	12	Ribosomal protein S8 (BS8)	350
rpsI	12	Ribosomal protein S9 (BS10)	79
rpsJ	12	Ribosomal protein S10 (BS13) (*tetA*)	438
rpsK	12	Ribosomal protein S11 (BS11)	79, 458*
rpsL	12	Ribosomal protein S12 (BS12) (*strA*); streptomycin resistance	79, 80, 152, 350
rpsM	11	Ribosomal protein S13	458*
rpsO	11	Ribosomal protein S15	T*
rpsP	145	Ribosomal protein S16 (BS17)	81
rpsQ	12	Ribosomal protein S17 (BS16)	350
rpsS	12	Ribosomal protein S19 (BS19)	350
rpsT	12	Ribosomal protein S20 (BS20)	348, 350
rrnA	0	rRNA operon	343*, 510
rrnB	275	rRNA operon	156*, 453, 487†
rrnD	67	rRNA operon	251†
rrnE	45	rRNA operon	251†
rrnF	NM	rRNA operon known to exist but not cloned or mapped	40†
rrnG	15	rRNA operon	40†
rrnH	15	rRNA operon	40†
rrnI	15	rRNA operon	40†
rrnJ	6	rRNA operon	503†
rrnO	0	rRNA operon	343*
rrnW	6	rRNA operon	503†
rvt		Mutations causing some phenotype as *rvtA* mutations but not mapping in the *rvtA* region	428
rvtA	217	Suppressor of sporulation defect in *spo0B*, *spo0E*, *spo0F*, and *spoIIA* mutants; allele of *spo0A*	428; N
sacA	330	β-Fructofuranosidase (sucrase)	121*, 263, 264

(*Continued on next page*)

Table 1—*Continued*

Gene symbol	Map position[a]	Phenotype, enzyme deficiency, or other characteristic	Reference[b]
sacB	296	Levansucrase	80†, 264, 450*
sacC	233	Levanase	285*, 419*
sacL	233	Increased expression of levanase	250
sacP	330	Sucrose transport; enzyme II of sucrose phosphotransferase system	121*, 122*, 263
sacQ	279	Hyperproduction of levansucrase and other extracellular enzymes; see *degQ*	249, 264
sacR	296	Constitutive for levansucrase	264, 430*, 452*
sacS	333	Regulation of α-fructofuranosidase; consists of two genes designated *sacX* and *sacY*	19†, 85†, 264
sacT	330	Constitutive α-fructofuranosidase production	264
sacU	306	Regulatory gene for levansucrase and other extracellular enzymes; see *degS,U*	249, 264, 266, 451
sacV	40	Regulatory gene for levansucrase	286*
sacX	333	Sucrase regulation	450*, 535*, CC*
sacY	333	Sucrase regulation; antiterminator	450*, 535*
sapA	114	Mutations overcome sporulation-phosphatase-negative phenotype of early blocked *spo* mutants	368, 373
sapB	56	Mutations overcome sporulation-phosphatase-negative phenotype of early blocked *spo* mutants	373
sas	211	Weak intragenic suppressor mutations of *spoIIA*	528
scoA	109	Sporulation control; protease and phosphatase overproduction; delayed spore formation	304
scoB	129	Sporulation control; protease and phosphatase overproduction; delayed spore formation	90, 304
scoC		Sporulation control; see *hpr*	304
scoD		Sporulation control; probably the same as *scoB*	304
sdhA	252	Flavoprotein subunit of succinate dehydrogenase; note that the *sdh* loci have been reorganized (206); originally designated *citF*	170†, 172, 365*
sdhB	252	Iron protein subunit of succinate dehydrogenase	170†, 172, 365*
sdhC	252	Cytochrome b_{558}; subunit of succinate dehydrogenase	170†, 172, 365*
secY	11	Secretion of proteins; homologous to *secY* of *E. coli*	T*
sen	70	Enhances protease production	493*
serA	207	Requirement for serine	196, 281
serC	257	Requirement for serine; defined by Tn*917* insertion	479
serR	25	Serine resistance	BB
sigA	224	Major RNA polymerase sigma factor, σ^A or σ^{43}; locus also called *rpoD*	150, 383†, 512*
sigB	40	RNA polymerase sigma factor, σ^B or σ^{37}	32*, 98*
sigC	NM	RNA polymerase sigma factor, σ^C or σ^{32}	225
sigD	UC	RNA polymerase sigma factor, σ^D or σ^{28}; involved in the synthesis of flagella and of autolytic enzymes	149, 173*
sigE	135	RNA polymerase sigma factor, σ^E or σ^{29}; locus the same as *spoIIGB*; see *spoIIG*	242*, 455*, 473*
sigF	211	RNA polymerase sigma factor, σ^F; locus the same as *spoIIAC*; see *spoIIA*	120*, 103*
sigG	135	RNA polymerase sigma factor, σ^G; locus the same as *spoIIIG*	288*; W*
sigH	11	RNA polymerase sigma factor, σ^H or σ^{30}; locus the same as *spo0H*; see *spo0H*	96*
sin	221	Sporulation inhibition at high copy number	141*, 142*
smo	295	Smooth/rough colony morphology	146†, 155, 229

(*Continued on next page*)

Table 1—Continued

Gene symbol	Map position[a]	Phenotype, enzyme deficiency, or other characteristic	Reference[b]
sof-1	217	Suppressor of sporulation defects in *spo0B*, *spo0E*, and *spo0F* mutants; mutation is an alteration in codon 12 of the *spo0A* gene	195*, 231
spcA		Spectinomycin resistance; see *rpsE*	46, 168, 216
spcB	146	Spectinomycin resistance	168, 276
spcD	2	Spectinomycin dependence; maps between *cysA* and *purA*	175
spdA	288	Sporulation derepressed; low pyruvate carboxylase activity	100
spe	NM	UV-sensitive spores (formerly *ssp-1*)	321
spg		Sporangiomycin resistance; 50S ribosomal alteration	28
spoCM	352	Stage 0 sporulation; allele of *spo0J*	44; C†, F†
spoL	227	"Decadent" sporulation	21
spo0A	217	Sporulation; mutants blocked at stage 0; mutants exhibit wide variety of pleiotropic phenotypes	115*, 192, 194, 246*, 299
spo0B	241	Sporulation; mutants blocked at stage 0; mutants have most phenotypes of mutants bearing *spo0A* mutations	36†, 42, 113†, 114*, 188†, 192
spo0C	217	Stage 0 sporulation; mutants with less pleiotropic phenotypes and known to have missense alterations in the *spo0A* gene product	195
spo0D		Stage 0 sporulation; single allele resulting in stage 0 block of sporulation; shown to be an allele of *spo0B*	211, 370; L
spo0E	120	Stage 0 sporulation; oligosporogenous mutations giving stage 0 block; less pleiotropic than *spo0A*, *spo0B*, or *spo0F* mutations	74, 359*, 370
spo0F	324	Stage 0 sporulation; DNA sequence contains single open reading frame for protein of 19,055 Da; inhibits sporulation when present in multiple copies	192, 232†, 471*, 522*
spo0G	223	Stage 0 sporulation; shown to be an allele of *spo0A*	211, 370; L
spo0H	11	Stage 0 sporulation; locus codes for RNA polymerase sigma factor, σ^H; also called *sigH*	96*, 192, 366, 370
spo0J	352	Stage 0 sporulation; the two alleles, 87 and 93, originally placed in the locus (370) are now thought to lie in distinct loci (R); there is disagreement as to the sporulation block caused by these mutations (69, 454)	104†, 202, 475
spo0K	104	Stage 0 sporulation; maps close to tryptophanyl-tRNA synthetase gene	74, 370
spo0L	106	Stage 0 sporulation; uncharacterized allele giving *spo0* phenotype; maps near *spo0K* but genetically distinct	L
spoIIA	211	Stage II sporulation; DNA sequence has three adjacent open reading frames, *spoIIAA*, *spoIIAB*, and *spoIIAC*, coding for proteins of 13, 16, and 29 kDa; the 29-kDa one has homology to *rpoD* and is designated σ^F; *spoIIAC* also called *sigF*	74, 103*, 120*, 211, 272†, 366
spoIIB	244	Stage II sporulation	74, 370
spoIIC		Stage II sporulation; shown to be an allele of *spoIID*	74, 370
spoIID	316	Stage II sporulation	13†, 275*, 366, 370, 402†
spoIIE	10	Stage II sporulation; DNA sequence has three adjacent open reading frames, *spoIIEA*, *spoIIEB*, and *spoIIEC*; *spoIIH* (527) is an allele of *spoIIE*	159*, 366, 370, 527†; Z†, AA*
spoIIF	120	Stage II sporulation; *spoIIJ* is an allele of *spoIIF*	202, 370; W*

(*Continued on next page*)

Table 1—*Continued*

Gene symbol	Map position[a]	Phenotype, enzyme deficiency, or other characteristic	Reference[b]
spoIIG	135	Stage II sporulation; DNA sequence has two adjacent open reading frames, *spoIIGA* and *spoIIGB*; *spoIIGB* (also called *sigE*) codes for an RNA polymerase sigma factor, σ^E	36†, 236, 242*, 366, 370, 454, 455*, 473*
spoIIH		See *spoIIE*	
spoIIJ		See *spoIIF*	
spoIIL	215	Defined by Tn*917* insertion; stage designation tentative pending characterization by electron microscopy	411
spoIIM	212	Defined by Tn*917* insertion; stage designation tentative pending characterization by electron microscopy; distinct from *spoIIA*	411
spoIIN	134	Defined by mutation *spo-279*; originally placed in *spoIIG*; maps near to, but separate from, *spoIIG*; *spoIIN279* may be an allele of *ftsA*	524, 525; N†
spoIIIA	220	Stage III sporulation; contains several genes	104†, 366, 370
spoIIIB	221	Stage III sporulation	366, 370
spoIIIC	227	Stage III sporulation, but very similar phenotype to *spoIVC*	105*, 211, 370
spoIIID	302	Stage III sporulation	211, 370; O*
spoIIIE	142	Stage III sporulation	50*, 211, 370
spoIIIF	239	Stage III sporulation; map order *nic-recB-spoIIIF-spoVB*	254
spoIIIG	135	Stage III sporulation; codes for RNA polymerase sigma factor, σ^G	288*; W*
spoIVA	204	Stage IV sporulation; linked to *trpC* by transformation	74, 281†, 366, 370
spoIVB	213	Stage IV sporulation; may be allele of *spo0A*	74, 370; L
spoIVC	227	Stage IV sporulation; contains at least two cistrons; linked to *aroD* by transformation	83, 104†, 126†, 247*, 370, 372
spoIVD	233	Stage IV sporulation	202, 370
spoIVE	234	Stage IV sporulation	211, 370
spoIVF	242	Stage IV sporulation; linked to *spo0B* by transformation	74, 104†, 254, 370
spoIVG	97	Stage IV sporulation	366, 370
spoVA	211	Stage V sporulation; contains five open reading frames, *spoVAA* through *spoVAE*	119*, 202, 370
spoVB	239	Stage V sporulation	104†, 202, 254
spoVC	7	Stage V sporulation	204*, 314†, 525
spoVD	133	Stage V sporulation; linked to *spoVE* by transformation	74, 202; F†
spoVE	133	Stage V sporulation	49*, 74, 202, 369†, 370, 514†
spoVF	148	Stage V sporulation; probably the same as *dpa*; mutants form octanol- and chloroform-resistant, heat-sensitive spores; form heat-resistant spores in presence of dipicolinic acid	20, 370, 372
spoVG	6	Stage V sporulation; previously called 0.4-kilobase gene; transcription turned on within 30 min of start of sporulation	161, 403, 420†
spoVH		Found to be an allele of *spoVA*	186; F
spoVJ	250	Stage V sporulation	104†, 186
spoVK	168	Defined by Tn*917* insertion	411
spoVL	240	Defined by Tn*917* insertion; considered to be distinct from *spoIVF* and *spoVB*	411
spoVM	145	Defined by Tn*917* insertion	411
spoVN	280	Defined by Tn*917* insertion; stage designation tentative pending electron microscopy	411

(*Continued on next page*)

Table 1—*Continued*

Gene symbol	Map position[a]	Phenotype, enzyme deficiency, or other characteristic	Reference[b]
spoVP	204	Defined by Tn917 insertion; relationship to *spoIVA* uncertain	411
spoVQ	213	Defined by Tn917 insertion; less closely linked to *lys* than is *spoVA*	411
spoVIA	255	Stage VI of sporulation; map order *argA-spoVIA-gerE-leuA*	221
spoVIB	247	Stage VI of sporulation	222
spoVIC	294	Stage VI of sporulation	220
sprA	NM	Derepression of homoserine kinase; homoserine dehydrogenase and the minor threonine dehydratase (*tdm*)	481
sprB		Partial suppression of isoleucine requirement; allows threonine dehydratase *sprA* mutants to grow in minimal medium; maps near *tdm* locus; see *tdm*	482
sprE		See *aprE*	
srf	29	Surfactin production	327†
srm	12	Modifies resistance of *spcA* strains	57
ssa	217	Alcohol-resistant sporulation; maps close to *spo0A*; *rvt* mutations have same phenotype	35, 428
sspA	266	One member of multigene family coding for small acid-soluble spore protein	71, 72*
sspB	65	Same as *sspA*	71, 72*
sspC	182	Same as *sspA*	71, 73*
sspD	121	Same as *sspA*	71, 72*
sspE	65	Same as *sspA*	160*
ssp-1		See *spe*	
std		Streptolydigin resistance; RNA polymerase; see *rpoC*	163, 446
strA		Streptomycin resistance; see *rpsL*	152
strB	130	Streptomycin resistance	447
strC	219	Streptomycin uptake deficient; possible cytochrome oxidase regulator	296, 447
suA20	145	Suppressor of *recH* mutations with increased ATP-dependent DNase	253
sub	NM	Structural gene for peptide antibiotic subtilin	23*
sul	9	Sulfanilamide resistance	210, 226, 295†
sup-3	29	Suppressor tRNA	180
sup-44	29	Suppressor tRNA	271
tag	309	Cell wall synthesis	43, 229, 230, 378
tal	240	Resistant to β-thienylalanine	70
tdm	285	Minor threonine dehydratase	482
ten	245	Constitutive transfection enhancement of SP82 DNA; transformation defective	J
terC	180	Terminus of replication	63*
tetB	3	Resistant to tetracycline	508
thfS	304	Tetrahydrofolate synthetase	N†
thiA	70	Thiamine requirement	233
thiB	105	Thiamine requirement	191
thiC	331	Thiamine requirement	BB
thrB	284	Threonine requirement; homoserine kinase; previously designated *thrA*	94, 355*, 435, 481
thrC	284	Threonine requirement; threonine synthase; previously called *thrB*	355*, 435
thyA	168	Thymidylate synthetase A	12, 333, 511
thyB	195	Thymidylate synthetase B	12, 217*, 333, 323†, 511

(*Continued on next page*)

Table 1—*Continued*

Gene symbol	Map position[a]	Phenotype, enzyme deficiency, or other characteristic	Reference[b]
til	225	Tilerone resistance	L
tkt	NM	Transketolase	414, 415
tmp	198	Trimethoprim resistance	490
tmrA	24	Tunicamycin resistance; hyperproductivity of extracellular α-amylase	337
tmrB	25	Tunicamycin resistance	337, 352†
tms-12	132	Temperature-sensitive cell division	75
tms-26	6	Temperature-sensitive cell division	75, 161†, 472
tolA		Tolerance to bacteriophage; see *abrB*	213, 214
tolB	NM	Tolerance to bacteriophage	213, 214
tre	62	Trehalase	265
trnA	0	Genes for Ile and Ala tRNAs located between 16S and 23S RNAs in the *rrnA* operon	343*, 486
trnB	275	Linked set of tRNA genes distal to *rrnB* that contains tRNAs for Val, Thr, Lys, Leu, Gly, Leu, Arg, Pro, Ala, Met, Ile, Ser, fMet, Asp, Phe, His, Gly, Ile/Met, Asn, Ser, and Glu; formerly *trnE*	156*, 486, 487†, 499*
trnI	15	Linked set of tRNA genes including tRNAs for Ala, Pro, Arg, Gly, Thr, and Asn located between *rrnH* and *rrnI*; formerly *trnH* and *trnB*	486, 498*
trnO	0	Same tRNA genes as *trnA* but located in *rrnO*	343*, 486
trnR	UC	Linked set of tRNA genes that contains tRNAs for Asn, Ser, Glu, Val, Met, Asp, Phe, Thr, Tyr, Trp, His, Gln, Gly, Cys, Leu, and Leu; formerly called *trnD*	486, 499*
trnY	NM	Linked set of tRNA genes for Lys, Glu, Asp, and Phe	486, 515*
trpA	203	Tryptophan synthase α	11, 179*, 197, 502
trpB	203	Tryptophan synthase β	11, 179*, 197, 502
trpC	203	Indol-3-glycerol-phosphate synthase	11, 179*, 197, 502
trpD	203	Anthranilate phosphoribosyltransferase	11, 22*, 197, 502
trpE	203	Anthranilate synthase	11, 22*, 197, 502
trpF	203	Phosphoribosylanthranilate isomerase	179*, 197, 502
trpS	104	Tryptophanyl-tRNA synthase	449; Y*
trpX	9	Glutamine-binding protein common to anthranilate synthase and *p*-aminobenzoate synthase	226
trx	252	Thioredoxin	Q*
ts-1	145	Temperature-sensitive division	55
ts-31	131	Temperature-sensitive division	306
ts-39	226	Temperature-sensitive synthesis of phosphatidylethanolamine	269
ts-341	312	Temperature-sensitive division	306
ts-526	3	Temperature-sensitive division	306
tscB, -G, -H	28	Temperature-sensitive vegetative growth	112†
tsi	48	Temperature-sensitive induction of all known SOS functions	277, 434
tsp		Thiostrepton resistance; 50S subunit; maps in ribosomal protein cluster	362
tsr	324	Temperature-sensitive RNA synthesis	470*
tuf	10	Elongation factor Tu	95
tyrA	202	Tyrosine requirement; prephenate dehydrogenase	178*, 328, 331, 480†
tzm	205	Tetrazolium reaction; probably an allele of *gerJ*	476, 495

(*Continued on next page*)

Table 1—*Continued*

Gene symbol	Map position[a]	Phenotype, enzyme deficiency, or other characteristic	Reference[b]
udk	NM	Uridine kinase; also lacks cytidine kinase; mutant resistant to fluorouridine	332
udp	NM	Uridine phosphorylase	292†
unc	NM	Uncoupler resistant	158
upp	NM	Uracil phosphoribosyltransferase	332
urg	NM	N-Glycosidase	283
urs		Uracil sensitivity; arginine-specific carbamoyl-phosphate synthase; see *cpa*	283
uvrA	305	Excision of UV light-induced pyrimidine dimers in DNA	189, 265, 321
uvrB	252	Excision of UV light-induced pyrimidine dimers in DNA; homologous to *uvrC* of *E. coli*	321; Q*
vas		Valine sensitivity; maps within threonine dehydratase locus; see *ilvA*	262
VS	12	Virginiamycin (VS component) resistance	400
xhi	112	Heat-inducible prophage PBSX	51, 53
xlt	112	Induced PBSX bacteriophage lack tails	138
xpaA	277	Hydrolysis of 5-bromo-4-chloroindoyl phosphate	M†
xpaB	114	Hydrolysis of 5-bromo-4-chloroindoyl phosphate	M†
xpaC	2	Hydrolysis of 5-bromo-4-chloroindoyl phosphate	M†
xpt	198	Xanthine phosphoribosyltransferase	417
xylA	48	Xylose isomerase	171†, 505†, 506*
xylB	48	Xylose kinase	171†, 505†
xylR	48	Regulation of xylose regulon	171†
xynA	48	Extracellular β-xylanase	31†, 131*, 171†, 354*, 401
xynB	48	Cell-associated β-xylosidase	171†, 401
xynC	48	Probably xyloside permease	171†

[a]NM, Not mapped; UC, map position not fully defined.

[b]†, Locus cloned; *, locus sequenced. References include (A) D. Dubnau, this volume, and unpublished data from (B) C. Anagnostopoulos, (C) M. Bramucci, (D) H. de Lencastre, (E) P. Dhaese, (F) J. Errington, (G) E. Freese, (H) A. Galizzi, (I) P. Gay, (J) D. Green, (K) L. Hederstedt, (L) J. A. Hoch, (M) M. Hulett, (N) T. Leighton, (O) R. Losick, (P) A. Moir, (Q) H. Paulus, (R) P. Piggot, (S) M. Polsinelli, (T) C. Price, (U) A. L. Sonenshein, (V) M. Steinmetz, (W) P. Stragier, (X) H. Taber, (Y) J. T.-F. Wong, (Z) M. Young, (AA) P. Youngman, (BB) S. Zahler, and (CC) M. Zukowski.

[c]kDa, Kilodalton(s).

[d]The nomenclature of the *pur* loci used here is that revised by Ebbole and Zalkin to conform to the *E. coli* system (99).

Acknowledgments. I thank the many colleagues who have given me advice about the map. In addition to those who have also given unpublished information and are listed in Table 1, I thank D. Dubnau, Y. Fujita, J. Hageman, D. Henner, J. Kooistra, J. Lutkenhaus, D. McConnell, G. Rapoport, J. Szulmajster, T. Tanaka, B. Vold, G. Wake, R. Yasbin, H. Zalkin, and P. Zuber.

This work was supported in part by Public Health Service grant RO1 AI23045 from the National Institutes of Health.

LITERATURE CITED

1. **Adams, A.** 1971. A class of poorly transformable multi-site mutations in *Bacillus subtilis*, p. 418–428. *In* L. Ledoux (ed.), *Informative Molecules in Biological Systems.* Elsevier, North Holland, Amsterdam.

2. **Adams, A.** 1972. Transformation and transduction of a large deletion mutation in *Bacillus subtilis. Mol. Gen. Genet.* **118**:311–322.

3. **Adams, A.** 1973. Transposition of the arsenate resistance locus of *Bacillus subtilis. Genetics* **74**: 197–213.

4. **Adams, A., and M. Oishi.** 1972. Genetic properties of arsenate sensitive mutants of *Bacillus subtilis*

168. *Mol. Gen. Genet.* **118:**295–310.

5. **Aharonowitz, Y., and E. Z. Ron.** 1972. A temperature sensitive mutant in *Bacillus subtilis* with an altered elongation factor G. *Mol. Gen. Genet.* **119:**131–138.

6. **Akamatsu, T., and J. Sekiguichi.** 1983. Properties of regeneration mutants of *Bacillus subtilis*. *FEMS Microbiol. Lett.* **20:**425–428.

7. **Albertini, A. M., M. L. Baldi, E. Ferrari, E. Isnenghi, M. T. Zambelli, and A. Galizzi.** 1979. Mutants of *Bacillus subtilis* affected in spore outgrowth. *J. Gen. Microbiol.* **110:**351–363.

8. **Albertini, A. M., T. Caramori, D. Henner, E. Ferrari, and A. Galizzi.** 1987. Nucleotide sequence of the *outB* locus of *Bacillus subtilis* and regulation of its expression. *J. Bacteriol.* **169:**1480–1484.

9. **Albertini, A. M., and A. Galizzi.** 1975. Mutant of *Bacillus subtilis* with a temperature-sensitive lesion in ribonucleic acid synthesis during germination. *J. Bacteriol.* **124:**14–25.

10. **Alonso, J. C., R. H. Tailor, and G. Lüder.** 1988. Characterization of recombination-deficient mutants of *Bacillus subtilis*. *J. Bacteriol.* **170:**3001–3007.

11. **Anagnostopoulos, C., and I. P. Crawford.** 1961. Transformation studies on the linkage of markers in the tryptophan pathway in *Bacillus subtilis*. *Proc. Natl. Acad. Sci. USA* **47:**378–390.

12. **Anagnostopoulos, C., and A. M. Schneider-Champagne.** 1966. Determinisme génétique de l'éxigence en thymine chez certains mutants de *Bacillus subtilis*. *C.R. Acad. Sci. Ser. D* **262:**1311–1314.

13. **Anaguchi, H., S. Fukui, H. Shimotsu, F. Kawamura, H. Saito, and Y. Kobayashi.** 1984. Cloning of sporulation gene *spoIIC* in *Bacillus subtilis*. *J. Gen. Microbiol.* **130:**757–760.

14. **Anderson, J. J., and A. T. Ganesan.** 1975. Temperature-sensitive deoxyribonucleic acid replication in a *dnaC* mutant of *Bacillus subtilis*. *J. Bacteriol.* **121:**173–183.

15. **Anderson, L. M., T. M. Henkin, G. H. Chambliss, and K. F. Bott.** 1984. New chloramphenicol resistance locus in *Bacillus subtilis*. *J. Bacteriol.* **158:**386–388.

16. **Andrup, L., T. Atlung, N. Ogasawara, H. Yoshikawa, and F. G. Hansen.** 1988. Interaction of the *Bacillus subtilis* DnaA-like protein with the *Escherichia coli* DnaA protein. *J. Bacteriol.* **170:**1333–1338.

17. **Attolini, C., G. Mazza, A. Fortunato, G. Ciarrochi, G. Mastromei, S. Riva, and A. Falaschi.** 1976. On the identity of *dnaP* and *dnaF* genes of *Bacillus subtilis*. *Mol. Gen. Genet.* **148:**9–17.

18. **Aubert, E., A. Klier, and G. Rapoport.** 1985. Cloning and expression in *Escherichia coli* of the regulatory *sacU* gene from *Bacillus subtilis*. *J. Bacteriol.* **161:**1182–1187.

19. **Aymerich, S., and M. Steinmetz.** 1987. Cloning and preliminary characterization of the *sacS* locus from *Bacillus subtilis* which controls the regulation of the exoenzyme levansucrase. *Mol. Gen. Genet.* **208:**114–120.

20. **Balassa, G., P. Milhaud, E. Raulet, M. T. Silva, and J. C. F. Sousa.** 1979. A *Bacillus subtilis* mutant requiring dipicolinic acid for the development of heat-resistant spores. *J. Gen. Microbiol.* **110:**365–379.

21. **Balassa, G., P. Milhaud, J. C. F. Sousa, and M. T. Silva.** 1979. Decadent sporulation mutants of *Bacillus subtilis*. *J. Gen. Microbiol.* **110:**381–392.

22. **Band, L., H. Shimotsu, and D. J. Henner.** 1984. Nucleotide sequence of the *Bacillus subtilis trpE* and *trpD* genes. *Gene* **27:**55–65.

23. **Banerjee, S., and J. N. Hansen.** 1988. Structure and expression of a gene encoding the precursor of subtilin, a small protein antibiotic. *J. Biol. Chem.* **263:**9508–9514.

24. **Barat, M., C. Anagnostopoulos, and A.-M. Schneider.** 1965. Linkage relationships of genes controlling isoleucine, valine, and leucine biosynthesis in *Bacillus subtilis*. *J. Bacteriol.* **90:**357–369.

25. **Baumberg, S., and A. Mountain.** 1984. *Bacillus subtilis* 168 mutants resistant to arginine hydroxamate in the presence of ornithine or citrulline. *J. Gen. Microbiol.* **130:**1247–1252.

26. **Bazill, G. W., and D. Karamata.** 1972. Temperature-sensitive mutants of *Bacillus subtilis* defective in deoxyribonucleotide synthesis. *Mol. Gen. Genet.* **117:**19–29.

27. **Bazzicalupo, M., E. Gallori, and M. Polsinelli.** 1980. Characterization of 5-fluoroindole and 5-fluorotryptophan resistant mutants in *Bacillus subtilis*. *Microbiologica* **3:**15–23.

28. **Bazzicalupo, M., B. Parisi, G. Pirali, M. Polsinelli, and F. Sala.** 1975. Genetic and biochemical characterization of a ribosomal mutant of *Bacillus subtilis* resistant to sporangiomycin. *Antimicrob. Agents Chemother.* **8:**651–656.

29. **Beall, B., M. Lowe, and J. Lutkenhaus.** 1988. Cloning and characterization of the *Bacillus subtilis* homologs of *Escherichia coli* cell division genes *ftsZ* and *ftsA*. *J. Bacteriol.* **170:**4855–4864.

30. **Berek, T., A. Miczak, and G. Ivanovics.** 1974. Mapping of the delta-aminolaevulinic acid dehydrase and porphobilinogen deaminase loci in *Bacillus subtilis*. *Mol. Gen. Genet.* **132:**233–239.

31. **Bernier, R., Jr., R. Driguez, and M. Desrochers.** 1983. Molecular cloning of a *Bacillus subtilis* xylanase gene. *Gene* **26:**59–65.

32. **Binnie, C., M. Lampe, and R. Losick.** 1986. Gene encoding the σ^{37} species of RNA polymerase σ factor from *Bacillus subtilis*. *Proc. Natl. Acad. Sci. USA* **83:**5943–5947.

33. **Bishop, P. E., and L. R. Brown.** 1973. Ethidium bromide-resistant mutant of *Bacillus subtilis*. *J. Bacteriol.* **115**:1077–1083.

34. **Bohannon, D. E., M. S. Rosenkrantz, and A. L. Sonenshein.** 1985. Regulation of *Bacillus subtilis* glutamate synthase genes by the nitrogen source. *J. Bacteriol.* **163**:957–964.

35. **Bohin, J.-P., and B. Lubochinsky.** 1982. Alcohol-resistant sporulation mutants of *Bacillus subtilis*. *J. Bacteriol.* **150**:944–955.

36. **Bonamy, C., and J. Szulmajster.** 1982. Cloning and expression of *Bacillus subtilis* spore genes. *Mol. Gen. Genet.* **188**:202–210.

37. **Borenstein, S., and E. Ephrati-Elizur.** 1969. Spontaneous release of DNA in sequential genetic order by *Bacillus subtilis*. *J. Mol. Biol.* **45**:137–152.

38. **Borriss, R., K.-H. Süss, M. Süss, R. Manteuffel, and J. Hofemeister.** 1986. Mapping and properties of *bgl* (β-glucanase) mutants of *Bacillus subtilis*. *J. Gen. Microbiol.* **132**:431–442.

39. **Bott, K. F., G. Reysset, J. Gregoire, D. Islert, and J.-P. Aubert.** 1977. Characterization of glutamine requiring mutants of *Bacillus subtilis*. *Biochem. Biophys. Res. Commun.* **79**:996–1003.

40. **Bott, K. F., G. C. Stewart, and A. G. Anderson.** 1984. Genetic mapping of cloned ribosomal genes, p. 19–34. *In* J. A. Hoch and A. T. Ganesan (ed.), *SYNTRO Conference on Genetics and Biotechnology of Bacilli*. Academic Press, Inc., New York.

41. **Boudreaux, D. P., E. Eisenstadt, T. Iijima, and E. Freese.** 1981. Biochemical and genetic characterization of an auxotroph of *Bacillus subtilis* altered in the acyl-CoA: acyl-carrier protein transacylase. *Eur. J. Biochem.* **115**:175–181.

42. **Bouvier, J., P. Stragier, C. Bonamy, and J. Szulmajster.** 1985. Nucleotide sequence of the *spo0B* gene of *Bacillus subtilis* and regulation of its expression. *Proc. Natl. Acad. Sci. USA* **81**:7012–7016.

43. **Boylan, R. J., N. H. Mendelsohn, D. Brooks, and F. E. Young.** 1972. Regulation of the bacterial cell wall: analysis of a mutant of *Bacillus subtilis* defective in biosynthesis of teichoic acid. *J. Bacteriol.* **110**:281–290.

44. **Bramucci, M. G., K. M. Keggins, and P. S. Lovett.** 1977. Bacteriophage PMB12 conversion of the sporulation defect in RNA polymerase mutants of *Bacillus subtilis*. *J. Virol.* **24**:194–200.

45. **Brandt, C., and D. Karamata.** 1987. Thermosensitive *Bacillus subtilis* mutants which lyse at the non-permissive temperature. *J. Gen. Microbiol.* **133**:1159–1170.

46. **Bresler, S. E., D. A. Perumov, A. P. Slivortzova, T. P. Chernik, and T. N. Sherchenka.** 1975. Study of the riboflavin operon in *Bacillus subtilis*. *Genetica* (USSR) **11**:95–100.

47. **Bron, S., and K. Murray.** 1975. Restriction and modification in *B. subtilis*. Nucleotide sequence recognized by restriction endonuclease R. *Bsu*R from strain R. *Mol. Gen. Genet.* **143**:25–33.

48. **Buchanan, C. E.** 1987. Absence of penicillin-binding protein 4 from an apparently normal strain of *Bacillus subtilis*. *J. Bacteriol.* **169**:5301–5303.

49. **Bugaichuk, U. D., and P. J. Piggot.** 1986. Nucleotide sequence of the *Bacillus subtilis* developmental gene *spoVE*. *J. Gen. Microbiol.* **132**:1883–1890.

50. **Butler, P. D., and J. Mandelstam.** 1987. Nucleotide sequence of the sporulation operon, *spoIIIE*, of *Bacillus subtilis*. *J. Gen. Microbiol.* **133**:2359–2370.

51. **Buxton, R. S.** 1976. Prophage mutation causing heat inducibility of defective *Bacillus subtilis* bacteriophage PBSX. *J. Virol.* **20**:22–28.

52. **Buxton, R. S.** 1978. A heat-sensitive lysis mutant of *Bacillus subtilis* 168 with a low activity of pyruvate carboxylase. *J. Gen. Microbiol.* **105**:175–185.

53. **Buxton, R. S.** 1980. Selection of *Bacillus subtilis* 168 mutants with deletions of PBSX prophage. *J. Gen. Virol.* **46**:427–437.

54. **Buxton, R. S., and J. B. Ward.** 1980. Heat-sensitive lysis mutants of *Bacillus subtilis* 168 blocked at three different stages of peptidoglycan synthesis. *J. Gen. Microbiol.* **120**:283–293.

55. **Callister, H., and R. G. Wake.** 1981. Characterization and mapping of temperature-sensitive division initiation mutations of *Bacillus subtilis*. *J. Bacteriol.* **145**:1042–1051.

56. **Cannon, J. G., and K. Bott.** 1979. Spectinomycin-resistant mutants of *Bacillus subtilis* with altered sporulation properties. *Mol. Gen. Genet.* **174**:149–162.

57. **Cannon, J. G., and K. F. Bott.** 1980. Mutation affecting expression of spectinomycin resistance in *Bacillus subtilis*. *J. Bacteriol.* **141**:409–412.

58. **Canosi, U., M. Nolli, E. Ferrari, R. Marinone, and G. Mazza.** 1979. Genetic mapping of caffeine resistant and sensitive mutants of *B. subtilis*. *Microbiologica* **2**:167–172.

59. **Canosi, U., A. G. Siccardi, A. Falaschi, and G. Mazza.** 1976. Effect of deoxyribonucleic acid replication inhibitors on bacterial recombination. *J. Bacteriol.* **126**:108–121.

60. **Cantwell, B., and D. J. McConnell.** 1983. Molecular cloning and expression of a *B. subtilis* glucanase gene in *Escherichia coli*. *Gene* **23**:211–219.

61. **Carlsson, P., and L. Hederstedt.** 1987. *Bacillus subtilis citM*, the structural gene for dihydrolipoamidetranssuccinylase: cloning and expression in *E. coli*. *Gene* **61**:217–224.

62. **Carrascosa, J. L., J. A. Garcia, and M. Salas.** 1982. A protein similar to *Escherichia coli groEL* is present in *Bacillus subtilis*. *J. Mol. Biol.* **148**:731–737.

63. **Carrigan, C. M., J. A. Haarsma, M. T. Smith, and R. G. Wake.** 1987. Sequence features of the replication terminus of the *Bacillus subtilis* chromo-

some. *Nucleic Acids Res.* **15**:8501–8509.

64. **Chalumeau, H., A. Delobbe, and P. Gay.** 1978. Biochemical and genetic study of D-glucitol transport and catabolism in *Bacillus subtilis. J. Bacteriol.* **134**:920–928.

65. **Champney, W. S., and R. A. Jensen.** 1969. D-Tyrosine as a metabolic inhibitor of *Bacillus subtilis. J. Bacteriol.* **98**:205–214.

66. **Chasin, L. A., and B. Magasanik.** 1968. Induction and repression of the histidine-degrading enzymes of *Bacillus subtilis. J. Biol. Chem.* **243**:5165–5178.

67. **Chaudhry, G. R., Y. S. Halpern, C. Saunders, N. Vasantha, B. J. Schmidt, and E. Freese.** 1984. Mapping of the glucose dehydrogenase gene in *Bacillus subtilis. J. Bacteriol.* **160**:607–611.

68. **Chen, N.-Y., F.-M. Hu, and H. Paulus.** 1987. Nucleotide sequence of the overlapping genes for the subunits of *Bacillus subtilis* aspartokinase II and their control regions. *J. Biol. Chem.* **262**:8787–8798.

69. **Clarke, S., and J. Mandelstam.** 1987. Regulation of Stage II of sporulation in *Bacillus subtilis. J. Gen. Microbiol.* **133**:2371–2380.

70. **Coats, J. H., and E. W. Nester.** 1967. Regulation reversal mutation: characterization of end product-activated mutants of *Bacillus subtilis. J. Biol. Chem.* **242**:4948–4955.

71. **Connors, M. J., S. Howard, J. Hoch, and P. Setlow.** 1986. Determination of chromosomal locations of four *Bacillus subtilis* genes which code for a family of small, acid-soluble spore proteins. *J. Bacteriol.* **166**:412–416.

72. **Connors, M. J., J. M. Mason, and P. Setlow.** 1986. Cloning and nucleotide sequencing of genes for three small, acid-soluble proteins from *Bacillus subtilis* spores. *J. Bacteriol.* **166**:417–425.

73. **Connors, M. J., and P. Setlow.** 1985. Cloning of a small, acid-soluble spore protein gene from *Bacillus subtilis* and determination of its complete nucleotide sequence. *J. Bacteriol.* **161**:333–339.

74. **Coote, J. G.** 1972. Sporulation in *Bacillus subtilis.* Characterization of oligosporogenous mutants and comparison of their phenotypes with those of asporogenous mutants. *J. Gen. Microbiol.* **71**:1–15.

75. **Copeland, J. C., and J. Marmur.** 1968. Identification of conserved genetic functions in *Bacillus* by use of temperature-sensitive mutants. *Bacteriol. Rev.* **32**:302–312.

76. **Cozzarelli, N. R., and R. L. Low.** 1973. Mutational alteration of *Bacillus subtilis* DNA polymerase III to hydroxy-phenylazopyrimidine resistance: polymerase III is necessary for DNA replication. *Biochem. Biophys. Res. Commun.* **51**:151–159.

77. **Cutting, S., and J. Mandelstam.** 1986. The nucleotide sequence and the transcription during sporulation of the *gerE* gene of *Bacillus subtilis. J. Gen. Microbiol.* **132**:3013–3124.

78. **Dabbs, E. R.** 1983. A pair of *Bacillus subtilis* ribosomal protein genes mapping outside the principal ribosomal protein cluster. *J. Bacteriol.* **156**:966–969.

79. **Dabbs, E. R.** 1983. Mapping of the genes for *Bacillus subtilis* ribosomal proteins S9, S11 and BL27 by means of antibiotic resistant mutants. *Mol. Gen. Genet.* **191**:295–300.

80. **Dabbs, E. R.** 1983. Arrangement of loci within the principal cluster of ribosomal protein genes of *Bacillus subtilis. Mol. Gen. Genet.* **192**:124–130.

81. **Dabbs, E. R.** 1983. Mapping of the genes for *Bacillus subtilis* ribosomal proteins S6 and S16: comparison of the chromosomal distribution of ribosomal protein genes in this bacterium with the distribution in *Escherichia coli. Mol. Gen. Genet.* **192**:386–390.

82. **Dabbs, E. R.** 1984. Order of ribosomal protein genes in the *rif* cluster of *Bacillus subtilis* is identical to that of *Escherichia coli. J. Bacteriol.* **159**:770–772.

83. **Dancer, B. N., and J. Mandelstam.** 1981. Complementation of sporulation mutations in fused protoplasts of *Bacillus subtilis. J. Gen. Microbiol.* **123**:17–26.

84. **Dean, D. R., J. A. Hoch, and A. I. Aronson.** 1977. Alteration of the *Bacillus subtilis* glutamine synthetase results in overproduction of the enzyme. *J. Bacteriol.* **131**:981–987.

85. **Débarbouillé, M., F. Kunst, A. Klier, and G. Rapoport.** 1987. Cloning of the *sacS* gene encoding a positive regulator of the sucrose regulon in *Bacillus subtilis. FEMS Microbiol. Lett.* **41**:137–140.

86. **Dedonder, R. A., J.-A. Lepesant, J. Lepesant-Kejzlarova, A. Billault, M. Steinmetz, and F. Kunst.** 1977. Construction of a kit of reference strains for rapid genetic mapping in *Bacillus subtilis* 168. *Appl. Environ. Microbiol.* **33**:989–993.

87. **deVos, W. M., and G. Venema.** 1983. Transformation of *Bacillus subtilis* competent cells: identification and regulation of the *recE* gene product. *Mol. Gen. Genet.* **190**:56–64.

88. **Diderichsen, B.** 1986. A genetic system for stabilization of cloned genes in *Bacillus subtilis*, p. 35–46. *In* A. T. Ganesan and J. A. Hoch (ed.), *Bacillus Molecular Genetics and Biotechnology Applications.* Academic Press, Inc., Orlando, Fla.

89. **Dingman, D. W., and A. L. Sonenshein.** 1987. Purification of aconitase from *Bacillus subtilis* and correlation of its N-terminal amino acid sequence with the sequence of the *citB* gene. *J. Bacteriol.* **169**:3062–3067.

90. **Dod, B., G. Balassa, E. Raulet, and V. Jeannoda.** 1978. Spore control (*Sco*) mutations in *Bacillus*

subtilis. II. Sporulation and the production of extracellular proteases and alpha-amylases by *Sco* mutants. *Mol. Gen. Genet.* **163**:45–56.

91. Doly, J., D. Le Roscouet, and C. Anagnostopoulos. 1974. ATP-dependent deoxyribonuclease in *Bacillus subtilis* and a mutant deficient in this activity. *Mutat. Res.* **22**:15–23.

92. Donovan, W., L. Zheng, K. Sandman, and R. Losick. 1987. Genes encoding spore coat polypeptides from *Bacillus subtilis*. *J. Mol. Biol.* **196**:1–10.

93. Dubnau, D., and C. Cirigliano. 1974. Genetic characterization of recombination-deficient mutants of *Bacillus subtilis*. *J. Bacteriol.* **117**:488–493.

94. Dubnau, D., C. Goldthwaite, I. Smith, and J. Marmur. 1967. Genetic mapping in *Bacillus subtilis*. *J. Mol. Biol.* **27**:163–185.

95. Dubnau, E., S. Pifko, A. Sloma, K. Cabane, and I. Smith. 1976. Conditional mutations in the translational apparatus of *Bacillus subtilis*. *Mol. Gen. Genet.* **147**:1–12.

96. Dubnau, E., J. Weir, G. Nair, L. Carter, III, C. Moran, Jr., and I. Smith. 1988. *Bacillus* sporulation gene *spo0H* codes for σ^{30} (σH). *J. Bacteriol.* **170**:1054–1062.

97. Dul, M. J., and F. E. Young. 1973. Genetic mapping of a mutant defective in D,L-alanine racemase in *Bacillus subtilis* 168. *J. Bacteriol.* **115**:1212–1214.

98. Duncan, M. L., S. S. Kalman, S. M. Thomas, and C. W. Price. 1987. Gene encoding the 37,000 dalton minor sigma factor of *Bacillus subtilis* RNA polymerase: isolation, nucleotide sequence, chromosomal locus, and cryptic function. *J. Bacteriol.* **169**:771–778.

99. Ebbole, D. J., and H. Zalkin. 1987. Cloning and characterization of a 12-gene cluster from *Bacillus subtilis* encoding nine enzymes for *de novo* purine nucleotide synthesis. *J. Biol. Chem.* **262**:8274–8287.

100. Endo, T., H. Ishikawa, and E. Freese. 1983. Properties of a *Bacillus subtilis* mutant able to sporulate continually during during growth in synthetic medium. *J. Gen. Microbiol.* **129**:17–30.

101. Endo, T., B. Uratani, and E. Freese. 1983. Purine salvage pathways of *Bacillus subtilis* and effect of guanine on growth of GMP reductase mutants. *J. Bacteriol.* **155**:169–179.

102. Ephrati-Elizur, E., P. R. Srinivasan, and S. Zamenhof. 1961. Genetic analysis by means of transformation of histidine linkage groups in *Bacillus subtilis*. *Proc. Natl. Acad. Sci. USA* **47**:56–63.

103. Errington, J., P. Fort, and J. Mandelstam. 1985. Duplicated sporulation genes in bacteria. *FEBS Lett.* **188**:184–188.

104. Errington, J., and D. Jones. 1987. Cloning in *Bacillus subtilis* by transfection with bacteriophage

vector φ105J27: isolation and preliminary characterization of transducing phages for 23 sporulation loci. *J. Gen. Microbiol.* **133**:493–502.

105. Errington, J., S. Rong, M. S. Rosenkrantz, and A. L. Sonenshein. 1988. Transcriptional regulation and structure of the *Bacillus subtilis* sporulation locus *spoIIIC*. *J. Bacteriol.* **170**:1162–1167.

106. Estrela, A. I., H. de Lencastre, and L. J. Archer. 1986. Resistance of a *Bacillus subtilis* mutant to a group of temperate bacteriophages. *J. Gen. Microbiol.* **132**:411–415.

107. Fani, R., G. Mastromei, and M. Polsinelli. 1984. Isolation and characterization of *Bacillus subtilis* mutants altered in competence. *J. Bacteriol.* **157**:152–157.

108. Feavers, I. M., J. S. Miles, and A. Moir. 1985. The nucleotide sequence of spore germination gene (*gerA*) of *Bacillus subtilis* 168. *Gene* **38**:95–102.

109. Fein, J. E., and H. J. Rogers. 1976. Autolytic enzyme-deficient mutants of *Bacillus subtilis* 168. *J. Bacteriol.* **127**:1427–1442.

110. Ferrari, E., D. J. Henner, and J. A. Hoch. 1981. Isolation of *Bacillus subtilis* genes from a Charon 4A library. *J. Bacteriol.* **146**:430–432.

111. Ferrari, E., D. J. Henner, and M. Y. Yang. 1985. Isolation of an alanine racemase gene from *Bacillus subtilis* and its use for plasmid maintenance in *B. subtilis*. *Bio/Technology* **3**:1003–1007.

112. Ferrari, E., F. Scoffone, G. Ciarrocchi, and A. Galizzi. 1985. Molecular cloning of a *Bacillus subtilis* gene involved in spore outgrowth. *J. Gen. Microbiol.* **131**:2831–2838.

113. Ferrari, F. A., D. Lang, E. Ferrari, and J. A. Hoch. 1982. Molecular cloning of the *spo0B* sporulation locus in bacteriophage lambda. *J. Bacteriol.* **152**:809–814.

114. Ferrari, F. A., K. Trach, and J. A. Hoch. 1985. Sequence analysis of the *spo0B* locus reveals a polycistronic transcription unit. *J. Bacteriol.* **161**:556–562.

115. Ferrari, F. A., K. Trach, D. LeCoq, J. Spence, E. Ferrari, and J. A. Hoch. 1985. Characterization of the *spo0A* locus and its product. *Proc. Natl. Acad. Sci. USA* **82**:2647–2651.

116. Fisher, S. H., and B. Magasanik. 1984. Synthesis of oxaloacetate in *Bacillus subtilis* mutants lacking the 2-ketoglutarate dehydrogenase enzymatic complex. *J. Bacteriol.* **158**:55–62.

117. Fisher, S. H., M. S. Rosenkrantz, and A. L. Sonenshein. 1984. Glutamine synthetase gene of *Bacillus subtilis*. *Gene* **32**:427–438.

118. Fisher, S. H., and A. L. Sonenshein. 1977. Glutamine-requiring mutants of *Bacillus subtilis*. *Biochem. Biophys. Res. Commun.* **79**:987–995.

119. Fort, P., and J. Errington. 1985. Nucleotide sequence and complementation analysis of a poly-

cistronic sporulation operon, *spoVA*, in *Bacillus subtilis*. *J. Gen. Microbiol.* **131**:1091–1105.

120. Fort, P., and P. J. Piggot. 1984. Nucleotide sequence of sporulation locus *spoIIA* in *Bacillus subtilis*. *J. Gen. Microbiol.* **130**:2147–2153.

121. Fouet, A., M. Arnaud, A. Klier, and G. Rapoport. 1987. *Bacillus subtilis* sucrose-specific enzyme II of the phosphotransferase system: expression in *Escherichia coli* and homology to enzymes II from enteric bacteria. *Proc. Natl. Acad. Sci. USA* **84**:8773–8777.

122. Fouet, A., A. Klier, and G. Rapoport. 1986. Nucleotide sequence of the sucrase gene of *Bacillus subtilis*. *Gene* **45**:221–225.

123. Freese, E., Y. K. Oh, E. B. Freese, M. D. Diesterhaft, and C. Prasad. 1972. Suppression of sporulation of *Bacillus subtilis*, p. 212–221. *In* H. O. Halvorson, R. Hanson, and L. L. Campbell (ed.), *Spores V*. American Society for Microbiology, Washington, D.C.

124. Freese, E., S. W. Park, and M. Cashel. 1964. The developmental significance of alanine dehydrogenase in *Bacillus subtilis*. *Proc. Natl. Acad. Sci. USA* **51**:1164–1172.

125. Freese, E. B., R. M. Cole, W. Klofat, and E. Freese. 1970. Growth, sporulation, and enzyme defects of glucosamine mutants of *Bacillus subtilis*. *J. Bacteriol.* **101**:1046–1062.

126. Fujita, M., and Y. Kobayashi. 1985. Cloning of sporulation gene *spoIVC* in *Bacillus subtilis*. *Mol. Gen. Genet.* **199**:471–475.

127. Fujita, Y., and T. Fujita. 1983. Genetic analysis of a pleiotropic deletion mutation (Δ*igf*) in *Bacillus subtilis*. *J. Bacteriol.* **154**:864–869.

128. Fujita, Y., and T. Fujita. 1987. The gluconate operon *gnt* of *Bacillus subtilis* encodes its own transcriptional negative regulator. *Proc. Natl. Acad. Sci. USA* **84**:4524–4528.

129. Fujita, Y., T. Fujita, Y. Miwa, J. Nihashi, and Y. Aratani. 1986. Organization and transcription of the gluconate operon. *J. Biol. Chem.* **261**:13744–13753.

130. Fujita, Y., J.-I. Nihashi, and T. Fujita. 1985. The organization and cloning of a gluconate (*gnt*) operon of *Bacillus subtilis*, p. 203–208. *In* J. A. Hoch and P. Setlow (ed.), *Molecular Biology of Microbial Differentiation*. American Society for Microbiology, Washington, D.C.

131. Fukusaki, E., W. Panbangred, A. Shinmyo, and H. Okada. 1984. The complete nucleotide sequence of the xylanase gene (*xynA*) of *Bacillus pumilus*. *FEBS Lett.* **171**:197–201.

132. Galizzi, A., A. M. Albertini, P. Plevan, and G. Cassani. 1976. Synthesis of RNA and protein in a mutant of *Bacillus subtilis* temperature sensitive during spore germination. *Mol. Gen. Genet.* **148**:159–164.

133. Galizzi, A., F. Gorrini, A. Rollier, and M. Polsinelli. 1973. Mutants of *Bacillus subtilis* temperature sensitive in the outgrowth phase of spore germination. *J. Bacteriol.* **113**:1482–1490.

134. Galizzi, A., A. G. Siccardi, A. M. Albertini, A. R. Amileni, G. Meneguzzi, and M. Polsinelli. 1975. Properties of *Bacillus subtilis* mutants temperature sensitive in germination. *J. Bacteriol.* **121**:450–454.

135. Gallori, E., M. Bazzicalupo, B. Parisi, G. Pedaggi, and M. Polsinelli. 1978. Resistance to (L)-azetidine-2-carboxylic acid in *Bacillus subtilis*. *Biochem. Biophys. Res. Commun.* **85**:1518–1525.

136. Gallori, E., and R. Fani. 1983. Characterization of D-cycloserine resistant mutants in Bacillus subtilis. *Microbiologica* **6**:19–26.

137. Gardner, A. L., and A. I. Aronson. 1984. Expression of the *Bacillus subtilis* glutamine synthetase gene in *Escherichia coli*. *J. Bacteriol.* **158**:967–971.

138. Garro, A. J., H. Leffert, and J. Marmur. 1970. Genetic mapping of a defective bacteriophage on the chromosome of *Bacillus subtilis* 168. *J. Virol.* **6**:340–343.

139. Garro, A. J., C. Sprouse, and J. G. Wetmur. 1976. Association of the recombination-deficient phenotype of *Bacillus subtilis recC* strains with the presence of an SPO2 prophage. *J. Bacteriol.* **126**:556–558.

140. Gass, K. B., and N. R. Cozzarelli. 1973. Further genetic and enzymological characterization of the three *Bacillus subtilis* deoxyribonucleic acid polymerases. *J. Biol. Chem.* **248**:7688–7700.

141. Gaur, N. K., K. Cabane, and I. Smith. 1988. Structure and expression of the *Bacillus subtilis sin* operon. *J. Bacteriol.* **170**:1046–1053.

142. Gaur, N. K., E. Dubnau, and I. Smith. 1986. Characterization of a cloned *Bacillus subtilis* gene that inhibits sporulation in multiple copies. *J. Bacteriol.* **169**:860–869.

143. Gay, P., H. Chalumeau, and M. Steinmetz. 1983. Chromosomal localizations of *gut*, *fruC*, and *pfk* mutations affecting genes involved in *Bacillus subtilis* D-glucitol catabolism. *J. Bacteriol.* **153**:1133–1137.

144. Gay, P., P. Cordier, M. Marquet, and A. Delobbe. 1973. Carbohydrate metabolism and transport in *Bacillus subtilis*. A study of *ctr* mutations. *Mol. Gen. Genet.* **121**:355–368.

145. Gay, P., and A. Delobbe. 1977. Fructose transport in *Bacillus subtilis*. *Eur. J. Biochem.* **79**:363–373.

146. Gay, P., D. LeCoq, M. Steinmetz, E. Ferrari, and J. Hoch. 1983. Cloning structural gene *sacB*, which codes for exoenzyme levansucrase of *Bacillus subtilis*: expression of the gene in *Escherichia coli*. *J. Bacteriol.* **153**:1424–1431.

147. Gianni, M., and A. Galizzi. 1986. Isolation of

genes preferentially expressed during *Bacillus subtilis* spore outgrowth. *J. Bacteriol.* **165**:123–132.

148. Gillespie, K., and R. E. Yasbin. 1987. Chromosomal locations of three *Bacillus subtilis din* genes. *J. Bacteriol.* **169**:3372–3374.

149. Gilman, M. Z., and M. J. Chamberlin. 1983. Developmental and genetic regulation of *Bacillus subtilis* genes transcribed by sigma-28-RNA polymerase. *Cell* **35**:285–293.

150. Gitt, M. A., L.-F. Wang, and R. H. Doi. 1985. A strong sequence homology exists between the major RNA polymerase σ factors of *Bacillus subtilis* and *Escherichia coli. J. Biol. Chem.* **260**:7178–7185.

151. Goldstein, B. J., and S. A. Zahler. 1976. Uptake of branched-chain alpha-keto acids in *Bacillus subtilis. J. Bacteriol.* **127**:667–670.

152. Goldthwaite, C., D. Dubnau, and I. Smith. 1970. Genetic mapping of antibiotic resistance markers in *Bacillus subtilis. Proc. Natl. Acad. Sci. USA* **65**:96–103.

153. Goldthwaite, C., and I. Smith. 1972. Genetic mapping of amino-glycoside and fusidic acid resistant mutations in *Bacillus subtilis. Mol. Gen. Genet.* **114**:181–189.

154. Gonzy-Tréboul, G., and M. Steinmetz. 1987. Phosphoenolpyruvate:sugar phosphotransferase system of *Bacillus subtilis*: cloning of the region containing the *pstH* and *pstI* genes and evidence for a *crr*-like gene. *J. Bacteriol.* **169**:2287–2290.

155. Grant, G. F., and M. I. Simon. 1969. Synthesis of bacterial flagella. II. PBS1 transduction of flagella-specific markers in *Bacillus subtilis. J. Bacteriol.* **99**:116–124.

156. Green, C. J., G. C. Stewart, M. A. Hollis, B. S. Vold, and K. F. Bott. 1985. Nucleotide sequence of the *Bacillus subtilis* ribosomal RNA operon *rrnB. Gene* **37**:261–266.

157. Guespin-Michel, J. F. 1971. Phenotypic reversion in some early blocked sporulation mutants of *Bacillus subtilis*. Genetic studies of polymyxin resistant partial revertants. *Mol. Gen. Genet.* **112**:243–254.

158. Guffanti, A. A., S. Clejan, L. H. Falk, D. B. Hicks, and T. A. Krulwich. 1987. Isolation and characterization of uncoupler-resistant mutants of *Bacillus subtilis. J. Bacteriol.* **169**:4469–4478.

159. Guzmán, P., J. Westpheling, and P. Youngman. 1988. Characterization of the promoter region of the *Bacillus subtilis spoIIE* operon. *J. Bacteriol.* **170**:1598–1609.

160. Hackett, R. H., and P. Setlow. 1987. Cloning, nucleotide sequencing, and genetic mapping of the gene for small, acid-soluble spore-protein γ of *Bacillus subtilis. J. Bacteriol.* **169**:1985–1992.

161. Haldenwang, W. B., C. D. B. Banner, J. F. Ollington, R. Losick, J. A. Hoch, M. B. O'Connor,

and A. L. Sonenshein. 1980. Mapping of a cloned gene under sporulation control by insertion of a drug resistance marker into the *Bacillus subtilis* chromosome. *J. Bacteriol.* **142**:90–98.

162. Halling, S. M., K. C. Burtis, and R. H. Doi. 1977. Reconstitution studies show that rifampicin resistance is determined by the largest polypeptide of *Bacillus subtilis* RNA polymerase. *J. Biol. Chem.* **252**:9024–9031.

163. Halling, S. M., K. C. Burtis, and R. H. Doi. 1978. Beta′ subunit of bacterial RNA polymerase is responsible for streptolydigin resistance in *Bacillus subtilis. Nature* (London) **272**:837–839.

164. Hammer-Jespersen, K. 1983. Nucleoside catabolism, p. 203–258. *In* A. Munch-Petersen (ed.), *Metabolism of Nucleotides, Nucleosides, and Nucleobases in Microorganisms*. Academic Press, Inc., New York.

165. Hara, H., and H. Yoshikawa. 1973. Asymmetric bidirectional replication of *Bacillus subtilis* chromosome. *Nature* (London) *New Biol.* **244**:200–203.

166. Harford, N. 1974. Genetic analysis of *rec* mutants of *Bacillus subtilis*. Evidence for at least six linkage groups. *Mol. Gen. Genet.* **129**:269–274.

167. Harford, N., J. Lepesant-Kejzlarova, J.-A. Lepesant, R. Hamers, and R. Dedonder. 1976. Genetic circularity and mapping of the replication origin region of the *Bacillus subtilis* chromosome, p. 28–34. *In* D. Schlessinger (ed.), *Microbiology—1976*. American Society for Microbiology, Washington, D.C.

168. Harford, N., and N. Sueoka. 1970. Chromosomal location of antibiotic resistance markers in *Bacillus subtilis. J. Mol. Biol.* **51**:267–286.

169. Harwood, C. R., and S. Baumberg. 1977. Arginine hydroxamate-resistant mutants of *Bacillus subtilis* with altered control of arginine metabolism. *J. Gen. Microbiol.* **100**:177–188.

170. Hasnain, S., R. Sammons, I. Roberts, and C. M. Thomas. 1985. Cloning and deletion analysis of a genomic segment of *Bacillus subtilis* coding for *sdhA, B, C* (succinate dehydrogenase) and *gerE* (spore germination) loci. *J. Gen. Microbiol.* **131**:2269–2279.

171. Hastrup, S. 1988. Analysis of the *Bacillus subtilis* xylose regulon, p. 79–83. *In* A. T. Ganesan and J. A. Hoch (ed.), *Genetics and Biotechnology of Bacilli*. Academic Press, Inc., San Diego, Calif.

172. Hederstedt, L., K. Magnusson, and L. Rutberg. 1982. Reconstitution of succinate dehydrogenase in *Bacillus subtilis* by protoplast fusion. *J. Bacteriol.* **152**:157–165.

173. Helmann, J. D., L. M. Márquez, and M. J. Chamberlin. 1988. Cloning, sequencing, and disruption of the *Bacillus subtilis* σ^{28} gene. *J. Bacteriol.* **170**:1568–1574.

174. Hemphill, E. H., I. Gage, S. A. Zahler, and R. Z. Korman. 1980. Prophage mediated production of bacteriocin-like substance by SP-beta lysogens of *Bacillus subtilis*. *Can. J. Microbiol.* **26:**1328–1333.

175. Henkin, T. M., K. M. Campbell, and G. H. Chambliss. 1979. Spectinomycin dependence in *Bacillus subtilis*. *J. Bacteriol.* **137:**1452–1455.

176. Henkin, T. M., and G. H. Chambliss. 1984. Genetic analysis of a streptomycin-resistant oligosporogenous *Bacillus subtilis* mutant. *J. Bacteriol.* **157:**202–210.

177. Henkin, T. M., and G. H. Chambliss. 1984. Genetic mapping of a mutation causing an alteration in *Bacillus subtilis* ribosomal protein S4. *Mol. Gen. Genet.* **193:**364–369.

178. Henner, D. J., L. Band, G. Flaggs, and E. Chen. 1986. The organization and nucleotide sequences of the *Bacillus subtilis hisH*, *tyrA*, and *aroE* genes. *Gene* **49:**147–152.

179. Henner, D. J., L. Band, and H. Shimotsu. 1985. Nucleotide sequence of the *Bacillus subtilis* tryptophan operon. *Gene* **34:**169–177.

180. Henner, D. J., and W. Steinberg. 1979. Genetic location of the *Bacillus subtilis sup-3* suppressor mutation. *J. Bacteriol.* **139:**668–670.

181. Henner, D. J., M. Yang, L. Band, and J. A. Wells. 1985. Characterization of cloned genes and their regulatory signals. Expression of cloned protease genes in *Bacillus subtilis*, p. 95–103. *In* J. A. Hoch and P. Setlow (ed.), *Molecular Biology of Microbial Differentiation*. American Society for Microbiology, Washington, D.C.

182. Henner, D. J., M. Yang, and E. Ferrari. 1988. Localization of *Bacillus subtilis sacU*(Hy) mutations to two linked genes with similarities to the conserved procaryotic family of two-component signalling systems. *J. Bacteriol.* **170:**5102–5109.

183. Higerd, T. B. 1977. Isolation of acetyl esterase mutants of *Bacillus subtilis* 168. *J. Bacteriol.* **129:**973–977.

184. Higerd, T. B., J. A. Hoch, and J. Spizizen. 1972. Hyperprotease-producing mutants of *Bacillus subtilis*. *J. Bacteriol.* **112:**1026–1028.

185. Higo, K.-I., E. Otaka, and S. Osawa. 1982. Purification and characterization of 30S ribosomal proteins from *Bacillus subtilis*: correlation to *Escherichia coli* 30S proteins. *Mol. Gen. Genet.* **185:**239–244.

186. Hill, S. H. A. 1983. *spoVH* and *spoVJ*: new sporulation loci in *Bacillus subtilis* 168. *J. Gen. Microbiol.* **129:**293–302.

187. Hilton, M. D., N. G. Alaeddinoglu, and A. L. Demain. 1988. *Bacillus subtilis* mutant deficient in the ability to produce the dipeptide antibiotic bacilysin: isolation and mapping of the mutation. *J. Bacteriol.* **170:**1018–1020.

188. Hirochika, H., Y. Kobayashi, F. Kawamura, and H. Saito. 1981. Cloning of sporulation gene *spo0B* of *Bacillus subtilis* and its genetic and biochemical analysis. *J. Bacteriol.* **146:**494–505.

189. Hoch, J. A., and C. Anagnostopoulos. 1970. Chromosomal location and properties of radiation sensitivity mutations in *Bacillus subtilis*. *J. Bacteriol.* **103:**295–301.

190. Hoch, J. A., and H. J. Coukoulis. 1978. Genetics of the alpha-ketoglutarate dehydrogenase complex of *Bacillus subtilis*. *J. Bacteriol.* **133:**265–269.

191. Hoch, J. A., and J. Mathews. 1972. Genetic studies in *Bacillus subtilis*, p. 113–116. *In* H. O. Halvorson, R. Hanson, and L. L. Campbell (ed.), *Spores V*. American Society for Microbiology, Washington, D.C.

192. Hoch, J. A., and J. L. Mathews. 1973. Chromosomal location of pleiotropic negative sporulation mutations in *Bacillus subtilis*. *Genetics* **73:**215–228.

193. Hoch, J. A., and E. W. Nester. 1973. Gene-enzyme relationships of aromatic acid biosynthesis in *Bacillus subtilis*. *J. Bacteriol.* **116:**59–66.

194. Hoch, J. A., and J. Spizizen. 1969. Genetic control of some early events in sporulation of *Bacillus subtilis* 168, p. 112–120. *In* L. L. Campbell (ed.), *Spores IV*. American Society for Microbiology, Washington, D.C.

195. Hoch, J. A., K. Trach, F. Kawamura, and H. Saito. 1985. Identification of the transcriptional suppressor *sof-1* as an alteration in the *spo0A* protein. *J. Bacteriol.* **161:**552–555.

196. Hoch, S. O. 1974. Mapping of the 5-methyltryptophan resistance locus in *Bacillus subtilis*. *J. Bacteriol.* **117:**315–317.

197. Hoch, S. O., C. Anagnostopoulos, and I. P. Crawford. 1969. Enzymes of the tryptophan operon of *Bacillus subtilis*. *Biochem. Biophys. Res. Commun.* **35:**838–844.

198. Hoch, S. O., C. W. Roth, I. P. Crawford, and E. W. Nester. 1971. Control of tryptophan biosynthesis by the methyltryptophan resistance gene in *Bacillus subtilis*. *J. Bacteriol.* **105:**38–45.

199. Hodgson, J. A., P. N. Lowe, and R. N. Perham. 1983. Wild-type and mutant forms of the pyruvate dehydrogenase multienzyme complex from *Bacillus subtilis*. *Biochem. J.* **211:**463–472.

200. Hofemeister, J., M. Israeli-Reches, and D. Dubnau. 1983. Integration of plasmid pE194 at multiple sites on the *Bacillus subtilis* chromosome. *Mol. Gen. Genet.* **189:**58–68.

201. Hoshino, T., T. McKenzie, S. Schmidt, T. Tanaka, and N. Sueoka. 1987. Nucleotide sequence of *Bacillus subtilis dnaB*: a gene essential for DNA replication, initiation and membrane attachment. *Proc. Natl. Acad. Sci. USA* **84:**653–667.

202. Hranueli, D., P. J. Piggot, and J. Mandelstam.

1974. Statistical estimate of the total number of operons specific for *Bacillus subtilis* sporulation. *J. Bacteriol.* **119**:684–690.

203. **Hutchison, K. W., and H. O. Halvorson.** 1980. Cloning of randomly sheared fragments from a phi-105 lysogen of *Bacillus subtilis*: identification of prophage-containing clones. *Gene* **8**:267–268.

204. **Igo, M., M. Lampe, and R. Losick.** 1988. Structure and regulation of a *Bacillus subtilis* gene that is transcribed by the Eσ^B form of RNA polymerase holoenzyme, p. 151–156. *In* A. T. Ganesan and J. A. Hoch (ed.), *Genetics and Biotechnology of Bacilli*, vol. 2. Academic Press, Inc., San Diego, Calif.

205. **Iijima, T., M. D. Diesterhaft, and E. Freese.** 1977. Sodium effect of growth on aspartate and genetic analysis of a *Bacillus subtilis* mutant with high aspartase activity. *J. Bacteriol.* **129**:1440–1447.

206. **Iijima, T., and Y. Ikeda.** 1970. Mutability of the phleomycin-resistant mutants of *Bacillus subtilis*. I. Isolation of genetically unstable mutants. *J. Gen. Appl. Microbiol.* **16**:419–427.

207. **Ikawa, S., T. Shibata, K. Matsumoto, T. Iijima, H. Saito, and T. Ando.** 1981. Chromosomal loci of genes controlling site-specific restriction endonucleases of *Bacillus subtilis*. *Mol. Gen. Genet.* **183**:1–6.

208. **Imada, S., L. E. Carroll, and N. Sueoka.** 1980. Genetic mapping of a group of temperature-sensitive *dna* initiation mutants in *Bacillus subtilis*. *Genetics* **94**:809–823.

209. **Imai, R., T. Sekiguchi, Y. Nosoh, and K. Tsuda.** 1987. The nucleotide sequence of 3-isopropylmalate dehydrogenase gene from *Bacillus subtilis*. *Nucleic Acids Res.* **15**:4988.

210. **Inselburg, J. W., T. Eremenko-Volpe, L. Greenwald, W. L. Meadow, and J. Marmur.** 1969. Physical and genetic mapping of the SPO2 prophage on the chromosome of *Bacillus subtilis* 168. *J. Virol.* **3**:627–628.

211. **Ionesco, H., J. Michel, B. Cami, and P. Schaeffer.** 1970. Genetics of sporulation in *Bacillus subtilis* Marburg. *J. Appl. Bacteriol.* **33**:13–24.

212. **Irie, R., T. Okamoto, and Y. Fujita.** 1982. A germination mutant of *Bacillus subtilis* deficient in response to glucose. *J. Gen. Appl. Microbiol.* **28**:345–354.

213. **Ito, J.** 1973. Pleiotropic nature of bacteriophage tolerant mutants obtained in early-blocked asporogenous mutants of *Bacillus subtilis* 168. *Mol. Gen. Genet.* **124**:97–106.

214. **Ito, J., G. Mildner, and J. Spizizen.** 1971. Early blocked asporogenous mutants of *Bacillus subtilis* 168. I. Isolation and characterization of mutants resistant to antibiotic(s) produced by sporulating *Bacillus subtilis* 168. *Mol. Gen. Genet.* **112**:104–109.

215. **Ito, J., and J. Spizizen.** 1973. Genetic studies of catabolite repression insensitive sporulation mutants of *Bacillus subtilis*. *Colloq. Int. C.N.R.S.* **227**:81–82.

216. **Itoh, T.** 1976. Amino acid replacement in the protein S5 from a spectinomycin resistant mutant of *Bacillus subtilis*. *Mol. Gen Genet.* **144**:39–42.

217. **Iwakura, M., M. Kawata, K. Tsuda, and T. Tanaka.** 1988. Nucleotide sequence of the thymidylate synthase B and dihydrofolate reductase genes contained in one *Bacillus subtilis* operon. *Gene* **64**: 9–20.

218. **Iyer, V. N.** 1966. Mutations determining mitomycin resistance in *Bacillus subtilis*. *J. Bacteriol.* **92**:1663–1669.

219. **James, W., and J. Mandelstam.** 1985. Protease production during sporulation of germination mutants of *Bacillus subtilis* and the cloning of a functional *gerE* gene. *J. Gen. Microbiol.* **131**: 2421–2430.

220. **James, W., and J. Mandelstam.** 1985. *spoVIC*, a new sporulation locus in *Bacillus subtilis* affecting spore coats, germination and the rate of sporulation. *J. Gen. Microbiol.* **131**:2409–2419.

221. **Jenkinson, H. F.** 1981. Germination and resistance defects in spores of *Bacillus subtilis* mutant lacking a coat polypeptide. *J. Gen. Microbiol.* **127**:81–91.

222. **Jenkinson, H. F.** 1983. Altered arrangement of proteins in the spore coat of a germination mutant of *Bacillus subtilis*. *J. Gen. Microbiol.* **129**:1945–1958.

223. **Jenkinson, H. F., and J. Mandelstam.** 1983. Cloning of the *Bacillus subtilis lys* and *spoIIIB* genes in phage ϕ105. *J. Gen. Microbiol.* **129**:2229–2240.

224. **Jentsch, S.** 1983. Restriction and modification in *Bacillus subtilis*: sequence specificities of restriction/modification systems *BsmM*, *BsuE*, and *BsuF*. *J. Bacteriol.* **156**:800–808.

225. **Johnson, W. C., C. P. Moran, Jr., and R. Losick.** 1983. Two RNA polymerase sigma factors from *Bacillus subtilis* discriminate between overlapping promoters for a developmentally regulated gene. *Nature* (London) **302**:800–804.

226. **Kane, J. F.** 1977. Regulation of a common aminotransferase subunit. *J. Bacteriol.* **132**:419–425.

227. **Kane, J. F., R. L. Goode, and J. Wainscott.** 1975. Multiple mutations in *cysA14* mutants of *Bacillus subtilis*. *J. Bacteriol.* **121**:204–211.

228. **Karamata, D., and J. D. Gross.** 1970. Isolation and genetic analysis of temperature-sensitive mutants of *Bacillus subtilis* defective in DNA synthesis. *Mol. Gen. Genet.* **108**:277–287.

229. **Karamata, D., M. McConnell, and H. J. Rogers.** 1972. Mapping of *rod* mutants of *Bacillus subtilis*. *J. Bacteriol.* **111**:73–79.

230. Karamata, D., H. M. Pooley, and M. Monod. 1987. Expression of heterologous genes for wall teichoic acid in *Bacillus subtilis* 168. *Mol. Gen. Genet.* 207:73–81.

231. Kawamura, F., and H. Saito. 1983. Isolation and mapping of a new suppressor mutation of an early sporulation gene *spo0F* mutation in *Bacillus subtilis*. *Mol. Gen. Genet.* 192:330–334.

232. Kawamura, F., H. Shimotsu, H. Saito, H. Hirochika, and Y. Kobayashi. 1981. Cloning of *spo0* genes with bacteriophage and plasmid vectors in *Bacillus subtilis*, p. 109–113. *In* H. S. Levinson, A. L. Sonenshein, and D. J. Tipper (ed.), *Sporulation and Germination*. American Society for Microbiology, Washington, D.C.

233. Kelly, M. S. 1967. Physical and mapping properties of distant linkages between genetic markers in transformation of *Bacillus subtilis*. *Mol. Gen. Genet.* 99:333–349.

234. Kelly, M. S. 1967. The causes of instability of linkage in transformation of *Bacillus subtilis*. *Mol. Gen. Genet.* 99:350–361.

235. Kelly, M. S., and R. H. Pritchard. 1963. Selection for linked loci in *Bacillus subtilis* by means of transformation. *Heredity* 17:598–603.

236. Kenney, T. J., and C. P. Moran, Jr. 1987. Organization and regulation of an operon that encodes a sporulation-essential sigma factor in *Bacillus subtilis*. *J. Bacteriol.* 169:3329–3339.

237. Kimhi, Y., and B. Magasanik. 1970. Genetic basis of histidine degradation in *Bacillus subtilis*. *J. Biol. Chem.* 245:3545–3548.

238. Kiss, A., G. Posfai, C. C. Keller, P. Venetianer, and R. J. Roberts. 1985. Nucleotide sequence of the *BsuR1* restriction modification system. *Nucleic Acids Res.* 13:6403–6421.

239. Kiss, I., I. Berek, and G. Ivanovics. 1971. Mapping the delta-aminolaevulinic acid synthetase locus in *Bacillus subtilis*. *J. Gen. Microbiol.* 66:153–159.

240. Klier, A. F., and G. Rapoport. 1988. Genetics and regulation of carbohydrate catabolism in *Bacillus*. *Annu. Rev. Microbiol.* 42:65–95.

241. Kobayashi, H., K. Kobayashi, and Y. Kobayashi. 1977. Isolation and characterization of fusidic acid-resistant, sporulation-defective mutants of *Bacillus subtilis*. *J. Bacteriol.* 132:262–269.

242. Kobayashi, Y., and H. Anaguchi. 1985. Cloning, amplification and characterization of sporulation genes, especially *spoIIG*, of *Bacillus subtilis*, p. 85–94. *In* J. A. Hoch and P. Setlow (ed.), *Molecular Biology of Microbial Differentiation*. American Society for Microbiology, Washington, D.C.

243. Koide, Y., A. Makamura, T. Uozumi, and T. Beppu. 1986. Cloning and sequencing of the major intracellular serine protease gene of *Bacillus subtilis*. *J. Bacteriol.* 167:110–116.

244. Kontinen, V. P., and M. Sarvas. 1988. Mutants of *Bacillus subtilis* defective in protein export. *J. Gen. Microbiol.* 134:2333–2344.

245. Kooistra, J., B. Vosman, and G. Venema. 1988. Cloning and characterization of a *Bacillus subtilis* transcription unit involved in ATP-dependent DNase synthesis. *J. Bacteriol.* 170:4791–4797.

246. Kudoh, J., T. Ikeuchi, and T. Kurahashi. 1985. Nucleotide sequence of the sporulation gene *spo0A* and its mutant genes of *Bacillus subtilis*. *Proc. Natl. Acad. Sci. USA* 82:2665–2668.

247. Kunkel, B., K. Sandman, S. Panzer, P. Youngman, and R. Losick. 1988. The promoter for a sporulation gene in the *spoIVC* locus of *Bacillus subtilis* and its use in studies of temporal and spatial control of gene expression. *J. Bacteriol.* 170:3513–3522.

248. Kunst, F., M. Debarbouille, T. Msadek, M. Young, C. Mauel, D. Karamata, A. Klier, G. Rapoport, and R. Dedonder. 1988. Deduced polypeptides encoded by the *Bacillus subtilis* sacU locus share homology with two-component sensor-regulator systems. *J. Bacteriol.* 170:5093–5101.

249. Kunst, F., M. Pascal, J. Lepesant-Kejzlarova, J.-A. Lepesant, A. Billault, and R. Dedonder. 1974. Pleiotropic mutations affecting sporulation conditions and the synthesis of extracellular enzymes in *Bacillus subtilis* 168. *Biochimie* 56:1481–1489.

250. Kunst, F., M. Steinmetz, J.-A. Lepesant, and R. Dedonder. 1977. Presence of a third sucrose hydrolyzing enzyme in *Bacillus subtilis* Marburg 168. *Biochimie* 59:287–292.

251. La Fauci, G., R. L. Widom, R. L. Eisner, E. D. Jarvis, and R. Rudner. 1986. Mapping of rRNA genes with integrable plasmids in *Bacillus subtilis*. *J. Bacteriol.* 165:204–214.

252. Laipis, P. J., and A. T. Ganesan. 1972. A deoxyribonucleic acid polymerase I-deficient mutant of *Bacillus subtilis*. *J. Biol. Chem.* 247:5867–5871.

253. Lakomova, N. M., T. S. Tsurikova, and A. A. Prozorov. 1980. Possible participation of RNA polymerase III in suppression of *recH* mutation of *Bacillus subtilis*. *Genetica* (USSR) 16:583–587.

254. Lamont, I. L., and J. Mandelstam. 1984. Identification of a new sporulation locus, *spoIIIF*, in *Bacillus subtilis*. *J. Gen. Microbiol.* 130:1253–1261.

255. Lampe, M., C. Binnie, R. Schmidt, and R. Losick. 1988. Cloned gene encoding the delta subunit of *Bacillus subtilis* RNA polymerase. *Gene* 67:13–19.

256. Lampe, M. F., and K. F. Bott. 1984. Cloning the *gyrA* gene of *Bacillus subtilis*. *Nucleic Acids Res.* 12:6307–6323.

257. Lampe, M. F., and K. F. Bott. 1985. Cloning of the *Bacillus subtilis recF* gene. *Gene* 38:139–144.

258. Lampe, M. F., and K. F. Bott. 1985. Genetic and physical organization of the cloned *gyrA* and *gyrB* genes of *Bacillus subtilis*. *J. Bacteriol.* 162:78–84.

259. Lampel, K. A., B. Uratani, G. R. Chaudhry, R. F. Ramaley, and S. Rudikoff. 1986. Characterization of the developmentally regulated *Bacillus subtilis* glucose dehydrogenase gene. *J. Bacteriol.* **166**:238–243.

260. LeHegarat, J.-C., and C. Anagnostopoulos. 1969. Localisation chromosomique d'un gene gouvernant la synthèse d'une phosphatase alcaline chez *Bacillus subtilis*. *C. R. Acad. Sci.* **269**:2048–2050.

261. LeHegarat, J.-C., and C. Anagnostopoulos. 1973. Purification, subunit structure and properties of two repressible phosphohydrolases of *Bacillus subtilis*. *Eur. J. Biochem.* **39**:525–539.

262. Leibovici, J., and C. Anagnostopoulos. 1969. Propriétés de la thréonine désaminase de la souche sauvage et d'un mutant sensible à la valine de *Bacillus subtilis*. *Bull. Soc. Chim. Biol.* **51**:691–707.

263. Lepesant, J.-A., A. Billault, J. Lepesant-Kejzlarova, M. Pascal, F. Kunst, and R. Dedonder. 1974. Identification of the structural gene for sucrase in *Bacillus subtilis* Marburg. *Biochimie* **56**:1465–1470.

264. Lepesant, J.-A., F. Kunst, J. Lepesant-Kejzlarova, and R. Dedonder. 1972. Chromosomal location of mutations affecting sucrose metabolism in *Bacillus subtilis* Marburg. *Mol. Gen. Genet.* **118**:135–160.

265. Lepesant-Kejzlarova, J., J.-A. Lepesant, J. Walle, A. Billault, and R. Dedonder. 1975. Revision of the linkage map of *Bacillus subtilis* 168: indications for circularity of the chromosome. *J. Bacteriol.* **121**:823–834.

266. Lepesant-Kejzlarova, J., J. Walle, A. Billault, F. Kunst, J.-A. Lepesant, and R. Dedonder. 1974. Etablissement de la carte génétique de *Bacillus subtilis*: reexamen de la localisation du segment chromosomique compris entre les marqueurs *sacQ36* et *gtaA12*. *C. R. Acad. Sci.* **278**:1911–1914.

267. Lerner, C. G., B. T. Stephenson, and R. L. Switzer. 1987. Structure of the *Bacillus subtilis* pyrimidine biosynthetic *(pyr)* gene cluster. *J. Bacteriol.* **169**:2202–2206.

268. Lerner, C. G., and R. L. Switzer. 1986. Cloning and structure of the *Bacillus subtilis* aspartate transcarbamylase gene *(pyrB)*. *J. Biol. Chem.* **261**:11156–11165.

269. Lindgren, V., E. Holmgren, and L. Rutberg.1977. *Bacillus subtilis* mutant with temperature-sensitive net synthesis of phosphatidylethanolamine. *J. Bacteriol.* **132**:473–484.

270. Lindgren, V., and L. Rutberg. 1974. Glycerol metabolism in *Bacillus subtilis*: gene-enzyme relationships. *J. Bacteriol.* **119**:431–442.

271. Lipsky, R. H., R. Rosenthal, and S. A. Zahler. 1981. Defective specialized SP-beta transducing bacteriophage of *Bacillus subtilis* that carry the *sup-3* or *sup-44* gene. *J. Bacteriol.* **148**:1012–1015.

272. Liu, H.-M., K.-F. Chak, and P. J. Piggot. 1982. Isolation and characterization of a recombinant plasmid carrying a functional part of the *Bacillus subtilis spoIIA* locus. *J. Gen. Microbiol.* **128**:2805–2812.

273. Loewen, P. C., and J. Switala. 1987. Genetic mapping of *katA*, a locus that affects catalase 1 levels in *Bacillus subtilis*. *J. Bacteriol.* **169**:5848–5851.

274. Lopez, M. E., F. A. Ferrari, A. G. Siccardi, G. Mazza, and M. Polsinelli. 1976. New purine markers in *Bacillus subtilis*. *J. Bacteriol.* **126**:533–535.

275. Lopez-Diaz, I., S. Clarke, and J. Mandelstam. 1986. *spoIID* operon of *Bacillus subtilis*: cloning and sequence. *J. Gen. Microbiol.* **132**:341–354.

276. Love, E., J. D'Ambrosio, N. C. Brown, and D. Dubnau. 1976. Mapping of the gene specifying DNA polymerase III of *Bacillus subtilis*. *Mol. Gen. Genet.* **144**:313–321.

277. Love, P. E., and R. E. Yasbin. 1984. Genetic characterization of the inducible SOS-like system of *Bacillus subtilis*. *J. Bacteriol.* **160**:910–920.

278. MacKay, R., A. Lo, G. Willick, M. Zuker, S. Baird, M. Dove, F. Moranelli, and V. Seligy. 1986. Structure of *Bacillus subtilis* endo-β-1,4-glucanase gene. *Nucleic Acids Res.* **14**:9159–9170.

279. Mackey, C. J., R. J. Warburg, H. O. Halvorson, and S. A. Zahler. 1984. Genetic and physical analysis of the *ilvBC-leu* region in *Bacillus subtilis*. *Gene* **32**:49–56.

280. Mackey, C. J., and S. A. Zahler. 1982. Insertion of bacteriophage SP-beta into the *citF* gene of *Bacillus subtilis* and specialized transduction of the *ilvBC-leu* genes. *J. Bacteriol.* **151**:1222–1229.

281. Mahler, I., R. Warburg, D. J. Tipper, and H. O. Halvorson. 1984. Cloning of an unstable *spoIIA-tyrA* fragment from *Bacillus subtilis*. *J. Gen. Microbiol.* **130**:411–421.

282. Makaroff, C. A., H. Zalkin, R. L. Switzer, and S. J. Vollmer. 1983. Cloning of the *Bacillus subtilis* glutamine phosphoribosylpyrophosphate amidotransferase gene in *Escherichia coli*. *J. Biol. Chem.* **258**:10586–10593.

283. Makino, F., and N. Munakata. 1977. Isolation and characterization of a *Bacillus subtilis* mutant with a defective N-glycosidase activity for uracil-containing deoxyribonucleic acid. *J. Bacteriol.* **131**:438–445.

284. Marrero, R., and R. E. Yasbin. 1988. Cloning of the *Bacillus subtilis recE*[+] gene and functional expression of *recE*[+] in *B. subtilis*. *J. Bacteriol.* **170**:335–344.

285. Martin, I., M. Débarbouillé, E. Ferrari, A. Klier, and G. Rapoport. 1987. Characterization of the levanase gene of *Bacillus subtilis* which shows homology to yeast invertase. *Mol. Gen. Genet.* **208**:177–184.

286. Martin, I., M. Débarbouillé, A. Klier, and G.

Rapoport. 1987. Identification of a new locus, *sacV*, involved in the regulation of levansucrase synthesis in *Bacillus subtilis*. *FEMS Microbiol. Lett.* **44**:39–43.

287. **Martinez, R. J., A. T. Ichiki, N. P. Lundh, and S. R. Tronick.** 1968. A single amino acid substitution responsible for altered flagellar morphology. *J. Mol. Biol.* **34**:559–564.

288. **Masuda, E. S., H. Anaguchi, K. Yamada, and Y. Kobayashi.** 1988. Two developmental genes encoding σ factor homologs are arranged in tandem in *Bacillus subtilis*. *Proc. Natl. Acad. Sci. USA* **85**:7637–7641.

289. **Matsuzaki, S., and Y. Kobayashi.** 1984. New mutation affecting the synthesis of some membrane proteins and sporulation in *Bacillus subtilis*. *J. Bacteriol.* **159**:228–232.

290. **Matsuzaki, S., and Y. Kobayashi.** 1985. Genetic heterogeneity in the *cysA-fus* region of the *Bacillus subtilis* chromosome: identification of the *hos* gene. *J. Bacteriol.* **163**:1336–1338.

291. **Mattioli, R., M. Bazzicalupo, G. Federici, E. Gallori, and M. Polsinelli.** 1979. Characterization of mutants of *Bacillus subtilis* resistant to S-(2-aminoethyl) cysteine. *J. Gen. Microbiol.* **114**:223–225.

292. **Maznitsa, I. I., V. V. Sukhodolets, and L. S. Ukhabotina.** 1983. Cloning of *Bacillus subtilis* 168 genes compensating the defect of mutations for thymidine phosphorylase and uridine phosphorylase in *Escherichia coli* cells. *Genetica* (USSR) **19**:881–887.

293. **Mazza, G., A. Fortunato, E. Ferrari, U. Canosi, A. Falaschi, and M. Polsinelli.** 1975. Genetic and enzymic studies on the recombination process in *Bacillus subtilis*. *Mol. Gen. Genet.* **136**:9–30.

294. **Mazza, G., and A. Galizzi.** 1975. The genetics of DNA replication, repair and recombination in *Bacillus subtilis*. *Microbiologica* **1**:111–135.

295. **McDonald, K. O., and W. F. Burke, Jr.** 1982. Cloning of the *Bacillus subtilis* sulfanilamide resistance gene in *Bacillus subtilis*. *J. Bacteriol.* **149**:391–394.

296. **McEnroe, A. S., and H. Taber.** 1984. Correlation between cytochrome aa_3 concentrations and streptomycin accumulation in *Bacillus subtilis*. *Antimicrob. Agents Chemother.* **26**:507–512.

297. **Meganathan, R., R. Bentley, and H. Taber.** 1981. Identification of *Bacillus subtilis men* mutants which lack O-succinylbenzoyl-coenzyme A synthetase and dihydroxynaphthoate. *J. Bacteriol.* **145**:328–332.

298. **Mendelson, N. H., and J. D. Gross.** 1967. Characterization of a temperature-sensitive mutant of *Bacillus subtilis* defective in deoxyribonucleic acid replication. *J. Bacteriol.* **94**:1603–1608.

299. **Michel, J. F., and B. Cami.** 1969. Selection de mutants de *Bacillus subtilis* bloqués au debut de la sporulation—nature des mutations selectionnées.

Ann. Inst. Pasteur Paris **116**:3–18.

300. **Miczak, A., I. Berek, and G. Ivanovics.** 1976. Mapping the uroporphyrinogen decarboxylase, coproporphyrinogen oxidase and ferrochelatase loci in *Bacillus subtilis*. *Mol. Gen. Genet.* **146**:85–87.

301. **Miczak, A., B. Pragai, and I. Berek.** 1979. Mapping the uroporphyrinogen III cosynthase locus in *Bacillus subtilis*. *Mol. Gen. Genet.* **174**:293–295.

302. **Miki, T., Z. Minimi, and Y. Ikeda.** 1965. The genetics of alkaline phosphatase formation in *Bacillus subtilis*. *Genetics* **52**:1093–1100.

303. **Miles, J. S., and J. R. Guest.** 1985. Complete nucleotide sequence of the fumarase gene *(citG)* of *Bacillus subtilis* 168. *Nucleic Acids Res.* **13**:131–140.

304. **Milhaud, P., G. Balassa, and J. Zucca.** 1978. Spore control (Sco) mutations in *Bacillus subtilis*. 1. Selection and genetic mapping of Sco mutations. *Mol. Gen. Genet.* **163**:35–44.

305. **Miller, P., A. Rabinowitz, and H. Taber.** 1988. Molecular cloning and preliminary genetic analysis of the *men* gene cluster of *Bacillus subtilis*. *J. Bacteriol.* **170**:2735–2741.

306. **Miyakawa, Y., and T. Komano.** 1981. Study on the cell cycle of *Bacillus subtilis* using temperature-sensitive mutants. I. Isolation and genetic analysis of the mutants defective in septum formation. *Mol. Gen. Genet.* **181**:207–214.

307. **Moir, A.** 1981. Germination properties of spore coat-defective mutant of *Bacillus subtilis*. *J. Bacteriol.* **146**:1106–1116.

308. **Moir, A.** 1983. The isolation of lambda transducing phage carrying the *citG* and *gerA* genes of *Bacillus subtilis*. *J. Gen. Microbiol.* **129**:303–310.

309. **Moir, A., I. M. Feavers, A. R. Zuberi, I. S. Sammons, I. S. Roberts, J. R. Yon, E. A. Wolf, and D. A. Smith.** 1985. Progress in the molecular genetics of spore germination in *Bacillus subtilis* 168, p. 35–46. *In* J. A. Hoch and P. Setlow (ed.), *Molecular Biology of Microbial Differentiation*. American Society for Microbiology, Washington, D.C.

310. **Moir, A., E. Lafferty, and D. A. Smith.** 1979. Genetic analysis of spore germination mutants of *Bacillus subtilis* 168: the correlation of phenotype with map location. *J. Gen. Microbiol.* **111**:165–180.

311. **Mollgaard, H.** 1980. Deoxyadenosine/deoxycytidine kinase from *Bacillus subtilis*. Purification, characterization, and physiological function. *J. Biol. Chem.* **255**:8216–8220.

312. **Mollgaard, H., and J. Neuhard.** 1978. Deoxycytidylate deaminase from *Bacillus subtilis*. Purification, characterization and physiological function. *J. Biol. Chem.* **253**:3536–3542.

313. **Monteiro, M. J., M. G. Sargent, and P. J. Piggot.** 1984. Characterization of the replication terminus of the *Bacillus subtilis* chromosome. *J. Gen. Microbiol.* **130:**2403–2414.

314. **Moran, C. P., Jr., R. Losick, and A. L. Sonenshein.** 1980. Identification of a sporulation locus in cloned *Bacillus subtilis* deoxyribonucleic acid. *J. Bacteriol.* **142:**331–334.

315. **Mordhoshi, F., and N. Munakata.** 1987. Multiple species of *Bacillus subtilis* DNA alkyltransferase involved in the adaptive response to simple alkylating agents. *J. Bacteriol.* **169:**587–592.

316. **Moriya, S., N. Ogasawara, and H. Yoshikawa.** 1985. Structure and function of the region of the replication origin of the *Bacillus subtilis* chromosome. III. Nucleotide sequence of some 10,000 base pairs in the origin region. *Nucleic Acids Res.* **13:**2251–2265.

317. **Mountain, A., and S. Baumberg.** 1980. Map locations of some mutations conferring resistance to arginine hydroxamate in *Bacillus subtilis* 168. *Mol. Gen. Genet.* **178:**691–701.

318. **Mountain, A., N. H. Mann, R. N. Munton, and S. Baumberg.** 1984. Cloning of *Bacillus subtilis* restriction fragment complementing auxotrophic mutants of eight *Escherichia coli* genes of arginine biosynthesis. *Mol. Gen. Genet.* **197:**82–89.

319. **Mountain, A., J. McChesney, M. C. M. Smith, and S. Baumberg.** 1986. Gene sequence encoding early enzymes of arginine synthesis within a cluster in *Bacillus subtilis*, as revealed by cloning in *Escherichia coli*. *J. Bacteriol.* **165:**1026–1028.

320. **Mueller, J. P., and H. W. Taber.** 1988. Genetic regulation of cytochrome *aa₃* in *Bacillus subtilis*, p. 91–95. *In* A. T. Ganesan and J. A. Hoch (ed.), *Genetics and Biotechnology of Bacilli*, vol. 2. Academic Press, Inc., San Diego, Calif.

321. **Munakata, H., and Y. Ikeda.** 1968. Mutant of *Bacillus subtilis* producing ultraviolet-sensitive spores. *Biochem. Biophys. Res. Commun.* **33:**469–475.

322. **Murphy, N., D. J. McConnell, and B. Cantwell.** 1984. DNA sequence of the gene and genetic control sites for the excreted *B. subtilis* β-glucanase. *Nucleic Acids Res.* **12:**5355–5367.

323. **Myoda, T. T., S. V. Lowther, V. L. Funange, and F. E. Young.** 1984. Cloning and mapping of the dihydrofolate reductase gene of *Bacillus subtilis*. *Gene* **29:**135–143.

324. **Nagahari, K., and K. Sakaguchi.** 1978. Cloning of *Bacillus subtilis* leucine A, B and C genes with *Escherichia coli* plasmids and expression of the *leuC* gene in *E. coli*. *Mol. Gen. Genet.* **158:**263–270.

325. **Nagami, Y., and T. Tanaka.** 1986. Molecular cloning and nucleotide sequence of a DNA fragment from *Bacillus natto* that enhances production of extracellular proteases and levansucrase in *Bacillus subtilis*. *J. Bacteriol.* **166:**20–28.

326. **Nakamura, A., T. Uozumi, and T. Beppu.** 1987. Nucleotide sequence of a cellulase gene of *Bacillus subtilis*. *Eur. J. Biochem.* **164:**317–320.

327. **Nakano, M. M., M. A. Marahiel, and P. Zuber.** 1988. Identification of a genetic locus required for biosynthesis of the lipopeptide antibiotic surfactin in *Bacillus subtilis*. *J. Bacteriol.* **170:**5662–5668.

328. **Nasser, D., and E. W. Nester.** 1967. Aromatic amino acid biosynthesis: gene-enzyme relationships in *Bacillus subtilis*. *J. Bacteriol.* **94:**1706–1714.

329. **Naumov, L. S., G. V. Savchenko, and A. Prozorov.** 1974. Mapping of *Bacillus subtilis* chromosomal region carrying the *rec342* mutation. *Genetika* **10:**126–131.

330. **Nester, E. W., and A. L. Montoya.** 1976. An enzyme common to histidine and aromatic amino acid biosynthesis in *Bacillus subtilis*. *J. Bacteriol.* **126:**699–705.

331. **Nester, E. W., M. Schafer, and J. Lederberg.** 1963. Gene linkage in DNA transfer: a cluster of genes concerned with aromatic biosynthesis in *Bacillus subtilis*. *Genetics* **48:**529–551.

332. **Neuhard, J.** 1983. Utilization of preformed pyrimidine bases and nucleosides, p. 95–148. *In* A. Munch-Petersen (ed.), *Metabolism of Nucleotides, Nucleosides, and Nucleobases in Microorganisms*. Academic Press, Inc., New York.

333. **Neuhard, J., A. R. Price, L. Schack, and E. Thomassen.** 1978. Two thymidylate synthetases in *Bacillus subtilis*. *Proc. Natl. Acad. Sci. USA* **75:**1194–1198.

334. **Niaudet, B., P. Gay, and R. Dedonder.** 1975. Identification of the structural gene of the PEP-phosphotransferase enzyme I in *Bacillus subtilis* Marburg. *Mol. Gen. Genet.* **136:**337–349.

335. **Nicholson, W. L., and G. H. Chambliss.** 1986. Molecular cloning of *cis*-acting regulatory alleles of the *Bacillus subtilis amyR* region by using gene conversion transformation. *J. Bacteriol.* **165:**663–670.

336. **Nilsson, D., and B. Hove-Jensen.** 1987. Phosphoribosylpyrophosphate synthetase of *Bacillus subtilis*. Cloning, characterization and chromosomal mapping. *Gene* **53:**247–255.

337. **Nomura, S., K. Yamane, T. Sasaki, M. Yamasaki, G. Tamura, and B. Maruo.** 1978. Tunicamycin-resistant mutants and chromosomal locations of mutational sites in *Bacillus subtilis*. *J. Bacteriol.* **136:**818–821.

338. **Nygaard, P., P. Duckert, and H. H. Saxild.** 1988. Purine gene organization and regulation in *Bacillus subtilis*, p. 57–61. *In* A. T. Ganesan and J. A. Hoch (ed.), *Genetics and Biotechnology of Bacilli*, vol. 2. Academic Press, Inc., San Diego, Calif.

339. **Oda, M., A. Sugishita, and K. Furukawa.** 1988.

Cloning and nucleotide sequences of histidase and regulatory genes in the *Bacillus subtilis hut* operon and positive regulation of the operon. *J. Bacteriol.* 170:3199–3205.

340. Ogasawara, N., S. Moriya, G. Mazza, and H. Yoshikawa. 1986. A *Bacillus subtilis dnaG* mutant harbours a mutation in a gene homologous to the *dnaN* gene of *Escherichia coli. Gene* 45:227–231.

341. Ogasawara, N., S. Moriya, G. Mazza, and H. Yoshikawa. 1986. Nucleotide sequence and organization of *dnaB* gene and neighbouring genes on the *Bacillus subtilis* chromosome. *Nucleic Acids Res.* 14:9989–9999.

342. Ogasawara, N., S. Moriya, K. von Meyenburg, F. G. Hansen, and H. Yoshikawa. 1985. Conservation of genes and their organization in the chromosomal replication origin region of *Bacillus subtilis* and *E. coli. EMBO J.* 4:3345–3350.

343. Ogasawara, N., S. Moriya, and H. Yoshikawa. 1983. Structure and organization of rRNA operons in the region of the replication origin of the *Bacillus subtilis* chromosome. *Nucleic Acids Res.* 11:6301–6318.

344. Oh, R. W., M. H. Barnes, N. C. Brown, and A. T. Ganesan. 1986. Cloning and characterization of the *polC* region of *Bacillus subtilis. J. Bacteriol.* 165:951–957.

345. O'Kane, C., B. A. Cantwell, and D. J. McConnell. 1985. Mapping of the gene for endo-β-1,3-1,4-glucanase of *Bacillus subtilis. FEMS Microbiol. Lett.* 29:135–139.

346. Ordal, G. W., D. P. Nettleton, and J. A. Hoch. 1983. Genetics of *Bacillus subtilis* chemotaxis: isolation and mapping of mutations and cloning of chemotaxis genes. *J. Bacteriol.* 154:1088–1097.

347. Ordal, G. W., H. M. Parker, and J. R. Kirby. 1985. Complementation and characterization of chemotaxis mutants of *Bacillus subtilis. J. Bacteriol.* 164:802–810.

348. Osawa, S. 1976. Gene locus of a 30S ribosomal protein S20 of *Bacillus subtilis. Mol. Gen. Genet.* 144:49–51.

349. Osawa, S., R. Takata, K. Tanaka, and M. Tamaki. 1973. Chloramphenicol resistant mutants of *Bacillus subtilis. Mol. Gen. Genet.* 127:163–173.

350. Osawa, S., A. Tokui, and H. Saito. 1978. Mapping by interspecies transformation experiments of several ribosomal protein genes near the replication origin of *Bacillus subtilis* chromosome. *Mol. Gen. Genet.* 164:113–129.

351. O'Sullivan, A., and N. Sueoka. 1967. Sequential replication of the *Bacillus subtilis* chromosome. IV. Genetic mapping by density transfer experiment. *J. Mol. Biol.* 27:349–368.

352. Otozai, K., Y. Takeichi, A. Nakayama, K. Yamane, T. Tanimoto, M. Yamasaki, G. Tamura, S. No-

mura, F. Kawamura, and H. Saito. 1984. Cloning of the *aroI*⁺ gene regions of *Bacillus subtilis* chromosomal DNAs by *B. subtilis* temperate phage p11 and *Escherichia coli* vector systems, and a comparison of physical maps of the gene regions. *J. Gen. Appl. Microbiol.* 30:15–25.

353. Pai, C. H. 1975. Genetics of biotin biosynthesis in *Bacillus subtilis. J. Bacteriol.* 121:1–8.

354. Paice, M. G., R. Bourbonnais, M. Desrochers, L. Jurasek, and M. Yaguchi. 1986. A xylanase gene from *Bacillus subtilis*: nucleotide sequence and comparison with *B. pumilus* gene. *Arch. Microbiol.* 144:201–206.

355. Parsot, C. 1986. Evolution of biosynthetic pathways: a common ancestor for threonine synthase, threonine dehydratase and D-serine dehydratase. *EMBO J.* 5:3013–3019.

356. Parsot, C., and G. N. Cohen. 1988. Cloning and nucleotide sequence of the *Bacillus subtilis hom* gene coding for homoserine dehydrogenases I and II. *J. Biol. Chem.* 263:14654–14660.

357. Paveia, M. H., and L. J. Archer. 1980. Location of genes for arabinose utilization in the *Bacillus subtilis* chromosome. *Brot. Genet.* (Lisbon) 1(86):169–176.

358. Perego, M., E. Ferrari, M. T. Bassi, A. Galizzi, and P. Mazza. 1987. Molecular cloning of *Bacillus subtilis* genes involved in DNA metabolism. *Mol. Gen. Genet.* 209:8–14.

359. Perego, M., and J. A. Hoch. 1987. The isolation and sequence of the *spo0E* gene. Its role in the initiation of sporulation in *Bacillus subtilis. Mol. Microbiol.* 1:125–132.

360. Perego, M., and J. A. Hoch. 1988. Sequence analysis and regulation of the *hpr* locus, a regulatory gene for protease production and sporulation in *Bacillus subtilis. J. Bacteriol.* 170:2560–2567.

361. Perego, M., G. B. Spiegleman, and J. A. Hoch. 1988. Structure of the gene for the transition state regulator *abrB*: regulator synthesis is controlled by the *spo0A* sporulation gene in *Bacillus subtilis. Mol. Microbiol.* 2:689–699.

362. Pestka, S., D. Weiss, R. Vince, B. Wienen, G. Stoffler, and I. Smith. 1976. Thiostrepton-resistant mutants of *Bacillus subtilis*: localization of resistance to 50S subunit. *Mol. Gen. Genet.* 144:235–241.

363. Peterson, A. M., and L. Rutberg. 1969. Linked transformation of bacterial and prophage markers in *Bacillus subtilis* 168 lysogenic for bacteriophage φ105. *J. Bacteriol.* 98:874–877.

364. Phang, C. H., and K. Jeyaseelan. 1988. Isolation and characterization of *citC* gene of *Bacillus subtilis*, p. 97–100. *In* A. T. Ganesan and J. A. Hoch (ed.), *Genetics and Biotechnology of Bacilli*, vol. 2. Academic Press, Inc., San Diego, Calif.

365. Phillips, M. K., L. Hederstedt, S. Hasnain, L. Rutberg, and J. R. Guest. 1987. Nucleotide sequence

encoding the flavoprotein and iron-sulfur protein subunits of the *Bacillus subtilis* PY79 succinate dehydrogenase complex. *J. Bacteriol.* **169**:864–873.

366. **Piggot, P. J.** 1973. Mapping of asporogenous mutations of *Bacillus subtilis*: a minimum estimate of the number of sporulation operons. *J. Bacteriol.* **114**:1241–1253.

367. **Piggot, P. J.** 1975. Characterization of a *cym* mutant of *Bacillus subtilis*. *J. Gen. Microbiol.* **89**:371–374.

368. **Piggot, P. J., and R. S. Buxton.** 1982. Bacteriophage PBSX-induced deletion mutants of *Bacillus subtilis* 168 constitutive for alkaline phosphatase. *J. Gen. Microbiol.* **128**:663–669.

369. **Piggot, P. J., K.-F. Chak, and U. D. Bugaichuk.** 1986. Isolation and characterization of a clone of the *spoVE* locus of *Bacillus subtilis*. *J. Gen. Microbiol.* **132**:1875–1881.

370. **Piggot, P. J., and J. G. Coote.** 1976. Genetic aspects of bacterial endospore formation. *Bacteriol. Rev.* **40**:908–962.

371. **Piggot, P. J., and J. A. Hoch.** 1985. Revised genetic linkage map of *Bacillus subtilis*. *Microbiol. Rev.* **49**:158–179.

372. **Piggot, P. J., A. Moir, and D. A. Smith.** 1981. Advances in the genetics of *Bacillus subtilis* differentiation, p. 29–39. *In* H. S. Levinson, A. L. Sonenshein, and D. J. Tipper (ed.), *Sporulation and Germination.* American Society for Microbiology, Washington, D.C.

373. **Piggot, P. J., and S. Y. Taylor.** 1977. New types of mutation affecting formation of alkaline phosphatase by *Bacillus subtilis* in sporulation conditions. *J. Gen. Microbiol.* **102**:69–80.

374. **Polsinelli, M.** 1965. Linkage relationship between genes for amino acid or nitrogenous base biosynthesis and genes controlling resistance to structurally correlated analogues. *Giorn. Microbiol.* **13**:99–110.

375. **Polucktova, E. V., and A. A. Prozorov.** 1981. The study of properties of the *Bacillus subtilis* rec149 mutant with an increased frequency of the defective phage induction. *Genetica* (USSR) **17**:1588–1592.

376. **Polvektova, E. U., N. M. Lakomova, T. S. Belova, and A. A. Prozorov.** 1984. Cloning of *purA16* locus in Rec⁺ cells of *Bacillus subtilis*. *Genetica* (USSR) **20**:943–948.

377. **Pooley, H. M., and D. Karamata.** 1984. Genetic analysis of autolysin-deficient and flagellaless mutants of *Bacillus subtilis*. *J. Bacteriol.* **160**:1123–1129.

378. **Pooley, H. M., D. Paschoud, and D. Karamata.** 1987. The *gtaB* marker in *Bacillus subtilis* 168 is associated with a deficiency in UDP glucose pyrophosphorylase. *J. Gen. Microbiol.* **133**:3481–

3493.

379. **Porter, A. C. G., and J. Mandelstam.** 1982. A mutant of *Bacillus subtilis* secreting a DNAase inhibitor during sporulation. *J. Gen. Microbiol.* **128**:1903–1914.

380. **Potvin, B. W., R. J. Kelleher, Jr., and H. Gooder.** 1975. Pyrimidine biosynthetic pathways of *Bacillus subtilis*. *J. Bacteriol.* **123**:604–615.

381. **Prasad, C., M. Diesterhaft, and E. Freese.** 1972. Initiation of spore germination in glycolytic mutants of *Bacillus subtilis*. *J. Bacteriol.* **110**:321–328.

382. **Price, C., S. Boylan, M. Duncan, S. Kalman, J.-W. Suh, S. Thomas, and B. Van Hoy.** 1988. Use of λgt11 and antibody probes to isolate genes encoding RNA polymerase subunits from *Bacillus subtilis*, p. 183–188. *In* A. T. Ganesan and J. A. Hoch (ed.), *Genetics and Biotechnology of Bacilli*, vol. 2. Academic Press, Inc., San Diego, Calif.

383. **Price, C. W., M. A. Gitt, and R. H. Doi.** 1983. Isolation and physical mapping of the gene encoding the major sigma-factor of *Bacillus subtilis* RNA polymerase. *Proc. Natl. Acad. Sci. USA* **80**:4074–4078.

384. **Price, V. L., and J. A. Gallant.** 1982. A new relaxed mutant of *Bacillus subtilis*. *J. Bacteriol.* **149**:635–641.

385. **Price, V. L., and J. A. Gallant.** 1983. *Bacillus subtilis relG* mutant: defect in glucose uptake. *J. Bacteriol.* **153**:270–273.

386. **Rabinovich, P. M., M. Y. Beburov, Z. K. Linevich, and A. I. Stepanov.** 1978. Amplification of *Bacillus subtilis* riboflavin operon genes in *Escherichia coli* cells. *Genetica* (USSR) **14**:1696–1705.

387. **Rabinovich, P. M., Y. V. Yomantas, M. Y. Haykinson, and A. I. Stepanov.** 1984. Cloning of genetic material in bacilli, p. 297–308. *In* A. T. Ganesan and J. A. Hoch (ed.), *Genetics and Biotechnology of Bacilli*. Academic Press, Inc., Orlando, Fla.

388. **Racine, F. M., and W. Steinberg.** 1974. Genetic location of two mutations affecting the lysyl-transfer ribonucleic acid synthetase of *Bacillus subtilis*. *J. Bacteriol.* **120**:384–389.

389. **Rapoport, G., A. Klier, A. Billault, F. Fargette, and R. Dedonder.** 1979. Construction of a colony bank of *E. coli* containing hybrid plasmids representative of the *Bacillus subtilis* 168 genome. Expression of functions harbored by the recombinant plasmids in *B. subtilis*. *Mol. Gen. Genet.* **176**:239–245.

390. **Raugei, G., M. Bazzicalupo, G. Federici, and E. Gallori.** 1981. Effect of a new pyrimidine analog on *Bacillus subtilis* growth. *J. Bacteriol.* **145**:1079–1081.

391. **Reeve, J. N., N. H. Mendelson, S. I. Coyne, L. L. Hallock, and R. M. Cole.** 1973. Minicells of *Ba-*

cillus subtilis. J. Bacteriol. **114**:860–873.

392. **Reich, C., K. J. Gardiner, G. J. Olsen, B. Pace, T. L. Marsh, and N. R. Pace.** 1986. The RNA component of the *Bacillus subtilis* RNaseP. Sequence, activity, and partial secondary structure. *J. Biol. Chem.* **261**:7888–7893.

393. **Reysset, G.** 1981. New class of *Bacillus subtilis* glutamine-requiring mutants. *J. Bacteriol.* **148**:653–658.

394. **Rima, B. K., and I. Takahashi.** 1978. Deoxyribonucleoside-requiring mutants of *Bacillus subtilis. J. Gen. Microbiol.* **107**:139–145.

395. **Rima, B. K., and I. Takahashi.** 1978. Synthesis of thymidine nucleotides in *Bacillus subtilis. Can. J. Biochem.* **56**:158–160.

396. **Riva, S., C. van Sluis, G. Mastromei, C. Attolini, G. Mazza, M. Polsinelli, and A. Falaschi.** 1975. A new mutant of *Bacillus subtilis* altered in the initiation of chromosome replication. *Mol. Gen. Genet.* **137**:185–202.

397. **Riva, S., G. Villani, G. Mastromei, and G. Mazza.** 1976. *Bacillus subtilis* mutant temperature sensitive in the synthesis of ribonucleic acid. *J. Bacteriol.* **127**:679–690.

398. **Robson, L. M., and G. H. Chambliss.** 1987. Endo-β-1,4-glucanase gene of *Bacillus subtilis* DLG. *J. Bacteriol.* **169**:2017–2025.

399. **Rogolsky, M.** 1968. Genetic mapping of a locus which regulates the production of pigment associated with spores of *Bacillus subtilis. J. Bacteriol.* **95**:2426–2427.

400. **Ron, E. Z., M.-P. de Bethune, and C. G. Cocito.** 1980. Mapping of virginiamycin S resistance in *Bacillus subtilis. Mol. Gen. Genet.* **180**:639–640.

401. **Roncero, M. I. G.** 1983. Genes controlling xylan utilization by *Bacillus subtilis. J. Bacteriol.* **156**:257–263.

402. **Rong, S., M. S. Rosenkrantz, and A. L. Sonenshein.** 1986. Transcriptional control of the *Bacillus subtilis spoIID* gene. *J. Bacteriol.* **165**:771–779.

403. **Rosenbluh, A., C. D. B. Banner, R. Losick, and P. C. Fitz-James.** 1981. Identification of a new developmental locus in *Bacillus subtilis* by construction of a deletion mutation in a cloned gene under sporulation control. *J. Bacteriol.* **148**:341–351.

404. **Rosenkrantz, M. S., D. W. Dingman, and A. L. Sonenshein.** 1985. *Bacillus subtilis citB* gene is regulated synergistically by glucose and glutamine. *J. Bacteriol.* **164**:155–164.

405. **Rumyantseva, E. V., V. V. Sukhodolets, and Y. V. Smirnov.** 1979. Isolation and characterization of mutants for genes of nucleoside catabolism in *Bacillus subtilis. Genetica* (USSR) **15**:594–604.

406. **Rutberg, B., and J. A. Hoch.** 1970. Citric acid cycle: gene-enzyme relationships in *Bacillus subtilis. J. Bacteriol.* **104**:826–833.

407. **Rutberg, L.** 1969. Mapping of a temperate bacteriophage active on *Bacillus subtilis. J. Virol.* **3**:38–44.

408. **Saito, H., T. Shibata, and T. Ando.** 1979. Mapping of genes determining nonpermissiveness and host-specific restriction to bacteriophage in *Bacillus subtilis* Marburg. *Mol. Gen. Genet.* **170**:117–122.

409. **Sammons, R. L., A. Moir, and D. A. Smith.** 1981. Isolation and properties of spore germination mutants of *Bacillus subtilis* 168 deficient in the initiation of germination. *J. Gen. Microbiol.* **124**:229–241.

410. **Sammons, R. L., G. M. Slynn, and D. A. Smith.** 1987. Genetical and molecular studies on *gerM*, a new developmental locus of *Bacillus subtilis. J. Gen. Microbiol.* **133**:3299–3312.

411. **Sandman, K., R. Losick, and P. Youngman.** 1987. Genetic analysis of *Bacillus subtilis spo* mutations generated by Tn917-mediated insertional mutagenesis. *Genetics* **117**:603–617.

412. **Sá-Nogueira, I., H. Paveia, and H. de Lancastre.** 1988. Isolation of constitutive mutants for L-arabinose utilization in *Bacillus subtilis. J. Bacteriol.* **170**:2855–2857.

413. **Santos, M. A., H. de Lancastre, and L. J. Archer.** 1983. *Bacillus subtilis* mutation blocking irreversible binding of bacteriophage SPP1. *J. Gen. Microbiol.* **129**:3499–3504.

414. **Sasajima, K., and T. Kumada.** 1981. Change in the regulation of enzyme synthesis under catabolite repression in *Bacillus subtilis* pleiotropic mutant lacking transketolase. *Agric. Biol. Chem.* **45**:2005–2012.

415. **Sasajima, K., and T. Kumada.** 1983. Deficiency of flagellation in *Bacillus subtilis* pleiotropic mutant lacking transketolase. *Agric. Biol. Chem.* **47**:1375–1376.

416. **Saxe, C. L., and N. H. Mendelson.** 1984. Identification of mutations associated with macrofiber formation in *Bacillus subtilis. Genetics* **107**:551–561.

417. **Saxild, H. H., and P. Nygaard.** 1987. Genetic and physiological characterization of *Bacillus subtilis* mutants resistant to purine analogs. *J. Bacteriol.* **169**:2977–2983.

418. **Saxild, H. H., and P. Nygaard.** 1988. Gene-enzyme relationships of the purine biosynthetic pathway in *Bacillus subtilis. Mol. Gen. Genet.* **211**:160–167.

419. **Schörgendorfer, K., H. Schwab, and R. M. Lafferty.** 1987. Nucleotide sequence of a cloned 2.5 kb *Pst*I-*Eco*R1 *Bacillus subtilis* DNA fragment coding for levanase. *Nucleic Acids Res.* **15**:9606.

420. **Segall, J., and R. Losick.** 1977. Cloned *Bacillus subtilis* DNA containing a gene that is activated

early during sporulation. *Cell* **11**:751–761.

421. **Seki, T., H. Yoshikawa, H. Takahashi, and H. Saito.** 1987. Cloning and nucleotide sequence of *phoP*, the regulatory gene for alkaline phosphatase and phosphodiesterase in *Bacillus subtilis. J. Bacteriol.* **169**:2913–2916.

422. **Seki, Y., H. Yoshikawa, H. Takahashi, and H. Saito.** 1988. Nucleotide sequence of the *Bacillus subtilis phoR* gene. *J. Bacteriol.* **170**:5935–5938.

423. **Sekiguchi, J., B. Ezaki, K. Kodana, and T. Akamatsu.** 1988. Molecular cloning of a gene affecting the autolysin level and flagellation in *Bacillus subtilis. J. Gen. Microbiol.* **134**:1611–1621.

424. **Sekiguchi, J., N. Takada, and H. Okada.** 1975. Genes affecting the productivity of alpha-amylase in *Bacillus subtilis* Marburg. *J. Bacteriol.* **121**:688–694.

425. **Sharrock, R. A., and T. Leighton.** 1981. Intergenic suppressors of temperature-sensitive sporulation in *Bacillus subtilis. Mol. Gen. Genet.* **183**:532–537.

426. **Sharrock, R. A., and T. Leighton.** 1982. Suppression of defective-sporulation phenotypes by the *Bacillus subtilis* mutation *rev4. Mol. Gen. Genet.* **186**:432–438.

427. **Sharrock, R. A., T. Leighton, and H. G. Wittmann.** 1981. Macrolide and aminoglycoside antibiotic resistance mutations in the *Bacillus subtilis* ribosome resulting in temperature-sensitive sporulation. *Mol. Gen. Genet.* **183**:538–543.

428. **Sharrock, R. A., S. Rubenstein, M. Chan, and T. Leighton.** 1984. Intergenic suppression of *spo0* phenotype by the *Bacillus subtilis* mutation *rvtA. Mol. Gen. Genet.* **194**:260–264.

429. **Shibata, T., S. Ikawa, Y. Komatsu, T. Ando, and H. Saito.** 1979. Introduction of host-controlled modification and restriction systems of *Bacillus subtilis* IAM 1247 into *Bacillus subtilis* 168. *J. Bacteriol.* **139**:308–310.

430. **Shimotsu, H., and D. J. Henner.** 1986. Modulation of *Bacillus subtilis* levansucrase gene expression by sucrose and regulation of the steady-state mRNA level by *sacU* and *sacQ* genes. *J. Bacteriol.* **168**:380–388.

431. **Shohayeb, M., and I. Chopra.** 1987. Mutations affecting penicillin-binding proteins 2a, 2b and 3 in *Bacillus subtilis* alter cell shape and peptidoglycan metabolism. *J. Gen. Microbiol.* **133**:1733–1742.

432. **Siccardi, A. G., E. Lanza, E. Nielsen, A. Galizzi, and G. Mazza.** 1975. Genetic and physiological studies on the site of action of distamycin A. *Antimicrob. Agents Chemother.* **8**:370–376.

433. **Siccardi, A. G., S. Ottolenghi, A. Fortunato, and G. Mazza.** 1976. Pleiotropic, extragenic suppresion of *dna* mutations in *Bacillus subtilis. J. Bac-*

teriol. **128**:174–181.

434. **Siegel, E. C., and J. Marmur.** 1969. Temperature-sensitive induction of bacteriophage in *Bacillus subtilis* 168. *J. Virol.* **4**:610–618.

435. **Skarstedt, M. T., and S. B. Greer.** 1973. Threonine synthetase of *Bacillus subtilis*. The nature of an associated dehydratase activity. *J. Biol. Chem.* **248**:1032–1044.

436. **Sloma, A., A. Ally, D. Ally, and J. Pero.** 1988. Gene encoding a minor extracellular protease in *Bacillus subtilis. J. Bacteriol.* **170**:5557–5563.

437. **Smith, D. A., A. Moir, and R. Sammons.** 1978. Progress in genetics of spore germination in *Bacillus subtilis*, p. 158–163. *In* G. Chambliss and J. C. Vary (ed.), *Spores VII*. American Society for Microbiology, Washington, D.C.

438. **Smith, I.** 1982. The translational apparatus of *Bacillus subtilis*, p. 111–145. *In* D. A. Dubnau (ed.), *Molecular Biology of the Bacilli*. Academic Press, Inc., New York.

439. **Smith, I., and P. Paress.** 1978. Genetic and biochemical characterization of kirromycin resistance mutations in *Bacillus subtilis. J. Bacteriol.* **135**:1101–1117.

440. **Smith, I., P. Paress, K. Cabane, and E. Dubnau.** 1980. Genetics and physiology of the *rel* system of *Bacillus subtilis. Mol. Gen. Genet.* **178**:271–279.

441. **Smith, I., and H. Smith.** 1973. Location of the SPO2 attachment site and the bryamycin resistance marker on the *Bacillus subtilis* chromosome. *J. Bacteriol.* **114**:1138–1142.

442. **Smith, I., D. Weiss, and S. Pestka.** 1976. A micrococcin-resistant mutant of *Bacillus subtilis*: localization of resistance to the 50S subunit. *Mol. Gen. Genet.* **144**:231–233.

443. **Smith, M. C. M., A. Mountain, and S. Baumberg.** 1986. Cloning in *Escherichia coli* of a *Bacillus subtilis* arginine repressor gene through its ability to confer structural stability on a fragment carrying genes of arginine biosynthesis. *Mol. Gen. Genet.* **205**:176–182.

444. **Sonenshein, A. L., and H. B. Alexander.** 1979. Initiation of transcription *in vitro* is inhibited by lipiarmycin. *J. Mol. Biol.* **127**:55–72.

445. **Sonenshein, A. L., H. B. Alexander, D. M. Rothstein, and S. H. Fisher.** 1977. Lipiarmycin-resistant ribonucleic acid polymerase mutants of *Bacillus subtilis. J. Bacteriol.* **132**:73–79.

446. **Sonenshein, A. L., B. Cami, J. Brevet, and R. Cote.** 1974. Isolation and characterization of rifampin-resistant and streptolydigin-resistant mutants of *Bacillus subtilis* with altered sporulation properties. *J. Bacteriol.* **120**:253–265.

447. **Staal, S. P., and J. A. Hoch.** 1972. Conditional dihydrostreptomycin resistance in *Bacillus subtilis. J. Bacteriol.* **110**:202–207.

448. Stahl, M. L., and E. Ferrari. 1984. Replacement of the *Bacillus subtilis* subtilisin structural gene with an in vitro-derived deletion. *J. Bacteriol.* 158:411–418.

449. Steinberg, W., and C. Anagnostopoulos. 1971. Biochemical and genetic characterization of a temperature-sensitive tryptophanyl-transfer ribonucleic acid synthetase mutant of *Bacillus subtilis. J. Bacteriol.* 105:6–19.

450. Steinmetz, M., S. Aymerich, G. Gonzy-Tréboul, and D. LeCoq. 1988. Levansucrase induction by sucrose in *Bacillus subtilis* involves an antiterminator. Homology with the *Escherichia coli bgl* operon, p. 11–15. *In* A. T. Ganesan and J. A. Hoch (ed.), *Genetics and Biotechnology of Bacilli*, vol. 2. Academic Press, Inc., San Diego, Calif.

451. Steinmetz, M., F. Kunst, and R. Dedonder. 1976. Mapping of mutations affecting synthesis of exocellular enzymes in *Bacillus*. Identity of *sacU, amyB* and *pap* mutations. *Mol. Gen. Genet.* 148:281–285.

452. Steinmetz, M., D. LeCoq, S. Aymerich, G. Gonzy-Tréboul, and P. Gay. 1985. The DNA sequence of the gene for the secreted *Bacillus subtilis* enzyme levansucrase and its genetic control sites. *Mol. Gen. Genet.* 200:220–228.

453. Stewart, G. C., and K. F. Bott. 1983. DNA sequence of the tandem ribosomal RNA promoter for *B. subtilis* operon *rrnB. Nucleic Acids Res.* 11:6289–6300.

454. Stragier, P., C. Bonamy, and C. Karmazyn-Campelli. 1988. Processing of a sporulation sigma factor in *Bacillus subtilis*: how morphological structure could control gene expression. *Cell* 52:697–704.

455. Stragier, P., J. Bouvier, C. Bonamy, and J. Szulmajster. 1984. A developmental gene product of *Bacillus subtilis* homologous to the sigma factor of *Escherichia coli. Nature* (London) 312:376–378.

456. Strauch, M. A., A. I. Aronson, S. W. Brown, H. J. Schreier, and A. L. Sonenshein. 1988. Sequence of the *Bacillus subtilis* glutamine synthetase gene region. *Gene* 71:257–265.

457. Struck, J. C. R., D. W. Vogel, N. Ulbrich, and V. A. Erdmann. 1988. A *dnaZX*-like open reading frame downstream from the *Bacillus subtilis* scRNA gene. *Nucleic Acids Res.* 16:2720.

458. Suh, J. W., S. A. Boylan, and C. W. Price. 1986. Gene for the alpha subunit of *Bacillus subtilis* RNA polymerase maps in the ribosomal protein gene cluster. *J. Bacteriol.* 168:65–71.

459. Sukhodolets, V. V., Y. V. Flyakh, and E. V. Rumyantseva. 1983. Mapping of mutations in genes for nucleoside catabolism on the *Bacillus subtilis* chromosome. *Genetica* (USSR) 19:221–226.

460. Sun, D., and I. Takahashi. 1982. Genetic mapping of catabolite-resistant mutants of *Bacillus subtilis. Can. J. Microbiol.* 28:1242–1251.

461. Sun, D., and I. Takahashi. 1984. A catabolite-resistance mutation is localized in the *rpo* operon of *Bacillus subtilis. Can. J. Microbiol.* 30:423–429.

462. Swanton, M., and G. Edlin. 1972. Isolation and characterization of an RNA relaxed mutant of *Bacillus subtilis. Biochem. Biophys. Res. Commun.* 46:583–588.

463. Taber, H. W., E. A. Dellers, and R. L. Lombardo. 1981. Menaquinone biosynthesis in *Bacillus subtilis*: isolation of *men* mutants and evidence for clustering of *men* genes. *J. Bacteriol.* 145:321–327.

464. Tanaka, K., M. Tamaki, S. Osawa, A. Kimura, and R. Takata. 1973. Erythromycin resistant mutants of *Bacillus subtilis. Mol. Gen. Genet.* 127:157–161.

465. Tanaka, T., and M. Kawata. 1988. Cloning and characterization of *Bacillus subtilis iep*, which has positive and negative effects on production of extracellular proteases. *J. Bacteriol.* 170:3593–3600.

466. Tipper, D. J., C. W. Johnson, C. L. Ginther, T. Leighton, and H. G. Wittmann. 1977. Erythromycin resistant mutations in *Bacillus subtilis* cause temperature sensitive sporulation. *Mol. Gen. Genet.* 150:147–159.

467. Todd, J. A., A. N. Roberts, K. Johnstone, P. J. Piggot, G. Winter, and D. J. Ellar. 1986. Reduced heat resistance of mutant spores after cloning and mutagenesis of the *Bacillus subtilis* gene encoding penicillin-binding protein 5. *J. Bacteriol.* 167:257–264.

468. Toma, S., M. Del Bue, A. Pirola, and G. Grandi. 1986. *nprR1* and *nprR2* regulatory regions for neutral protease expression in *Bacillus subtilis. J. Bacteriol.* 167:740–743.

469. Tominaga, A., and Y. Kobayashi. 1978. Kasugamycin-resistant mutations of *Bacillus subtilis. J. Bacteriol.* 135:1149–1150.

470. Trach, K., J. W. Chapman, P. Piggot, D. LeCoq, and J. A. Hoch. 1988. Complete sequence and transcriptional analysis of the *spo0F* region of the *Bacillus subtilis* chromosome. *J. Bacteriol.* 170:4194–4208.

471. Trach, K. A., J. W. Chapman, P. J. Piggot, and J. A. Hoch. 1985. Deduced product of the stage 0 sporulation gene *spo0F* shares homology with the Spo0A, OmpR, and SfrA proteins. *Proc. Natl. Acad. Sci. USA* 82:7260–7264.

472. Trautner, T. A., B. Pawlek, S. Bron, and C. Anagnostopoulos. 1974. Restriction and modification in *Bacillus subtilis*. Biological aspects. *Mol. Gen. Genet.* 131:181–191.

473. Trempy, J. E., C. Bonamy, J. Szulmajster, and

W. G. Haldenwang. 1985. *Bacillus subtilis* sigma factor σ²⁹ is the product of sporulation gene *spoIIG*. *Proc. Natl. Acad. Sci. USA* **82**:4189–4192.

474. **Trowsdale, J., S. M. H. Chen, and J. A. Hoch.** 1978. Genetic analysis of phenotype revertants of *spo0A* mutants in *Bacillus subtilis*: a new cluster of ribosomal genes, p. 131–135. *In* G. Chambliss and J. C. Vary (ed.), *Spores VII*. American Society for Microbiology, Washington, D.C.

475. **Trowsdale, J., S. M. H. Chen, and J. A. Hoch.** 1979. Genetic analysis of a class of polymyxin resistant partial revertants of stage 0 sporulation mutants of *Bacillus subtilis*: map of the chromosome region near the origin of replication. *Mol. Gen. Genet.* **173**:61–70.

476. **Trowsdale, J., and D. A. Smith.** 1975. Isolation, characterization, and mapping of *Bacillus subtilis* 168 germination mutants. *J. Bacteriol.* **123**:85–95.

477. **Uehara, H., K. Yamane, and B. Maruo.** 1979. Thermosensitive, extracellular neutral proteases in *Bacillus subtilis*. *J. Bacteriol.* **139**:583–590.

478. **Van Alstyne, D., and M. I. Simon.** 1971. Division mutants of *Bacillus subtilis*: isolation and PBS1 transduction of division-specific markers. *J. Bacteriol.* **108**:1366–1379.

479. **Vandeyar, M. A., and S. A. Zahler.** 1986. Chromosomal insertions of Tn917 in *Bacillus subtilis*. *J. Bacteriol.* **167**:530–534.

480. **van Randen, J., and G. Venema.** 1984. Direct plasmid transfer from replica plated *E. coli* colonies to competent *B. subtilis* cells. *Mol. Gen. Genet.* **195**:57–61.

481. **Vapnek, D., and S. Greer.** 1971. Suppression by derepression in threonine dehydratase-deficient mutants of *Bacillus subtilis*. *J. Bacteriol.* **106**:615–625.

482. **Vapnek, D., and S. Greer.** 1971. Minor threonine dehydratase encoded within the threonine synthetic region of *Bacillus subtilis*. *J. Bacteriol.* **106**:983–993.

483. **Vasantha, N., L. D. Thompson, C. Rhodes, C. Banner, J. Nagle, and D. Filpula.** 1984. Genes for alkaline protease and neutral protease from *Bacillus amyloliquefaciens* containing a large open reading frame between the regions coding for signal sequence and mature proteins. *J. Bacteriol.* **159**:811–819.

484. **Vasantha, N., B. Uratani, R. F. Ramaley, and E. Freese.** 1983. Isolation of a developmental gene of *Bacillus subtilis* and its expression in *Escherichia coli*. *Proc. Natl. Acad. Sci. USA* **80**:785–789.

485. **Vazquez-Ramos, J. M., and J. Mandelstam.** 1981. Oxolinic acid-resistant mutants of *Bacillus subtilis*. *J. Gen. Microbiol.* **127**:1–9.

486. **Vold, B. S.** 1985. Structure and organization of genes for transfer RNA in *Bacillus subtilis*. *Microbiol. Rev.* **49**:71–80.

487. **Vold, B. S., and C. J. Green.** 1985. Transcription, processing, and expression of tRNA genes from *Bacillus subtilis*, p. 135–141. *In* J. A. Hoch and P. Setlow (ed.), *Molecular Biology of Microbial Differentiation*. American Society for Microbiology, Washington, D.C.

488. **Vosman, B., G. Kuiken, and G. Venema.** 1988. Transformation in *Bacillus subtilis*: involvement of the 17-kilodalton DNA-entry nuclease and the competence-specific 18-kilodalton protein. *J. Bacteriol.* **170**:3703–3710.

489. **Wabiko, H., K. Ochi, D. M. Nguyen, E. R. Allen, and E. Freese.** 1988. Genetic mapping and physiological consequences of *metE* mutations of *Bacillus subtilis*. *J. Bacteriol.* **170**:2705–2710.

490. **Wainscott, V. J., and J. F. Kane.** 1976. Dihydrofolate reductase in *Bacillus subtilis*, p. 208–213. *In* D. Schlessinger (ed.), *Microbiology—1975*. American Society for Microbiology, Washington, D.C.

491. **Walton, D. A., A. Moir, R. Morse, I. Roberts, and D. A. Smith.** 1984. The isolation of λ phage carrying DNA from the histidine and isoleucine-valine regions of the *Bacillus subtilis* chromosome. *J. Gen. Microbiol.* **130**:1577–1586.

492. **Wang, L.-F., C. W. Price, and R. H. Doi.** 1985. *Bacillus subtilis dnaE* encodes a protein homologous to DNA primase of *Escherichia coli*. *J. Biol. Chem.* **260**:3368–3372.

493. **Wang, L.-F., S.-L. Wong, S.-S. Park, and R. H. Doi.** 1988. Isolation and characterization of a novel *Bacillus subtilis* and *Bacillus natto* gene that enhances protease production, p. 45–50. *In* A. T. Ganesan and J. A. Hoch (ed.), *Genetics and Biotechnology of Bacilli*, vol. 2. Academic Press, Inc., San Diego, Calif.

494. **Warburg, R. J., I. Mahler, D. J. Tipper, and H. O. Halvorson.** 1984. Cloning the *Bacillus subtilis* 168 *aroC* gene encoding dehydroquinase. *Gene* **32**:57–66.

495. **Warburg, R. J., and A. Moir.** 1981. Properties of a mutant of *Bacillus subtilis* 168 in which spore germination is blocked at a late stage. *J. Gen. Microbiol.* **124**:243–253.

496. **Ward, J. B., Jr., and S. A. Zahler.** 1973. Genetic studies of leucine biosynthesis in *Bacillus subtilis*. *J. Bacteriol.* **116**:719–726.

497. **Ward, J. B., Jr., and S. A. Zahler.** 1973. Regulation of leucine biosynthesis in *Bacillus subtilis*. *J. Bacteriol.* **116**:727–735.

498. **Wawrousek, E. F., and J. N. Hansen.** 1983. Structure and organization of a cluster of six tRNA genes in the space between tandem ribosomal RNA gene sets in *Bacillus subtilis*. *J. Biol. Chem.* **258**:291–298.

499. Wawrousek, E. F., N. Narasimhan, and J. N. Hansen. 1984. Two large clusters with thirty-seven transfer RNA genes adjacent to ribosomal RNA gene sets in *Bacillus subtilis*: sequence and organization of *trnD* and *trnB* gene clusters. *J. Biol. Chem.* **259**:3694–3702.

500. Whiteman, P., C. Marks, and E. Freese. 1980. The sodium effect of *Bacillus subtilis* growth on aspartate. *J. Gen. Microbiol.* **119**:493–504.

501. Whiteman, P. A., T. Iijima, M. D. Diesterhaft, and E. Freese. 1978. Evidence for a low affinity but high velocity aspartate transport system needed for rapid growth of *Bacillus subtilis* on aspartate as sole carbon source. *J. Gen. Microbiol.* **107**:297–307.

502. Whitt, D. D., and B. C. Carlton. 1968. Characterization of mutants with single and multiple defects in the tryptophan biosynthetic pathway in *Bacillus subtilis*. *J. Bacteriol.* **96**:1273–1280.

503. Widom, R. L., E. D. Jarvis, G. LaFauci, and R. Rudner. 1988. Instability of rRNA operons in *Bacillus subtilis*. *J. Bacteriol.* **170**:605–610. (Author's correction, **170**:2003.)

504. Wienen, B., R. Erlich, M. Stoffler-Meilicke, G. Stoffler, I. Smith, D. Weiss, R. Vince, and S. Pestka. 1979. Ribosomal protein alterations in thiostrepton- and micrococcin-resistant mutants of *Bacillus subtilis*. *J. Biol. Chem.* **254**:8031–8041.

505. Wilhelm, M., and C. P. Hollenberg. 1984. Selective cloning of *Bacillus subtilis* xylose isomerase and xylulokinase in *E. coli* genes by IS5-mediated expression. *EMBO J.* **3**:2555–2560.

506. Wilhelm, M., and C. P. Hollenberg. 1985. Nucleotide sequence of the *Bacillus subtilis* xylose isomerase gene: extensive homology between the *Bacillus* and *E. coli* enzyme. *Nucleic Acids Res.* **13**:5717–5722.

507. Willecke, E., and A. B. Pardee. 1971. Fatty acid-requiring mutant of *Bacillus subtilis* defective in branched chain alpha-keto acid dehydrogenase. *J. Biol. Chem.* **146**:5264–5272.

508. Williams, G., and I. Smith. 1979. Chromosomal mutants causing resistance to tetracycline in *Bacillus subtilis*. *Mol. Gen. Genet.* **177**:23–29.

509. Williams, M. T., and F. E. Young. 1977. Temperate *Bacillus subtilis* bacteriophage φ3T: chromosomal attachment site and comparison with temperate bacteriophage φ105 and SPO2. *J. Virol.* **21**:522–529.

510. Wilson, F. E., J. A. Hoch, and K. Bott. 1981. Genetic mapping of a linked cluster of ribosomal ribonucleic acid genes in *Bacillus subtilis*. *J. Bacteriol.* **148**:624–628.

511. Wilson, M. C., J. L. Farmer, and F. Rothman. 1966. Thymidylate synthesis and aminopterin resistance in *Bacillus subtilis*. *J. Bacteriol.* **92**:186–196.

512. Wong, L. F., and R. H. Doi. 1986. Nucleotide sequence and organization of *Bacillus subtilis* RNA polymerase major sigma factor (σ^{43}). *Nucleic Acids Res.* **14**:4293–4307.

513. Wong, S.-L., C. W. Price, D. S. Goldfarb, and R. H. Doi. 1984. The subtilisin E gene of *Bacillus subtilis* is transcribed from a sigma-37 promoter *in vivo*. *Proc. Natl. Acad. Sci. USA* **81**:1184–1188.

514. Yamada, H., H. Anaguchi, and Y. Kobayashi. 1983. Cloning of the sporulation gene *spoVE* in *Bacillus subtilis*. *J. Gen. Appl. Microbiol.* **29**:477–486.

515. Yamada, Y., M. Ohki, and H. Ishikura. 1983. The nucleotide sequence of *Bacillus subtilis* tRNA genes. *Nucleic Acids Res.* **11**:3037–3045.

516. Yamane, K., and B. Maruo. 1978. Alkaline phosphatase possessing alkaline phosphodiesterase activity and other phosphodiesterases in *Bacillus subtilis*. *J. Bacteriol.* **134**:108–114.

517. Yang, M., E. Ferrari, E. Chen, and D. J. Henner. 1986. Identification of pleiotropic *sacQ* gene of *Bacillus subtilis*. *J. Bacteriol.* **166**:113–119.

518. Yang, M., A. Galizzi, and D. Henner. 1983. Nucleotide sequence of the amylase gene from *Bacillus subtilis*. *Nucleic Acids Res.* **11**:237–249.

519. Yang, M., H. Shimotsu, E. Ferrari, and D. J. Henner. 1987. Characterization and mapping of the *Bacillus subtilis prtR* gene. *J. Bacteriol.* **169**:434–437.

520. Yang, M. Y., E. Ferrari, and D. Henner. 1984. Cloning of the neutral protease gene of *Bacillus subtilis* and the use of the cloned gene to create an in vitro-derived deletion mutation. *J. Bacteriol.* **160**:15–21.

521. Yoneda, Y., and B. Maruo. 1975. Mutation of *Bacillus subtilis* causing hyperproduction of alpha-amylase and protease, and its synergistic effect. *J. Bacteriol.* **124**:48–54.

522. Yoshikawa, H., J. Kazami, S. Yamashita, T. Chibazakura, H. Sone, F. Kawamura, M. Oda, M. Isaka, Y. Kobayashi, and H. Saito. 1986. Revised assignment for the *Bacillus subtilis spoOF* gene and its homology with *spoOA* and two *Escherichia coli* genes. *Nucleic Acids Res.* **14**:1063–1072.

523. Young, F. E., D. Smith, and B. E. Reilly. 1969. Chromosomal location of genes regulating resistance to bacteriophage in *Bacillus subtilis*. *J. Bacteriol.* **98**:1087–1097.

524. Young, M. 1975. Genetic mapping of sporulation operons in *Bacillus subtilis* using a thermosensitive sporulation mutant. *J. Bacteriol.* **122**:1109–1116.

525. Young, M. 1976. The use of temperature-sensitive mutants to study gene expression during sporulation in *Bacillus subtilis*. *J. Bacteriol.* **126**:928–938.

526. Young, M. 1983. The mechanism of insertion of

a segment of heterologous DNA into the chromosome of *Bacillus subtilis*. *J. Gen. Microbiol.* **129:**1497–1512.

527. **Youngman, P., J. B. Perkins, and R. Losick.** 1984. A novel method for the rapid cloning in *Escherichia coli* of *Bacillus subtilis* chromosomal DNA adjacent to Tn*917* inserts. *Mol. Gen. Genet.* **195:**424–433.

528. **Yudkin, M. D., and L. Turley.** 1980. Suppression of asporogeny in *Bacillus subtilis*. Allele-specific suppression of a mutation in the *spoIIA* locus. *J. Gen. Microbiol.* **121:**69–78.

529. **Yuki, S.** 1975. The chromosomal location of the structural gene for amylase in *Bacillus subtilis. Jpn. J. Genet.* **50:**155–157.

530. **Zahler, S. A.** 1978. An adenine-thiamin auxotrophic mutant of *Bacillus subtilis. J. Gen. Microbiol.* **107:**199–201.

531. **Zahler, S. A., L. G. Benjamin, B. S. Glatz, P. F. Winter, and B. J. Goldstein.** 1976. Genetic mapping of *alsA*, *alsR*, *thyA*, *kauA*, and *citD* markers in *Bacillus subtilis*, p. 35–43. *In* D. Schlessinger (ed.), *Microbiology—1976*. American Society for Microbiology, Washington, D.C.

532. **Zahler, S. A., R. Z. Korman, R. Rosenthal, and H. E. Hemphill.** 1977. *Bacillus subtilis* bacteriophage SP-beta: localization of the prophage attachment site, and specialized transduction. *J. Bacteriol.* **129:**556–558.

533. **Zeigler, D. R., B. E. Burke, R. M. Pfister, and D. H. Dean.** 1987. Genetic mapping of cadmium resistance mutations in *Bacillus subtilis. Curr. Microbiol.* **16:**163–165.

534. **Zeigler, D. R., and D. H. Dean.** 1985. Revised genetic map of *Bacillus subtilis* 168. *FEMS Microbiol. Rev.* **32:**101–134.

535. **Zukowski, M., L. Miller, P. Cogswell, and K. Chen.** 1988. Inducible expression system based on sucrose metabolism genes of *Bacillus subtilis*, p. 17–22. *In* A. T. Ganesan and J. A. Hoch (ed.), *Genetics and Biotechnology of Bacilli*, vol. 2. Academic Press, Inc., San Diego, Calif.

Regulation of Procaryotic Development
Edited by Issar Smith, Ralph A. Slepecky, and Peter Setlow
© 1989 American Society for Microbiology, Washington, DC 20006

Chapter 2

Spore Thermoresistance Mechanisms

Philipp Gerhardt and Robert E. Marquis

In considering mechanisms of resistance in bacterial spores, we recognize that resistance to one type of environmental stress may involve a mechanism quite different from that to another stress. For example, the mechanism (or mechanisms) for resistance against heat is different from that against radiation. This article deals only with thermoresistance. Further, we distinguish between the physicochemical determinants responsible for stabilizing essential macromolecules and the physiological and molecular processes responsible for attaining this stabilization during sporulation and maintaining it during dormancy. For example, a low content or activity of water in the spore protoplast (core) may partly account for stabilization against heat; transport, synthesis, and bonding of ions and small molecules in high concentration, causing loss of water from the protoplast, may account for attainment of this state; and integrity of cortex tension exerting inward pressure on the protoplast may account for maintenance of this state. Furthermore, we recognize that several determinants and processes may prevail. For example, mineralization may account for thermostabilization by processes quite different from those for dehydration. Each determinant is examined as an independent variable and is quantitatively correlated with change in resistance as the determinant is varied in amount. In examining these determinants and processes, we avoid focusing on spores of a single model species, especially the ones that are convenient or well studied in other respects but are relatively low in heat resistance. Instead, various species of spores spanning a wide range of heat resistance are examined, especially those having high heat resistance and practical importance.

Resistance mechanisms have been a significant component of research reported in the literature and at international spore symposia. Recent reviews on spore resistance mechanisms include those of Gombas (37), Gould (38, 39), Warth (91), Lindsay et al. (55), Gerhardt (34), and Murrell (68, 69). The last cited review also provides a concise current description of the structure and composition of spores and of the biophysical state of the spore protoplast.

In this article we seek to review critically the experimental reports on heat resistance mechanisms of bacterial spores, keeping in mind the aforementioned caveats. Mostly we focus on recent advances, yet provide historical perspective where necessary. Unsubstantiated theories have too often colored discussions in research reports and reviews of this field, so we concentrate on the experimental evidence that is directly correlated with

Philipp Gerhardt • Department of Microbiology and Public Health, Michigan State University, East Lansing, Michigan 48824. *Robert E. Marquis* • Department of Microbiology and Immunology and Department of Dental Research, University of Rochester, Rochester, New York 14642.

heat resistance in vivo. Thermoresistance of spores is now attributable to at least three physicochemical determinants that affect the vital protoplast: dehydration, mineralization, and thermal adaptation. We develop the evidence for this assertion and also consider evidence on the physiological and molecular processes for attaining and maintaining heat resistance.

DEHYDRATION

Background

Dehydration can account for resistance and dormancy in a wide range of plants and animals and has generally been accepted as the doctrinaire explanation for the extraordinary heat resistance of bacterial spores. The total water content of spores, which varies from about 35 to 65% (wet-weight basis), is clearly too great for dehydration of the entire spore to account for high thermoresistance, although there is a loose correlation between total water content and heat resistance (11). The total water content would have to be distributed unequally, with more in the integument and less in the protoplast. The experimental evidence for the presumption of a dehydrated protoplast, however, had been limited until recently to the historical light-microscopic image of a dormant spore with a brightly refractile center and a darkly non-refractile periphery. Efforts to quantify the unequal distribution of water content had been thwarted because the protoplast could not be isolated and maintained without loss of the essential resistance properties that exist in the intact spore. The few predictions of protoplast water content have varied from 7 to 28% (wet-weight basis) if dehydration alone were to account for sporal heat resistance (see discussion by Nakashio and Gerhardt [72]).

The dehydration explanation was contradicted in 1981 by Bradbury et al. (19) on the basis of proton nuclear magnetic resonance studies of water mobility in structures isolated from spores. However, the interpre-

tation of these data was apparently based on the assumption that such isolated structures have water mobility and sorption properties similar to those within intact spores. In actuality, spore structures and especially the protoplast change quickly and dramatically upon isolation. In reviewing the evidence of Bradbury et al. and others, Lindsay et al. (55) developed an alternative theory based on general molecular stabilization. They concluded that "better evidence of the degree of dehydration of the protoplast in the intact spore, when immersed in water, is required to substantiate the partial dehydration theory." Gould (39) similarly concluded that "more direct measurement of water levels in the various spore compartments is urgently needed."

Measurement in the Protoplast

Beaman et al. (12) first successfully quantified the water content of the protoplast in situ within the fully hydrated dormant spore by the use of spores in which the complex of coat and outer membrane was genetically defective or was chemically disrupted or stripped. This was evidenced by susceptibility of the cortex to digestion by lysozyme and therefore also to permeation by smaller molecules. Consequently, the compartmental water contents became determinable in lysozyme-sensitive spores by differential permeability measurements with ^3H-labeled water for the entire spore, with a ^{14}C-labeled sugar for the periprotoplast integument, and by difference for the protoplast.

Later on, Lindsay et al. (52) established a correlation between the permeability-measured protoplast-water content of diverse types of lysozyme-sensitive dormant spores and their protoplast wet density, which was obtained by the use of a density medium (Metrizamide or Nycodenz) that permeates through the periprotoplast integument to the protoplast membrane, as shown by Tisa et al. (87). The determination of protoplast wet density by gradient centrifugation can be accomplished with small, impure samples and is precise and easy.

The conversion from lysozyme resistance in native spores to lysozyme susceptibility in chemically stripped spores is usually accompanied by a partial reduction in heat resistance, but this difference is small in comparison with the large difference between the stripped spores and the corresponding germinated spores and vegetative cells (48). Lysozyme-sensitive mutants are more like native or revertant spores in heat resistance and are used, when obtainable, in addition to chemically stripped spores.

The appearances of lysozyme-resistant and lysozyme-sensitive spores are exemplified in electron micrographs of stained sections of *Bacillus megaterium* spores. Figure 1 displays several lysozyme-resistant morphotype spores. Figure 1A shows a native spore of the commonly used strain QM-B1551 (ATCC 12872), which contains all of the usual structural components, including a prominent exosporium. Figure 1B shows the usual appearance of an exosporiumless variant strain (ATCC 33729). Figure 1C shows an unusual appearance of this variant strain in which the coat is separated and wrinkled so as to reveal the outer pericortex membrane, which usually is not detectable.

Figure 2 displays several lysozyme-sensitive morphotype spores of *B. megaterium*. Figure 2A shows a spore of variant strain NCIB 8291, in which the apical openings in the exosporium and defects in the coat-outer membrane complex provide entrance to the cortex for lysozyme. Figure 2B shows a spore of strain QM-B1551 after chemical treatment with alkaline sodium dodecyl sulfate and dithiothreitol, which removes the coat and outer membrane. Figure 2C shows a spore of variant strain 33729 after the same chemical treatment, which results in a cortex-encased protoplast. All such lysozyme-sensitive morphotype spores, in this and other species, are permeable to smaller probing molecules penetrating to the protoplast membrane and can be used for the determination of water content within the protoplast in situ.

Correlation with Heat Resistance

In the initial experiments with the differential permeability method (12), the extent of protoplast hydration in three different lysozyme-sensitive morphotype spores of *B. megaterium* was found to be 27 to 29% on a wet-weight basis (i.e., 27 to 29 g of water per 100 g of wet protoplast), which in itself seems sufficient to account for their very low level of moist-heat resistance (D_{100}, 0.03 to 0.17 min). The water content in the integument of the three spore types is much higher (59 to 74%) than that in the protoplast, whereas the water content in the entire spore is intermediate (37 to 49%).

Extension of the differential permeability method to four other spore species by Nakashio and Gerhardt (72) showed that their protoplast water content varies from 30 to 55% and correlates with heat resistance quite separately from *B. megaterium*, and indicated that the other spore species depend on factors in addition to protoplast dehydration for their much greater heat resistance (D_{100}, 2.80 to 238 min).

A generalized correlation between moist-heat resistance and protoplast water content of spores was subsequently established in a comprehensive study by Beaman and Gerhardt (10), using the density method (52). Twenty-eight types of lysozyme-sensitive spores of seven *Bacillus* species (other than *B. megaterium*) were examined, representing thermophiles, mesophiles, and psychrophiles spanning a 3,000-fold range in moist-heat resistance (measured as a D_{100} value) (Fig. 3). Protoplast water contents correlate linearly with the log of D_{100} values over much, but not all, of the range. There is a maximum of about 57% and a minimum of about 28% in protoplast water content in the lysozyme-sensitive spores and presumably also in lysozyme-resistant spores. Beyond these limits, determinants other than protoplast dehydra-

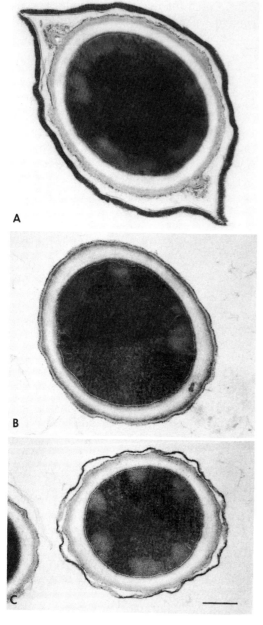

tion (see the sections Mineralization and Thermal Adaptation, below) independently cause changes in heat resistance. Recently, spores of an eighth species, *Bacillus sphaericus*, were found to behave consistently with the foregoing correlation (14).

Differing markedly from this generalized relationship, four spore types of *B. megaterium* are at the minimum limit of protoplast water content (about 28%), yet have extremely low heat resistance (D_{100}, 0.03 to 0.38 min) (Fig. 4). Spores of *B. stearothermophilus* with this same protoplast water content (and similar mineralization) have D_{100} values at least 3 logs higher. The only parameter yet observed to account for the anomalously low heat resistance of *B. megaterium* spores is the low volume ratio of protoplast to protoplast-plus-cortex (48) (Fig. 5; see Role of the Cortex, below). At lethal temperature, the relatively small volume of cortex may be inadequate to maintain protective dehydration in the protoplast.

Refractility

The relative dehydration in a central region has, from first observation, been indicated by the light-microscopic image of the bacterial spore with a refractile center and a nonrefractile periphery. This qualitative image has been refined by measuring the refractive indices of spores with microscopic (50, 77), photometric (35), and laser diffractometric (88, 97) techniques. Even though the refractive index of the entire spore mainly reflects the spore protoplast, an accurate conversion to protoplast water content has not been possible because of inability to obtain an experimentally determined value of the refractive index increment of the protoplast

Figure 1. Native lysozyme-resistant spores of *B. megaterium* strains in stained sections, shown by transmission electron microscopy (48). Bar, 100 nm. (A) Strain QM-B1551 (ATCC 12872). This morphotype spore contains all of the usual structural components: exosporium, complex of coat and outer pericortex membrane, complex of cortex and primordial cell wall, inner periprotoplast membrane, DNA-containing nucleoplasm, and ribosome-containing cytoplasm. The protoplast consists of the cytoplasm and nucleoplasm enclosed by the inner membrane; the integument consists of everything outside of the protoplast. See the original publication (48) for further identification of structural components. (B) Variant strain ATCC 33729, which lacks exosporium. (C) Unusual appearance of the variant strain with the coat wrinkled and separated, revealing the outer membrane around the cortex.

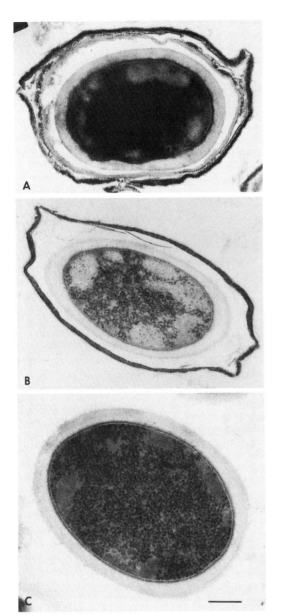

Figure 2. Genetically defective or chemically treated, lysozyme-sensitive spores of *B. megaterium* strains. Bar, 200 μm. (A) Variant strain NCIB 8291, which contains genetically defective coat-outer membrane complex (12). (B) Strain QM-B1551 after chemical treatment that removed the coat-outer membrane complex, resulting in a cortex-encased protoplast with surrounding exosporium (48). (C) Exosporiumless variant strain ATCC 33729 after chemical treatment that removed the coat-outer membrane complex, resulting in a cortex-encased protoplast (48).

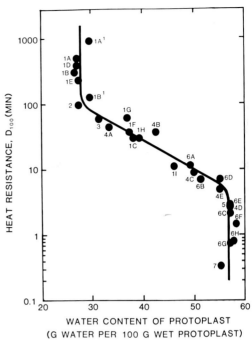

Figure 3. Heat resistance correlated with protoplast water content of lysozyme-sensitive spore types from seven *Bacillus* species which vary in thermal adaptation and in mineralization: 1, *B. stearothermophilus*; 2, "*B. caldolyticus*"; 3, *B. coagulans*; 4, *B. subtilis*; 5, *B. thuringiensis*; 6, *B. cereus*; 7, *B. macquariensis*. For further identification, see Table 1 in the original publication (10).

itself. However, Gerhardt et al. (35) have correlated the average apparent refractive indices for the sporoplast (the structures within the outer pericortex membrane) and the entire spore with the moist-heat resistance of five spore types, spanning 3 decades of D_{100} values. The results indicated that the protoplast is yet more dehydrated than the sporoplast and that its heat resistance can be correlated with dehydration.

Water Activity

Water content does not directly reflect the water activity (a_w) within spore protoplasts, since the two parameters are not equal. Water activity may be more important than water content for molecular stabilization. From examining the dehydration necessary

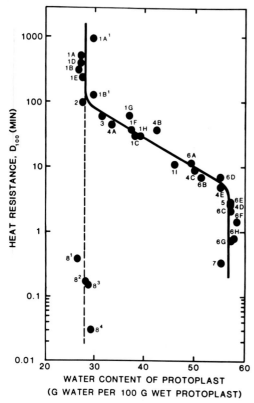

Figure 4. Heat resistance correlated with protoplast water content of four lysozyme-sensitive spore types of *B. megaterium*, in comparison with the generalized relationship shown in Fig. 3. The identity numbers 8[1], 8[2], 8[3], and 8[4] for *B. megaterium* correspond to 10, 12, 13, and 14, respectively, in the original publication (72).

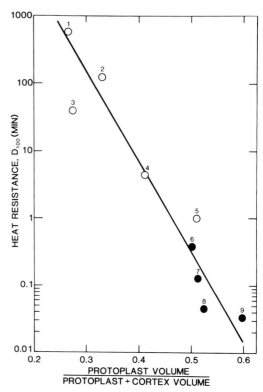

Figure 5. Heat resistance of *B. megaterium* morphotype spores (●) and various other types of spores (○) correlated with volume ratio of protoplast to protoplast-plus-cortex. The identity numbers correspond to those in the original publication (48): 1, *B. stearothermophilus* smooth; 2, *B. stearothermophilus* rough; 3, *B. subtilis* subsp. *niger*; 4, *B. cereus* T, high calcium; 5, *B. cereus* T, low calcium; 6, *B. megaterium* native parent; 7, *B. megaterium* decoated parent; 8, *B. megaterium* exosporiumless variant; 9, *B. megaterium* decoated, exosporiumless variant.

in vitro to stabilize a spore enzyme, Warth (91) found that an a_w of 0.73 (corresponding to a water content of 20%, dry weight basis) is required to stabilize proteins within the spore protoplast. For a spore in water, this reduced a_w must be maintained by high pressure exerted inward by elastic cortex layers. Algie (6) calculated from several parameters that the effective a_w of protoplasts of *B. stearothermophilus* spores is about 0.83. On the other hand, Bradbury et al. (19) deduced from nuclear magnetic resonance studies that the internal water is relatively free (see Background, above). If so, there could be an a_w near 1.0.

Although the turgor pressure within bacterial spores has never been determined

directly, there have been estimates based on predicted water activities within spores that the turgor pressure may be as great as 50 MPa (59, 60, 91). In fact, one of the proposed functions for the thick network of cortical peptidoglycan is to maintain the high turgor pressure and associated low water activity predicted within the spore. Turgor pressure can be related to differences in water activity between the inside and the outside of the cell by means of the equation used previously by Marquis (60): $p^{in} - p^{out}$ = turgor pressure = $(2.30\ RT/\overline{V})\ (\log a_w^{out}/a_w^{in})$, where p is the hydrostatic pressure, R is the gas constant, T is the Kelvin temperature, $\overline{V}$ is the molar volume of water, a_w^{out} is the water activity of the medium outside of the spore (technically, outside of the inner membrane), and a_w^{in} is the water activity within the protoplast.

Unfortunately, a method has not yet been devised to measure water activity in situ within the protoplast of fully hydrated spores. Attempts to do so involved a dew-point hygrometer probe and gram pellets of lysozyme-sensitized *B. stearothermophilus* and *B. megaterium* spores equilibrated in a 50% glucose solution and in water within a closed container (T. C. Beaman and P. Gerhardt, personal communication). The results gave a_w values with the spore pellets that were the same as those with only the suspending medium, whether glucose solution or water. These experimental findings are consistent with the alternative prediction that if surrounding water freely permeates throughout the spore, as has been shown (17, 67), at equilibrium in a closed system the a_w within the spore protoplast must be the same as the a_w of the surrounding medium and proportional to the relative humidity of the surrounding atmosphere. The experimental findings may be inconclusive, however, because of (i) the large volume fraction and high content of environmental water in the interstitium and integument relative to the small volume fraction and low content of cytoplasmic water within the protoplast; (ii) the poor sensitivity of the instrument at high a_w values; and (iii) the likely high a_w value of the cytoplasmic water. A different approach seems necessary, e.g., the use of differential scanning calorimetry to measure freezing-point depression in the surrounding water and the cytoplasmic water of spores.

Resistance after Further Dehydration

The water content of spores as a function of atmospheric relative humidity is shown in the typically shaped sorption isotherms obtained by various investigators (58, 64, 73,

89, 93). This relationship takes on added significance if resistance to dry heat is considered. For moist-heat resistance, the spores are fully hydrated in an environment of liquid water or steam vapor (as in an autoclave). For dry-heat resistance, the spores are usually first dried to equilibrium with air at ambient temperature and moderate humidity and then are further dried (as in a hot-air oven) to equilibrium with air at high temperature and low humidity. Consequently, spores generally are more resistant to dry heat than to moist heat because they become drier. However, dried vegetative cells do not attain the same degree of heat resistance as do wet or dry spores.

The qualitative generalization that spores become more resistant as they become drier was quantified in a classic study by Murrell and Scott (70) in which sporal water activity (equating this term with atmospheric relative humidity) was controlled at the temperature of heating. For six spore species that vary widely in moist-heat resistance (a 10^5-fold range in D_{110} values), heat resistance increases as a_w is decreased. Maximum resistance occurs at a_w values of 0.2 to 0.4, where the originally wide differences in D_{110} values are reduced to only about a 10-fold range. The corresponding water contents (dry weight basis) are about 5 to 10% (65). Although the distribution of water within the spore was not considered in these early studies, the water content of the protoplasts must similarly have been extremely low at the low water content of the entire spores.

Role of the Cortex

The physiological process (or processes) by which the spore protoplast becomes dehydrated during sporogenesis has been subjected to much theorizing, which has been amply dealt with in previous reviews (36, 55, 91). These theories mainly have focused on mechanical processes by which the protoplast may become dehydrated by the cortex through its contraction (51), osmoregulation (40), or anisotropic expansion (91).

Heat resistance, and presumably protoplast water content also, can be manipulated osmotically. Coat-modified (and thus lysozyme-sensitized) dormant spores of *Bacillus cereus* T that are heat sensitized by 3 M CaCl$_2$ regain heat resistance if >2 M sucrose is added, as shown by Gould and Dring (40). Furthermore, just-germinated spores of this species regain the thermoresistance level of the original dormant spores if heated in >1.5 M sucrose (27).

All of the proposals for the possible role of the cortex in dehydration and heat resistance are based on a view that the peptidoglycan of the cortex has a high charge density and contains large numbers of mineral cations to balance the excess negative charge on the polymer matrix. This view is reasonable in terms of the structure described by Warth and Strominger (92) and by Tipper and Gauthier (86) for cortical peptidoglycans. They found cortical peptidoglycans to have a low degree of peptide cross-linking, many muramyl residues substituted with only L-alanine, and many muramyl residues having delta-lactam structures with no substituent peptides. Their proposed structure indicates a high net negative charge for cortical peptidoglycans, as compared with vegetative cell wall peptidoglycans.

Assessments of the dielectric properties of a number of types of spores, including those with coats removed chemically, allowed for estimations of the conductivities, and therefore charge densities, of cortical peptidoglycans in situ (22–24). In general, the conductivities are considerably lower than those predicted on the basis of the prevailing view of the structure of spore peptidoglycan. Spores of *B. cereus* T are notable for their extremely low conductivity, less than 0.03 S/m even when a current of 50 MHz is used so that the membranes of the cells are capacitatively short circuited. Thus, the cortex and the protoplast both have low conductivity and are in a state of extreme electrostasis. Clearly, the dielectric picture and the prevailing picture of cortical peptidoglycan structure are not compatible.

Consequently, the properties of cortical peptidoglycans from spores of *B. cereus* T, *B. megaterium* ATCC 19213, and other *Bacillus* spores were reinvestigated (G. R. Bender and R. E. Marquis, personal communication). Initial isolations of cortex were made from mechanically disrupted spores, basically according to the mechanical isolation procedures used previously by others. The product from *B. megaterium* ATCC 19213, for example, was found to be a lysozyme-sensitive peptidoglycan with high net negative charge and a highly expandable structure. The latter was indicated by a value for dextran-impermeable volume of 3.8 ml/g at a pH value of 2, which increased to 13.3 ml/g when the structures were titrated to a pH value of 10. Clearly, such cortical fragments could change from a contracted state when the carboxyl groups are protonated to a swollen state when the groups are negatively charged at high pH values. The structures behaved as one would expect from the theories of a mechanically active cortex.

A different approach then was taken with isolation of cortical peptidoglycans from decoated spores by means of chemical extraction basically according to the fractionation procedure of Park and Hancock (75). The structures isolated in this way were very different from those isolated from mechanically disrupted spores. The peptidoglycans isolated by means of chemical extraction were sensitive to lysozyme, and addition of the enzyme to suspensions of the structures isolated from *B. cereus* T or *B. megaterium* ATCC 19213 resulted in reductions in optical density of up to 90%. However, the chemically isolated structures had relatively few free amino groups able to react with fluorodinitrobenzene. For example, cortical structures isolated from chemically extracted *B. megaterium* ATCC 19213 spores had an average of only 23 nmol of free amino groups per mg (dry weight), and this value was in sharp contrast to values of 980 nmol/g for cell walls isolated from mechanically disrupted *Micrococcus luteus* cells or of 1,190 nmol/g for cortical peptidoglycans isolated from mechanically disrupted *B. megaterium* spores. The protein contents of the cortical structures obtained by chemical extraction were only about 50 μg/mg (dry weight) when protein was assayed by means of the Folin reaction with bovine serum albumin as the standard. The phosphate content of the acid-hydrolyzed cortical structure was less than 1 μmol/mg (dry weight). That the cortical structures are peptidoglycan was indicated by sensitivity to lysozyme. That they are highly cross-linked was indicated by the low levels of free amino groups. Moreover, when the cortical stuctures were titrated with acid or base, they did not swell and shrink in the way that loosely cross-linked peptidoglycans have been found to do.

The dielectric properties of spores and the nature of the cortical peptidoglycan isolated by means of chemical extraction suggest a very different view of cortical structure and function from that previously held. It now seems that the major function of the cortex is to act as a mechanical restraining structure to maintain dehydration of the protoplast. In keeping with this function, the cortical peptidoglycan is highly cross-linked. The previous findings of sparsely cross-linked, very open structures for cortical peptidoglycans probably reflect damage done to the structures during isolation. In fact, cortical peptidoglycans isolated by chemical extraction of decoated spores can be converted to open, loosely cross-linked structures by mechanical shearing (for example, by sonication in the presence of glass microbeads).

The cortex in itself is clearly capable of and necessary for maintaining, if not attaining, the dehydrated resistant state of the protoplast. The lack of a major role for coat layers was indicated early on by the isolation of a coatless mutant of *Clostridium perfringens* that retains refractility and heat resistance (25) and was further indicated in highly and poorly resistant spores by the divestment of their coat without the loss of spore-characteristic heat resistance (48, 49). The necessity of cortex synthesis for heat resistance was also shown by Imae and Strominger (44), using mutants defective in a cortex-synthesizing enzyme. The isolation of free spore protoplasts devoid of cortex but retaining heat resistance has not been accomplished, despite efforts to do so (32, 48).

The essential role of the cortex in maintaining the resistant state of the protoplast is also indicated by the correlation between the volume ratio of protoplast to protoplast-plus-cortex and the heat resistance of various spores (Fig. 5). This correlation significantly includes *B. megaterium* spores, in which the low protoplast water content should provide high heat resistance but the relatively small cortex volume apparently causes low heat resistance. The results in Fig. 5 are consistent with similar findings by various investigators in a wide range of species (5, 11; J. E. Algie and L. S. Tisa, *Spore Newsl.* 7:20–21, 1981; A. D. Hitchins and R. A. Slepecky, *Spore Newsl.* 7:103–104, 1981). Apparently, the larger the relative volume of cortex (or the smaller the relative volume of protoplast), the greater the heat resistance.

Role of the Membranes

Marquis (61) has reviewed the concept that protoplast dehydration initially occurs by osmotic loss of water concomitant with the movement of potassium out of the forespore. These transfers result from acidification of the cortical space between the two spore membranes. ATP is hydrolyzed by the unique inversion of the F_1 components of the outer membrane. The cortex is then laid down and serves to maintain, rather than to attain, the initial state of dehydration in the protoplast. This concept is further detailed below (Uptake during Sporulation).

Molecular Mechanisms

The molecular processes by which the spore protoplast may become dehydrated are exemplified by the in vitro studies of Lindsay and Murrell (53, 54) on the interaction of nucleic acids with calcium dipicolinate and dipicolinic acid (DPA). These spore-specific chelate compounds bind strongly to DNA in vitro so as to increase its wet and dry density and its thermal stability. Calcium dipicolinate is apparently intercalated within the helical structure of DNA, thus displacing intramolecular water. Calcium dipicolinate is also bound to different RNA species, although the mechanism of binding is not known. Because of the primary role of Ca and other minerals and the dispensable role of DPA in determining heat resistance in vivo (see the section on DPA, below), it will be desirable to examine the interaction of minerals alone with nucleic acids.

At the molecular level, thermostability of the dehydrated spore protoplast may not result from simple volume exclusion of water, but instead (or additionally) may be due to the formation of covalent or noncovalent bonds between reactive groups on essential macromolecules and between counterions to form an insoluble, gelled, high-polymer matrix. The gel concept was formulated early on by Black and Gerhardt (17) from experimental evidence in the literature on disulfide and calcium cross-linking in thermostable proteins. The bond concept was developed more recently by Bradbury et al. (19) from experimental evidence in the literature on enzymes immobilized in solid supports, in which stabilization results from noncovalent electrostatic interactions and hydrogen bonds between negatively charged carboxyl groups of a polymer and positively charged groups on the surface of an enzyme. As water is removed and close molecular packing is increased, there is more and more opportunity for specific and nonspecific intermolecular bonding, especially with minerals. The result is a protoplast in which small molecules, macromolecules, and supramolecular structures are tightly packed, bonded, dehydrated, and immobilized and thus are thermostabilized. Evidence for this physical condition within the dormant spore protoplast is the almost complete absence of electrical conductivity even at high current frequencies, as revealed in the dielectric studies of Carstensen et al. (22–24).

MINERALIZATION

Background

The view that mineralization, particularly calcification, is associated with heat resistance of spores has an intuitive basis. One feels that a calcified cell should be dormant and protected against thermal damage, in essence that it should be a living fossil. However, any such intuitive feelings need to be translated into specific molecular biological terms. Currently, we have a detailed picture of the mineral contents of spores, the physical states of spore minerals, and the contributions of specific minerals to heat resistance. In addition, we have a reasonably good view of the physiology of mineral uptake and deposition during sporogenesis, although molecular and genetic details of the uptake systems are lacking. However, we have only speculative perceptions of the mechanisms by which minerals act to protect spore components against heat damage, and only a clouded view of the nature of heat damage and the major targets of heat damage. Moreover, as indicated in the previous section, increased thermoresistance associated with increased mineralization appears to involve changes in water content of spores between defined limits, so that mineralization and dehydration are not always separable in terms of their contributions to resistance.

Correlation with Heat Resistance

Murrell and Warth (71) carried out an extensive set of experiments with 19 types of *Bacillus* spores and correlated increased heat resistance with increased Ca content. They found, on average, a 30-fold increase in D_{100} value for each percentage increase in Ca content over the range from 1.2 to 3.2% (dry weight). Heat resistance was positively related also to DPA content and to cortex content assessed as diaminopimelate content. Resistance was inversely related to Mg content. Similar associations between heat resistance and mineralization have been found in studies of the spores of single species with variations in mineral content induced by changes in the growth and sporulation medium (84), by endotrophic sporulation in water (18, 33), or by ion exchange (2).

Although Ca is the major mineral associated with resistance, spores contain other minerals. Mn is prominent in many spores, as is Mg. Na usually is present at up to about 10% of the mineral content, while minerals such as K, Zn, Fe, Cu, Co, and Ni are present at lower levels. Spores may also contain silicates, as indicated by the results of X-ray microanalyses (80, 85).

The recent findings of Beaman and Gerhardt (10) indicate that mineralization of spores is accompanied by reduction in protoplast water content over much of the range, so that at least part of the increased heat resistance associated with mineralization is due to dehydration. At the upper and lower limits of protoplast water content in fully hydrated spores, however, mineralization acts independently.

Minerals have been found to increase resistance to dry heat as well as to moist heat (3). An example is *Bacillus subtilis* subsp. *niger* spores which were heated after being dried (Fig. 6) (Bender and Marquis, personal communication). The native form was clearly more resistant to dry-heat killing than the protonated H form. However, when the H-form spores were converted to the Ca form, the native level of resistance to dry-heat killing was largely restored.

Minerals have also been shown to decrease spore thermoresistance. Multivalent cations (Sr^{2+}, Ca^{2+}, Mg^{2+}, and La^{3+}, in order

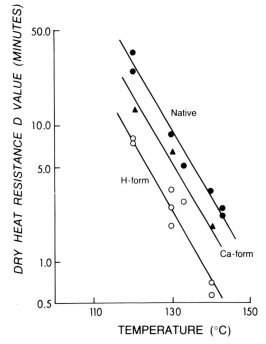

Figure 6. Dry heat resistance of native mineralized (●), H-form demineralized (○), and Ca-form remineralized (▲) spores of *B. subtilis* subsp. *niger*.

of increasing effectiveness) at high concentrations (e.g., 4 M) sensitize various species of spores, as shown by Gould and Dring (40). Fe-loaded spores of *Clostridium botulinum* are highly sensitized to thermal killing (E. A. Johnson, personal communication).

Localization

Most of the minerals in spores appear to be localized in the protoplast, as indicated by X-ray microanalysis of spores of *B. cereus* (79, 80, 85), *B. megaterium* (45, 74, 79), and *C. perfringens* (7). Mn (and probably other minerals) can be deposited on the surfaces of certain spores (78). However, the minerals involved in heat resistance are likely to be those associated with the protoplast. A possible exception was described by Ando and Tsuzuki (8), who found that the cortex-lytic enzyme of *C. perfringens* NCTC 8238 is damaged by heat and can be protected by minerals. Heat-damaged spores of this organism can be superdormant, but they can be made to germinate after decoating by means of lysozyme action. This situation seems to be unusual. In general, heat-killed spores cannot be recovered by decoating and lysozyme treatment.

Uptake during Sporulation

The sequence of cytological and biochemical events in sporulation has been defined in some detail. Early in the process, sporulating cells undergo asymmetric division after which the smaller cell is engulfed by the larger cell in a type of procaryotic phagocytosis. The smaller, forespore compartment is then contained within the larger, sporangial compartment.

The forespore is unusual in that it has two membranes, an inner right-side-out membrane and an outer inside-out membrane, so that the F_1 sectors of the F_1F_0 ATPases of the outer forespore membrane are in the sporangial cytoplasm. Forespores isolated from disrupted sporulating cells can hydrolyze exogenously added ATP, whereas vegetative protoplasts of sporulating cells cannot (94). The contra-workings of the two forespore membranes result in loss of K from forespores. Isolated forespores have very low levels of K and other minerals (63). Potassium loss is associated with osmotic loss of water, which is reflected by the darkened image of the forespore in the phase-contrast microscope and by the low values for dextran-impermeable volume per gram (dry weight) of isolated forespores.

For aerobic species, forespores would have restricted access to oxygen from the environment, which would pass first through the sporangial membrane and then through the outer forespore membrane before arriving within the forespore. Detailed analyses of the levels of metabolites and enzymes in forespores and sporangial cells carried out by Singh et al. (83) indicated that forespores have basically an anaerobic metabolism. Levels of tricarboxylic acid-cycle enzymes are very low in forespores, while levels of glycolytic en-

zymes are not reduced. The forespores accumulate 3-phosphoglycerate, even though phosphoglycerate mutase, enolase, and pyruvate kinase are present in high levels. The low levels of K in forespores would be expected to be inhibitory for pyruvate kinase, which requires K at millimolar levels for activity. The enzymes of both the inner and outer forespore membranes would act to acidify the cortical space between the two membranes. Acidification of the cortical space would allow for ATP synthesis by the forespore inner membrane as protons diffuse into the cell through the F_1F_0 ATPase acting in the synthase mode. However, the net result would be acidification of the forespore, and this acidification would in part account for the very low levels of K in the cells.

Some 3 h before uptake of DPA by forespores, ATP levels in the forespores were found to be high but to decline thereafter to the low values characteristic of dormant spores (83). Moreover, the level of NADH and the NADH/NAD ratio drop sharply in the forespores, but not in the sporangial cells. Thus, it appears that forespores become deficient in ATP and other energy currency and are then unable to maintain proton currents across the cell membrane. The initial osmotic demineralization and dehydration and the acidification of the forespore set the stage for subsequent mineralization.

The uptake of Ca by sporulating cells and forespores has been studied in some detail, especially by Ellar (29). The molecular specifics of the transport systems are still unknown, and there is no knowledge of how the systems are regulated. Moreover, sporulation-specific transport systems for minerals (such as Mn, Mg, Zn, Fe, and possibly even K) have been identified only in very preliminary ways, certainly not in terms of specific genes and proteins. At about stage IV of sporulation, the sporangial cell develops a transport system for uptake of Ca, first identified by Eisenstadt and Silver (28) and later by Bronner et al. (21). The flow of Ca into the spore involves initial concentration by the sporangial cell via systems sensitive to respiratory inhibitors and subsequent movement of the mineral into the forespore by passive mechanisms (82).

Physical States

The high levels of mineralization of spores and their dehydrated states result in the immobilization of minerals, possibly because of precipitation. Since DPA and minerals can be extracted separately from spores, it seems that anionic groups other than or in addition to those of DPA can be involved in mineral salt formation in spores (see DPA, below). Mineral precipitates or complexes within spores may involve proteins, anionic lipids, and nucleic acids. The conclusion that mineral ions are immobilized within spores is based on findings with a variety of techniques, including dielectric techniques (22–24) and techniques for assessing electron paramagnetic resonance (46, 96). The immobilization may be energetically important in mineral uptake by forespores in allowing for passive flow of mineral ions from the sporangial cell into the forespore.

Exchange and Effectiveness of Specific Cations

Alderton and Snell (2) found early on that the mineral contents of spores can be altered by means of ion exchange. The basic procedure involves first titrating spores with acid to displace minerals and then back-titrating with mineral base solutions to obtain remineralized forms. Alderton and colleagues (1, 3, 4) and others (8, 76) have used the exchange procedure to obtain spores with altered mineral contents. The general finding is that demineralized spores are much more sensitive to wet or dry heat than are native spores or spores that have been remineralized. Resistance to radiation and to disinfectants does not appear to be greatly affected by demineralization. However, resistance to the germinating action and subsequent

lethality of hydrostatic pressure is actually increased by demineralization (15).

In most of the early studies of ion exchange, assays of mineral contents of spores were not carried out, and so it is difficult to know how effective the exchange was. Subsequently, Marquis and Bender (16, 62) used atomic absorption spectrophotometric assays to monitor changes in mineral composition. They found that spores of *B. megaterium* ATCC 19213, *B. subtilis* subsp. *niger*, and *B. stearothermophilus* ATCC 7953 can be demineralized completely without loss of viability in populations by means of controlled acidification and heating. Acid sensitivities vary greatly among various types of spores, and specific procedures must be developed for each specific type. As expected, the demineralized spores are more sensitive to heat than are the native spores. For example, killing of native spores of *B. stearothermophilus* ATCC 7953 is characterized by a D value of 1.0 min at a temperature of 123°C. This same D value is obtained by heating demineralized spores at only 102°C. However, demineralized spores remain greatly more heat resistant than vegetative cells.

Spore remineralization with Ca produces cells with the full heat resistance of native spores, but other minerals also are effective in restoring resistance. A hierarchy for effectiveness is defined, with Ca more effective than Mn, Mg, or K. Na is entirely ineffective. Minerals were found (16) to be more effective in protecting spores against lower killing temperature than against higher temperatures. As an example (Fig. 7), spores of *B. subtilis* subsp. *niger* were demineralized by acid extraction, remineralized with incremental amounts of Ca, and then used for determinations of D values at 90 and 95°C. Demineralized spores had D values of less than 1 min at both temperatures. The D value at 90°C increased progressively as the cells took up more and more Ca, to a maximum value of about 20 min for spores containing 0.6 μmol of Ca per mg (dry weight). The D

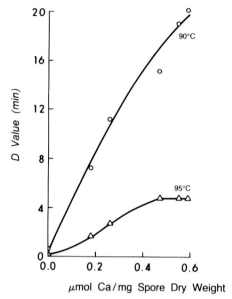

Figure 7. Increase in heat resistance (*D* value) at two temperatures, associated with remineralization by calcium from H-form spores of *B. subtilis* subsp. *niger* (16).

value at 95°C could be increased only to about 4 min, when spores took up 0.45 μmol of Ca per mg and further calcification did not increase resistance. The subject of mineral content and spore resistance has been reviewed recently (61).

THERMAL ADAPTATION

Correlation with Heat Resistance

Evolutionary selection for heat stability has long been known to affect thermoresistance in bacterial spores, just as with vegetative cells. That is, the spores of thermophilic species inherently are more resistant than those of mesophilic or psychrophilic species. This generalization has been quantified by Warth (90). In each species, the spore is resistant by about 40°C more than the vegetative cell.

In addition to this inherent (or intrinsic or constitutive) thermal adaptation, however, there is an imposed (or extrinsic or inducible) component. That is, the spores of a given species grown at maximum tempera-

ture generally are more resistant than those grown at optimum or minimum temperature, as exemplified by the early experiments of Williams and Robertson (95) and recent experiments by Khoury et al. (47). An extreme instance of imposed thermal adaptation was accomplished by Heinen and Lauwers (43), who gradually adapted "*Bacillus caldolyticus*" to grow and form some spores at 105°C.

Both the inherent and the imposed components of thermal adaptation were shown by Beaman and Gerhardt (10) to share a common linear correlation with heat resistance over a wide range (Fig. 8). For example, *B. stearothermophilus* spores themselves vary more than 10-fold in D_{100} value, depending on the growth temperature, and differ more than 1,000-fold from the psychrophilic spores of *Bacillus macquariensis*. Yet even this psychrophilic spore is more heat resistant (D_{100}, 0.33 min; 10) than the thermophilic vegetative cell of *B. stearothermophilus* (D_{100}, 0.067 min; 13) when compared at their optimum growth temperatures.

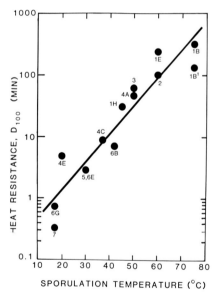

Figure 8. Heat resistance correlated with sporulation temperature of lysozyme-sensitive native spores of seven species. The identity numbers correspond to those in Fig. 3. For further identification, see Table 1 in the original publication (10).

Inherent thermal adaptation in vegetative cells is manifested in various molecular modifications; for example, thermophiles are distinctive for the primary structure of their proteins and the incorporation of high-melting-point fatty acids in the lipid of their membranes (20). In spores, however, both the inherent and imposed components of thermal adaptation are reflected in the protoplast water content over much, but not all, of the temperature range (10). Increased sporulation temperature is correlated with decreased protoplast water content between limits of about 57 and 28%. Above and below these limits, however, sporulation temperature and protoplast water content are independent (Fig. 9). The relationships shown in Fig. 8 and 9 are reflected in the generalized correlation between thermoresistance and protoplast water content shown in Fig. 3.

Resistance after Sublethal Heating

Increased temperature during sporulation can cause increased heat resistance, but enhancement of resistance to lethal heating may also result from prior sublethal heating of dormant spores, which reportedly can sometimes induce an increase in dormancy (31). Increased heat resistance of certain vegetative cells can be induced by sublethal heating before lethal heating (56). With bacterial spores, sublethal heating ("heat shock" or "heat activation") before inoculation for viable counting is commonly used to overcome the inability, or to stimulate the ability, of dormant spores to germinate and grow in otherwise favorable conditions.

The effect of sublethal heating on thermoresistance and other properties of B. stearothermophilus ATCC 7953 spores was recently investigated by Beaman et al. (13). The preheated spores were shown to be separable from a heterogeneous population into two distinct fractions by the use of buoyant density centrifugation with an appropriate (Nycodenz) gradient. One band has a density of 1.240 g/ml and the other has a density of 1.340 g/ml, the same as that of the original dormant spores. The spores in both bands are germinable faster than the original dormant spores and thus are considered to be activated. The proportion of 1.240-g/ml spores becomes larger with longer time of preheating. The 1.240-g/ml spores are considered to be fully activated and permeabilized at the outer membrane in that they are fully depleted of DPA and partly depleted of minerals, become susceptible to lysozyme, have a different appearance by electron microscopy, and lose some of their heat resistance. The 1.34-g/ml spores, which are considered to be partly activated, show no evidence of permeabilization and are much more heat resistant (D_{100}, 1,960 min) than the original dormant spores (D_{100}, 760 min). This phenomenon of enhanced resistance may involve either in situ induction or selection of a preexisting subpopulation.

Etoa and Michiels (30) similarly reported that heat resistance of B. stearothermophilus CNCH 5781 spores increases after heat activation. Preheating the spores at 100°C for 30 min in water increases the D_{121} value by 29%, and the same preheating in nutrient medium increases the D_{121} value by

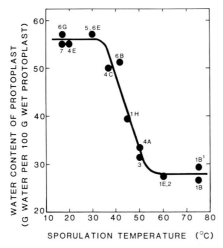

Figure 9. Protoplast water content correlated with sporulation temperature of lysozyme-sensitive native spores of seven species. The identity numbers correspond to those in Fig. 3. For further identification, see Table 1 in the original publication (10).

62%. Further heating in water or medium gradually decreases the heat resistance. The spore population was apparently not analyzed by buoyant density centrifugation, and the relatively small increase in D values may be explained by a mixed population of more resistant, partly activated spores and less resistant, fully activated spores, as distinguished by Beaman et al. (13).

The increased heat resistance induced by prior sublethal heating raises the question, does an increase in heat resistance occur during the process of heating up spores to lethal temperature? In certain bacterial vegetative cells, heat resistance is increased not only by prior sublethal heating (56) but also during the process of heating the cells up to lethal temperature, with the extent of increase depending on the slowness of temperature rise (57). This same phenomenon had been reported for *B. megaterium* spores early on by Alderton et al. (4). However, their test was made with acid-demineralized spores in the presence of Ca ions at elevated pH, so that the increase in heat resistance can now be explained by heat-enhanced remineralization rather than by thermal adaptation per se. This may also explain the greater increase in heat resistance of *B. stearothermophilus* spores caused by sublethal heating in nutrient medium rather than in water, as observed by Etoa and Michiels (30). Experiments are still needed to learn whether increased heat resistance of spores can be induced, independently of mineral gain, during the process of slowly heating native spores up to lethal temperature.

OTHER POSSIBLE DETERMINANTS

DPA

DPA is unique to bacterial spores, is concentrated at extraordinarily high levels within the protoplast, and apparently is chelated with Ca or other cations. In the older literature, correlations were presented repeatedly indicating that DPA makes a major contribution to and is correlated with heat resistance, yet contradictory data were also published.

A turning point occurred in 1972 when Hanson et al. (42) isolated and characterized a *B. cereus* T revertant spore that was devoid of DPA (less than 0.01% [dry weight]) and low in Ca and Mn content, yet was as heat resistant as the parent spore. This evidence appeared to rule out the direct involvement of DPA in heat resistance.

In 1981, furthermore, this genetic dissociation of DPA from heat resistance was repeated in confirming experiments (G. J. Dring and G. W. Gould, *Spore Newsl.* 7:130–131, 1981). Using the same DPA-negative but heat-sensitive *B. cereus* T mutant spore that was originally isolated by Halvorson and Swanson (41) and employed by Hanson et al. (42), Dring and Gould isolated DPA-negative heat-resistant revertant strain 47. The D_{80} value (70 min) of this spore strain was indistinguishable from that of the original wild-type strain. DPA was not detected by two separate assays, with which 0.05% of spore dry weight would have been readily detectable in strain 47 and with which 10% was found in the wild-type strain. One might argue that DPA is required in only relatively low concentration because *B. cereus* T spores are relatively low in heat resistance. The 0.05% detectability limit, however, is far below the limits at which the DPA contents of *B. cereus* T spores are seemingly correlated with heat resistance (18, 26).

Zytkovicz and Halvorson (98) reported the isolation of a DPA-minus mutant spore in another species, *B. subtilis* 168, which also retains the same heat resistance as the wild strain. However, Balassa et al. (9) reported the isolation of a DPA-minus mutant spore of the same organism that requires the incorporation of added DPA for the development of heat resistance. The spores of *B. subtilis* 168, like those of *B. cereus* T, are relatively low in heat resistance.

The dissociation of DPA from heat resistance in a highly resistant species and with a different type of evidence was recently shown in a study of heat shock in *B. stearothermophilus* ATCC 7953 spores by Beaman et al. (13). Fully heat-activated spores, which were separated by buoyant density centrifugation into a band at 1.240 g/ml, are virtually devoid of DPA (0.5 nmol/mg [dry weight] of spores, or 0.01%) yet retain a high level of heat resistance (D_{100}, 453 min).

Another line of evidence in additional organisms was mentioned by Marquis (61) and is further reported here. Spores of *B. subtilis* subsp. *niger*, a relatively resistant type, were demineralized to yield protonated spores, which were nearly devoid of minerals but retained essentially all of the original DPA (306 nmol/mg [dry weight] of spores, or 5.1%). During Na remineralization of some of the protonated spores, there was a major loss of DPA to a final content of only 0.5 nmol/mg (dry weight) of spores, or 0.01%. These spores were found to have about the same heat resistance (D_{100}, 0.11 min) as the H-form spores, which was less than that for the original dormant spores (D_{100}, 1.0 min) but still greatly more than for vegetative cells (extrapolated D_{100} value, about 10^{-6} min). Furthermore, conversion of the Na-form spores to the Ca form resulted in regain of heat resistance to approximately the same level (D_{100}, 1.1 min) as in the original native spores, yet with the DPA content remaining at the same negligible level as in the Na-form spores.

Taken together, these several different lines of evidence with several different types of spores lead to the conclusion that DPA is probably not an essential determinant of heat resistance. Hanson et al. (42) have suggested that DPA may instead function in the maintenance of resistance in that heat-resistant, DPA-free revertant spores are unstable to storage in water and in normal germination. Hanson et al. and others (55) have considered that other small molecules (e.g., phosphoglyceric acid, sulfolactic acid, glutamic acid) that occur in the spore in significant (but far lower) levels may have functionally replaced DPA in DPA-free thermoresistant spores, but there is no evidence to support this speculation. The underlying problem in attempting to correlate heat resistance with DPA content seems to have been not separating the effects of DPA from those of Ca and other minerals, i.e., of not dealing with DPA independently.

SASPs

Mason and Setlow (66) have used mutants of *B. subtilis* 168 with deletions in the *sspA* and *sspB* genes to implicate the small, acid-soluble proteins (SASPs) coded for by these genes in the resistance of spores to UV irradiation. However, these proteins appear to have little, if any, role in heat resistance. SASPα$^-$ and SASPα$^-$β$^-$ spores have slightly decreased resistance to heating at 85°C, but this decrease could have been the result of alterations other than the lack of SASPs in the spores. The general nature of these unique proteins has been recently reviewed by Setlow (81).

CONCLUSIONS

Experimental evidence enables us now to define quantitatively the main physicochemical determinants of heat resistance and to envisage an increasingly clear view of the physiological processes by which thermoresistance is attained during sporogenesis and maintained during dormancy of bacterial spores. Spores are resistant by about 40°C more than their corresponding vegetative cells. Dehydration of the protoplast is the only determinant necessary and sufficient in itself for the elevated level of heat resistance characteristic of spores. The protoplast water content of fully hydrated spores varies between limits of about 57 to 28% on a wet-weight basis in lysozyme-sensitive spores and presumably also in lysozyme-resistant spores. The processes for attaining protoplast dehydration are not fully understood, but we propose a novel concept of potassium-coupled osmotic egress of water at an early stage of sporogenesis. The maintenance of protoplast dehydration requires an intact cortex, but not coat nor exosporium. The extent of protoplast dehydration depends on the relative amount of cortical peptidoglycan, which is now shown to be tightly cross-linked.

A second determinant of spore heat resistance is protoplast mineralization, mostly by calcium. Mineralization affects heat resistance independently at the upper and lower limits of protoplast dehydration, but between them is reflected in changed water content. Mineralization is apparently attained by active transport through the sporangial membrane and passive flow into the developing spore. Although often associated with minerals, DPA is apparently not necessary for attaining heat resistance, though may function in retaining it.

A third determinant of spore heat resistance is thermal adaptation, which mostly is genetically inherent. However, increasing the temperature of sporulation can impose additional heat resistance, with or without affecting protoplast dehydration. Heat resistance is increased temporarily in spores partly activated by sublethal heat shock, but is decreased in fully activated spores.

At the molecular level, as evidenced especially by dielectric measurements, the spore protoplast contains complexes of small molecules, macromolecules, and supramolecular structures so compacted, bonded, and dehydrated that they become immobilized and thus thermostabilized.

Acknowledgments. Our efforts were supported by contract DAAL03-86-K-0075 from the Biological Sciences Program of the U.S. Army Research Office. This is journal article no. 12982 from the Michigan Agricultural Experiment Station.

LITERATURE CITED

1. Alderton, G., J. K. Chen, and K. A. Ito. 1980. Heat resistance of the chemical resistance forms of *Clostridium botulinum* 62A spores over the water activity range of 0 to 0.9. *Appl. Environ. Microbiol.* 40:511–515.

2. Alderton, G., and N. Snell. 1963. Base exchange and heat resistance in bacterial spores. *Biochem. Biophys. Res. Commun.* 10:139–143.

3. Alderton, G., and N. Snell. 1969. Bacterial spores: chemical sensitization to heat. *Science* 163:1212–1213.

4. Alderton, G., P. A. Thompson, and N. Snell. 1964. Heat adaptation and ion exchange in *Bacillus megaterium* spores. *Science* 143:141–143.

5. Algie, J. E. 1983. The heat resistance of bacterial spores and its relationship to the contraction of the forespore protoplasm during sporulation. *Curr. Microbiol.* 9:173–175.

6. Algie, J. E. 1984. Effect of the internal water activity of bacterial spores on their heat resistance in water. *Curr. Microbiol.* 11:293–296.

7. Ando, Y. 1976. Some properties of ionic forms of spores of a *Clostridium perfringens* strain. *Jpn. J. Bacteriol.* 31:713–717.

8. Ando, Y., and T. Tsuzuki. 1983. Mechanism of chemical manipulation of heat resistance of *Clostridium perfringens* spores. *J. Appl. Bacteriol.* 54:197–202.

9. Balassa, G., P. Milhaud, E. Raulet, M. T. Silva, and J. C. F. Sousa. 1979. A *Bacillus subtilis* mutant requiring dipicolinic acid for the development of heat-resistant spores. *J. Gen. Microbiol.* 110:365–379.

10. Beaman, T. C., and P. Gerhardt. 1986. Heat resistance of bacterial spores correlated with protoplast dehydration, mineralization, and thermal adaptation. *Appl. Environ. Microbiol.* 52:1242–1246.

11. Beaman, T. C., J. T. Greenamyre, T. R. Corner, H. S. Pankratz, and P. Gerhardt. 1982. Bacterial spore heat resistance correlated with water content, wet density, and protoplast/sporoplast volume ratio. *J. Bacteriol.* 150:870–877.

12. Beaman, T. C., T. Koshikawa, H. S. Pankratz, and P. Gerhardt. 1984. Dehydration partitioned within core protoplast accounts for heat resistance of bacterial spores. *FEMS Microbiol. Lett.* 24:47–51.

13. Beaman, T. C., H. S. Pankratz, and P. Gerhardt. 1988. Heat shock affects permeability and resistance of *Bacillus stearothermophilus* spores. *Appl. Environ. Microbiol.* 54:2515–2520.

14. Beaman, T. C., H. S. Pankratz, and P. Gerhardt. 1989. Low heat resistance of *Bacillus sphaericus* spores correlated with high protoplast water content. *FEMS Microbiol. Lett.* 58:1–4.

15. Bender, G. R., and R. E. Marquis. 1982. Sensitivity of various salt forms of *Bacillus megaterium* spores to the germinating action of hydrostatic pressure. *Can. J. Microbiol.* 28:643–649.

16. Bender, G. R., and R. E. Marquis. 1985. Spore heat resistance and specific mineralization. *Appl. Environ. Microbiol.* 50:1414–1421.

17. Black, S. H., and P. Gerhardt. 1962. Permeability

of bacterial spores. IV. Water content, uptake, and distribution. *J. Bacteriol.* 83:960–967.

18. **Black, S. H., T. Hashimoto, and P. Gerhardt. 1960.** Calcium reversal of the heat susceptibility and dipicolinate deficiency of spores formed "endotrophically" in water. *Can. J. Microbiol.* 6:213–224.

19. **Bradbury, J. H., J. R. Foster, B. Hammer, J. Lindsay, and W. G. Murrell. 1981.** The source of the heat resistance of bacterial spores; study of water in spores by NMR. *Biochim. Biophys. Acta* 678:157–164.

20. **Brock, T. D. 1985.** Life at high temperatures. *Science* 230:132–138.

21. **Bronner, F., W. C. Nash, and E. E. Golub. 1975.** Calcium transport in *Bacillus megaterium*, p. 356–361. *In* P. Gerhardt, R. N. Costilow, and H. L. Sadoff (ed.), *Spores VI.* American Society for Microbiology, Washington, D.C.

22. **Carstensen, E. L., and R. E. Marquis. 1975.** Dielectric and electrochemical properties of bacterial cells, p. 563–571. *In* P. Gerhardt, R. N. Costilow, and H. L. Sadoff (ed.), *Spores VI.* American Society for Microbiology, Washington, D.C.

23. **Carstensen, E. L., R. E. Marquis, S. Z. Child, and G. R. Bender. 1979.** Dielectric properties of native and decoated spores of *Bacillus megaterium. J. Bacteriol.* 140:917–928.

24. **Carstensen, E. L., R. E. Marquis, and P. Gerhardt. 1971.** Dielectric study of the physical state of electrolytes and water within *Bacillus cereus* spores. *J. Bacteriol.* 107:106–113.

25. **Cassier, M., and A. Ryter. 1971.** Sur un mutant de *Clostridium perfringens* donnant des spores sans tuniques à germination lysozyme-dépendante. *Ann. Inst. Pasteur* (Paris) 121:717–732.

26. **Church, B. D., and H. Halvorson. 1959.** Dependence of heat resistance of bacterial endospores on the dipicolinic acid content. *Nature* (London) 183:124–125.

27. **Dring, G. J., and G. W. Gould. 1975.** Reimposition of heat-resistance on germinated spores of *Bacillus cereus* by osmotic manipulation. *Biochem. Biophys. Res. Commun.* 66:202–208.

28. **Eisenstadt, E., and S. Silver. 1972.** Calcium transport during sporulation in *Bacillus subtilis*, p. 425–433. *In* H. O. Halvorson, R. Hanson, and L. L. Campbell (ed.), *Spores V.* American Society for Microbiology, Washington, D.C.

29. **Ellar, D. J. 1978.** Spore specific structures and their function, p. 295–325. *In* R. Y. Stanier, H. J. Rogers, and J. B. Ward (ed.), *Relations between Structure and Function in the Prokaryotic Cell.* Cambridge University Press, Cambridge.

30. **Etoa, F. X., and L. Michiels. 1988.** Heat-induced resistance of *Bacillus stearothermophilus* spores. *Lett. Appl. Microbiol.* 6:43–45.

31. **Finley, N., and M. L. Fields. 1962.** Heat activation

and heat-induced dormancy of *Bacillus stearothermophilus* spores. *Appl. Microbiol.* 10:231–236.

32. **Fitz-James, P. C. 1971.** Formation of protoplasts from resting spores. *J. Bacteriol.* 78:755–764.

33. **Foerster, H. F., and J. W. Foster. 1966.** Endotrophic calcium, strontium, and barium spores of *Bacillus megaterium* and *Bacillus cereus. J. Bacteriol.* 91:1333–1345.

34. **Gerhardt, P. 1988.** The refractory homeostasis of bacterial spores, p. 41–49. *In* R. Whittenbury, J. G. Banks, G. W. Gould, and R. G. Board (ed.), *Homeostatic Mechanisms in Microorganisms.* Bath University Press, Bath, United Kingdom.

35. **Gerhardt, P., T. C. Beaman, T. R. Corner, J. T. Greenamyre, and L. S. Tisa. 1982.** Photometric immersion refractometry of bacterial spores. *J. Bacteriol.* 150:643–648.

36. **Gerhardt, P., and W. G. Murrell. 1978.** Basis and mechanisms of spore resistance: a brief preview, p. 18–20. *In* G. H. Chambliss and J. C. Vary (ed.), *Spores VII.* American Society for Microbiology, Washington, D.C.

37. **Gombas, D. E. 1983.** Bacterial spore resistance to heat. *Food Technol.* 37:105–110.

38. **Gould, G. W. 1983.** Mechanisms of resistance and dormancy, p. 173–209. *In* A. Hurst and G. W. Gould (ed.), *The Bacterial Spore*, vol. 2. Academic Press, Inc. (London), Ltd., London.

39. **Gould, G. W. 1986.** Water and survival of bacterial spores, p. 143–156. *In* A. C. Leopold (ed.), *Membranes, Metabolism and Dry Organisms.* Cornell University Press, Ithaca, N.Y.

40. **Gould, G. W., and G. J. Dring. 1975.** Heat resistance of bacterial endospores and concept of an expanded osmoregulatory cortex. *Nature* (London) 258:401–405.

41. **Halvorson, H. O., and A. Swanson. 1969.** Role of dipicolinic acid in the physiology of bacterial spores, p. 121–132. *In* L. L. Campbell (ed.), *Spores IV.* American Society for Microbiology, Bethesda, Md.

42. **Hanson, R. S., M. V. Curry, J. V. Garner, and H. O. Halvorson. 1972.** Mutants of *Bacillus cereus* strain T that produce thermoresistant spores lacking dipicolinate and have low levels of calcium. *Can. J. Microbiol.* 18:1139–1143.

43. **Heinen, W., and A. M. Lauwers. 1981.** Growth of bacteria at 100 C and beyond. *Arch. Microbiol.* 129:127–128.

44. **Imae, Y., and J. L. Strominger. 1976.** Cortex content of asporogenous mutants of *Bacillus subtilis. J. Bacteriol.* 126:914–918.

45. **Johnstone, K., D. J. Ellar, and T. C. Appleton. 1980.** Location of metal ions in *Bacillus megaterium* spores by high resolution electron probe X-ray microanalysis. *FEMS Microbiol. Lett.* 7:97–101.

46. **Johnstone, K., G. S. A. B. Stewart, M. D. Barratt,**

and D. J. Ellar. 1982. An electron paramagnetic resonance study of the manganese environment within dormant spores of *Bacillus megaterium* KM. *Biochim. Biophys. Acta* **714**:379–381.

47. Khoury, P. H., S. J. Lombardi, and R. A. Slepecky. 1987. Perturbation of the heat resistance of bacterial spores by sporulation temperature and ethanol. *Curr. Microbiol.* **15**:15–19.

48. Koshikawa, T., T. C. Beaman, H. S. Pankratz, S. Nakashio, T. R. Corner, and P. Gerhardt. 1984. Resistance, germination, and permeability correlates of *Bacillus megaterium* spores successively divested of integument layers. *J. Bacteriol.* **159**:624–632.

49. Labbe, R. G., R. R. Reich, and C. L. Duncan. 1978. Alteration in ultrastructure and germination of *Clostridium perfringens* type A spores following extraction of spore coats. *Can. J. Microbiol.* **24**:1526–1536.

50. Leman, A. 1973. Interference microscopic determination of bacterial dry weight during germination and sporulation. *Jena Rev.* **5**:263–270.

51. Lewis, J. C., N. S. Snell, and H. K. Burr. 1960. Water permeability of bacterial spores and the concept of a contractile cortex. *Science* **132**:544–545.

52. Lindsay, J. A., T. C. Beaman, and P. Gerhardt. 1985. Protoplast water content of bacterial spores determined by buoyant density sedimentation. *J. Bacteriol.* **163**:735–737.

53. Lindsay, J. A., and W. G. Murrell. 1985. Changes in density of DNA after interaction with dipicolinic acid and its possible role in heat resistance. *Curr. Microbiol.* **12**:329–334.

54. Lindsay, J. A., and W. G. Murrell. 1986. Solution spectroscopy of dipicolinic interaction with nucleic acids: role in spore heat resistance. *Curr. Microbiol.* **13**:255–259.

55. Lindsay, J. A., W. G. Murrell, and A. D. Warth. 1985. Spore resistance and the basic mechanism of heat resistance, p. 162–186. *In* L. E. Harris and A. J. Skopek (ed.), *Sterilization of Medical Products*, vol. 3. Johnson & Johnson Pty. Ltd., Botany, N.S.W., Australia.

56. Mackey, B. M., and C. M. Derrick. 1986. Elevation of the heat resistance of *Salmonella typhimurium* by sublethal heat shock. *J. Appl. Bacteriol.* **61**:389–393.

57. Mackey, B. M., and C. M. Derrick. 1987. Changes in heat resistance of *Salmonella typhimurium* during heating at rising temperatures. *Lett. Appl. Microbiol.* **4**:13–16.

58. Maeda, Y., T. Fujita, T. Sigiura, and S. Koga. 1968. Physical properties of water in spores of *Bacillus megaterium*. *J. Gen. Appl. Microbiol.* **14**:217–226.

59. Marquis, R. E. 1984. Reversible actions of hydrostatic pressure and compressed gases in microorganisms, p. 273–301. *In* A. Hurst and A. Nasim

(ed.), *Repairable Lesions in Microorganisms*. Academic Press, Inc. (London), Ltd., London.

60. Marquis, R. E. 1988. Turgor pressure, sporulation, and the physical properties of cell walls, p. 21–32. *In* P. Actor, L. Daneo-Moore, M. L. Higgins, M. R. J. Salton, and G. D. Shockman (ed.), *Antibiotic Inhibition of Bacterial Surface Assembly and Function*. American Society for Microbiology, Washington, D.C.

61. Marquis, R. E. 1989. Minerals and bacterial spores, p. 247–274. *In* T. J. Beveridge and R. J. Doyle (ed.), *Metal Ions and Bacteria*. J. Wiley & Sons, Inc., New York.

62. Marquis, R. E., and G. R. Bender. 1985. Mineralization and heat resistance of bacterial spores. *J. Bacteriol.* **161**:789–791.

63. Marquis, R. E., G. R. Bender, E. L. Carstensen, and S. Z. Child. 1983. Dielectric characterization of forespores isolated from *Bacillus megaterium*. *J. Bacteriol.* **153**:436–442.

64. Marshall, B. J., and W. G. Murell. 1970. Biophysical analysis of the bacterial spore. *J. Appl. Bacteriol.* **33**:103–129.

65. Marshall, B. J., W. G. Murrell, and W. J. Scott. 1963. The effect of water activity, solutes and temperature on the viability and heat resistance of freeze-dried bacterial spores. *J. Gen. Microbiol.* **31**:451–460.

66. Mason, J. M., and P. Setlow. 1986. Essential role of small, acid-soluble spore proteins in resistance of *Bacillus subtilis* spores to UV light. *J. Bacteriol.* **167**:174–178.

67. Murrell, W. G. 1961. Discussion (on permeability of bacterial spores), p. 229–236. *In* H. O. Halvorson (ed.), *Spores II*. Burgess Publishing Company, Minneapolis.

68. Murrell, W. G. 1981. Biophysical studies on the molecular mechanisms of spore heat resistance and dormancy, p. 64–77. *In* H. S. Levinson, A. L. Sonenshein, and D. J. Tipper (ed.), *Sporulation and Germination*. American Society for Microbiology, Washington, D.C.

69. Murrell, W. G. 1988. Bacterial spores—nature's ultimate survival package, p. 311–346. *In* W. G. Murrell and I. R. Kennedy (ed.), *Microbiology in Action*. John Wiley & Sons, Inc., New York.

70. Murrell, W. G., and W. J. Scott. 1966. The heat resistance of bacterial spores at various water activities. *J. Gen. Microbiol.* **43**:411–425.

71. Murrell, W. G., and A. D. Warth. 1965. Composition and heat resistance of bacterial spores, p. 1–24. *In* L. L. Campbell and H. O. Halvorson (ed.), *Spores III*. American Society for Microbiology, Ann Arbor, Mich.

72. Nakashio, S., and P. Gerhardt. 1985. Protoplast dehydration correlated with heat resistance of bacterial spores. *J. Bacteriol.* **162**:571–578.

73. Neihof, R., J. K. Thompson, and V. R. Deitz. 1967. Sorption of water vapour and nitrogen gas by bacterial spores. *Nature* (London) **216**:1304–1306.

74. Nishihara, T., T. Ichikawa, and M. Kondo. 1980. Location of elements in ashed spores of *Bacillus megaterium*. *Microbiol. Immunol.* **24**:495–506.

75. Park, J. T., and R. Hancock. 1960. A fractionation procedure for studies of the synthesis of cell-wall mucopeptide and of other polymers in cells of *Staphylococcus aureus*. *J. Gen. Microbiol.* **22**:249–258.

76. Rode, L. J., and J. W. Foster. 1966. Quantitative aspects of exchangeable calcium in spores of *Bacillus megaterium*. *J. Bacteriol.* **91**:1589–1593.

77. Ross, K. F. A., and E. Billing. 1957. The water and solid content of living bacterial spores and vegetative cells as indicated by refractive index measurements. *J. Gen. Microbiol.* **16**:418–425.

78. Rosson, R. A., and K. H. Nealson. 1982. Manganese binding and oxidation by spores of a marine *Bacillus*. *J. Bacteriol.* **151**:1027–1034.

79. Scherrer, R., and P. Gerhardt. 1972. Location of calcium within *Bacillus* spores by electron probe X-ray microanalysis. *J. Bacteriol.* **112**:559–568.

80. Scherrer, R., and V. E. Shull. 1987. Microincineration and elemental X-ray microanalysis of single *Bacillus cereus* T spores. *Can. J. Microbiol.* **33**:304–313.

81. Setlow, P. 1988. Small acid-soluble spore proteins of *Bacillus* species: structure, synthesis, function and degradation. *Annu. Rev. Microbiol.* **42**:319–338.

82. Seto-Young, D. L. T., and D. J. Ellar. 1981. Studies of calcium transport during growth and sporulation. *Microbios* **30**:191–208.

83. Singh, R. P., B. Setlow, and P. Setlow. 1977. Levels of small molecules and enzymes in the mother cell compartment and the forespore of sporulating *Bacillus megaterium*. *J. Bacteriol.* **130**:1130–1138.

84. Slepecky, R., and J. W. Foster. 1959. Alterations in metal content of spores of *Bacillus megaterium* and the effects on some spore properties. *J. Bacteriol.* **78**:117–123.

85. Stewart, M., A. P. Somylo, A. V. Somylo, H. Shuman, J. A. Lindsay, and W. G. Murrell. 1980. Distribution of calcium and other elements in cryosectioned *Bacillus cereus* T spores determined by high-resolution scanning electron probe X-ray microanalysis. *J. Bacteriol.* **143**:481–491.

86. Tipper, D. J., and J. J. Gauthier. 1972. Structure of the bacterial endospore, p. 3–12. *In* H. O. Halvorson, R. Hanson, and L. L. Campbell (ed.), *Spores V*. American Society for Microbiology, Washington, D.C.

87. Tisa, L. S., T. Koshikawa, and P. Gerhardt. 1982. Wet and dry bacterial spore densities determined by buoyant sedimentation. *Appl. Environ. Microbiol.* **43**:1307–1310.

88. Ulanowski, Z., I. K. Ludlow, and W. M. Waites. 1987. Water content and size of spore components determined by laser diffractometry. *FEMS Microbiol. Lett.* **43**:229–232.

89. Waldham, D. G., and H. O. Halvorson. 1954. Studies on the relationship between equilibrium vapor pressure and moisture content of bacterial endospores. *Appl. Microbiol.* **2**:333–338.

90. Warth, A. D. 1978. Relationship between the heat resistance of spores and the optimum and maximum growth temperatures of *Bacillus* species. *J. Bacteriol.* **134**:699–705.

91. Warth, A. D. 1985. Mechanisms of heat resistance, p. 209–225. *In* G. J. Dring, D. J. Ellar, and G. W. Gould (ed.), *Fundamental and Applied Aspects of Bacterial Spores*. Academic Press, Inc. (London), Ltd., London.

92. Warth, A. D., and J. L. Strominger. 1972. Structure of the peptidoglycan from spores of *Bacillus subtilis*. *Biochemistry* **11**:1389–1396.

93. Watt, I. C. 1981. Water vapor adsorption by *Bacillus stearothermophilus* endospores, p. 253–255. *In* H. S. Levinson, A. L. Sonenshein, and D. J. Tipper (ed.), *Sporulation and Germination*. American Society for Microbiology, Washington, D.C.

94. Wilkinson, B. J., J. A. Deans, and D. J. Ellar. 1975. Biochemical evidence for the reversed polarity of the outer membrane of the bacterial forespore. *Biochem. J.* **152**:561–569.

95. Williams, O. B., and W. J. Robertson. 1954. Studies on heat resistance. VI. Effect of temperature of incubation at which formed on heat resistance of aerobic thermophilic spores. *J. Bacteriol.* **67**:377–378.

96. Windle, J. J., and L. E. Sacks. 1963. Electron paramagnetic resonance of manganese (II) and copper (II) in spores. *Biochim. Biophys. Acta* **66**:173–179.

97. Wyatt, P. J. 1975. Observations on the structure of spores. *J. Appl. Bacteriol.* **38**:47–51.

98. Zytkovicz, T. H., and H. O. Halvorson. 1972. Some characteristics of dipicolinic acid-less mutant spores of *Bacillus cereus*, *Bacillus megaterium*, and *Bacillus subtilis*, p. 49–52. *In* H. O. Halvorson, R. Hanson, and L. L. Campbell (ed.), *Spores V*. American Society for Microbiology, Washington, D.C.

Regulation of Procaryotic Development
Edited by Issar Smith, Ralph A. Slepecky, and Peter Setlow
© 1989 American Society for Microbiology, Washington, DC 20006

Chapter 3

Methods for Genetic Manipulation, Cloning, and Functional Analysis of Sporulation Genes in *Bacillus subtilis*

Philip Youngman, Harald Poth, Brian Green, Karen York, Gabriela Olmedo, and Kerry Smith

Over the past decade, revolutionary advances have been made in the methodology available for investigating mechanisms of gene regulation in *Bacillus subtilis*, much of the innovation coming from work directed at understanding the complex changes in gene expression that occur during sporulation. In some cases, these advances have consisted simply in adapting to *B. subtilis* techniques pioneered in other organisms, chiefly *Escherichia coli*. In other cases, however, distinctive biological properties of *B. subtilis*, in particular the highly recombinogenic nature of DNA uptake by competent cells, have encouraged the development of novel strategies. This review will be an effort to summarize many of the current strategies and several new vectors devised to facilitate the identification and functional analysis of regulated genes in *B. subtilis*, with particular emphasis given to approaches that exploit special characteristics of the organism (which therefore have no counterparts in *E. coli*), approaches developed very recently (which therefore may not be widely appreciated), and approaches designed expressly for investi-

gating the regulation of sporulation-specific gene expression. We also take advantage of this opportunity to describe several new vectors which have not appeared in previous publications, and we speculate about the potential utility in *B. subtilis* of approaches that have not to our knowledge actually been explored by other investigators. These include the use of "shuttle mutagenesis" for *B. subtilis* and the use of *pcnB* mutants of *E. coli* as cloning hosts for ColE1-derived vectors containing inserts of *B. subtilis* chromosomal DNA.

IDENTIFICATION OF REGULATED GENES

Conventional Mutagenesis To Generate *spo* Mutations

Mutations that block sporulation at discrete stages have identified approximately 45 to 50 distinct genetic loci (35). Early efforts to clone genes comprising these loci emphasized (but were not restricted to) the screening of *E. coli* phage and plasmid libraries of

Philip Youngman, Brian Green, Karen York, Gabriela Olmedo, and Kerry Smith • Department of Microbiology, University of Pennsylvania School of Medicine, Philadelphia, Pennsylvania 19104. *Harald Poth* • Department of Microbiology, Gesellschaft für Biotechnologie Forschung mbH, Mascheroder Weg, D-3300 Braunschweig, Federal Republic of Germany.

B. subtilis genomic DNA for clones that could correct or complement *spo* mutations upon reintroduction into *B. subtilis* by transformation (e.g., reference 17). More recently, *B. subtilis* temperate phage cloning vectors based on φ105 (12, 26) and SPβ (42) have provided a more effective way to clone the wild-type copies of *spo* genes by selection for complementation (see below).

A somewhat surprising discovery which has emerged from the DNA sequence analysis of cloned *spo* genes from various sources is that a disproportionate percentage of genes in which mutations can cause a Spo phenotype appear to encode proteins that perform functions obviously regulatory in nature. Striking examples include the observations that *spo0A* and *spo0F* encode proteins homologous to responder elements of two-component signal transduction systems in bacteria (54) and that *spo0H* (10), *spoIIG* (55), *spoIIA* (49), *spoIIIC* (14), *spoIVC* (51), and *spoIIIG* (27, 37) all encode RNA polymerase σ factors (or portions of σ factors). In contrast, mutations in sporulation-specific genes cloned by "reverse genetics," even with probes derived from proteins that represent major structural components of the spore coat or core, do not generally produce a Spo colonial morphology phenotype (47, 68). These observations suggest that genes in which *spo* mutations can occur must represent a select and relatively small subset of genes whose expression is sporulation specific. One implication of this is that the cloning and sequence analysis of *spo* genes is justified as likely to reveal a specific kind of regulatory function for the gene product simply through inspection of homologies with sequenced genes in existing data bases. Another implication is that sporulation-specific genes whose products do not perform regulatory functions (genes perhaps more representative of typical sporulation-specific regulons) must usually be identified by means other than screening for Spo mutants. Examples of screens based strictly on expression patterns will be discussed in a later section.

Insertional Mutagenesis with Tn917 Derivatives

Transposon Tn917, originally identified in an isolate of *Enterococcus faecalis* (53), has proved to be a highly effective agent for insertional mutagenesis in *B. subtilis* and has been exploited for the isolation of a large collection of Spo mutants (43). As discussed below, an important advantage of producing *spo* mutations with Tn917 is the ease with which the mutated gene can be cloned. In addition, it should be noted that recent improvements in vectors available for recovering chromosomal insertions of Tn917 have significantly facilitated the use of the transposon for insertional mutagenesis, and recent information concerning the randomness of Tn917 insertions has alleviated earlier concerns that the utility of the transposon might be compromised by the presence of "hot spots" in the chromosome where Tn917 insertions are strongly preferred (62). Thus, random "libraries" of Tn917 insertions likely to include disruptions of any nonessential gene are now relatively straightforward to obtain.

In earlier work with Tn917 in *B. subtilis* (63), methods used to recover populations of bacteria containing chromosomal insertions of the transposon were based on the use of temperature-sensitive vectors derived from a small *Staphylococcus aureus* plasmid called pE194 (24). Bacteria containing these vectors could be grown at a low temperature (<32°C) and then diluted into fresh medium at a temperature nonpermissive for plasmid replication (>45°C), maintaining a selection for the erythromycin resistance (Em^r) gene associated with the transposon. Extensive outgrowth at the high temperature eventually resulted in a strong enrichment for bacteria containing chromosomal insertions. However, practical complications arose from the fact that the thermosensitivity of replication in the pE194-derived vectors was incomplete. Replication was severely impaired at high temperatures, but not abolished, even at temperatures above 48°C. As a consequence, it was necessary to make a relatively extreme dilution of 32°C cultures into broth pre-warmed to at least 48°C to enrich effectively for bacteria that contained chromosomal insertions. The recommended procedure called for a 1:100 dilution of a 32°C culture at mid-log phase into at least 20 liters of 48°C broth (62). This clearly made it inconvenient for many laboratories to generate comprehensive insertion libraries.

New Tn917 delivery vectors are based on the use of a mutant derivative of pE194, isolated by Sandra Gruss in Richard Novick's laboratory (Gruss and Novick, personal communication), which displays an extremely tight replication block at temperatures above 37°C, yet maintains essentially wild-type copy number at temperatures below 32°C. The replication properties of this highly useful mutant, which has been referred to as pE194Ts, were characterized in detail by Dubnau and colleagues (56) and were found to be the result of a simple point mutation (or perhaps a pair of closely spaced point mutations) that can easily be transferred by recombination to other pE194-derived replicons. It was possible, in fact, to transfer the *ts* mutation readily to several pE194-derived

vectors containing derivatives of Tn917 (Fig. 1), including pTV1 (62), pTV32 (62), and pTV51 (A. Camilli and D. Portnoy, personal communication). These "pTVTs" vectors offer several advantages over their wild-type counterparts. First, because the block to replication is absolutely tight at the nonpermissive temperature, less extreme dilutions are required to enrich effectively for the products of transposition. Second, because the non-permissive temperature is lower, the vectors are useful for generating transpositions in a much broader range of bacterial species. Vectors such as pTV32Ts have been introduced successfully into such diverse species as *S. aureus* (P. Pattee, personal communication), *E. faecalis* (57), *Streptococcus mutans* (L. Daneo-Moore, personal communication), *Clostridium acetobutylicum* (M. Young, personal communication), and *Listeria monocytogenes* (Camilli and Portnoy, personal communication) and have proved useful for generating insertional mutations. Third, because of reduced background growth of bacteria in which transpositions have not occurred, populations of bacteria containing vectors like pTV32Ts, grown at 32°C, can be plated directly on solid media and incubated over-

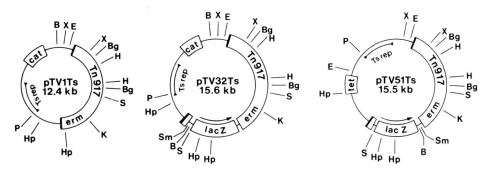

Figure 1. New, highly temperature-sensitive vectors for recovering transpositions of Tn917 and its derivatives. Ts rep, Replication functions derived from pE194Ts (56); cat, a chloramphenicol acetyltransferase gene derived from pC194 (24); tet, a *tetL* tetracycline resistance determinant derived from pBC16 (see reference 62); erm, an inducible erythromycin resistance gene naturally associated with Tn917 (53); lacZ, a promoterless copy of the *E. coli lacZ* gene, substituted with a ribosome-binding site derived from *B. subtilis* gene *spoVG* (see reference 65); B, BamHI recognition sites; Bg, BglII sites; E, EcoRI sites; H, HindIII sites; Hp, HpaI sites; K, KpnI sites; P, PstI sites, S, SalI sites; Sm, SmaI sites; X, XbaI sites. The origins of pTV1Ts and pTV32Ts were described previously (62); pTV51Ts was obtained as the result of a plasmid cross between pTV51 (65) and pBD95Ts (62) (Camilli and Portnoy, personal communication).

night at 42°C to recover isolated single colonies produced by insert-containing bacteria.

Insertional Mutagenesis Using Integrational Vectors

It has long been appreciated that fragments of B. subtilis chromosomal DNA cloned into E. coli plasmids (which cannot replicate autonomously in B. subtilis) can mediate the efficient recombinational integration of the plasmid into the chromosome of a transformation recipient by a "Campbell-like" mechanism (11). Integrational vectors in general use for this purpose usually consist of ColE1-derived replicons that carry a cat gene of gram-positive origin (other kinds of integrational vectors will be described in a later section). Among the first to recognize the potential utility of integrational vectors for insertional mutagenesis were Ehrlich and colleagues (39) and Hoch and colleagues (18). Recombinational integration is mutagenic, of course, whenever the chromosomal fragment cloned into the vector comes entirely from within a chromosomal transcription unit. Thus, any method that generates a relatively random population of relatively small fragments of genomic DNA can be used to prepare a plasmid library in E. coli that will generate insertional mutations when re-turned to B. subtilis by transformation. An important advantage of this approach is that disrupted genes are adjacent to the E. coli replicon in the B. subtilis chromosome, which then allows for the recovery of DNA, including all or portions of the mutated gene, by digestion of chromosomal DNA with appropriate restriction enzymes, followed by ligation and transformation of an E. coli strain. Some of the earlier vectors developed in part for such applications are shown in Fig. 2. In recent years, several laboratories have developed their own general-purpose integrational vectors, which include such features as polylinkers, detection of inserts by disruption of lac alpha-complementation, origins of replication for coliphages M13 or f1 to produce single-stranded DNA with helper phages, and T7 or SP6 promoter sites for generating in vitro transcripts through cloned inserts. Three such vectors in use by our laboratory are shown in Fig. 3.

Shuttle Mutagenesis

"Shuttle mutagenesis" is a term first used by Heffron and colleagues (45) to describe a method of indirect transposon-mediated insertional mutagenesis in which E. coli transposons were used to create insertions into foreign DNA cloned into E. coli cosmid vec-

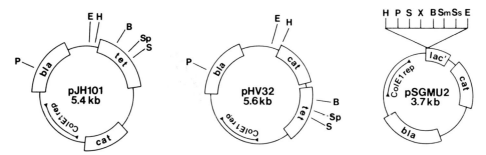

Figure 2. Integrational vectors commonly used in *B. subtilis*. ColE1 rep, Replication functions derived from pBR322 (4)((pJH101 and pHV32) or pUC13 (see reference 59); bla, a beta-lactamase gene derived from pBR322; tet, a tetracycline resistance gene derived from pBR322; cat, a chloramphenicol resistance gene derived from pC194 (24); lac', a fragment of the *E. coli lac* operon capable of alpha complementation (see reference 59); Sp, *Sph*I recognition sites; Ss, *Sst*I sites. All other restriction sites are designated as in Fig. 1. The origins of pJH101 (18), pHV32 (39), and pSGMU2 (15) were described previously.

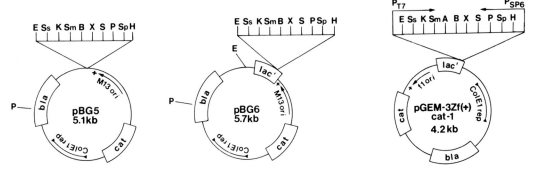

Figure 3. Integrational vectors that incorporate polylinker cloning sites and M13 or f1 origins of replication. M13 ori, Single-stranded phage origin of replication derived from the M13 constructions of Messing and colleagues (see reference 59); f1 ori, single-stranded phage origin of replication derived from Promega phagemid pGEM-3Zf(+); P_{T7}, promoter for *E. coli* phage T7; P_{SP6}, promoter for *Salmonella typhimurium* phage SP6; ColE1 rep, lac', cat, bla, see legend for Fig. 2; A, *Ava*I restriction sites; all other restriction sites are as designated in the legends of Fig. 1 and 2. Vector pBG6 was constructed by inserting a *Cla*I-*Bal*I fragment from M13mp19 (59) that contains M13 ori, lac', and the polylinker sites into the *Cla*I-*Bal*I backbone of pJH101 (Fig. 2; 18); vector pBG5 was derived from pBG6 by deleting the small *Eco*RI fragment containing lac'; and vector pGEM-3Zf(+) *cat* was constructed by inserting a 1.0-kb *Ava*II fragment from M13mp18 *cat* containing the *cat* gene into the unique *Nde*I site in pGEM-3Zf(+), after filling in *Ava*II- and *Nde*I-generated ends with Klenow.

tors. The transposons were engineered to carry a genetic marker selectable in the organism from which the clone library was prepared, so that insert-containing DNA could be "shuttled" back into that organism by transformation. A selection for the transposon-associated marker forced transfer of the insert into the genome of the transformation recipient by homologous recombination. Although such an approach has not to our knowledge been attempted using a genomic library of *B. subtilis* DNA, the very high efficiency of transformation in this organism would appear to make it particularly attractive. The advantage of this approach over insertional mutagenesis with Tn917 would be the possibility of using certain kinds of insertion elements that are already available in *E. coli* but which would be difficult to construct from existing derivatives of Tn917. Examples would include mini-Mu derivatives that mediate translational *lacZ* fusions (6), derivatives of Tn5 that mediate translational fusions to *phoA* (36), and derivatives of Tn5 that contain an isopropyl-β-D-thiogalacto-pyranoside (IPTG)-inducible promoter ori-

ented to direct transcription outward from the transposon (D. Berg, personal communication). Clearly, it would be far easier, for example, to insert an *erm* gene into Tn5-*phoA* than to engineer Tn917 to mediate translational *phoA* fusions. Although problems have been encountered in earlier attempts to establish a fully representative genomic library of *B. subtilis* DNA in *E. coli*, these were probably a consequence of using multicopy plasmid or phasmid cloning vectors. A good genomic library of *B. subtilis* DNA could almost certainly be obtained with a unit copy number cosmid vector.

Use of *lac* Fusions To Identify Sporulation-Specific Genes

As discussed above, identifying sporulation-specific genes by isolation of Spo mutants has the advantage that the mutated genes are likely to be ones whose products play important regulatory roles, but also has the disadvantage that the great majority of sporulation-specific genes would escape detection. Certain classes of sporulation regulons might be missed altogether. More general methods for identifying sporulation-specific transcription units have been devised which are based on direct detection of sporulation-specific gene expression. One approach involves the use of *lacZ* fusions in conjunction with the fluorogenic substrate 4-methylumbelliferyl-β-D-galactopyranoside (MUG) (66). When this substrate is hydrolyzed by β-galactosidase, it yields a methylumbelliferone that is highly fluorescent in response to long-wavelength UV light. A population of bacteria containing *lacZ* fusions to random chromosomal promoters can be plated for single colonies on an agar medium that supports sporulation and then can be sprayed with a solution of MUG well after colonies have reached their full size (about 36 h at 37°C). At this time, bacteria in the colonies would have been in stationary phase for a substantial period of time, allowing for turnover of proteins produced exclusively during growth, and many of the bacteria would be in the process of sporulating. Thus, colonies exhibiting significant fluorescence are likely to be ones formed by bacteria containing a *lac* fusion to a chromosomal promoter active during sporulation, or at least during the stationary phase. Further examination can quickly reveal whether the fusion is exclusively active postexponentially. For this, and for other applications discussed below, there are several approaches for generating relatively random *lac* fusions to chromosomal promoters. Three will be considered here, as follows.

Tn917 derivatives

There are several derivatives of Tn917 that mediate transcriptional *lac* fusions as an automatic consequence of gene disruption. With the availability of highly temperature-sensitive selection vectors such as pTV32Ts and pTV51Ts (Fig. 1), one advantage of this approach is that it is very easy to carry out. For some applications, the fact that the process of generating a fusion is mutagenic

is also an advantage. In addition, once *lac* fusions have been obtained that display an expression pattern of interest, DNA including the promoter responsible for driving expression of the fusion can readily be cloned into *E. coli* by using vectors such as pTV21*del*2 (64), as discussed in a later section.

Integrational vectors

There are also integrational vectors that can mediate *lac* fusions as an automatic consequence of integrative recombination. One advantage of this approach (for some applications) is that obtaining *lac* fusions does not necessarily result in gene disruption. Thus, unlike transposon-mediated fusions, integrational vector-mediated fusions may be used to monitor the expression patterns of essential genes. Integrational fusion vectors are typically derivatives of ColE1 that contain a *cat* gene and a promoterless copy of the *E. coli lacZ* coding sequence, preceded by multiple cloning sites. Because the natural Shine-Dalgarno sequence of *lacZ* is relatively ineffective as a ribosome-binding site in *B. subtilis*, integrational vectors designed to generate transcriptional *lacZ* fusions contain versions of *lacZ* substituted with ribosome-binding sites derived from other genes. Some are based on constructions of Donnelly and Sonenshein (9) which utilize translation initiation signals for *lacZ* derived from the *E. coli trpA* gene, and some are based on constructions of Errington (13) which utilize translation initiation signals derived from the *B. subtilis spoIIAA* gene. Another alternative is a family of vectors similar to pGV34 (Fig. 4) (21), which was derived from constructions of Zuber and Losick (69). In these constructions, a ribosome-binding site from the *spoVG* gene of *B. subtilis* provides the translation initiation signal for *lacZ*. An advantage of vectors like pGV34 is that they can be used either for Campbell-like recombinational integration or for recombinational substitution into an SPβ prophage, as described below.

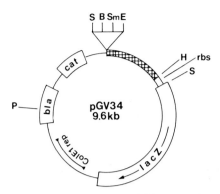

Figure 4. Integrational vector useful for generating transcriptional *lacZ* fusions. The *lacZ* gene in pGV34 is a promoterless, truncated version of the *E. coli lacZ* coding sequence, preceded by the ribosome-binding site (rbs) of the *B. subtilis spoVG* gene plus the first few codons of *spoVG* fused in frame with the *lacZ* coding sequence (see reference 69). The cross-hatched portion of pGV34 is a "stuffer" fragment consisting of *B. subtilis* chromosomal DNA just upstream from the *spoVG* promoter. Thus, pGV34 is suitable for receiving promoter-containing *Eco*RI-*Hind*III, *Sma*I-*Hind*III, or *Bam*HI-*Hind*III fragments. All other features of pGV34 are as described for Fig. 1 and 2. The construction of pGV34 was described previously (21).

Integration into SPβ

A highly useful method for integrating *lacZ* fusions into the *B. subtilis* temperate phage SPβ has been described by Zuber and Losick (70). In essence, transcriptional fusions are first created in an *E. coli* vector and then transferred into a special SPβ prophage by homologous recombination. Among the *E. coli* vectors whose structures are appropriate for this purpose is pGV34 (Fig. 4). The kind of recombinational event that would transfer a *lac* fusion from pGV34 to the SPβ prophage is diagrammed in Fig. 5. Such an approach is suitable either for obtaining an SPβ-associated *lac* fusion to a specific cloned gene or for generating a relatively random library of transcriptional fusions (P. Zuber, personal communication). One important advantage of this approach is that SPβ-associated fusions are highly portable. Although they are stably maintained in the cell

when integrated into the prophage, they can easily be cured from the cell by streaking for single colonies at high temperature (a mutant prophage is used that has a thermolabile repressor), and they can easily be transferred from strain to strain by transduction to examine the behavior of the fusions in various mutant backgrounds. An advantage of this approach for generating random libraries of fusions for the purpose of identifying promoters with special expression properties is that the fusions are adjacent to pBR322-derived sequences in the SPβ phages and thus may readily be excised and "rescued" back into *E. coli*.

Use of *lac-cat* Fusions To Screen for Genes Expressed Exclusively during Sporulation

Although simple *lac* fusions are useful for identifying many different kinds of regulated genes, vectors that generate simultaneous transcriptional fusions to *lac* and *cat* genes offer additional advantages. One such vector, pTV53 (Fig. 6), generates such simultaneous fusions by transposition (65). It contains a derivative of Tn917 in which promoterless copies of a *lac* and a *cat* gene have been inserted in tandem near one end of the transposon, with no transcription terminator between them. Thus, gene disruptions can simultaneously place expression of *lac* and *cat* under control of the promoter for the disrupted gene. One use for such a transposon is to provide a means to identify transcription units expressed exclusively during sporulation. Bacteria containing random insertions of Tn917-*lac-cat* can be plated for single colonies on a sporulation-inducing medium and screened with MUG after colonies have developed, as described above. Bacteria from colonies that exhibit fluorescence can then be screened for chloramphenicol resistance or sensitivity (Cm^r, Cm^s) simply by picking onto plates containing selective levels of the drug. Fluorescing bacteria that express a *lac-cat* fusion during the vegetative phase will grow in the presence of chloramphenicol,

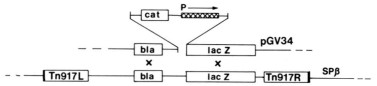

Figure 5. Schematic diagram of the kind of recombination event that can transfer a transcriptional *lacZ* fusion from pGV34 into an SPβ prophage. The cross-hatched portion of pGV34 represents a promoter-containing fragment cloned into pGV34 in a way that replaces the stuffer fragment (see legend to Fig. 4). Tn917L and Tn917R refer to sequences derived from Tn*917*. The prophage depicted here has the same structure as that present in lysogen ZB307 described by Zuber and Losick (70).

perhaps indicating that the fusion is expressed during both growth and sporulation. Bacteria that produce fluorescent colonies when sprayed with MUG (or blue colonies on X-Gal [5-bromo-4-chloro-3-indolyl-β-D-galactopyranoside] plates) but are Cm^s must be expressing the *lac-cat* fusion exclusively during sporulation. Another use for *lac-cat* fusions is to provide a selection for recovering mutants that derepress certain regulated promoters.

Use of the SPAC Expression Cassette To Identify Genes Whose Expression Depends upon Specific Positive Regulatory Factors

A family of expression vectors was described in 1984 by Yansura and Henner (60) which were based on the concept of using the *E. coli lac* repressor to control the transcription of recombinant genes in *B. subtilis*. Plasmid-borne expression cassettes were constructed which contained the structural gene for the *lac* repressor (the *lacI* gene), engineered for constitutive expression in *B. subtilis*, together with expression promoters that consisted of a *lac* operator sequence placed adjacent to otherwise constitutive promoters. One such expression promoter, p_{spac}, was derived from *B. subtilis* phage SPO1. When promoterless foreign genes were inserted downstream from expression promoters such as p_{spac}, it was found that transcription of the genes became inducible by IPTG. More recently, several investigators have modified

these so-called "SPAC" expression cassettes and devised new strategies for using them to investigate the network of regulatory mechanisms that control sporulation gene expression. One very useful construction that incorporates the SPAC cassette into a ColE1-derived integrational vector was described recently by Grossman and colleagues (K. J. Jaacks, J. Healy, R. Losick, and A. D. Grossman, submitted for publication) (Fig. 7). In this vector, pAG58, several unique restriction sites are available for insertion of promoterless genes to be placed under p_{spac} control. The resulting constructions can then be in-

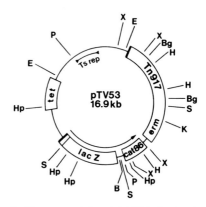

Figure 6. Vector containing a Tn*917* derivative that generates simultaneous transcriptional fusions to promoterless copies of the *E. coli lacZ* gene and the *Bacillus pumilus cat-86* gene. Restriction sites and other features of pTV53 are as described for Fig. 1, except that Ts rep refers to replication functions derived from pE194 rather than pE194Ts. The construction of pTV53 was described previously (65).

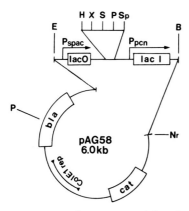

Figure 7. Integrational vector containing the SPAC expression cassette devised by Yansura and Henner (60). P$_{spac}$, A promoter derived from *B. subtilis* phage SPO1; P$_{pcn}$, a promoter derived from the penicillinase gene of *Bacillus licheniformis*; *lacI*, the *E. coli lac* repressor gene; *lacO*, the *E. coli lac* operator; Nr, NruI recognition sites; all other restriction sites and features of pAG58 as described for Fig. 2. The construction of pAG58 was described by Jaacks et al. (submitted).

tegrated into the *B. subtilis* chromosome by using either homology to the promoterless gene or homology to ColE1-derived sequences, selecting for Cm^r. Other forms of the SPAC cassette will be described in a later section.

The specific purpose for which pAG58 was constructed serves well to illustrate one important kind of application for the SPAC cassette: the identification of genes whose expression depends upon the presence of a *trans*-acting positive regulatory factor. In the work of Grossman and colleagues, a truncated copy of *spo0H*, the gene that codes for σ^H, was inserted into the SPAC cassette. Subsequent recombinational integration into the chromosome via a crossover within *spo0H* sequences resulted in placing an intact chromosomal copy of *spo0H* under *p*$_{spac}$ control. A library of random *lacZ* fusions to chromosomal transcription units was then obtained using Tn*917-lac*, which made it possible to screen for transcription units activated in the presence of σ^H (phenotype, blue on X-Gal in the presence of IPTG, less blue or white in the absence of IPTG). This is a general

strategy, of course, which is currently being explored by other investigators to identify genes controlled by other sigma species. Other kinds of applications for the SPAC cassette will be discussed in a later section.

CLONING STRATEGIES

As reviewed recently by Errington (15), the cloning of *B. subtilis* DNA in *B. subtilis* plasmid vectors has presented many problems, most of which have proved difficult to ovecome completely. First, DNA is fragmented upon uptake by naturally competent *B. subtilis* cells and is converted into a single-stranded form that is very recombinationally interactive with homologous sequences already present in the cell. Thus, the recovery of cloned sequences by selection for complementation using plasmid vectors requires the use of recombination-deficient protoplasts. Second, it has become clear in recent years that the plasmids which have been developed for molecular cloning in *B. subtilis* replicate through a single-stranded intermediate (52), which tends to promote the deletion of cloned sequences. Third, maintaining cloned sequences in multiple copies is often deleterious to the host or interferes with sporulation (3, 30, 41). Although moderately low-copy-number vectors that do not replicate through single-stranded intermediates have recently been developed (L. Jannière and S. D. Ehrlich, personal communication), these are not yet in wide use and are relatively untested as general-purpose cloning vehicles. Temperate phage vectors based on φ105 or SPβ have proved to be much more effective than plasmid vectors for the recovery of *B. subtilis* DNA fragments by selection for complementation in a *B. subtilis* host, and a description of the currently available cloning systems of this kind will be emphasized here. In addition, two different kinds of approaches will be described for taking advantage of Tn*917* insertions for cloning *B. subtilis* DNA in *E. coli*.

Cloning of DNA Adjacent to Existing Tn917 Insertions

As discussed above, derivatives of Tn917 or Tn917-lac have been in use for several years to identify regulated genes of many kinds by insertional disruption, and an effective system has been described for the cloning of chromosomal DNA adjacent to such insertions (64). In essence, this system consists of a collection of plasmid vectors that make it possible to integrate different variants of a ColE1-derived replicon into chromosomal copies of Tn917 or Tn917-lac by homologous recombination. Three such vectors are shown in Fig. 8. An advantage of this approach is that substitution of different vectors by recombination provides virtually limitless possibilities for the kinds and configurations of restriction sites that might be introduced. A disadvantage is that the approach is difficult to use in B. subtilis strains that cannot be made competent for transformation (e.g., com mutants, spo0A mutants, or rec mutants) and cannot be adapted easily for use in other species of gram-positive bacteria that are not naturally transformable. It was in part to overcome these limitations that the vector described in the following section was developed. Another disadvantage of this cloning strategy, as it was utilized in the past, was the fact that many segments of B. subtilis chromosomal DNA are toxic to E. coli when cloned into muticopy plasmids such as

pBR322 and its derivatives (64). Mutations in a new E. coli locus, pcnB, were recently described that reduce the copy number of ColE1-derivatives by a factor of 10 to 15, but which have no detectable effect on the general viability of E. coli strains (33, 34). Use of pcnB mutants as cloning hosts should in most cases remove any barriers to the cloning of B. subtilis DNA that are due to high-level expression of B. subtilis gene products. To take advantage of this, pcnB mutations have been introduced by transduction into hsdR E. coli strains useful as general-purpose cloning hosts (K. Smith and P. Youngman, unpublished data).

Transposition-Proficient Derivatives of Tn917 That Contain E. coli Cloning Vectors

To take full advantage of the observation that temperature-sensitive vectors such as pTV32Ts (Fig. 1) can be used to recover chromosomal insertions of Tn917 in L. monocytogenes, a gram-positive intracellular pathogen that is not naturally transformable and which is completely undeveloped as a genetic system, Portnoy and colleagues have recently constructed derivatives of the transposon that contain both a promoterless lacZ gene and a pBR322-derived cloning vector inserted into nonessential transposon DNA (Fig. 9) (Camilli and Portnoy, personal communication). Despite the fact that these con-

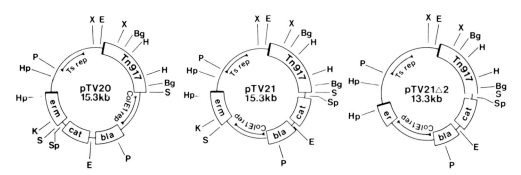

Figure 8. Vectors useful for cloning chromosomal sequences adjacent to Tn917 insertions into E. coli. Restriction sites and other features of pTV20, pTV21, and pTV21del2 are as described for Fig. 1 and 2. The construction of these vectors was described previously (64).

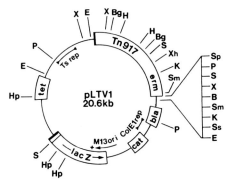

Figure 9. pLTV1, a transposition-proficient, *lac* fusion-generating derivative of Tn*917* that contains a pBR322-derived cloning vector. Restriction sites and other features of pLTV1 are as described in the legends of Fig. 1, 2, and 3. This vector was constructed by ligating *Hin*dIII-cut pBG5 (Fig. 3) into the *Bam*HI site in pTV51Ts (Fig. 1), after filling in *Hin*dIII- and *Bam*HI-generated ends with Klenow fragment (Camilli and Portnoy, personal communication).

structions have added more than 8 kilobases (kb) of DNA to Tn*917*, they are fully proficient for transposition. Thus, like the transposon derivatives in pTV32Ts and pTV51Ts (Fig. 1), they can generate *lacZ* fusions by insertional disruption of chromosomal genes. In addition, however, they contain sequences derived from the integrational vector pBG5 (Fig. 3), inserted immediately behind the *lacZ*

gene. This vector (Fig. 9) includes a *cat* gene derived from pC194 (24), a coliphage M13 origin of replication, and a polylinker derived from M13mp19 (59). The pBG5 sequences are oriented in the transposon such that all of the polylinker sites are distal to *lacZ*. Thus, when insertions of these transposon derivatives create *lacZ* fusions, several restriction sites are available for rescuing into *E. coli* chromosomal DNA that might contain the promoter driving expression of *lacZ* (Fig. 10).

The φ105 Cloning System

φ105 is a relatively small temperate phage of *B. subtilis* (39.2 kb) which has proved to be a very effective cloning vector. It is well characterized physically (5) and is apparently not prone to delete or rearrange cloned inserts, providing they are small enough not to strain the packaging limits of the phage (26). Although workable methods have been developed for cloning by "prophage transformation" with φ105 (44), the protoplast transfection approach developed by Errington and colleagues (12, 16) (Fig. 11) would appear to be a far superior way to make use of the phage for cloning. The most useful unique cloning site in currently available vectors is a *Bam*HI site. Thus, genomic libraries

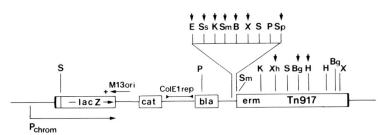

Figure 10. Schematic depiction of the structure of a transposition-mediated *lacZ* fusion generated by the transposon derivative in pLTV1. P_{chrom}, Chromosomal promoter driving expression of the *lac* fusion; Xh, *Xho*I recognition sites. All other features of the diagram are identified as in the legends to Fig. 1, 2, and 3. Restriction sites marked by vertical arrowheads are all useful for rescuing chromosomal DNA adjacent to the site of insertion on the p_{chrom}-proximal side of the transposon. Digestion of chromosomal DNA with any enzyme that recognizes these sites will generate a ColE1 *rep*-containing fragment extending outward into chromosomal DNA in the direction of p_{chrom} and can be recovered after ligation by transformation of an *E. coli* strain selecting for Ap^r.

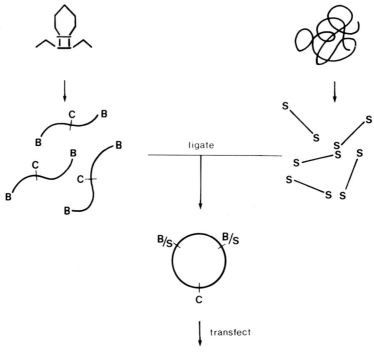

Figure 11. Schematic diagram illustrating the use of φ105 cloning vectors. B, *Bam*HI-generated cohesive ends; S, *Sau*3A-generated cohesive ends; C, cohesive ends of phage genome joined by ligation; B/S, *Bam*HI-generated and *Sau*3A-generated cohesive ends joined by ligation. As represented at the upper left, phage genomic DNA is first ligated together at high DNA concentration to produce linear concatemers, followed by digestion with *Bam*HI. As represented at the upper right, chromosomal DNA is subjected to a partial digestion with *Sau*3A to produce fragments approximately 3 to 4 kb in size. *Bam*HI-cut phage DNA is then ligated in the presence of a molar excess of *Sau*3A-cut chromosomal DNA, at low DNA concentration, to favor circularization of the phage DNA with the inclusion of a chromosomal fragment. Ligated DNA is then used to transfect *B. subtilis* protoplasts, and infectious centers are plated on a lawn of φ105-sensitive bacteria to produce plaques.

are generated by first ligating the cohesive ends of the phage together, followed by *Bam*HI digestion. *Bam*HI-cut phage DNA is then ligated at a dilute DNA concentration in the presence of a molar excess of chromosomal fragments that have cohesive ends that are compatible with *Bam*HI-generated ends. Chromosomal fragments are best prepared by partial digestion with *Sau*3A, followed by purification of fragments approximately 3 to 4 kb in size or by random shear to produce fragments 3 to 4 kb in size, followed by attachment of *Bam*HI linkers. The ligation mix is then used to transfect *B. sub-*

tilis protoplasts in the presence of polyethylene glycol. Because the BamHI site in vectors like φ105J27 is in nonessential DNA, phage containing inserts less than 4 kb in size produce plaques on a lawn of φ105-sensitive bacteria (26). This approach takes advantage of the fact that polyethylene glycol-mediated protoplast transfection can be highly efficient, capable of producing >10⁶ plaques per μg of phage DNA (12). One disadvantage of the approach, however, is that only fragments less than 4 kb in size can be accommodated without impairing propagation of the phage. Another disadvantage is that there is no simple way with existing vectors to force inclusion of an insert and to exclude insertion of more than one insert. A further disadvantage is that inserts must be subcloned into muticopy plasmid vectors by conventional methods before the genes they contain can conveniently be characterized physically and functionally. This can be a significant technical obstacle if an inconvenient configuration of restriction sites is present within the insert.

The SPβ Cloning System

At first glance, SPβ would appear to be a poor choice as a cloning vector. First, it is a very large phage, almost 120 kb in size (19). The naive expectation would be that cloned inserts would be more difficult to recover from such a phage than from the much smaller φ105 derivatives. Second, genomic fragments cloned into SPβ (42) or the related phage ρ11 (28) are generally not as stable with respect to deletion as fragments cloned into φ105. However, an important advantage of SPβ is that its large size allows for modifications that circumvent its apparent disadvantages,

modifications that would not be possible with φ105 derivatives. In fact, newly developed derivatives of SPβ offer several important features not available in existing φ105 vectors. It is possible to clone much larger inserts into these SPβ derivatives than into φ105, and easier to subclone them from the recombinant phages. Moreover, these operations can be carried out in a way that makes the issue of insert stability of little concern.

The approaches that have been developed recently to exploit SPβ as a cloning vector are based on the concept of "prophage transformation," first introduced by Kawamura and colleagues (29), and they utilize derivatives of SPβ that contain an insertion of Tn*917*, a deletion (*del2*) of approximately 10 kb of nonessential phage DNA, and a mutation (*c2*) that makes the phage repressor themolabile and thus allows heat induction to be used to prepare a phage lysate (see reference 42). These phage derivatives are used in conjunction with plasmid vectors pCV1 and pCV2 (Fig. 12), which contain a *cat* gene selectable in *B. subtilis*, together with sequences derived from pBR322 (4) that include a ColE1 replication origin and a *bla*

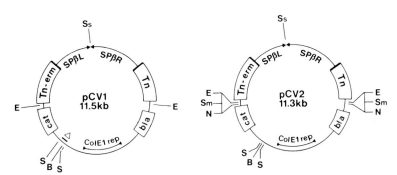

Figure 12. Plasmid vectors designed for SPβ-mediated cloning of chromosomal DNA fragments by prophage transformation. SPβL, SPβ sequences adjacent to the *erm*-proximal end of Tn*917* in the SPβ*c2del2*::Tn*917* prophage (see reference 42); SPβR, SPβ sequences adjacent to the *erm*-distal end of Tn*917* in the prophage; Δ, a portion of pCV1 deleted in the course of constructing pCV2; N, *Not*I recognition sites; all other features of pCV1 and pCV2 are as indicated in the legends to Fig. 1 and 2. The construction of pCV1 was described previously (42). To obtain pCV2, pCV1 was first digested with *Sal*I and recircularized at low DNA concentration to delete the small *Sal*I-*Bam*HI-*Sal*I fragment. A *Bam*HI site was restored by the insertion of a synthetic *Sal*I-*Bam*HI-*Sal*I adapter. *Not*I and *Sma*I sites were then added by inserting another synthetic adapter into both *Eco*RI sites in pCV1.

gene, all bracketed by the "left" and "right" arms of Tn917 and flanked by additional SPβ DNA. Vector pCV1 was described previously (42). Vector pCV2 was derived from pCV1 by the addition of linkers containing NotI and SmaI sites directly adjacent to the EcoRI sites in pCV1 and by the deletion of sequences from pCV1 indicated in Fig. 12.

In both pCV1 and pCV2, either the SalI or BamHI sites are available for the insertion of fragments of chromosomal DNA with appropriate cohesive ends. As explained above in the context of using φ105J27, the BamHI site may be used for the insertion of size-selected fragments generated by partial digestion of genomic DNA with Sau3A. In the case of SPβ, however, fragments as large as 8 to 10 kb may be used (42). The SalI sites offer an additional kind of option which is extremely useful. As first pointed out by Zabrovsky and Allikments (67), SalI-generated cohesive ends (single-stranded extension 3'-AGCT-5') can be made compatible with Sau3A-generated cohesive ends (single-stranded extension 3'-CTAG-5') if the SalI-generated ends are partially filled in with dTTP and dCTP and the Sau3A-generated

ends are partially filled in with dGTP and dATP. Moreover, the partial filling-in prevents self-ligation of either vector or insert. Thus, a genomic library can be generated with either pCV1 or pCV2 (Fig. 13). The plasmid vector is first cut with SalI and SstI and then incubated with Klenow fragment plus dTTP and dCTP. Chromosomal DNA is subjected to a light partial digestion with Sau3A, and fragments in the size range of 8 to 10 kb are isolated by electrophoresis or fractionation in a sucrose gradient. These are then incubated with Klenow fragment plus dGTP and dATP. Partially filled-in SalI and SstI vector fragments are then ligated with partially filled-in chromosomal fragments at high DNA concentration to generate long, linear concatemers which will frequently include joining of vector and insert fragments in the appropriate orientation to permit the integrative recombination event depicted in Fig. 13, when the ligation mix is used to transform a lysogen containing the prophage SPβc2del2::Tn917, selecting for Cmʳ.

When partially filled-in vector fragments are ligated to partially filled-in chromosomal fragments, Cmʳ transformants cannot arise

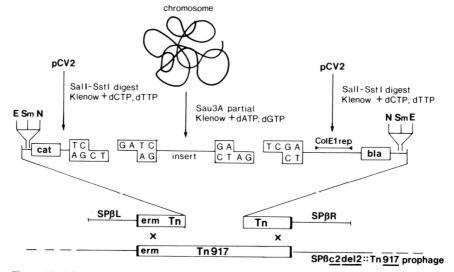

Figure 13. Schematic diagram illustrating how a genomic library can be obtained using pCV2. Restriction sites and other features of pCV2 fragments and the SPβc2del2::Tn917 prophage are as indicated in the legends to other figures.

except as the result of a recombination event that produces a recombinant prophage, and because inserts are relatively large, as few as 10^3 transformants should produce a clone library fully representative of genomic sequences. Phages carrying inserts that include genes capable of complementing specific mutations can be recovered by heat induction of a pooled population of primary Cm^r transformants, followed by infection of strains containing the mutations of interest. In the case of auxotrophic mutations or *spo* mutations, recombinant phages that carry complementing inserts can be recovered by direct selection. Because only a few thousand phage would be sufficient to cover the entire genome, it should even be feasible to screen for nonselectable complementing phenotypes.

A further advantage of generating libraries with pCV2 is the presence of *Not*I sites bracketing the portion of the cloning vector that includes the *Sal*I and *Bam*HI sites,

plus the *cat* gene and pBR322-derived sequences (Fig. 14). This provides a simple way to subclone inserts from the phage. If the insert is relatively stable with respect to deletion, as is typical with pCV2-generated recombinant phages (H. Poth and D. Heitmann, unpublished data), this can be accomplished by preparing a small amount of phage DNA, cutting with *Not*I, ligating at dilute DNA concentration, and transforming an *E. coli* strain with a selection for ampicillin resistance (Ap^r). If the insert is relatively unstable, insert-containing sequences can be rescued directly from *Not*I digests of chromosomal DNA. As noted above, it is not uncommon for *B. subtilis* chromosomal DNA fragments carried on ColE1-derived vectors to be deleterious or even lethal to an *E. coli* host. For example, in previous work (42) it proved impossible to recover a cloned *B. subtilis* fragment containing the *recE* gene from a recombinant SPβ phage by using methods

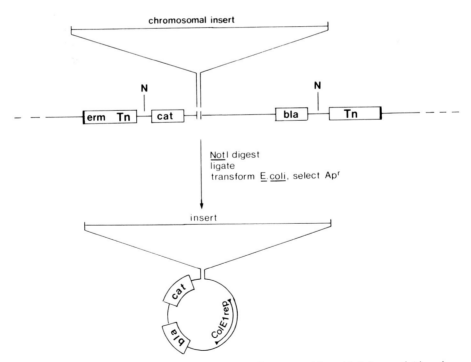

Figure 14. Diagram illustrating how cloned inserts can be recovered from *Not*I digests of either phage DNA or chromosomal DNA harboring a recombinant prophage. All features of the recombinant molecules are as indicated in other figure legends.

similar to those described here. The use of a pcnB mutant of E. coli as a cloning host might be extremely helpful in such cases.

FUNCTIONAL ANALYSIS OF CLONED GENES

Efforts to facilitate the physical and functional characterization of cloned genes have led to the development of numerous innovative methods for manipulating cloned DNA fragments and reintroducing them into the B. subtilis chromosome. Again, most of this methodology was designed initially to investigate the structure and regulation of sporulation genes.

Integrative Gene Disruption To Define the Functional Boundaries of Transcription Units

Having cloned a gene or operon of interest, the first order of business is generally to define its functional boundaries. These would consist, at the 5' extremity, of sequences required for normal expression and regulation of the promoter and, at the 3' extremity, of sequences encoding the carboxy terminus of the most promoter-distal gene product (or, conceivably, of sequences required for stability of the transcript). Because of the efficiency with which integrational vectors can be used in B. subtilis, initial characterization of cloned sequences usually involves subcloning of portions of the cloned DNAs into vectors such as those shown in Fig. 2 and 3, followed by transformation into a wild-type B. subtilis recipient. The underlying principle, of course, is that integration of vectors containing either boundary of the transcription unit will regenerate an intact copy of the transcription unit as one of the products of recombination, whereas integration of vectors containing fragments internal to the transcription unit will cause gene disruption, producing a mutant phenotype. One of the first systematic applications of this approach was by Piggot and colleagues (40) to define the functional boundaries of the spoIIA operon of B. subtilis.

High-Resolution Gene-Disruption Analysis

Once the approximate functional boundaries of a transcription unit have been established, it is often desirable to define them at higher resolution and to generate DNA sequence information. This is particularly relevant when the objective is to investigate the structure and regulation of the promoter for the cloned gene. The expression phenotypes of a series of deletions toward the promoter from the 5' direction help to identify sequences that function in cis to determine the activation or regulation of the promoter. DNA sequence information in the region of the promoter is necessary for designing "primer-extension" probes (38) to establish the apparent in vivo start site(s) for transcription.

One approach for generating unidirectional deletions toward a promoter of interest is the strategy of exonuclease III-mediated resection described by Henikoff (23). A promoter-containing restriction fragment is subcloned into one of the polylinker sites in an integrational vector such as pGEM-3Zf(+) cat-1 (B. Green and P. Youngman, unpublished) (Fig. 3), oriented such that the end of the fragment upstream from the promoter is preceded by at least one recognition site for a restriction enzyme that leaves a blunt end or a cohesive end with a 5' overhang, which in turn is preceded by at least one recognition site for an enzyme that leaves a 3' overhang, preferably one with a four-base extension (such as SstI, KpnI, or SphI). Because cohesive ends with 3' overhangs are insensitive to the action of exonuclease III, double-cut vector can produce a substrate for unidirectional deletions toward the promoter. Subsequent treatment with S1 nuclease and T4 DNA polymerase will allow the two blunt ends of the molecule to be rejoined with ligase. Carrying out such operations with a vector like pGEM-3Zf(+)cat-1, which has an f1 origin of replication, also facilitates DNA sequence analysis of the deletion products.

An easier and more rapid method for generating unidirectional deletions in DNA cloned into M13 derivatives was described by Dale and colleagues (8). This approach is based on the discovery that when a short oligonucleotide is annealed to M13mp19 virion DNA at a site that overlaps the *Eco*RI-proximal end of the polylinker, it becomes possible to linearize the single-stranded DNA by digestion with *Eco*RI. Unidirectional deletions can then be generated into DNA cloned into the polylinker by using the 3′-to-5′ exonuclease activity of T4 DNA polymerase. After addition of a short poly(dG) tail with terminal transferase, the two ends of the deleted molecule can be held together for ligation by annealing the same oligonucleotide used in the linearization reaction. This method is very efficient, easy to control, and not technically demanding. In part to take advantage of this approach to facilitate the characterization of *B. subtilis* promoters, a new family of integrational vectors has been constructed simply by inserting a *cat* gene into various M13 derivatives (22). One such vector is shown in Fig. 15.

Reintroduction of Mutationally Altered DNA into the Chromosome

In the course of analyzing *cis*-acting sequences that participate in determining the regulation of a promoter, it is frequently necessary to introduce mutational changes in vitro and return the mutant promoters to the chromosome, in single copy, in a way that makes it possible to assess the effects of the mutational changes on transcription. Normally, this cannot be accomplished using simple integrational vectors to return the mutant promoters to their normal chromosomal locations, because gene-conversion events that accompany vector integration, or reciprocal recombination itself, may correct the mutations or separate them from the reporter gene with which they are associated in the

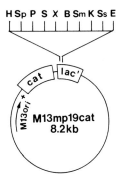

Figure 15. An integrational vector derived from single-stranded phage cloning vector M13mp19 (59). Restriction sites and other features of M13mp19*cat* are as indicated in the legends to other figures. The construction of this vector was described previously (22).

vector. One solution is to use the integration system devised by Zuber and Losick (70) (Fig. 5). For example, alterations might be introduced by oligonucleotide-directed mutagenesis into a promoter-containing fragment cloned into M13 and then transferred to a vector such as pGV34 (Fig. 4) to create a *lacZ* fusion. When subsequently integrated into an SPβ prophage as shown in Fig. 5, there is no opportunity for the mutant promoter to interact with its wild-type counterpart in the chromosome in a way that might correct the mutation. An advantage of this approach is that the fusion construction is now easily transferred to other genetic backgrounds by SPβ-mediated transduction.

An alternative approach is possible based on vectors developed by Shimotsu and Henner (48). With these vectors, *lacZ* fusions may be constructed in a way that places them adjacent to a *cat* gene selectable in *B. subtilis* and bracketed by segments of DNA containing, respectively, the "front" and "back" halves of the *amyE* gene. Thus, integration of the expression cassette can occur by a gene replacement-type double-crossover event that inserts the expression cassette into the chromosomal copy of the *amyE* gene. This kind of recombination event can be distinguished from other events that might produce a Cm^r transformant, because disruption of the *amyE* gene produces an easily recognized phenotype on starch plates.

Yet another approach was described recently by Lewandoski and Smith (32), which makes it possible to analyze the behavior of a *lacZ* translational fusion either in multicopy or inserted into the chromosome either at the normal chromosomal location of the promoter or at a heterologous locus defined by integrated pE194 sequences. The vector constructed for this purpose, pIS112, is a chimera of a temperature-sensitive, high-copy, *cat*-containing pE194 derivative and a ColE1 derivative described by Zuber and Losick (69) which contains the *lacZ* coding sequence preceded by polylinker sites that permit the construction of in-frame *lac* fusions. Although pE194 *rep* functions have been inactivated in pIS112, they can be complemented in *trans* by a chromosomal insertion of pE194 to allow replication in *B. subtilis* as a multicopy plasmid. Thus, when a fragment of chromosomal DNA has been inserted into pIS112 to produce a fusion and the fusion-containing construction is used to transform a *B. subtilis* strain, Cm^r transformants can arise in three ways: (i) if the recipient contains a chromosomal pE194 insertion, a selection for Cm^r at low temperature will establish the construction as a high-copy-number autonomous replicon; (ii) if the recipient contains a chromosomal pE194 insertion, a selection for Cm^r at high temperature can result from integration of the construction at the site of the insertion; and (iii) if the recipient does not contain a pE194 insertion, Cm^r transformants can only arise by integration at the homologous chromosomal locus.

Creation of Permanent Chromosomal Deletions by Using Cloned DNAs

Several methods are available in *B. subtilis* for disrupting the chromosomal copy of a cloned gene. One, described above, involves the use of integrational vectors that include sequences internal to the gene. A disadvantage of this approach is that the disruption is easily revertable by "Campbell excision" and would require continuous antibiotic selection to be maintained. Another disadvantage is that it "uses up" a drug resistance gene and thus limits the kinds of additional manipulations that can be carried out in the disruption mutant. Alternatively, antibiotic resistance genes may be inserted into cloned genes in vitro and then transferred to the chromosome by marker-replacement recombination. In this case, the disruption is stable, but the antibiotic resistance gene again remains permanently in the chromosome. These disadvantages are overcome with an approach described by Yang and colleagues (58). A permanent deletion was introduced into the chromosomal copy of a protease-encoding gene by first constructing, in vitro, a deletion internal to the gene cloned into the integrational vector pJH101 (Fig. 2). The deletion-containing vector was then inserted into the chromosome by homologous recombination within the protease gene, selecting for Cm^r. This kind of integrational event may or may not itself be mutational, depending upon whether the promoter for the protease gene is present in the vector and depending upon where within the cloned sequences the integrative crossover occurs. Either way, Campbell excision, which can be recognized as producing Cm^s segregants, may leave behind a permanent deletion of the gene. A better way to carry out such a procedure would be to use a shuttle vector that included *rep* functions derived from pE194Ts (56). After introduction into *B. subtilis*, integration into the chromosome could be forced by raising the temperature above 37°C while maintaining a selection for the vector-associated drug resistance. When the temperature is returned to 32°C, integrants are highly unstable, and antibiotic-sensitive segregants would be much easier to obtain than with vectors such as pJH101.

Copy Number Manipulation

It has long been recognized that maintaining sporulation genes (30) or even the promoters for sporulation genes (3) on multicopy vectors in *B. subtilis* can inhibit sporulation, and this is usually attributed to titration of DNA-binding proteins or other regulatory factors that act at the posttranslational level. The discovery that DNA sequences cloned into integrational vectors can be stably amplified in the chromosome of *B. subtilis* in a stepwise and controlled fashion (1, 25, 61) has led to the development of systematic procedures for evaluating the effects of copy number on gene activity. For example, Piggot and colleagues (41) were able to demonstrate that as few as four copies of the *spo0F* gene of *B. subtilis* can inhibit sporulation. Manipulating the copy number of *lacZ* fusions has proved useful for determining whether regulation of specific promoters is accomplished by positive or negative control mechanisms (P. Piggot, personal communication), and "overexpression" of genes maintained on multicopy plasmids is providing insights into regulatory interactions of gene products that apparently play a role in the control of initiation of sporulation (I. Smith, personal communication).

Use of Vectors That Confer Phleomycin Resistance

As genetic manipulations of *B. subtilis* become ever more elaborate, it becomes increasingly useful to have available additional drug resistance genes that can be selected for in single copy. Until recently, the only drug resistance determinants of this kind in wide use were *cat* and *erm* genes. Although it has been common knowledge for several years that the *tet* genes associated with Tn*916* and its relatives are selectable in *B. subtilis* in single copy (e.g., reference 7), these genes have not yet been characterized physically and functionally in sufficient detail to allow their easy incorporation into plasmid, transposon, or integrational vectors. It was recently determined that *S. aureus* plasmid pUB110 (24) contains, in addition to its kanamycin resistance gene, a gene that confers resistance to

antibiotics such as bleomycin (Bm) and phleomycin (Pm) in both *E. coli* and *B. subtilis* (46). It has further emerged, through the experience of investigators in several laboratories, that this gene confers cleanly selectable Pm^r in single copy in *B. subtilis* (J. Alonso, personal communication; D. Dubnau, personal communication). A vector that should be very useful for constructing translational *lacZ* fusions which incorporates the Pm^r gene (*ble*) has already been described by Alonso (2), and a small DNA fragment derived from pUB110 that contains the *ble* gene has been inserted into the polylinker of pUC18 (H. Poth and P. Youngman, unpublished data) (Fig. 16a), which should facilitate its incorporation into other constructions. Additional vectors that make use of the *ble* gene are described below.

New SPAC Cassettes

As mentioned above, various versions of the SPAC expression cassette developed by Henner and colleagues (e.g., reference 60) have proved useful for the identification of classes of genes whose expression depends upon certain positive regulatory factors (e.g., Jaacks et al., submitted). Numerous other applications for this expression cassette have been explored by other investigators. For example, Stragier and colleagues (50) have used strategies involving the SPAC cassette to help define the genetic and physiological requirements for the developmentally regulated proteolytic processing of pro-σ^E. In recent work from our laboratory (K. York and P. Youngman, unpublished), carried out in collaboration with Moran and colleagues at Emory University (T. Kenney and C. P. Moran, personal communication), the SPAC cassette was used in yet another way, and this involved the construction of new vectors which may be of use to other investigators. The objective of the work was to establish whether σ^A-associated RNA polymerase is responsible for transcription of the *spoIIE* and *spoIIG* operons of *B. subtilis*, both of which have

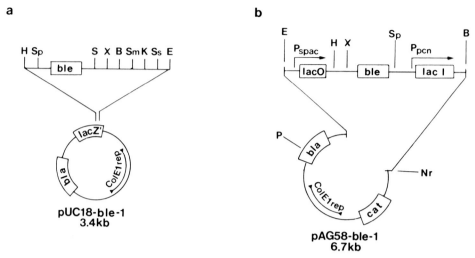

Figure 16. (a) pUC18-*ble*-1 vector, which can serve as the source of a *ble* gene flanked by convenient restriction sites that may facilitate its tansfer to other vectors. This vector was constructed by inserting a 0.7-kb *Hae*III-*Dra*I fragment from pUB110 (24) that contains the *ble* gene into the *Pst*I site of pUC18 (59) after making the *Pst*I-generated ends flush by treatment with T4 DNA polymerase. (b) pAG58-*ble*-1 vector, which can be used to integrate a SPAC expression cassette into the chromosome with a selection for Pm[r]. This vector was constructed by inserting a *ble*-containing *Sal*I-*Sph*I fragment from pUC18-*ble*-1 (panel a) into the *Sal*I-*Sph*I backbone of pAG58 (Fig. 7). Restriction sites and other features of pUC18-*ble*-1 and pAG58-*ble*-1 are as described in the legends to other figures.

very unusual promoters that contain inappropriately spaced −10 and −35 consensus sequences (21, 31). Our approach was to create mutational changes within the −10 hexamers of the *spoIIE* and *spoIIG* promoters that abolished transcription and then ask whether certain amino acid substitutions in σ[A] that might be predicted to change its recognition specificity could restore activity of the mutant promoters. We expected, however, and later confirmed, that such change-of-specificity mutations in *sigA* would be dominant lethals. This required placing expression of the mutant *sigA* genes under control of an inducible promoter such as p_{spac}. Existing versions of the SPAC cassette, which utilized *cat* as the selectable marker for integration, were not suitable for this purpose, because our strains that contained *lacZ* fusions to the mutant *spoIIE* and *spoIIG* promoters were already Cm[r] and Em[r] (they were constructed as diagrammed in Fig. 5). Thus, it was necessary to construct the vector shown in Fig. 16b, which contains a *ble* gene in-

serted within the SPAC cassette. We further needed to insert the SPAC cassette into the chromosome at a location other than SPβ (into which we had already inserted the mutant p_{spoIIE}-*lac* and p_{spoIIG}-*lac* fusions). This was accomplished by constructing a vector in which the SPAC-*ble* cassette that was engineered to express mutant versions of *sigA* was bracketed by sequences from the left and right arms of Tn917. This permitted the transfer of the expression cassettes by homologous recombination, selecting for Pm[r], into phenotypically silent chromosomal insertions of Tn917. This series of experiments was highly successful, yielding results which suggest strongly that σ[A] is utilized in some unusual way, well into sporulation, for the transcription of important sporulation-specific genes.

SUMMARY

B. subtilis has long been regarded as a useful organism in which to investigate

mechanisms of developmental gene regulation, but as recently as a decade ago it was
also regarded as relatively primitive, by
comparison with *E. coli*, with respect to the
genetic tools available for gene regulation
studies. Innovative efforts by many investigators have not only narrowed this methodology gap between *B. subtilis* and
E. coli, but have actually resulted in the
development of strategies for gene manipulation in *B. subtilis* that have no counterpart in *E. coli*. In any event, ease of genetic
manipulation no longer represents a significant barrier to progress in understanding
the regulation of sporulation genes, and
it can safely be said that methodological
innovation has been an important factor
contributing to the recent and remarkable acceleration of research activity in
this area.

Acknowledgments. We thank C. P.
Moran, P. Zuber, A. Grossman, P. Piggot, I. Smith, P. Pattee, D. Dubnau, D. Portnoy, J. Alonso, R. Novick, M. Young, L.
Daneo-Moore, and colleagues in their
laboratories for many helpful discussions and for the communication of
data before publication. We thank D. Heitmann for technical assistance and help with
preparation of figures for the manuscript.

Work from our laboratory was supported by Public Health Service grant
GM35495 from the National Institutes of
Health; by a postdoctoral fellowship to H.P.
from the Deutscher Akademischer Aus-
tauschdienst Sonderprogramm Gentechnol-
ogie; and by a predoctoral fellowship to G.O.
from Merck Sharp & Dohme Research Laboratories.

LITERATURE CITED

1. Albertini, A. M., and A. Galizzi. 1985. Amplification of a chromosomal region in *Bacillus subtilis*.
J. Bacteriol. **162**:1203–1211.
2. Alonso, J. C. 1988. New plasmid vectors for the
construction of translational gene fusions in *Bacillus
subtilis*. *Gene* **65**:325–328.
3. Banner, C. D. B., C. P. Moran, Jr., and R. Losick.
1983. Deletion analysis of a complex promoter for
a developmentally regulated gene from *Bacillus subtilis*. *J. Mol. Biol.* **168**:351–365.
4. Bolivar, F., R. L. Rodriguez, P. J. Greene, M. C.
Betlach, H. L. Heynecker, H. W. Boyer, J. H. Crosa,
and S. Falkow. 1977. Construction and characterization of new cloning vehicles. II. A multipurpose
cloning system. *Gene* **2**:95–113.
5. Bugaichuk, U.D., M. E. Deadman, J. Errington, and
D. Savva. 1984. Restriction enzyme analysis of *Bacillus subtilis* bacteriophage φ105 DNA. *J. Gen.
Microbiol.* **130**:2165–2167.
6. Castilho, B. A., P. Olfson, and M. J. Casadaban.
1984. Plasmid insertion mutagenesis and *lac* fusion
with mini-Mu bacteriophage transposons. *J. Bacteriol.* **158**:488–495.
7. Christie, P. J., R. Z. Korman, S. A. Zahler, J. C.
Adsit, and G. M. Dunny. 1987. Two conjugation
systems associated with *Streptococcus faecalis* plasmid pCF10: identification of a conjugative transposon that transfers between *S. faecalis* and *Bacillus
subtilis*. *J. Bacteriol.* **169**:2529–2536.
8. Dale, R. M. K., B. A. McClure, and J. P. Houchins.
1985. A rapid single-stranded cloning strategy for
producing a sequential series of overlapping clones
for use in DNA sequencing: application to sequencing in the corn mitochondrial 18S rDNA. *Plasmid*
13:31–40.
9. Donnelly, C. E., and A. L. Sonenshein. 1984. Promoter-probe plasmid for *Bacillus subtilis*. *J. Bacteriol.* **157**:965–967.
10. Dubnau, E., J. Weir, G. Nair, L. Carter III, C. P.
Moran, Jr., and I. Smith. 1988. *Bacillus* spor-
ulation gene *spo0H* codes for σ^{30} (σ^H). *J. Bacteriol.*
170:1054–1062.
11. Duncan, C. H., G. A. Wilson, and F. E. Young.
1978. Mechanism of integrating foreign DNA during transformation of *Bacillus subtilis*. *Proc. Natl.
Acad. Sci. USA* **75**:3664–3665.
12. Errington, J. 1984. Efficient *Bacillus subtilis* cloning
system using bacteriophage vector φ105J9. *J. Gen.
Microbiol.* **130**:2615–2628.
13. Errington, J. 1986. A general method for fusion of
the *Escherichia coli lacZ* gene to chromosomal genes
in *Bacillus subtilis*. *J. Gen Microbiol.* **132**:2953–
2966.
14. Errington, J. 1987. Two separable functional domains in the σ subunit of RNA polymerase in *Bacillus subtilis*. *FEBS Lett.* **224**:257–260.
15. Errington, J. 1988. Generalized cloning vectors for
Bacillus subtilis, p. 345–362. *In* R. L. Rodriguez
and D. T. Denhardt (ed.), *Vectors: a Survey of Molecular Cloning Vectors and Their Uses.* Butter-
worths, Boston.
16. Errington, J., and D. Jones. 1987. Cloning in *Bacillus subtilis* by transfection with bacteriophage
vector φ105J27: isolation and preliminary charac-

terization of transducing phages for 23 sporulation loci. *J. Gen. Microbiol.* **133**:493–502.

17. Ferrari, E., D. J. Henner, and J. A. Hoch. 1981. Isolation of *Bacillus subtilis* genes from a charon 4A library. *J. Bacteriol.* **146**:430–432.

18. Ferrari, F. A., A. Nguyen, D. Land, and J. A. Hoch. 1983. Construction and properties of an integrable plasmid for *Bacillus subtilis*. *J. Bacteriol.* **154**:1513–1515.

19. Fink, P. S., and S. A. Zahler. 1982. Restriction fragment maps of the genome of *Bacillus subtilis* bacteriophage SPβ. *Gene* **19**:235–238.

20. Fort, P., and J. Errington. 1985. Nucleotide sequence and complementation analysis of a polycistronic sporulation operon, *spoVA*, in *Bacillus subtilis*. *J. Gen. Microbiol.* **131**:1091–1105.

21. Guzmán, P., J. Westpheling, and P. Youngman. 1988. Characterization of the promoter region of the *Bacillus subtilis spoIIE* operon. *J. Bacteriol.* **170**:1598–1609.

22. Guzmán, P., and P. Youngman. 1988. Novel integrational vectors for *Bacillus subtilis* based on coliphage M13 and their use for the analysis of regulated promoters, p. 299–303. *In* A. T. Ganesan and J. A. Hoch (ed.), *Genetics and Biotechnology of Bacilli*, vol. 2. Academic Press, Inc., San Diego, Calif.

23. Henikoff, S. 1984. Unidirectional digestion with exonuclease III creates targeted breakpoints for DNA sequencing. *Gene* **28**:351–359.

24. Iordanescu, S. 1976. Three distinct plasmids originating in the same *Staphylococcus aureus* strain. *Arch. Roum. Pathol. Exp. Microbiol.* **35**:111–118.

25. Jannière, L., B. Michel, E. Pierre, and S. D. Ehrlich. 1985. Stable gene amplification in the chromosome of *Bacillus subtilis. Gene* **40**:47–55.

26. Jones, D., and J. Errington. 1987. Construction of improved φ105 vectors for cloning by transfection in *Bacillus subtilis. J. Gen. Microbiol.* **133**:483–492.

27. Karmazyn-Campelli, C., C. Bonamy, B. Savelli, and P. Stragier. 1989. Tandem genes encoding σ-factors for consecutive steps in development in *Bacillus subtilis. Genes Dev.* **3**:150–157.

28. Kawamura, F., T. Mizukami, H. Anzai, and H. Saito. 1981. Frequent deletion of *Bacillus subtilis* chromosomal fragments in artificially constructed phage ρ11*phisA*⁺. *FEBS Lett.* **136**:244–246.

29. Kawamura, F., H. Saito, and Y. Ikeda. 1979. A method for construction of specialized transducing phage ρ11 of *Bacillus subtilis. Gene* **5**:87–91.

30. Kawamura, F., H. Shimotsu, H. Saito, H. Hirochka, and Y. Koboyashi. 1981. Cloning of *spo0* genes with bacteriophage and plasmid vectors in *Bacillus subtilis*, p. 109–113. *In* H. S. Levinson, A. L. Sonenshein, and D. J. Tipper (ed.), *Sporulation and Germination*. American Society for Microbiology, Washington, D.C.

31. Kenney, T. J., and C. P. Moran, Jr. 1987. Organization and regulation of an operon that encodes a sporulation-essential sigma factor in *Bacillus subtilis. J. Bacteriol.* **169**:3329–3339.

32. Lewandoski, M., and I. Smith. 1989. Use of a versatile *lacZ* vector to analyze the upstream region of the *Bacillus subtilis spo0F* gene. *Plasmid* **20**:148–154.

33. Liu, J., and J. S. Parkinson. 1989. Genetics and sequence analysis of the *pcnB* locus, an *Escherichia coli* gene involved in plasmid copy number control. *J. Bacteriol.* **171**:1254–1261.

34. Lopilato, J., S. Bortner, and J. Beckwith. 1986. Mutations in a new chromosomal gene of *Escherichia coli* K-12, *pcnB*, reduce plasmid copy number of pBR322 and its derivatives. *Mol. Gen. Genet.* **205**:285–290.

35. Losick, R., P. Youngman, and P. J. Piggot. 1986. Genetics of endospore formation in *Bacillus subtilis. Annu. Rev. Genet.* **20**:625–669.

36. Manoil, C., and J. Beckwith. 1985. Tn5-*phoA*: a transposon probe for protein exports signals. *Proc. Natl. Acad. Sci. USA* **82**:8129–8133.

37. Masuda, E. S., H. Anaguchi, K. Yamada, and Y. Koboyashi. 1988. Two developmental genes encoding sigma factor homologs are arranged in tandem in *Bacillus subtilis. Proc. Natl. Acad. Sci. USA* **85**:7637–7641.

38. McNight, S. L., and R. Kingsbury. 1982. Transcription control signals of a eukaryotic protein-coding gene. *Science* **217**:316–324.

39. Niaudet, B., A. Goze, and S. D. Ehrlich. 1982. Insertional mutagenesis in *Bacillus subtilis*: mechanism and use in gene cloning. *Gene* **19**:277–284.

40. Piggot, P. J., C. A. M. Curtis, and H. De Lencastre. 1984. Use of integrational plasmid vectors to demonstrate the polycistronic nature of a transcription unit (*spoIIA*) required for sporulation of *Bacillus subtilis. J. Gen. Microbiol.* **130**:2123–2136.

41. Piggot, P. J., J. J. Wu, C. A. M. Curtis, and J. W. Chapman. 1988. Manipulation of gene copy number in *Bacillus subtilis* using integrative plasmids, p. 141–145. *In* A. T. Ganesan and J. A. Hoch (ed.), *Genetics and Biotechnology of Bacilli*, vol. 2. Academic Press, Inc., San Diego, Calif.

42. Poth, H., and P. Youngman. 1988. A new gene cloning system for *Bacillus subtilis* comprising elements of phage, plasmid and transposon vectors. *Gene* **73**:215–226.

43. Sandman, K., R. Losick, and P. Youngman. 1987. Genetic analysis of *Bacillus subtilis spo* mutations generated by Tn917-mediated insertional mutagenesis. *Genetics* **117**:603–617.

44. Savva, D., and J. Mandelstam. 1984. Cloning of the

Bacillus subtilis *spoIIA* and *spoVA* loci in phage φ105DI:1t. *J. Gen. Microbiol.* **130**:2137–2145.

45. Seifert, H. S., E. Y. Chen, M. So, and F. Heffron. 1986. Shuttle mutagenesis: a method of transposon mutagenesis for *Saccharomyces cerevisiae*. *Proc. Natl. Acad. Sci. USA* **83**:735–739.

46. Semon, D., N. R. Movva, T. F. Smith, N. El Alama, and J. Davies. 1987. Plasmid-determined bleomycin resistance in *Staphylococcus aureus*. *Plasmid* **17**:46 –53.

47. Setlow, P. 1988. Small, acid-soluble spore proteins of *Bacillus* species: structure, synthesis, genetics, function and degradation. *Annu. Rev. Microbiol.* **42**:319–338.

48. Shimotsu, H., and D. J. Henner. 1986. Construction of a single-copy integration vector and its use in the analysis of regulation of the *trp* operon of *Bacillus subtilis*. *Gene* **43**:85–94.

49. Stragier, P. 1986. Comment on "Duplicated Sporulation Genes in Bacteria" by J. Errington, P. Fort, and J. Mandelstam. *FEBS Lett.* **195**:9–11.

50. Stragier, P., C. Bonamy, and C. Karmazyn-Campelli. 1988. Processing of a sporulation sigma factor in *Bacillus subtilis*: how morphological structure could control gene expression. *Cell* **52**:697–704.

51. Stragier, P., B. Kunkel, L. Kroos, and R. Losick. 1989. Chromosomal rearrangement generating a composite gene for a developmental transcription factor. *Science* **243**:507–512.

52. te Riele, H., B. Michel, and S. D. Ehrlich. 1986. Single-stranded plasmid DNA in *Bacillus subtilis* and *Staphylococcus aureus*. *Proc. Natl. Acad. Sci. USA* **83**:2541–2545.

53. Tomich, P. K., F. Y. An, and D. B. Clewell. 1980. Properties of erythromycin-inducible transposon Tn917. *J. Bacteriol.* **141**:1366–1374.

54. Trach, K., J. W. Chapman, P. J. Piggot, and J. A. Hoch. 1985. Deduced product of the stage 0 sporulation gene *spo0F* shares homology with Spo0A, OmpR and SfrA proteins. *Proc. Natl. Acad. Sci. USA* **82**:7260–7264.

55. Trempy, J. E., C. Bonamy, J. Szulmajster, and W. G. Haldenwang. 1985. *Bacillus subtilis* sigma factor σ^{29} is the product of sporulation gene *spoIIG*. *Proc. Natl. Acad. Sci. USA* **82**:4189–4192.

56. Villafane, R., D. H. Bechhofer, C. S. Narayanan, and D. Dubnau. 1987. Replicaton control genes of plasmid pE194. *J. Bacteriol.* **169**:4822–4829.

57. Weaver, K. E., and D. B. Clewell. 1988. Regulation of the pAD1 sex pheromone response in *Enterococcus faecalis*: construction and characterization of *lacZ* transcription fusions in a key control region of the plasmid. *J. Bacteriol.* **170**:4343–4352.

58. Yang, M. Y., E. Ferrari, and D. J. Henner. 1984. Cloning of the neutral protease gene of *Bacillus subtilis* and the use of the cloned gene to create an in vitro-derived deletion mutation. *J. Bacteriol.* **160**:15–21.

59. Yanisch-Perron, C., J. Vieira, and J. Messing. 1985. Improved M13 phage cloning vectors and host strains: nucleotide sequences of the M13mp18 and pUC19 vectors. *Gene* **33**:103–119.

60. Yansura, D. G., and D. J. Henner. 1984. Use of the *Escherichia coli lac* repressor and operator to control gene expression in *Bacillus subtilis*. *Proc. Natl. Acad. Sci. USA* **81**:439–443.

61. Young, M. 1984. Gene amplification in *Bacillus subtilis*. *J. Gen. Microbiol.* **130**:1613–1621.

62. Youngman, P. 1987. Plasmid vectors for recovering and exploiting Tn917 transpositions in *Bacillus* and other gram-positive bacteria, p. 79–103. *In* K. Hardy (ed.), *Plasmids: a Practical Approach*. IRL Press, Oxford.

63. Youngman, P. J., J. B. Perkins, and R. Losick. 1983. Genetic transposition and insertional mutagenesis in *Bacillus subtilis* with *Streptococcus faecalis* transposon Tn917. *Proc. Natl. Acad. Sci. USA* **80**:2305–2309.

64. Youngman, P., J. B. Perkins, and R. Losick. 1984. A novel method for the rapid cloning in *Escherichia coli* of *Bacillus subtilis* chromosomal DNA adjacent to Tn917 insertions. *Mol. Gen. Genet.* **195**:424–433.

65. Youngman, P., J. B. Perkins, and K. Sandman. 1985. Use of Tn917-mediated transcriptional fusions to *lacZ* and *cat-86* for the identification and study of regulated genes in the *Bacillus subtilis* chromosome, p. 47–54. *In* J. A. Hoch and P. Setlow (ed.), *Molecular Biology of Microbial Differentiation*. American Society for Microbiology, Washington, D.C.

66. Youngman, P., P. Zuber, J. B. Perkins, K. Sandman, M. Igo, and R. Losick. 1985. New ways to study developmental genes in spore-forming bacteria. *Science* **228**:285–291.

67. Zabarovsky, E. R., and R. L. Allikments. 1986. An improved technique for the efficient construction of gene libraries by partial filling-in of cohesive ends. *Gene* **42**:119–123.

68. Zheng, L., W. P. Donovan, P. C. Fitz-James, and R. Losick. 1988. Gene encoding a morphogenetic protein required in the assembly of the outer coat of the *Bacillus subtilis* endospore. *Genes Dev.* **2**:1047–1054.

69. Zuber, P., and R. Losick. 1983. Use of a *lac* fusion to study developmental regulation by the *spo0* genes of *Bacillus subtilis*. *Cell* **35**:275–283.

70. Zuber, P., and R. Losick. 1987. Role of AbrB in Spo0A- and Spo0B-dependent utilization of a sporulation promoter in *Bacillus subtilis*. *J. Bacteriol.* **169**:2223–2230.

Regulation of Procaryotic Development
Edited by Issar Smith, Ralph A. Slepecky, and Peter Setlow
© 1989 American Society for Microbiology, Washington, DC 20006

Chapter 4

The Trigger Mechanism of Bacterial Spore Germination

Simon J. Foster and Keith Johnstone

The germination response of bacterial spores represents an environmentally controlled developmental system. Paradoxically, although bacterial spores are metabolically dormant (25), they retain an alert sensory mechanism which allows them to monitor the external environment and to trigger germination in the presence of specific germinants (39). Intriguingly, the components responsible for germination triggering must escape the mechanisms of stabilization which result in spore resistance and dormancy; they must, however, retain the intrinsic resistance properties of the whole spore. This germination response results in the differentiation of the dormant spore into a metabolically active vegetative cell. Fundamental to this process are the mechanisms of germinant recognition and initiation of germination.

The purpose of this article is to review the approaches which have been employed to analyze the mechanism of germination triggering, to interrelate the findings from these approaches, and to ask whether there is evidence for a common mechanism of spore germination. To this end, only selected references which address fundamental concepts will be cited. No attempt will be made to review comprehensively the wealth of existing information about spore structure, function, and germination; for this purpose the reader is referred to previous articles (25, 38, 65, 66, 85, 86, 120, 122, 123, 144, 158). Due to the extensive genetical and biochemical studies of germination triggering in *Bacillus subtilis* and *Bacillus megaterium*, respectively, we inevitably concentrate on these organisms as model systems and refer to other organisms where relevant.

The definition of germination is somewhat confused (65). In this article, the term "germination" is used to describe the irreversible degradative biochemical reactions which occur when spores are exposed to germinants and which result in the breaking of spore dormancy and the loss of typical spore properties. Commitment to germination may be defined as the time a spore suspension must be exposed to germinants to allow the normal progress of germination events to take place when the germinants are subsequently removed (139). During commitment, initiation of germination takes place, which involves interaction of the germinant with its receptor. This results subsequently in irreversible commitment to germination. The initial germinant-mediated response constitutes the trigger reaction (65).

Simon J. Foster and Keith Johnstone • Department of Botany, University of Cambridge, Downing Street, Cambridge CB2 3EA, United Kingdom.

GERMINATION

Activation

Dormant spores will not germinate optimally unless they have been previously activated. Activation is a reversible process and involves "conditioning" the spore population such that the rate and extent of germination are increased (66). Since activation also increases the rate of commitment to germination (139), the germinant receptor mechanism must be stimulated by this process. Activation is most commonly achieved by sublethal heat treatment, although several other treatments have been used, e.g., extremes of pH and reducing agents (31, 66, 108). Heat activation causes loss of protein from spore coats and a change in their ultrastructure (48, 135). Also, release of small amounts of dipicolinic acid (DPA) and Zn^{2+}, which are located in the core and coats, respectively, occurs during heat activation (62, 66). In addition, heat activation results in a small increase in inner membrane fluidity (57, 138). The activation energy for heat activation is approximately 50 kJ mol^{-1} (73) and is within the range for both enzyme-catalyzed reactions (45) and protein denaturation (73). How this stimulation of germination occurs is unknown, although it is clearly worthy of further study since activation is of fundamental importance to our understanding of the regulation of spore germination.

Germinants

A wide range of bacterial spore germinants has been described, which may be divided into nutrient, nonnutrient (chemical), enzymatic, and physical (including mechanical) germinants (39, 40). Although these classes of germinants may initiate germination by different mechanisms, a common model for nutrient-induced germination must be able to take account of the wide variety of nutrient germinants. L-Alanine is the most common nutrient germinant (39, 47, 164); others include several amino acids (105), purine ribosides (129), and sugars (99, 143). Na$^+$ or K$^+$ ions are also generally required for L-alanine-induced germination (39). Some spores, particularly those of *Clostridium* spp., will only germinate in the presence of a combination of germinants (38). Spores of several species can, however, be germinated ionically; e.g., spores of *Clostridium perfringens* and *B. megaterium* QMB1551 germinate in 50 mM KCl (2, 4, 39, 109) and 40 mM KBr (32), respectively.

Nonnutrient germinants include surfactants and calcium dipicolinate, but the mechanisms whereby they cause germination remain to be elucidated (40, 102, 103). It has been suggested that the alkyl amines may cause displacement of calcium from spore membranes to initiate germination (26). Several peptidoglycan-lytic enzymes will cause germination by breakdown of the spore cortex in spores which have had their coats chemically extracted (36, 41, 43). Given the central role of the spore cortex in maintaining spore dormancy (25), the action of these peptidoglycan-lytic enzymes is to be expected. Also, subtilisin has been found to initiate germination, possibly by the liberation of L-alanine from the spore coats (98). Mechanical methods of spore germination include abrasion and hydrostatic pressure (16, 40, 44, 87). Germination under high hydrostatic pressure may occur due to an increase in permeability of an unknown spore component (44). Alternatively, high pressures could result in an increase in ionization of certain electrolytes in the spore (17), which triggers the endogenous germination mechanism. The fact that pressure germination is potentiated by L-alanine (44) supports this view.

The rest of this article will focus on the nutrient germinants which have been most studied and which clearly represent the physiological spore germination pathway. Spores of *B. megaterium* KM have a single L-alanine-dependent germination pathway (139), whereas *B. subtilis* spores may be germinated by L-alanine and by a mixture of D-glucose, fructose, L-asparagine, and KCl (GFAK) (84).

Physiological and Biochemical Events

During germination, the spore sequentially loses many of its characteristic properties (38). This process results in a decrease in spore refractility, the associated microscopic change from phase-white to phase-dark, loss of heat and chemical resistance, an increase in stainability, and the breaking of spore dormancy.

The biochemical events which accompany these physiological changes include release of ions, e.g., Ca^{2+}, Zn^{2+}, and DPA; limited cortex hydrolysis; release of soluble hexosamine; proteolysis of spore proteins; and the onset of general spore metabolism (63, 113, 115, 147). Early metabolic changes include increases in pools of ATP, NADH, glycolytic intermediates, and tricarboxylic acid cycle intermediates (113, 117–119).

When analyzing *B. megaterium* KM spore germination, these events can be divided into two categories: first, those early events which can be detected within 1 min of germination triggering and which may therefore be associated with the germination-triggering event, and second, those events which are initiated at a later time (S. J. Foster, Ph.D. Thesis, University of Cambridge, Cambridge, United Kingdom, 1986). After 1 min of germination, 11, 5, and 0.8% of the 40-min-germinated values of commitment, loss of heat resistance, and DPA release, respectively, have occurred (Fig. 1). Loss of heat resistance is the first event detected during germination of spores of *B. megaterium* QMB1551, *B. subtilis*, and *Bacillus cereus* (22, 23, 49, 72). In contrast, selective cortex hydrolysis, loss of soluble hexosamine, decrease in A_{600}, net ATP synthesis, and the onset of general spore metabolism cannot be detected until at least 2 min after the addition of germinants (Fig. 2). Thus these later germination events appear to be dependent on the initial triggering and commitment reactions.

FUNCTIONAL APPROACHES

Understanding the mechanisms of germinant recognition and commitment during nutrient-induced germination has remained a primary research goal for several decades.

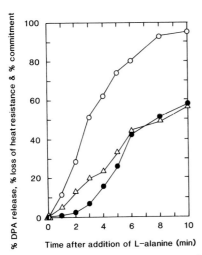

Figure 1. Early events during germination triggering. Percent commitment (○), percent DPA release (●), and percent loss of heat resistance (△) during triggering of heat-shocked spores of *B. megaterium* KM by L-alanine. Data taken from Foster (thesis).

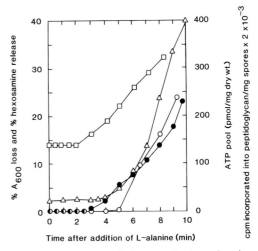

Figure 2. Late events during germination triggering. Percent hexosamine release (○), percent A_{600} loss (●), ATP pool (△), and reducing termini in cortical peptidoglycan determined by reduction with NaB³H₄ (□) during triggering of heat-shocked spores of *B. megaterium* KM by L-alanine. Data taken from reference 61 and Foster (thesis).

Several functional approaches towards elucidating these mechanisms have emerged, as discussed below.

Biochemical Analysis

The germinant receptor

Identification and localization of the germinant receptor still remains a major objective. In many species, D-alanine acts as a competitive inhibitor of L-alanine-induced germination (29, 97, 139, 160), possibly by binding to the same site as L-alanine (164). These stereospecific properties of the germinant receptor demonstrate that this component of the trigger mechanism must be a protein which is activated allosterically (162). The inhibitory effects of alcohols on germination of *B. subtilis* spores lead to the interpretation that the L-alanine receptor site is located close to a hydrophobic region (165). Kinetic studies of L-alanine-induced germination of *B. subtilis* have suggested that three molecules of L-alanine bind to each receptor molecule to trigger germination (54).

There is some evidence which suggests an inner membrane location for the germinant receptor. It is possible to produce refractile spore protoplasts having no spore coats or cortex but which can be germinated under suitable conditions (30). Also, spores can be germinated after extraction of coat proteins, even when the outer membrane has been disrupted by the extraction process (88, 148). Thus the germinant receptor does not appear to be located in the spore coats or outer membrane. More direct evidence comes from work with *B. megaterium* QMB1551, in which the germinant affinity analog L-proline chloromethyl ketone and the specific germination inhibitor acetic anhydride were both reported to label a 10.2-kilodalton (kDa) inner membrane protein (100, 106). Furthermore, the germinant L-proline causes a change in fluidity (57) and anisotropy (130) of isolated inner membranes of *B. megaterium* QMB1551. Also, the germination rate temperature profile correlates well with the inner

membrane transition temperature of spores produced in different media (151). No changes in inner membrane fluidity, however, could be detected in vivo in spores of *B. megaterium* KM until after germination triggering (138). A mutant of *B. megaterium* QMB1551 unable to germinate in L-proline showed decreased labeling of the 10.2-kDa inner membrane band by L-proline chloromethyl ketone (106), and membranes isolated from this mutant showed no anisotropy changes in response to L-proline (130). Also, a 10.5-kDa inner membrane protein covalently labeled by the germination inhibitor $^{203}Hg^{2+}$ has been identified in spores of *B. megaterium* KM (Foster, thesis). Finally, procaine, a membrane-bulking agent which inhibits membrane protein processing in *Escherichia coli* (71), has been shown to inhibit germination of spores of *B. megaterium* KM at a precommitment site (33). Triggering of spore germination of *B. megaterium* QMB1551 is altered by growing spores under conditions which modify the lipid composition of spore membranes (131). Thus, although definitive proof that the 10.2-kDa protein is the proline receptor remains to be established, part of all of the commitment process appears to be membrane associated.

The evidence to date, therefore, suggests that in spores of *B. megaterium* the germinant receptor is a protein which may be located in the spore inner membrane, although its functional role remains to be identified.

Role of metabolism

There has been much controversy as to whether metabolism of the germinant and/or internal spore metabolism is essential for initiation of spore germination.

Germinant metabolism. Three lines of evidence have demonstrated that germination triggering in spores of *B. megaterium* and *B. subtilis* is independent of metabolism of the germinant. First, no incorporation of label from radiolabeled L-alanine or D-glucose into other compounds was detected during germination triggering in spores of *B. megaterium* KM (114) and *B. megaterium* QMB1551 (128), respectively. Second, nonmetabolizable germinant analogs including L-proline chloromethyl ketone, L-alanine chloromethyl ketone, and 6-deoxymethyl-α-D-glucose are capable of initiating germination (106, 128, 149; Foster, thesis). Third, mutant spores lacking enzymes potentially involved in germinant metabolism germinate normally. Thus mutant spores of *B. subtilis* lacking L-alanine dehydrogenase (95) and glycolytic enzymes (96, 128) do not have germination defects. Although it was suggested that glucose dehydrogenase may be required for glucose-triggered germination of spores of *B. subtilis* (96, 141, 161), it has recently been demonstrated that a plasmid insertion into the reading frame of glucose dehydrogenase abolished the production of this enzyme during sporulation, but had no effect on growth or on germination of the mutant spores in L-alanine or GFAK (Y. Nakatani, K.-D. Neitzke, and E. Freese, personal communication) or on Luria-Bertani agar (101).

However, uricase action is necessary for uric acid-induced germination of spores of *Bacillus fastidiosus* (110). It cannot be assumed, therefore, that germination triggering occurs in the absence of germinant metabolism in all organisms.

General spore metabolism. The possibility that germinants might activate metabolic processes within the spore as part of germination triggering has also been intensively studied. No significant changes in metabolite pools including ATP, NADH, and glycolytic and tricarboxylic acid cycle intermediates were detected in the first minutes of *B. megaterium* KM spore germination, during which time a substantial proportion of the spore population becomes committed to germination (63, 113). Furthermore, no incorporation of ^3H from ^3H_2O into spore metabolites was detected during the same time period, indicating that the major metabolic pathways, including glycolysis, the pentose phosphate pathway, and the tricarboxylic acid cycle, are inoperative during germination triggering (114). In addition, a wide range of metabolic inhibitors including potassium fluoride, potassium cyanide, and carbonylcyanide *m*-chlorophenyl hydrazone have no effect on germination in *B. megaterium* (20, 107, 116). Recent analysis of the effects of sodium, potassium, and calcium channel blockers on nutrient-induced germination of spores of *B. megaterium* QMB1551 suggests that ion channels may be necessary for germination in this organism (79).

Hydrolytic reactions, including the activity of proteases, phosphatases, lipases, and amidases, would not have been detected by the techniques described above and therefore remain strong candidates for the trigger event.

Sequencing events

Determination of the sequential interrelationships between the early germination events is essential for an understanding of the trigger reaction in relation to the overall germination process. However, the inherent asynchrony of germination in the spore population and the differential sensitivities of biochemical assays make it difficult to identify the sequence of events by direct measurement as described above. Inhibitors have been widely used to probe the germination pathway (134, 152). Mercuric (Hg^{2+}) ions act as a potent inhibitor of spore germination (72, 93), and work with *B. megaterium* QMB1551 suggested the presence of two Hg^{2+}-sensitive sites during germination (51, 104, 105). These two sites of Hg^{2+} inhibition have been characterized during germination of spores of *B. megaterium* KM (33). The first site (referred to as site I) represents the L-alanine-binding site since it can be protected from the effects of Hg^{2+} by D-alanine. The second site (referred to as site II) is probably due to inhibition of a cortex lytic enzyme. Using the differential Hg^{2+} sensitivities of these two sites, loss of heat resistance was identified as the only pre- or cocommitment event. This may be due to the creation of a heat-sensitive active enzyme by the action of L-alanine as part of the trigger reaction. Loss of spore A_{600}, hexosamine, DPA, and Zn^{2+}, as well as selective cortex hydrolysis and loss of UV resistance, were all shown to be postcommitment events (33).

Specific inhibitor studies have suggested the involvement of a trypsinlike proteinase in spore germination which is activated early in the germination sequence of *B. cereus* T (6, 7). Also, a *B. cereus* protease mutant which produces spores defective in an early stage of germination has been isolated (13). Whether or not this protease is directly involved in the germination response remains to be established. Using spores of *B. megaterium* KM, we have shown that certain proteinase inhibitors act at a precommitment site and thus the L-alanine receptor protein may itself have proteolytic activity (33). These results must, however, be interpreted with caution, since all the effective proteinase inhibitors displayed atypical reversibility.

In spores of *B. megaterium* KM the germination pathway may therefore comprise the following steps: (i) interaction of the receptor with L-alanine; (ii) activation of proteolytic activity; (iii) activation of a cortex lytic enzyme; (iv) selective cortex hydrolysis and early germination events, e.g., DPA release; and (v) late germination events, e.g., ATP synthesis and the onset of general spore metabolism. It is not known whether commitment is represented by steps (ii) or (iii).

Cortex lytic enzymes

The three-dimensional configuration of the spore cortex has been extensively implicated in the maintenance but not the establishment of spore dormancy (25). It is thus central to several models of spore dormancy and resistance, including the anisotropically expanded cortex hypothesis originally proposed by Alderton and Snell (1) and modified by Warth (158). According to this model, the cortical peptidoglycan is synthesized in such

a manner that it swells, exerting an inward radial pressure against the spore protoplast. In this way, the spore cortex prevents core expansion and water uptake, thus maintaining the dormancy and certain of the resistance properties of the spore (86).

Activation of a lytic enzyme as a primary event in spore germination was originally suggested by Powell and Strange (94). Given the central role of the cortex in the maintenance of spore dormancy, cortex lytic enzymes are prime candidates for such lytic activity. In *C. perfringens*, deficiencies in, or damage to, cortex lytic enzymes can result in superdormant spores (24). Although such superdormant spores are viable, they cannot germinate and outgrow without an exogenous source of a lytic enzyme such as lysozyme.

The activity of lytic enzymes would not be detected by the metabolic studies described above. Although release of cortex peptidoglycan fragments due to extensive hydrolysis of the cortex is a relatively late germination event (51, 52; Fig. 2), selective hydrolysis of the cortex takes place early in the germination pathway (61, 63; Fig. 2). New cortical reducing groups can be detected within 2 min after the addition of germinants to spores of *B. megaterium* KM, and these arise due to an increase in the number of cortical muramic acid δ-lactam residues which are a unique component of spore peptidoglycan (159).

A lytic enzyme capable of lysing spore cortex was first isolated from spores of *B. cereus* by Strange and Dark (140). Further studies by Gould and co-workers (42, 43) demonstrated that a partially purified preparation of lytic enzyme from *B. cereus* would induce phase darkening and loss of refractility in permeabilized spores. These studies were unable to determine which of the several cortex-lytic enzymes associated with the spore was responsible for the germinationlike changes observed.

More recently, a number of lytic enzymes which can be divided into two major categories have been isolated from bacterial spores (8). First, "core" enzymes, which cause geminationlike changes in permeabilized spores, have been isolated from germinating and disrupted spores (5, 10–12, 34, 36). Second, "surface-bound" enzymes, which are unable to cause "germination" of permeabilized spores but can hydrolyze isolated spore cortex, can be extracted from dormant spores (9, 11, 34, 135). The surface-bound enzymes from both *B. megaterium* KM (34) and *B. cereus* T (11) are probably equivalent in that they both hydrolyze isolated spore cortex but are unable to germinate coat-stripped (permeabilized) spores or lyse vegetative cell walls. The surface-bound enzyme from *B. cereus* T is stimulated by the presence of germinants (11). The surface-bound enzymes do not, however, appear to be essential for germination, since after extraction of these enzymes, spore germination still occurs (34). These enzymes may therefore be sporulation-specific products with a role in cortex synthesis or mother cell lysis.

A germination-specific cortex-lytic enzyme (GSLE) has been purified from germinating spores of *B. megaterium* KM (34) and has been strongly implicated as an essential component of the germination mechanism of this organism for the following reasons.

Substrate specificity. GSLE shows a narrow substrate range for activity and only hydrolyzes the structurally intact peptidoglycan of permeabilized spores. Binding-site recognition may depend on the presence of spore-specific muramic acid δ-lactam residues (159), but less specific interactions must also be involved since the enzyme binds very tightly to, but does not hydrolyze, vegetative cell peptidoglycan. This requirement for an intact cortical structure for enzyme activity suggests that a correctly stressed three-dimensional peptidoglycan substrate is essential for GSLE activity. Autolysins have previously been proposed to regulate vegetative cell wall expansion by hydrolyzing highly stressed peptidoglycan (67). A similar substrate specific-

ity has been demonstrated for the α-spore lytic enzyme extracted from spores of *C. perfringens* (37). Approximately 20% of the spore peptidoglycan is autolysin resistant during germination and differs from the rest of the peptidoglycan in terms of its degree of cross-linking and its chemical composition (15). The high degree of substrate specificity displayed by the GSLE may therefore be important in recognition of these sensitive and resistant peptidoglycans in vivo. This would allow selective breakdown of the cortex, in the absence of primordial cell wall hydrolysis, which would otherwise result in lysis of the germinated spore.

Activation during germination. Active GSLE cannot be extracted from dormant spores. During germination, extractable GSLE activity coincides with initiation of absorbance loss. Since germination occurs in the absence of de novo protein synthesis (65), GSLE must be present in an inactive form in the dormant spore. Activation of an unpurified lytic enzyme also occurs during germination of spores of *C. perfringens* (3).

Inhibitor profile. There is a high degree of correlation between the effects of a wide range of inhibitors, including $HgCl_2$, on GSLE activity and postcommitment absorbance loss.

Mode of action. Johnstone and Ellar (61) showed that selective hydrolysis of the cortex takes place early in germination, which results in an increase in the number of reducing groups due to the creation of new muramic acid δ-lactam residues. When GSLE is added to permeabilized spores of *B. megaterium* KM, a similar increase in the muramic acid δ-lactam content of the spore cortex is observed. These new muramic acid δ-lactam residues may arise due to amidase activity which breaks cross-links in the cortical peptidoglycan (61).

These results therefore suggest that GSLE activity represents the second $HgCl_2$-sensitive site (site II) observed during *B. megaterium* KM spore germination and that GSLE is responsible for catalyzing core rehydration and loss of spore refractility.

Mechanisms of autolysin activation

Both the location and mechanism of activation of GSLE are central to our understanding of the mechanism of germination triggering. Enzyme activation by release from an inactive, bound form was proposed for the spore lytic enzyme isolated from *B. cereus* T (43). Subsequently an "enhancing factor" was extracted from germinating spores of the same organisms; this factor increased lytic enzyme activity by a direct effect on the enzyme, possibly by proteolytic activation (78). Activation of an autolytic enzyme during germination of *C. perfringens* spore germination has been shown to be an energy-dependent process, although the mechanism of activation is unknown (5). It has previously been recognized that the autolytic muramidase of *Streptococcus faecium* is proteolytically activated (64, 91). The observation that protease inhibitors inhibit the germination process at a precommitment site suggests that such proteolytic activation of GSLE may be an essential step in germination of spores of *B. megaterium* KM.

Anti-GSLE serum was used to establish the molecular mechanism of GSLE activation during *B. megaterium* KM spore germination (35). Active GSLE has a molecular size of 29 to 30 kDa (35). Western blotting (immunoblotting) of dormant spore fractions demonstrated that >95% of the GSLE-cross-reactive protein is present as a 63-kDa proform which is apparently covalently bound to the spore cortex peptidoglycan and can only be visualized after lysozyme solubilization of the cortex (35). During germination in the presence of chloramphenicol, active GSLE appears, with a concomitant disappearance of the GSLE proform (Fig. 3). Although active GSLE cannot be extracted from dormant spores, a 30-kDa protein is present in sodium dodecyl sulfate (SDS) extracts of dormant spores. This protein may represent a remnant

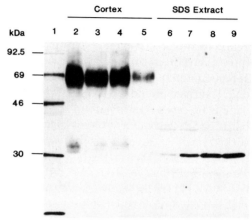

Figure 3. GSLE activation during germination. Cortex was purified from germinating spores after SDS extraction and spore breakage, and samples from 2.5 mg (dry weight) of spores were analyzed by SDS-polyacrylamide gel electrophoresis and Western blotting (35). Lane 1, Molecular weight standards; lanes 2 through 5, lysozyme digest of cortex from germinating spores 0, 15, 30, and 60 min, respectively, after the addition of the germinant L-alanine; lanes 6 through 9, trichloroacetic acid precipitate of SDS extracts from germinating spores 0, 15, 30, and 60 min, respectively, after addition of 1 mM L-alanine. Reproduced, with permission, from reference 35.

of GSLE activated during sporulation for cortical peptidoglycan synthesis.

To relate GSLE activation to the germination sequence, its relationship to the commitment event was established. When spores of *B. megaterium* KM were treated with 1 mM L-alanine in the presence of 1 mM $HgCl_2$, 25 to 30% of the spore population became committed to germination, whereas <5% germination, as measured by A_{600} loss, hexosamine release, and DPA release, occurred (33). Triggering of germination was >95% inhibited by the solely precommitment protease inhibitor phenylmethylsulfonyl fluoride (3.33 mM) (33). Thus, commitment-associated events can be identified as L-alanine-dependent reactions which take place in the presence of 1 mM $HgCl_2$ and which are inhibited by 3.33 mM phenylmethylsulfonyl fluoride. Approximately 76% of the germination control GSLE was activated in the presence of 1 mM HgCl$_2$ (Fig. 4, lane 3), whereas GSLE activation was inhibited by >90% in the presence of 3.33 mM phenylmethylsulfonyl fluoride (lane 6). In the presence of 4 mM HgCl$_2$, which completely inhibits commitment, GSLE activation was also >90% inhibited (Fig. 4, lane 5). These observations suggest that proteolytic cleavage of the GSLE proform constitutes part of the commitment reaction.

Western blotting of dormant spore fractions of other species after separation by SDS-polyacrylamide gel electrophoresis has revealed that antigenically cross-reactive proteins are present in spores of *B. cereus* T, *B.*

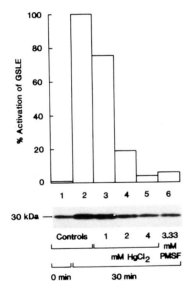

Figure 4. Effect of inhibitors on GSLE activation. Spores were germinated in the presence of 0 to 4 mM HgCl$_2$ or 3.33 mM phenylmethylsulfonyl fluoride at 0.5 mg (dry weight) ml^{-1} and extracted with SDS and Western blotting carried out following SDS-polyacrylamide gel electrophoresis as described in the legend of Fig. 3. Lanes 1 and 2, SDS extracts from 0- and 30-min-germinated spores; lanes 3 through 5, SDS extracts from spores germinated for 30 min in the presence of 1, 2, and 4 mM HgCl$_2$, respectively; lane 6, SDS extract from spores germinated for 30 min in the presence of 3.33 mM phenylmethylsulfonyl fluoride. The corresponding bar graphs show the counts determined by scintillation counting in each track, expressed as the percent of control 30-min stimulation. Reproduced, with permission, from reference 35.

subtilis 168, and *Clostridium bifermentans* M 86b (35). These proteins range in molecular size from 30 to 65 kDa and are present in spore fractions of all species examined. The presence of GSLE-cross-reactive proteins, together with the proposed involvement of autolysins and proteases in germination of spores of other organisms (6, 37), suggests that autolysin activation by proteolytic cleavage may represent a common mechanism for spore germination. It should be noted, however, that in spores of *B. fastidiosus* (110) and *B. megaterium* QMB1551 (89) no increase in cortical muramic acid δ-lactam residues has been detected during germination; glucosaminidase activity is responsible for cortex degradation in *B. megaterium* QMB1551 spore germination (89). Also, the presence of GSLE-cross-reactive proteins does not lead to the suggestion that a cortex-bound GSLE proform is present in spores of other organisms. In these organisms, therefore, an autolysin of different specificity may be activated. Alternatively, GSLE may be an important component of the germination mechanism, but its activity may be regulated by different mechanisms in these organisms. Since muramic acid δ-lactam is a common cortical structural component in all endospore formers (144), the presence of GSLE-cross-reactive proteins implies that these proteins may play a role during sporulation of these organisms.

Degradation of other spore components

The pitted layer of the spore coats disappears during germination (48), although relatively little change occurs in the protein profile of spore integuments (18). The pitted layer of the spore coat is also necessary for rapid germination of *Bacillus stearothermophilus* spores (14). It has been reported that in *B. megaterium* KM breakdown of a substantial proportion of inner membrane protein occurs during germination (127). The subsequent recognition that *B. megaterium* QMB1551 spore inner membranes contain two membrane fractions led to the suggestion that significant inner membrane protein degradation does not occur during germination (142). Lipases, however, are activated after 5 min of germination (68). These degradative events have not been detected within the germination-triggering period.

The low-molecular-weight, acid-soluble core proteins (SASPs) are also degraded during germination (124, 126). A specific endoprotease has been isolated from spores of *B. megaterium* QMB1551 which is proteolytically activated early in germination and hydrolyzes the spore SASPs (46, 74, 75, 121, 125). Neither the α and β SASPs nor the endoprotease appears to be essential for germination, since mutants with deletions of α and β SASP genes (77), or with decreased levels of endoprotease activity (92), germinate normally.

Genetical Analysis

Molecular genetics represents a powerful tool for analysis of the germination pathway. Mutants altered in spore germination properties have been isolated in a number of species of *Bacillus* and *Clostridium* (13, 146, 150, 163). The organism most extensively used for studies of germination mutants is *B. subtilis* 168. More than 100 *B. subtilis* mutants whose spores germinate abnormally have been isolated, and progress in their study has been reviewed frequently (70, 81–84, 90, 132, 133).

Mutations which block germination have been designated as *ger* mutations. It must, however, be recognized that since germination will take place in the absence of de novo protein synthesis, expression of *ger* genes occurs only during sporulation. Thus *ger* genes may be regarded as a subclass of *spo* loci that are required for a normal germination phenotype and which may be subject to the complexity of regulatory control demonstrated for other *spo* loci (76). As such, *ger* genes may be divided into a number of classes, as follows.

(i) Class I genes include structural genes for components of the germination mechanism per se. These gene products may include germinant receptor(s), cortex lytic enzymes, etc. Assuming that these gene products are not necessary for sporulation, deletion of their structural genes will only affect the spore germination response, and temperature-sensitive (*ts*) mutations in these genes will only be temperature sensitive for germination.

(ii) Class II genes are regulatory genes whose products affect expression of germination mechanism components during sporulation. If these genes solely regulate class I genes, then deletion will give rise to a defective germination phenotype, but *ts* mutations would be expected to be temperature sensitive for sporulation with respect to germination phenotype. Alternatively, if these genes in addition regulate expression of other *spo* loci, then a pleotropic phenotype would be expected.

(iii) Class III genes encode essential gene products for posttranslational processing and/or assembly of the germination mechanism.

(iv) Class IV genes encode essential gene products for synthesizing the correct dormant spore structure(s) which is a substrate for the germination mechanism, e.g., the correct cortex peptidoglycan substrate.

Although knowledge of all four classes of gene is necessary for a complete picture of the germination process, it is clear that class I mutants, which can be identified due to the *ts* germination phenotype, will be of most use in identifying functional components of the germination trigger mechanism.

Mutants blocked in the trigger response

Studies of *ger* mutants have demonstrated that there are two different nutrient-induced germination pathways in *B. subtilis*. Mutants which will germinate in GFAK but not in L-alanine alone have been isolated and termed *gerA* and *gerC* (111, 146). Mutants temperature sensitive in their L-alanine germination response map at the *gerA* locus, and thus the *gerA* gene product(s) may act as the L-alanine receptor. Conversely, mutations in *gerB*, *fruB*, and *gerK* germinate normally in L-alanine but not in GFAK (53, 83, 96, 143). Although GFAK *ts* germination mutants have not been isolated, it has been suggested that *gerK* may encode for the glucose receptor in this system (53).

Mutants blocked after triggering

The *gerD*, *gerE*, *gerF*, *gerH*, *gerJ*, *gerM*, *spoIVB*, *spoVA*, and *spoVIC* mutants comprise a heterogeneous group whose response to both the L-alanine and GFAK germinant systems is altered (50, 55, 56, 58, 59, 84, 112, 153, 154, 156, 157). In addition, many of these mutations are pleiotropic and show altered sporulation characteristics. Genes in which these mutations occur clearly have an important role in the successful establishment or functioning of the germination apparatus. Their products may act either directly, as common elements of the different triggering systems, or indirectly, as regulators of expression or assembly of other germination-associated proteins during sporulation. Thus these mutations may fall into any of the four functional classes outlined above. A model of the germination pathway of *B. subtilis* based on this genetical analysis is shown in Fig. 5.

Structural defects have been identified in spores of *gerE*, *spoVIB*, and *spoVIC* mutants. The integument structure is altered in all of these mutants (28, 56, 59, 60), and this may be due, at least in part, to lack of processing by a sporulation-associated proteinase (55, 59). Germination mutants of *B. cer-*

eus have also been observed to have altered coat structure (136, 137), and one of these coat mutants also is defective in proteinase synthesis (13). Recently it has been demonstrated that insertional inactivation of the *B. subtilis cotA*, *cotB*, and *cotC* genes has no effect on germination phenotype, whereas insertional inactivation of *cotD* results in spores with a slower germination response to L-alanine (21). There is also biochemical evidence for involvement of the spore coat in germination of *B. cereus* T spores (69). It is not known whether the defects in spore coat assembly and proteinase production in these mutants are directly or indirectly related to their germination phenotype. For example, mutations which affect the permeability of the spore coats to germinants might cause changes in germination characteristics.

All the mutants affected in both germinant systems lose some spore properties before germination is halted; e.g., *gerD* and *gerJ* spores become heat sensitive, but are blocked at the phase grey stage (80, 84, 156). No coat structure alterations have been observed in either mutant, but spores of *gerJ* are abnormally heat sensitive. The observation that spores of *B. subtilis* with altered cross-linking of spore cortex due to mutation of the PBP 5a gene are also abnormally heat sensitive (145) suggested that *gerJ* mutants may have a modified spore cortex structure. It has recently been demonstrated that *gerJ* spores have an altered cortex structure and are abnormally heat sensitive, and the appearance of PBP 5* during sporulation is delayed (154). These findings demonstrate that the interpretation of *B. subtilis* mutants defective in both germinant pathways as evidence for common functional components in both pathways may not be correct. For example, assembly of the germination receptor complexes may require common components, or alternatively, the two pathways may activate different cortex-lytic enymes both of which require a correctly formed cortical peptidoglycan substrate for activity.

The *gerD* mutants show no structural

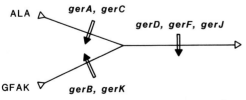

Figure 5. Relationships between *ger* mutations in the *B. subtilis* germination pathway.

defects, and germination of these mutants in the L-alanine system could be stimulated by the presence of monovalent cations (157). As no structural or resistance defects were observed, this mutant may be altered enzymatically, possibly in a cortex-lytic enzyme.

Cloning *ger* genes

Recently *gerA*, *gerE*, and *gerJ* loci have been cloned to identify their gene products and to elucidate their function in the germination pathway (19, 50, 81, 155). The *gerA* locus consists of three complementation units (166) coding for three proteins of proposed molecular sizes 53.5, 41.2, and 42.3 kDa (27, 167). All three *gerA* genes must be expressed during sporulation to allow a normal germination phenotype. These proteins differ in their hydropathy profiles, and it has been suggested that they may form a membrane-bound multisubunit receptor complex (167). Since a mutation in the 41.2-kDa protein results in a *ts* germination phenotype, this protein represents a class I gene product and must play a functional role in the *B. subtilis* L-alanine germination response (166); a functional role of the two other *gerA* gene products during germination has yet to be established.

In contrast, the *gerE* locus has been shown to code for a single 8.5-kDa protein which may have a regulatory function, controlling the transcription of several proteins during sporulation (19). This locus therefore represents a class II *ger* mutation.

MODEL FOR GERMINATION TRIGGERING

The model shown in Fig. 6 for the germination pathway of *B. megaterium* KM was proposed (33), based on previous germination models and the results described above (35, 63, 139).

Heat shock activates the L-alanine receptor (R → R* in Fig. 6), which becomes triggered by L-alanine (R* → R**). The L-alanine receptor has been identified as the first site of $HgCl_2$ inhibition (site I). R** has proteolytic activity, which converts a proenzyme (L) to an active heat-sensitive cortex-lytic enzyme (L*), which is also $HgCl_2$ sensitive (site II). Commitment may represent either the R* → R** or L → L* reactions, the loss of heat resistance prior to commitment being due to the heat sensitivity of either R** or L*. Thus the pre- or cocommitment

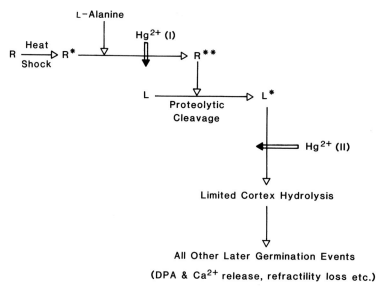

Figure 6. Model for germination triggering in *B. megaterium* KM.

loss of heat resistance may be associated with the heat sensitivity of the activated enzyme and is not associated with loss of CaDPA or spore rehydration. Cortex hydrolysis then allows uptake of water and the onset of all other downstream germination events, including loss of Ca^{2+} and DPA, release of soluble hexosamine, onset of core metabolism, and the loss of spore refractility. Very low levels of cortex hydrolysis, which cannot be detected by existing techniques, may allow the release of small amounts of DPA from the spore core early in germination (61). The cortex-lytic enzyme thus regulates the rehydration of the spore core during germination. Once water enters the spore core, it may solubilize and release CaDPA as well as rehydrating core proteins and initiating the onset of general spore metabolism. More extensive cortex hydrolysis will allow the release of peptidoglycan fragments, which occurs as a late germination event (22, 51, 52).

CONCLUSIONS AND FUTURE PROSPECTS

Although the physiological changes which occur during the germination response have been well characterized, the molecular mechanism which controls the ordered cascade of events consequential upon interaction of germinants with the spore is still unknown. Recently, however, both biochemical and genetical studies have produced evidence which may not only result in the identification of essential components of the germination pathway, but may also give an insight into the functional interactions of these components during the germination response.

Studies of spore germination of many species, and strains, of *Bacillus* and *Clostridium* have led to a fragmentary knowledge of the germination process in a large number of organisms. For a unifying model for amino acid-induced germination to be proposed, it must first be established whether there is a common germination mechanism. The overall germination response, which results in the characteristic change from a phase-bright to a phase-dark spore, is essentially similar in all organisms examined. In addition, in all endospores studied, there appear to be common mechanisms of spore dormancy. It therefore might be suggested that the same fundamental processes are responsible for breaking of spore dormancy and the onset of germination in bacterial endospores. Triggering of germination results in commitment, loss of heat resistance, release of DPA, Ca^{2+}, and hexosamine-containing fragments, the onset of spore metabolism, etc. In those organisms which have been studied intensively, metabolism has been shown not to be necessary for the triggering of germination. One notable exception is *B. fastidiosus*, in which metabolism of the germinant uric acid is essential for the initiation of germination. The nature of the trigger mechanism in all organisms is unknown, but both proteases and cortex-lytic enzymes have been strongly implicated in this process.

If a single model can be proposed to encompass the germination response of most organisms, the components involved in that response, together with their molecular and sequential interrelationships, must first be identified. GSLE, a crucial component of the above model proposed for the germination response of *B. megaterium* KM, may be the key to unraveling the germination mechanism in this organism. Alternatively, identification of the function of cloned *ger* genes may allow similar analysis of *B. subtilis* spore germination. Questions concerning the location and precise function of the germinant receptor protein still remain unanswered. The number of components involved in activating GSLE is unknown. If the germinant receptor is membrane associated, then activation of an immobilized cortex-lytic enzyme may require one or more steps.

Several experimental approaches are currently being employed to attempt to answer these fundamental questions, as follows.

(i) By forward genetics, the *gerA* locus

of *B. subtilis* has been cloned and sequenced. Three open reading frames have been identified in this locus, although the role of the encoded proteins in germination is unknown. For the function of these proteins to be established, L-alanine-dependent effects on other spore components to control the germination response need to be identified.

(ii) In biochemical studies, GSLE is the substrate for an activating factor, which may be the L-alanine receptor itself. An apparent in situ requirement for activation, however, has so far hampered attempts to isolate this factor from spores of *B. megaterium*. Analysis of *B. subtilis* germination, using techniques similar to those used to study *B. megaterium*, may also allow the germination mechanisms of these two organisms to be compared.

(iii) By reverse genetics, cloning, sequencing, and mapping of the GSLE structural gene, together with the search for gene homologs in *B. subtilis*, may allow relationships between GSLE and known *spo* and *ger* loci to be established. Generation of mutants in which the GSLE structural gene has been insertionally inactivated is necessary to unequivocally prove the involvement of GSLE in the germination response of spores of *B. megaterium* KM.

We are therefore at an exciting time in the analysis of the trigger mechanism of bacterial spore germination. Within the next decade, forward and reverse genetics, together with biochemical analysis, may converge to identify the molecular basis of this unique biological process.

Acknowledgments. We thank the Managers of the Broodbank Fund (S.J.F.) and the Science Research Council (grant GR/E 68594) for financial support.

LITERATURE CITED

1. **Alderton, G., and N. S. Snell.** 1963. Base exchange and heat resistance in bacterial spores. *Biochem. Biophys. Res. Commun.* **10:**139–143.
2. **Ando, Y.** 1978. Biochemical mechanism of ionic germination of spores of *Clostridium perfringens* type A, p. 75–79. *In* G. Chambliss and J. C. Vary (ed.), *Spores VII.* American Society for Microbiology, Washington, D.C.
3. **Ando, Y.** 1979. Spore lytic enzyme released from *Clostridium perfringens* spores during germination. *J. Bacteriol.* **140:**59–64.
4. **Ando, Y., and T. Tsuzuki.** 1984. The role of surface charge in ionic germination of *Clostridium perfringens* spores. *J. Gen. Microbiol.* **130:**267–273.
5. **Ando, Y., and T. Tsuzuki.** 1984. Energy-dependent activation of spore-lytic enzyme precursor by germinated spores of *Clostridium perfringens*. *Biochem. Biophys. Res. Commun.* **123:**463–467.
6. **Boschwitz, H., H. O. Halvorson, A. Keynan, and Y. Milner.** 1985. Trypsin-like enzymes from dormant and germinated spores of *Bacillus cereus* T and their possible involvement in germination. *J. Bacteriol.* **164:**302–309.
7. **Boschwitz, H., Y. Milner, A. Keynan, H. O. Halvorson, and W. Troll.** 1983. Effect of inhibitors of trypsin-like enzymes on *Bacillus cereus* T spore germination. *J. Bacteriol.* **153:**700–708.
8. **Brown, W. C.** 1977. Autolysins in *Bacillus subtilis*, p. 75–84. *In* D. Schlessinger (ed.), *Microbiology—1977.* American Society for Microbiology, Washington, D.C.
9. **Brown, W. C., and R. L. Cuhel.** 1975. Surface-localized cortex-lytic enzyme in spores of *Bacillus cereus* T. *J. Gen. Microbiol.* **91:**429–432.
10. **Brown, W. C., R. L. Cuhel, and C. Greer.** 1977. Isolation and properties of a surface-bound cortex-lytic enzyme from spores of *Bacillus cereus* T, p. 336–349. *In* A. N. Barker, G. J. Dring, D. J. Ellar, G. W. Gould, and J. Wolf (ed.), *Spore Research 1976.* Academic Press Inc. (London), Ltd., London.
11. **Brown, W. C., D. Vellom, I. Ho, N. Mitchell, and P. McVay.** 1982. Interaction between *Bacillus* spore hexosaminidase and specific germinants. *J. Gen. Microbiol.* **149:**969–976.
12. **Brown, W. C., D. Vellom, E. Schnepf, and C. Greer.** 1978. Purification of a surface-bound hexosaminidase from spores of *Bacillus cereus* T. *FEMS Microbiol. Lett.* **3:**247–251.
13. **Cheng, Y. E., P. C. Fitz-James, and A. I. Aronson.** 1978. Characterization of a *Bacillus cereus* protease mutant defective in an early stage of spore germination. *J. Bacteriol.* **133:**336–344.
14. **Cheung, H.-Y., and M. R. W. Brown.** 1985. Coat structure and morphogenesis of bacterial spores in relation to the initiation of spore germination, p. 317–327. *In* G. J. Dring, D. J. Ellar, and G. W. Gould (ed.), *Fundamental and Applied Aspects of Bacterial Spores.* Academic Press, Inc. (London), Ltd., London.
15. **Cleveland, E. F., and C. Gilvarg.** 1975. Selective degradation of peptidoglycan from *Bacillus meg-*

aterium spores during germination, p. 458–464. *In* P. Gerhardt, R. N. Costilow, and H. L. Sadoff (ed.), *Spores VI.* American Society for Microbiology, Washington, D.C.

16. Clouston, J. G., and P. A. Wills. 1969. Initiation of germination and inactivation of *Bacillus pumilus* spores by hydrostatic pressure. *J. Bacteriol.* 97:684–690.

17. Clouston, J. G., and P. A. Wills. 1970. Kinetics of initiation of germination of *Bacillus pumilus* spores by hydrostatic pressure. *J. Bacteriol.* 103:140–143.

18. Crafts-Lighty, A., and D. J. Ellar. 1980. The structure and function of the outer membrane in dormant and germinated spores of *Bacillus megaterium.* *J. Appl. Bacteriol.* 48:135–145.

19. Cutting, S., and J. Mandelstam. 1986. The nucleotide sequence and the transcription of the *gerE* gene of *Bacillus subtilis. J. Gen. Microbiol.* 132:3013–3024.

20. Dills, S. S., and J. C. Vary. 1978. An evaluation of respiration chain-associated functions during initiation of germination of *Bacillus megaterium* spores. *Biochim. Biophys. Acta* 541:301–311.

21. Donovan, W., L. Zheng, K. Sandman, and R. Losick. 1987. Genes encoding spore coat polypeptides from *Bacillus subtilis. J. Mol. Biol.* 196:1–10.

22. Dring, G. J., and G. W. Gould. 1971. Sequence of events during rapid germination of spores of *Bacillus cereus. J. Gen. Microbiol.* 65:101–104.

23. Dring, G. J., and G. W. Gould. 1971. Movement of potassium ions during L-alanine initiated germination of *Bacillus subtilis* spores, p. 133–142. *In* A. N. Barker, G. W. Gould, and J. Wolf (ed.), *Spore Research 1971.* Academic Press, Inc. (London), Ltd., London.

24. Duncan, C. L., R. G. Labbe, and R. R. Reich. 1972. Germination of heat- and alkali-altered spores of *Clostridium perfringens* type A by lysozyme and an initiation protein. *J. Bacteriol.* 109:550–559.

25. Ellar, D. J. 1978. Spore specific structures and their function. *Symp. Soc. Gen. Microbiol.* 28:295–324.

26. Ellar, D. J., M. W. Eaton, and J. A. Posgate. 1974. Calcium release and germination of bacterial spores. *Biochem. Soc. Trans.* 2:947–948.

27. Feavers, I. M., J. S. Miles, and A. Moir. 1985. The nucleotide sequence of a spore-germination gene (*gerA*) of *Bacillus subtilis* 168. *Gene* 38:95–102.

28. Feng, P., and A. I. Aronson. 1986. Characterization of a *Bacillus subtilis* germination mutant with pleiotropic alterations in spore coat structure. *Curr. Microbiol.* 13:221–226.

29. Fey, G., G. W. Gould, and A. D. Hitchins. 1964. Identification of D-alanine as the auto-inhibitor of germination of *Bacillus globigii* spores. *J. Gen. Microbiol.* 35:229–236.

30. Fitz-James, P. C. 1971. Formation of protoplasts from resting spores. *J. Bacteriol.* 105:1119–1136.

31. Foerster, H. F. 1985. The effects of alteration in

the suspending medium on low temperature activation of spores of *Bacillus stearothermophilus.* *Arch. Microbiol.* 142:185–189.

32. Foerster, H. F., and J. W. Foster. 1966. Response of *Bacillus* spores to combinations of germinative compounds. *J. Bacteriol.* 91:1168–1177.

33. Foster, S. J., and K. Johnstone. 1986. The use of inhibitors to identify early events during *Bacillus megaterium* KM spore germination. *Biochem. J.* 237:865–870.

34. Foster, S. J., and K. Johnstone. 1987. Purification and properties of a germination-specific cortex-lytic enzyme from spores of *Bacillus megaterium* KM. *Biochem. J.* 242:573–579.

35. Foster, S. J., and K. Johnstone. 1988. Germination-specific cortex-lytic enzyme is activated during triggering of *Bacillus megaterium* KM spore germination. *Mol. Microbiol.* 2:727–733.

36. Gombas, D. E., and R. G. Labbe. 1981. Extraction of spore-lytic enzyme from *Clostridium perfringens* spores. *J. Gen. Microbiol.* 126:37–44.

37. Gombas, D. E., and R. G. Labbe. 1985. Purification and properties of spore-lytic enzymes from *Clostridium perfringens* type A spores. *J. Gen. Microbiol.* 131:1487–1496.

38. Gould, G. W. 1969. Germination, p. 397–444. *In* G. W. Gould and A. Hurst (ed.), *The Bacterial Spore.* Academic Press, Inc. (London), Ltd., London.

39. Gould, G. W. 1970. Germination and the problem of dormancy. *J. Appl. Bacteriol.* 33:34–49.

40. Gould, G. W., and G. J. Dring. 1972. Biochemical mechanisms of spore germination, p. 401–408. *In* H. O. Halvorson, R. Hanson, and L. L. Campbell (ed.), *Spores V.* American Society for Microbiology, Washington, D.C.

41. Gould, G. W., and A. D. Hitchins. 1963. Sensitization of bacterial spores to lysozyme and to hydrogen peroxide with agents which rupture disulphide bonds. *J. Gen. Microbiol.* 33:413–423.

42. Gould, G. W., and A. D. Hitchins. 1965. Germination of spores with Strange and Dark's spore lytic enzyme, p. 213–221. *In* L. L. Campbell and H. O. Halvorson (ed.), *Spores III.* American Society for Microbiology, Ann Arbor, Mich.

43. Gould, G. W., A. D. Hitchins, and K. L. King. 1966. Function and location of a "germination enzyme" in spores of *Bacillus cereus. J. Gen. Microbiol.* 44:293–302.

44. Gould, G. W., and A. J. H. Sale. 1970. Initiation of germination of bacterial spores by hydrostatic pressure. *J. Gen. Microbiol.* 60:335–346.

45. Gutfreund, H. 1972. *Enzymes: Physical Principles,* p. 161–162. Wiley Interscience, New York.

46. Hackett, R. H., and P. Setlow. 1983. Enzymatic activity of precursors of *Bacillus megaterium* spore protease. *J. Bacteriol.* 153:375–378.

47. Harrell, W. R., and H. O. Halvorson. 1955. Stud-

ies on the role of L-alanine in the germination of spores of *Bacillus terminalis*. *J. Bacteriol.* **69**:275–279.

48. Hashimoto, T., and S. F. Conti. 1971. Ultrastructural changes associated with activation and germination of *Bacillus cereus* T spores. *J. Bacteriol.* **105**:361–368.

49. Hashimoto, T., W. R. Frieben, and S. F. Conti. 1969. Germination of single bacterial spores. *J. Bacteriol.* **100**:1385–1392.

50. Hasnain, S., R. Sammons, I. Roberts, and C. M. Thomas. 1985. Cloning and deletion analysis of a genomic segment of *Bacillus subtilis* coding for the *sdh A, B, C* (succinate dehydrogenase) and *gerE* (spore germination) loci. *J. Gen. Microbiol.* **131**:2269–2279.

51. Hsieh, L. K., and J. C. Vary. 1975. Germination and peptidoglycan solubilization in *Bacillus megaterium* spores. *J. Bacteriol.* **123**:463–470.

52. Hsieh, L. K., and J. C. Vary. 1975. Peptidoglycan autolysis during initiation of spore germination in *Bacillus megaterium*, p. 465–471. *In* P. Gerhardt, R. N. Costilow, and H. L. Sadoff (ed.), *Spores VI*. American Society for Microbiology, Washington, D.C.

53. Irie, R., T. Okamoto, and Y. Fujita. 1982. A germination mutant of *Bacillus subtilis* deficient in response to glucose. *J. Gen. Appl. Microbiol.* **28**:345–354.

54. Irie, R., T. Okamoto, and Y. Fujita. 1984. Kinetics of spore germination of *Bacillus subtilis* in low concentrations of L-alanine. *J. Gen. Appl. Microbiol.* **30**:109–113.

55. James, W., and J. Mandelstam. 1985. *spoVIC*, a new sporulation locus in *Bacillus subtilis* affecting spore coats, germination and the rate of sporulation. *J. Gen. Microbiol.* **131**:2409–2419.

56. James, W., and J. Mandelstam. 1985. Protease production during sporulation of germination mutants of *Bacillus subtilis* and the cloning of a functional *gerE* gene. *J. Gen. Microbiol.* **131**:2421–2430.

57. Janoff, A. S. R., T. Coughlin, F. M. Racine, E. J. McGroarty, and J. C. Vary. 1979. Use of electron spin resonance to study *Bacillus megaterium* spore membranes. *Biochem. Biophys. Res. Commun.* **89**:569–570.

58. Jenkinson, H. F. 1981. Germination and resistance defects in spores of a *Bacillus subtilis* mutant lacking a coat polypeptide. *J. Gen. Microbiol.* **127**:81–91.

59. Jenkinson, H. F., and H. Lord. 1983. Protease deficiency and its association with defects in spore coat structure, germination and resistance properties in a mutant of *Bacillus subtilis*. *J. Gen. Microbiol.* **129**:2727–2737.

60. Jenkinson, H. F., and J. Mandelstam. 1985. Two new sporulation loci affecting coat assembly and

late properties of *Bacillus subtilis* spores, p. 129–137. *In* G. J. Dring, D. J. Ellar, and G. W. Gould (ed.), *Fundamental and Applied Aspects of Bacterial Spores*. Academic Press, Inc. (London), Ltd., London.

61. Johnstone, K., and D. J. Ellar. 1982. The role of cortex hydrolysis in the triggering of germination of *Bacillus megaterium* KM endospores. *Biochim. Biophys. Acta* **714**:185–191.

62. Johnstone, K., D. J. Ellar, and T. C. Appleton. 1980. Location of metal ions in *Bacillus megaterium* spores by high resolution electron probe x-ray microanalysis. *FEMS Microbiol. Lett.* **7**:97–101.

63. Johnstone, K., G. S. A. B. Stewart, I. R. Scott, and D. J. Ellar. 1982. Zinc release and the sequence of biochemical events during triggering of *Bacillus megaterium* KM spore germination. *Biochem. J.* **208**:407–411.

64. Kawamura, T., and G. D. Shockman. 1983. Purification and some properties of the endogenous, autolytic N-acetylmuramoyl-hydrolase of *Streptococcus faecium*, a bacterial glycoenzyme. *J. Biol. Chem.* **258**:9514–9521.

65 Keynan, A. 1978. Spore structure and its relation to resistance, dormancy, and germination, p. 43–53. *In* G. Chambliss and J. C. Vary (ed.), *Spores VII*. American Society for Microbiology, Washington, D.C.

66. Keynan, A., and Z. Evenchik. 1969. Activation, p. 359–395. *In* G. W. Gould and A. Hurst (ed.). *The Bacterial Spore*. Academic Press, Inc. (London), Ltd., London.

67. Koch, A. L. 1985. Bacterial wall growth and division or life without actin. *Trends Biochem. Sci.* **10**:11–14.

68. Koncewicz, M. A., D. J. Ellar, and J. A. Posgate. 1977. Metabolism of membrane lipids during bacterial spore germination and outgrowth. *Biochem. Soc. Trans.* **5**:118–119.

69. Kutima, P. M., and P. M. Foegeding. 1987. Involvement of the spore coat in the germination of *Bacillus cereus* T spores. *Appl. Environ. Microbiol.* **53**:47–52.

70. Lafferty, E., and A. Moir. 1977. Further studies on conditional germination mutants of *Bacillus subtilis* 168, p. 87–105. *In* A. N. Barker, J. Wolf, D. J. Ellar, G. J. Dring, and G. W. Gould (ed.), *Spore Research 1976*. Academic Press, Inc. (London), Ltd., London.

71. Lazdunski, C., D. Baty, and J. M. Pages. 1979. Procaine, a local anaesthetic interacting with the cell membrane, inhibits the processing of precursor forms of periplasmic proteins in *Escherichia coli*. *Eur. J. Biochem.* **96**:49–57.

72. Levinson, H. S., and M. T. Hyatt. 1966. Sequence of events during *Bacillus megaterium* spore germination. *J. Bacteriol.* **91**:1811–1818.

73. **Levinson, H. S., and M. T. Hyatt.** 1969. Heat activation kinetics of *Bacillus megaterium* spores. *Biochem. Biophys. Res. Commun.* **37**:909–916.

74. **Loshon, C. A., and P. Setlow.** 1982. *Bacillus megaterium* spore protease: purification, radioimmunoassay, and analysis of antigen level and localization during growth, sporulation, and spore germination. *J. Bacteriol.* **150**:303–311.

75. **Loshon, C. A., B. M. Swerdlow, and P. Setlow.** 1982. *Bacillus megaterium* spore protease: synthesis and processing of precursor forms. *J. Biol. Chem.* **257**:10838–10845.

76. **Losick, R., and P. Youngman.** 1985. Endospore formation in *Bacillus*, p. 63–88. *In* R. Losick and L. Shapiro (ed.), *Microbial Development*. Cold Spring Harbor Laboratory, Cold Spring Harbor, N.Y.

77. **Mason, J. M., and P. Setlow.** 1986. Essential role of small acid-soluble spore proteins in resistance of *Bacillus subtilis* spores to UV light. *J. Bacteriol.* **167**:174–178.

78. **Mencher, J. R., and L. C. Blankenship.** 1971. Enhancement of *Bacillus cereus* spore lytic enzyme by a heat labile non-dialysable factor in spore extracts. *Biochim. Biophys. Acta* **230**:646–648.

79. **Mitchell, C., J. F. Skomurski, and J. C. Vary.** 1986. Effect of ion channel blockers on germination of *Bacillus megaterium* spores. *FEMS Microbiol. Lett.* **34**:211–214.

80. **Moir, A.** 1981. Germination properties of a spore coat-defective mutant of *Bacillus subtilis*. *J. Bacteriol.* **146**:1106–1116.

81. **Moir, A., I. M. Feavers, and A. R. Zuberi.** 1986. A spore germination operon in *Bacillus subtilis* 168, p. 183–194. *In* A. T. Ganesan and J. A. Hoch (ed.), *Bacillus Molecular Genetics and Biotechnology Applications*. Academic Press, Inc. (London), Ltd., London.

82. **Moir, A., I. M. Feavers, A. R. Zuberi, R. L. Sammons, I. S. Roberts, J. R. Yon, E. A. Wolff, and D. A. Smith.** 1985. Progress in the molecular genetics of spore germination in *Bacillus subtilis* 168, p. 35–46. *In* J. A. Hoch and P. Setlow (ed.), *Molecular Biology of Microbial Differentiation*. American Society for Microbiology, Washington, D.C.

83. **Moir, A., E. Lafferty, and D. A. Smith.** 1979. Genetic analysis of spore germination mutants of *Bacillus subtilis* 168: the correlation of phenotype with map location. *J. Gen. Microbiol.* **111**:165–180.

84. **Moir, A., and D. A. Smith.** 1985. The genetics of spore germination in *Bacillus subtilis*, p. 89–100. *In* G. J. Dring, D. J. Ellar, and G. W. Gould (ed.), *Fundmental and Applied Aspects of Bacterial Spores*. Academic Press, Inc. (London), Ltd., London.

85. **Murrell, W. G.** 1981. Biophysical studies on the molecular mechanisms of spore heat resistance and dormancy, p. 64–77. *In* H. L. Levinson, A. L. Sonenshein, and D. J. Tipper (ed.), *Sporulation and Germination*. American Society for Microbiology, Washington, D.C.

86. **Murrell, W. G.** 1988. Bacterial spores: nature's ultimate survival package, p. 311–348. *In* W. G. Murrell and I. R. Kennedy (ed.), *Microbiology in Action*. Research Studies Press Ltd., Letchworth, England.

87. **Murrell, W. G., and P. A. Wills.** 1977. Initiation of *Bacillus* spore germination by hydrostatic pressure: effect of temperature. *J. Bacteriol.* **129**:1272–1280.

88. **Nakatani, Y., M. Imagawa, Y. Takubo, J. Nishikawa, T. Nishihara, and M. Kondo.** 1985. Germination of the decoated spores of *Bacillus megaterium*. *Microbiol. Immunol.* **29**:1139–1149.

89. **Nakatani, Y., I. Tanida, T. Koshikawa, M. Imagawa, T. Nishihara, and M. Kondo.** 1985. Collapse of cortex expansion during germination of *Bacillus megaterium* spores. *Microbiol. Immunol.* **29**:689–699.

90. **Piggot, P. J., A. Moir, and D. A. Smith.** 1981. Advances in the genetics of *Bacillus subtilis* differentiation, p. 29–390. *In* H. S. Levinson, A. L. Sonenshein, and D. J. Tipper (ed.), *Sporulation and Germination*. American Society for Microbiology, Washington, D.C.

91. **Pooley, H. M., and G. D. Shockman.** 1969. Relationship between the latent form and the active form of the autolytic enzyme of *Streptococcus faecalis*. *J. Bacteriol.* **100**:617–624.

92. **Postemsky, C. J., S. S. Dignam, and P. Setlow.** 1978. Isolation and characterization of mutants of *Bacillus megaterium* having decreased levels of spore protease. *J. Bacteriol.* **135**:841–850.

93. **Powell, J. F.** 1950. Factors affecting the germination of thick suspensions of *Bacillus subtilis* spores. *J. Gen. Microbiol.* **4**:330–339.

94. **Powell, J. F., and R. E. Strange.** 1956. Biochemical changes occurring during sporulation in *Bacillus* species. *Biochem. J.* **63**:661–668.

95. **Prasad, C.** 1974. Initiation of spore germination in *Bacillus subtilis*: relationship to inhibition of L-alanine metabolism. *J. Bacteriol.* **119**:805–810.

96. **Prasad, C., M. Diesterhaft, and E. Freese.** 1972. Initiation of spore germination in glycolytic mutants of *Bacillus subtilis*. *J. Bacteriol.* **110**:321–328.

97. **Preston, R. A., and H. A. Douthit.** 1984. Germination of *Bacillus cereus* spores: critical control by DL-alanine racemase. *J. Gen. Microbiol.* **130**:3123–3133.

98. **Quesnel, L. B., J. A. Owers, V. E. Farmer, and D. Coupes.** 1977. Subtilisin induced germination of *Bacillus cereus* PX spores and the effects of dimethylsulphoxide, p. 753–770. *In* A. N. Barker,

J. Wolf, and D. J. Ellar (ed.), *Spore Research 1976.* Academic Press, Inc. (London), Ltd., London.

99. Racine, F. M., S. S. Dills, and J. C. Vary. 1979. Glucose-triggered germination of *Bacillus megaterium* spores. *J. Bacteriol.* **138**:442–445.

100. Racine, F. M., J. F. Skomurski, and J. C. Vary. 1981. Alterations in *Bacillus megaterium* QMB1551 spore membranes with acetic anhydride and L-proline, p. 224–227. *In* H. S. Levinson, A. L. Sonenshein, and D. J. Tipper (ed.), *Sporulation and Germination.* American Society for Microbiology, Washington, D.C.

101. Rather, P. N., and C. P. Moran. 1988. Compartment-specific transcription in *Bacillus subtilis*: idenification of the promoter for *gdh. J. Bacteriol.* **170**:5086–5092.

102. Rieman, H., and Z. J. Ordal. 1961. Germination of bacterial endospores with calcium and dipicolinic acid. *Science* **133**:1703–1704.

103. Rode, L. J., and J. W. Foster. 1961. Germination of bacterial spore with alkyl primary amines. *J. Bacteriol.* **81**:768–779.

104. Rossignol, D. P., and J. C. Vary. 1977. A unique method for studying the initiation of *Bacillus megaterium* spore germination. *Biochem. Biophys. Res. Commun.* **79**:1098–1103.

105. Rossignol, D. P., and J. C. Vary. 1978. L-Proline-initiated germination in *Bacillus megaterium* spores, p. 90–94. *In* G. Chambliss and J. C. Vary (ed.), *Spores VII.* American Society for Microbiology, Washington, D.C.

106. Rossignol, D. P., and J. C. Vary. 1979. L-Proline site for triggering *Bacillus megaterium* spore germination. *Biochem. Biophys. Res. Commun.* **89**:547–551.

107. Rossignol, D. P., and J. C. Vary. 1979. Biochemistry of L-proline-triggered germination of *Bacillus megaterium* spores. *J. Bacteriol.* **138**:431–441.

108. Russell, A. D. 1982. *The Destruction of Bacterial Spores,* p. 1–29. Academic Press, Inc. (London), Ltd., London.

109. Sacks, L. E. 1981. Influence of cations on lysozyme-induced germination of coatless spores of *Clostridium perfringens* 8.6. *Biochim. Biophys. Acta* **674**:118–127.

110. Salas, H. A., K. Johnstone, and D. J. Ellar. 1985. Role of uricase in the triggering of germination of *Bacillus fastidiosus* spores. *Biochem. J.* **229**:241–249.

111. Sammons, R. L., A. Moir, and D. A. Smith. 1981. Isolation and properties of spore germination mutants of *Bacillus subtilis* 168 deficient in the initiation of germination. *J. Gen. Microbiol.* **124**:229–241.

112. Sammons, R. L., G. M. Slynn, and D. A. Smith. 1987. Genetical and molecular studies on *gerM*, a new developmental locus of *Bacillus subtilis. J. Gen. Microbiol.* **133**:3299–3312.

113. Scott, I. R., and D. J. Ellar. 1978. Metabolism and the triggering of germination of *Bacillus megaterium*: concentrations of amino acids, organic acids, adenine nucleotides and nicotinamide nucleotides during germination. *Biochem. J.* **174**:627–634.

114. Scott, I. R., and D. J. Ellar. 1978. Metabolism and the triggering of germination of *Bacillus megaterium*: use of L-[^{3}H]alanine and tritiated water to detect metabolism. *Biochem. J.* **174**:635–640.

115. Scott, I. R., and D. J. Ellar. 1978. Study of calcium dipicolinate release during bacterial spore germination by using a new, sensitive assay for dipicolinate. *J. Bacteriol.* **1315**:133–137.

116. Scott, I. R., G. S. A. B. Stewart, M. A. Koncewicz, D. J. Ellar, and A. Crafts-Lighty. 1978. Sequence of biochemical events during germination of *Bacillus megaterium* spores, p. 95–103. *In* G. Chambliss and J. C. Vary (ed.), *Spores VII.* American Society for Microbiology, Washington, D.C.

117. Setlow, B., and P. Setlow. 1977. Levels of oxidized and reduced pyridine nucleotides in dormant spores and during growth, sporulation, and spore germination of *Bacillus megaterium. J. Bacteriol.* **129**:857–865.

118. Setlow, B., and P. Setlow. 1977. Levels of acetyl coenzyme A, reduced and oxidized coenzyme A, and coenzyme A in disulfide linkage to protein in dormant and germinating spores and sporulating cells of *Bacillus megaterium. J. Bacteriol.* **132**:444–452.

119. Setlow, B., L. K. Shay, J. C. Vary, and P. Setlow. 1977. Production of large amounts of acetate during germination of *Bacillus megaterium* spores in the absence of exogenous carbon sources. *J. Bacteriol.* **132**:744–746.

120. Setlow, P. 1975. Energy and small-molecule metabolism during germination of *Bacillus* spores, p. 443–450. *In* P. Gerhardt, R. N. Costilow, and H. L. Sadoff (ed.), *Spores VI.* American Society for Microbiology, Washington, D.C.

121. Setlow, P. 1976. Purification and properties of a specific proteolytic enzyme present in spores of *Bacillus megaterium. J. Biol. Chem.* **251**:7853–7862.

122. Setlow, P. 1981. Biochemistry of bacterial forespore development and spore germination, p. 13–28. *In* H. S. Levinson, A. L. Sonenshein, and D. J. Tipper (ed.), *Sporulation and Germination.* American Society for Microbiology, Washington, D.C.

123. Setlow, P. 1983. Germination and outgrowth, p. 211–254. *In* A. Hurst and G. W. Gould (ed.), *The Bacterial Spore,* vol. 2. Academic Press, Inc. (London), Ltd., London.

124. Setlow, P. 1985. Protein degradation during bacterial spore germination, p. 285–296. *In* G. J. Dring, D. J. Ellar, and G. W. Gould (ed.), *Fundamental and Applied Aspects of Bacterial Spores.* Academic

Press, Inc. (London), Ltd., London.

125. Setlow, P., C. Gerard, and J. Ozols. 1980. The amino acid sequence specificity of a protease from spores of *Bacillus megaterium. J. Biol. Chem.* 255:3624–3628.

126. Setlow, P., and G. Primus. 1975. Protein degradation and amino acid metabolism during germination of *Bacillus megaterium* spores, p. 451–457. *In* P. Gerhardt, R. N. Costilow, and H. L. Sadoff (ed.), *Spores VI.* American Society for Microbiology, Washington, D.C.

127. Seto-Young, D. L. T., and D. J. Ellar. 1979. Membrane changes during germination of *Bacillus megaterium* KM spores. *Microbios* 26:7–15.

128. Shay, L. K., and J. C. Vary. 1978. Biochemical studies on glucose-initiated germination in *Bacillus megaterium. J. Bacteriol.* 117:126–132.

129. Shibata, H., N. Ohnishi, K. Takeda, H. Fukunaga, K. Shimamura, E. Yasunobu, I. Tani, and T. Hashimoto. 1986. Germination of *Bacillus cereus* spores induced by purine ribosides and their analogues: effects of modification of base and sugar moieties of purine nucleosides on germination-inducing activity. *Can. J. Microbiol.* 32:186–189.

130. Skomurski, J. F., F. M. Racine, and J. C. Vary. 1983. Steady state anistropy changes of 1,6-diphenyl 1,3,5-hexatriene in membranes from *Bacillus megaterium. Biochim. Biophys. Acta* 731:428–436.

131. Skomurski, J. F., and J. C. Vary. 1982. Fatty acid and phospholipid composition of *Bacillus megaterium* spores with altered germination properties. *Lipids* 17:914–923.

132. Smith, D. A., A. Moir, and E. Lafferty. 1977. Spore germination genetics in *Bacillus subtilis*, p. 69–85. *In* A. N. Barker, J. Wolf, D. J. Ellar, G. J. Dring, and G. W. Gould (ed.), *Spore Research 1976.* Academic Press, Inc. (London), Ltd., London.

133. Smith, D. A., A. Moir, and R. Sammons. 1978. Progress in genetics of spore germination in *Bacillus subtilis*, p. 158–163. *In* G. Chambliss and J. C. Vary (ed.), *Spores VII.* American Society for Microbiology, Washington, D.C.

134. Smoot, L. A., and M. D. Pierson. 1982. Mechanisms of sorbate inhibition of *Bacillus cereus* T and *Clostridium botulinum* 62A sprore germination. *Appl. Environ. Microbiol.* 42:477–483.

135. Srivastava, O. P., and P. C. Fitz-James. 1981. Alteration by heat activation of enzymes localised in spore coats of *Bacillus cereus. Can. J. Microbiol.* 27:408–416.

136. Stelma, G. N., A. I. Aronson, and P. C. Fitz-James. 1978. Properties of *Bacillus cereus* temperature-sensitive mutants altered in spore coat formation. *J. Bacteriol.* 134:1157–1170.

137. Stelma, G. N., A. I. Aronson, and P. C. Fitz-James. 1980. A *Bacillus cereus* mutant defective in spore coat deposition. *J. Gen. Microbiol.* 116:173–185.

138. Stewart, G. S. A. B., M. W. Eaton, K. Johnstone, M. D. Barrett, and D. J. Ellar. 1980. An investigation of membrane fluidity change during sporulation and germination of *Bacillus megaterium* KM measured by electron spin and nuclear magnetic resonance spectroscopy. *Biochim. Biophys. Acta* 600:270–290.

139. Stewart, G. S. A. B., K. Johnstone, E. Hagelberg, and D. J. Ellar. 1981. Commitment of bacterial spores to germinate: a measure of the trigger reaction. *Biochem. J.* 198:101–106.

140. Strange, R. E., and F. A. Dark. 1957. A cell-wall lytic enzyme associated with spores of *Bacillus* species. *J. Gen. Microbiol.* 16:236–249.

141. Strauss, N. 1983. Role of glucose dehydrogenase in germination of *Bacillus subtilis* spores. *FEMS Microbiol. Lett.* 20:379–384.

142. Swerdlow, R. D., and P. Setlow. 1984. Isolation and characterization of two distinct fractions from the inner membrane of dormant *Bacillus megaterium* spores. *J. Bacteriol.* 158:9–15.

143. Thibeault, D., and G. M. Lefebvre. 1984. Triggering in unactivated *Bacillus megaterium* spores. *Can. J. Microbiol.* 30:997–1000.

144. Tipper, D. J., and J. J. Gauthier. 1972. Structure of the bacterial endospore, p. 3–12. *In* H. O. Halvorson, R. Hanson, and L. L. Campbell (ed.), *Spores V.* American Society for Microbiology, Washington, D.C.

145. Todd, J. A., A. N. Roberts, K. Johnstone, P. J. Piggot, G. Winter, and D. J. Ellar. 1986. Reduced heat resistance of mutant spores after cloning and mutagenesis of the *Bacillus subtilis* gene encoding penicillin-binding protein 5. *J. Bacteriol.* 167:257–264.

146. Trowsdale, J., and D. A. Smith. 1975. Isolation, characterization, and mapping of *Bacillus subtilis* 168 germination mutants. *J. Bacteriol.* 123:83–95.

147. Uehara, M., and H. A. Frank. 1967. Sequence of events during germination of putrefactive anaerobe 3679 spores. *J. Bacteriol.* 94:506–511.

148. Vary, J. C. 1973. Germination of *Bacillus megaterium* spores after various extraction procedures. *J. Bacteriol.* 116:797–802.

149. Vary, J. C. 1978. Glucose-initiated germination in *Bacillus megaterium* spores, p. 104–108. *In* G. Chambliss and J. C. Vary (ed.), *Spores VII.* American Society for Microbiology, Washington, D.C.

150. Vary, J. C., and A. Kornberg. 1970. Biochemical studies of bacterial sporulation and germination. XXI. Temperature-sensitive mutants for initiation of germination. *J. Bacteriol.* 101:327–329.

151. Vary, J. C., J. F. Skomurski, and B. A. Cornell. 1984. Differential scanning calorimetry of membranes isolated from *Bacillus megaterium* spores. *Can. J. Microbiol.* 30:854–856.

152. Vinter, V. 1970. Germination and outgrowth: ef-

fect of inhibitors. *J. Appl. Bacteriol.* **33**:50–59.

153. **Warburg, R. J.** 1981. Defective sporulation of a spore germination mutant of *Bacillus subtilis* 168, p. 98–100. *In* H. S. Levinson, A. L. Sonenshein, and D. J. Tipper (ed.), *Sporulation and Germination.* American Society for Microbiology, Washington, D.C.

154. **Warburg, R. J., C. E. Buchanan, K. Parent, and H. O. Halvorson.** 1986. A detailed study of *gerJ* mutants of *Bacillus subtilis. J. Gen. Microbiol.* **132**:2309–2319.

155. **Warburg, R. J., M. P. Davis, I. Mahler, D. J. Tipper, and H. O. Halvorson.** 1985. Cloning of *gerA* and other genes from the *spoIIA-tyrA* region of *Bacillus subtilis*, p. 67–70. *In* J. A. Hoch and P. Setlow (ed.), *Molecular Biology of Microbial Differentiation.* American Society for Microbiology, Washington, D.C.

156. **Warburg, R. J., and A. Moir.** 1981. Properties of a mutant of *Bacillus subtilis* 168 in which spore germination is blocked at a late stage. *J. Gen. Microbiol.* **124**:243–253.

157. **Warburg, R. J., A. Moir, and D. A. Smith.** 1985. Influence of alkali cations on the germination of spores of wild-type and *gerD* mutants of *Bacillus subtilis. J. Gen. Microbiol.* **131**:221–230.

158. **Warth, A. D.** 1978. Molecular structure of the bacterial spore. *Adv. Microb. Physiol.* **17**:1–47.

159. **Warth, A. D., and J. L. Strominger.** 1969. Structure of the peptidoglycan of bacterial spores: occurrence of the lactam of muramic acid. *Proc. Natl. Acad. Sci. USA* **64**:528–535.

160. **Watabe, K., K. Sano, M. Otani, Y. Okada, Y. Kak-**

iuchi, and M. Kondo. 1979. Effect of alanine-containing dipeptides on germination of *Bacillus thiaminolyticus* spores. *J. Bacteriol.* **139**:126–131.

161. **Wax, R., and E. Freese.** 1968. Initiation of the germination of *Bacillus subtilis* spores by a combination of compounds in place of L-alanine. *J. Bacteriol.* **95**:433–438.

162. **Wolgamott, G. D., and N. N. Durham.** 1971. Initiation of spore germination in *Bacillus cereus*: a proposed allosteric receptor. *Can. J. Microbiol.* **17**:1043–1048.

163. **Wyatt, L. R., and W. M. Waites.** 1971. Spores of *Clostridium bifermentans*: comparison of germination mutants, p. 123–131. *In* A. N. Barker, G. W. Gould, and J. Wolf (ed.), *Spore Research 1971.* Academic Press, Inc. (London), Ltd., London.

164. **Yasuda, Y., and K. Tochikubo.** 1985. Germination initiation and inhibitory activities of L- and D-alanine analogues for *Bacillus subtilis* spores. *Microbiol. Immunol.* **29**:229–241.

165. **Yasuda-Yasaki, Y., S. Namiki-Kanie, and Y. Hachisuka.** 1978. Inhibition of *Bacillus subtilis* spore germination by various hydrophobic compounds: demonstration of hydrophobic character of the L-alanine receptor site. *J. Bacteriol.* **136**:484–490.

166. **Zuberi, A. R., I. M. Feavers, and A. Moir.** 1985. Identification of three complementation units in the *gerA* spore germination locus of *Bacillus subtilis. J. Bacteriol.* **162**:756–762.

167. **Zuberi, A. R., A. Moir, and I. M. Feavers.** 1987. The nucleotide sequence and gene organization of the *gerA* spore germination operon of *Bacillus subtilis* 168. *Gene* **51**:1–11.

Regulation of Procaryotic Development
Edited by Issar Smith, Ralph A. Slepecky, and Peter Setlow
© 1989 American Society for Microbiology, Washington, DC 20006

Chapter 5

Metabolic Regulation of Sporulation and Other Stationary-Phase Phenomena

Abraham L. Sonenshein

In 1965, Schaeffer et al. (76) published a seminal paper in which they concluded that sporulation of *Bacillus subtilis* is regulated by a form of catabolic repression that involves metabolites containing carbon or nitrogen or both. This notion was reinforced by the carefully constructed demonstration by Dawes and Mandelstam (10) that sporulation frequency in a chemostat culture is inversely proportional to growth rate. Ever since the publication of Schaeffer's work, the key metabolites and the regulatory proteins with which they are presumed to interact have been the Holy Grail for *Bacillus* sporulation physiologists. This search has led more often than not to blind alleys or, at best, to interesting red herrings. The purpose of this review is not to recount these disappointments and diversions (see instead references 26 and 86) but rather to question basic assumptions, to summarize the current state of knowledge in the area of metabolic regulation of sporulation, and to describe approaches that seem likely to be fruitful.

SPORULATION: A STATIONARY-PHASE PHENOMENON

For a process as complex as spore formation, the temptation to simplify and generalize is all but overwhelming. This has led to the simplifying idea that a signal indicating carbon or nitrogen (or phosphorus) depletion induces cells in a direct sense to activate a set of genes whose products are necessary for and are specific to sporulation. The assumption that what is important for sporulation is necessarily specific to sporulation is clearly wrong (the sigma factor coded for by *spo0H* is required for transcription of some genes whose direct functions are unrelated to sporulation); the implication that the response to the environment leads immediately to a sporulation-committed pathway may be misleading; and the notion that the only important phenomena for sporulation are those that are absolutely required for sporulation is too limiting (see below). A more complex view is that sporulation is a confluence of pathways that is chosen by a cell that has already perceived an environmental change and has called into play a multitude of regulatory responses. In the best-studied laboratory conditions, a population of cells that exhausts a complex medium (e.g., nutrient broth) will activate a mixture of responses as it passes from balanced growth to unbalanced growth to stationary phase. These responses include stringency, derepression of some biosynthetic pathways, induction of some transport

Abraham L. Sonenshein • Department of Molecular Biology and Microbiology, Tufts University Health Sciences Campus, 136 Harrison Avenue, Boston, Massachusetts 02111.

systems and degradative pathways for alternative nutrient sources, induction of chemotaxis and motility, induction of alternative energy-generating systems, induction of macromolecule-degrading enzymes (and secretion of some of them), and induction of competence for transformation. The relevant questions are: within this Babel of gene expression, which messages, if any, are necessary for sporulation, and how and when does the cell make the choice to give up on growth and opt irreversibly for sporulation? These are not new questions, but it is only recently that it has become possible to see some of the levels at which these various regulatory responses overlap and interrelate.

In the soil, the natural habitat for *B. subtilis*, growth, when it occurs at all, is slow. Under such conditions, a large fraction of the population should become committed to sporulate during each round of cell division (10, 76). One view of this phenomenon is that *B. subtilis* is geared to sporulate rather than to adapt to poor growth conditions. This view is supported by the fact that this organism, compared to enteric bacteria, has the capacity to utilize only a limited number of sources of carbon and nitrogen. An alternative (but not incompatible) view is that a slowly growing bacterium in nature spends most of its time at the transition point between balanced growth and stationary phase. It therefore might be expressing all the time many of the responses that a rapidly growing cell would turn on only when it depletes a rich medium. In either case, a likely explanation for the phenomenon is that in each cell there is a certain probability that the pool of one or more key metabolites will drop below a critical, threshold level and thereby signal the onset of sporulation; this probability would be inversely proportional to growth rate.

The various phenomena that occur during stationary phase (three of which, competence for transformation by DNA, secretion of extracellular enzymes, and sporulation, were arbitrarily chosen for the illustration in Fig. 1) may be viewed as arising from (I) multiple, totally independent pathways; (II) pathways with shared intermediates; or (III) a common main pathway with exclusive branches.

It is already clear that scheme I is not correct; the products of the *spo0A* and *spo0H* genes are required, directly or indirectly, for expression of most competence (*com*) genes, *aprE* and *amyE* (coding for secreted proteins), and all *spo* genes tested. The model most compatible with current knowledge has elements of schemes II and III (Fig. 2). It is admittedly flawed, since events that transpire at the onset of stationary phase cannot be summarized in a single, two-dimensional pathway. One caveat is that cells that exhaust one or more essential nutrients will always enter stationary phase, but they may or may not form spores. This is reflected in the fact that expression of several stationary-phase genes is known to vary considerably depending on the nature of the medium used for growth and the mode of sporulation induction. That is, the phenotype of a cell induced to sporulate in minimal-glucose medium is not identical to that of a cell that has

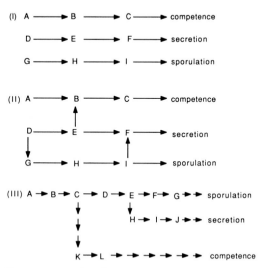

Figure 1. Examples of possible regulatory schemes that might control sporulation and sporulation-associated events. The capital letters could represent metabolites or regulatory macromolecules.

exhausted a nutrient broth medium (for example, see the discussion of the *citG* gene below). Moreover, with the exception of a subset of *com* genes, no two genes expressed in stationary phase have turned out to be regulated in exactly the same way.

ROLE OF GTP CONCENTRATION IN SIGNALLING NUTRIENT DEPLETION

The most influential discovery in the long search for the key metabolite that might monitor the availability of sources of carbon, nitrogen, and phosphorus is the finding of Freese and colleagues that the intracellular concentrations of GTP and GDP drop significantly whenever cells are induced to sporulate (50). Moreover, treatments that force the GTP pool to drop, such as addition of a partially inhibitory concentration of decoyinine, a drug that inhibits GMP synthetase, cause cells in rapid, balanced growth to slow their growth and initiate sporulation (27, 28). This remarkable phenomenon has both practical and theoretical implications. It allows measurements of very early sporulation events by defining an arbitrary zero time (T_0). This is in contrast to induction of sporulation by nutrient depletion, for which a T_0 time point is not precisely predictable and not uniform for the entire population. Similarly, induction of sporulation by suspension in an incomplete medium does not have a precise T_0, since the time required for depletion of intracellular pools is uncertain. The fact that changes in gene expression can be measured as early as 4 min after addition of decoyinine (C. Mathiopoulos and A. L. Sonenshein, *Mol. Microbiol.*, in press) argues that the drug acts very rapidly.

Decoyinine fools cells in rapid, balanced growth into behaving as though they had exhausted an essential nutrient, but this effect is not necessarily as simple as it appears. Addition of guanine prevents induction of sporulation by decoyinine (29); this argues that the drug acts by a mechanism that depends on its effect on guanine metabolism. At least one additional factor is necessary for decoyinine action; the cells must be at a density of 10^8/ml or higher. This effect has been attributed to cell-cell signalling by an extracellular factor which is produced during growth and accumulates in the medium (35). Synthesis of this factor is regulated by the *spo0A-abrB* system (see below). The factor appears to be an oligopeptide; its exact nature remains to be determined (35).

While addition of decoyinine mimics exhaustion of a complete medium in its ability to induce sporulation, the two situations are not really equivalent physiological conditions. First, the mechanism of inhibition of GTP accumulation is different. During exhaustion of a nutrient broth medium, the stringent response is activated (50) and seems to be responsible for reducing GTP synthesis (through inhibition of IMP dehydrogenase) and for conversion of GTP to other compounds (although the extent of decrease in intracellular GTP cannot be accounted for entirely by increases in ppGpp or pppGpp [27]). Decoyinine interrupts the flow of purines to GTP by inhibiting GMP synthetase, but the possibility that decoyinine has other activities has not been excluded. This point could be tested by isolating GMP synthetase mutants that are resistant to decoyinine and asking whether they can be induced to sporulate by decoyinine. Second, as noted below, not all genes that are expressed during nutrient exhaustion or required for sporulation under that condition are induced by decoyinine. In some cases this may be instructive with respect to sporulation-specific events, but in other cases it is clearly not.

The effect of inhibition of GTP synthesis on sporulation gene expression suggests that one or more regulatory proteins monitor the intracellular GTP concentration and adjust gene expression accordingly. Thus, it should be fruitful to search for GTP-binding proteins in the hope that one or more of them might prove to regulate transcription of some target gene. It should be noted, however, that the putative GTP-binding protein need not in-

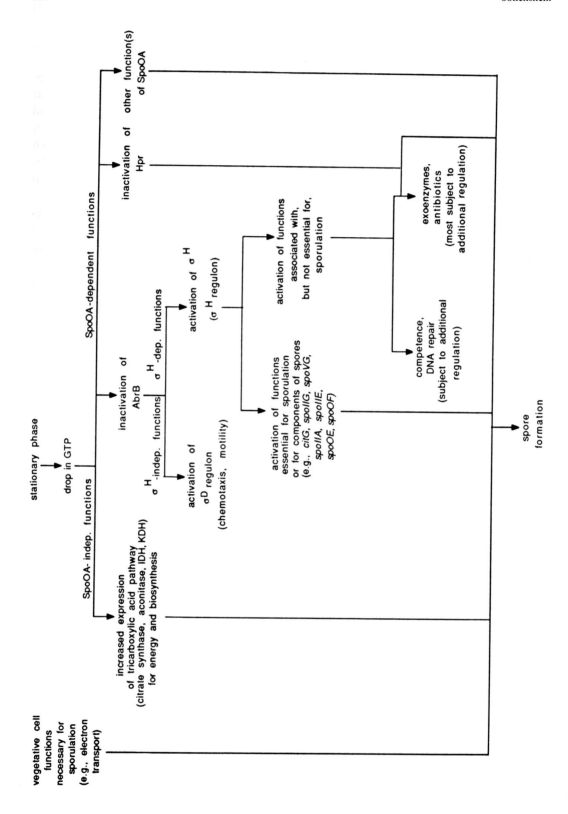

teract with DNA directly, but may act instead by modulating the function of a separate protein that is the direct regulator. Such GTP-dependent cascades of regulatory proteins are common in eucaryotic cells (32). J. Vary and C. Mitchell (personal communication) have found at least a dozen GTP-binding proteins in extracts of vegetative and early sporulating cells of *B. subtilis*. Some of these polypeptides vary in abundance in cells at different stages of the developmental cycle. Moreover, three proteins, one of which is specific to sporulating cells, react with antibody to a subunit of transducin, a eucaryotic GTP-binding protein. Trach and Hoch (91) have discovered that the open reading frame immediately downstream from *spo0B* (and possibly in the same transcription unit) codes for a polypeptide with substantial sequence similarity to known GTP-binding proteins. The role of this polypeptide remains to be established.

ROLE OF THE *spo0A* PRODUCT

The *spo0A* gene was first identified by mutations that cause the most severe and pleiotropic block in sporulation-associated events (42, 59). The original notion was that these were hyperrepressed mutants, that is, mutants unable to sense or respond to nutrient depletion (58). That idea is close to currently accepted thinking. As discussed in greater detail by I. Smith elsewhere in this volume, the *spo0A* product (Spo0A) has striking similarity to a class of regulatory proteins in gram-negative bacteria each of which interacts with a specific sensory protein (21).

The *spo0A* product is made during growth and at an increased rate during stationary phase (98), but seems to act predominantly in stationary phase (70, 71). It may regulate its own synthesis (98). Part of its function is to repress the *abrB* (also known

as *cpsX*) gene (72; P. Zuber, personal communication) or inactivate its product (99) or both. The product of *abrB* appears to act as a repressor of many genes normally expressed during stationary phase. Thus, a *spo0A* mutant does not sporulate and does not express most stationary-phase genes; a *spo0A abrB* double mutant regains expression of these genes but still does not sporulate (36, 92). This has been taken to mean that the *spo0A* product has functions in addition to its effect on *abrB*. An *abrB* single mutant has no defect in sporulation (i.e., sporulation is not constitutive). Some genes that are regulated by *abrB* (e.g., *tycA* [56], σ^D-dependent genes [D. Arnosti and M. Chamberlin, personal communication]) become constitutive in its absence, but others (e.g., *spoVG* [99]) still require some stationary-phase signal for full induction. This can be understood in the following way. Some genes are regulated only by the negative regulator coded for by *abrB*. These genes can be turned on by inactivation of AbrB either by mutation or by activation of its antagonist, Spo0A, at the end of growth. Other genes are regulated by both AbrB and a second regulator (either positive or negative) which does not depend for its activity on Spo0A but does depend on some signal indicating nutrient depletion.

The product of *spo0A* is also required for regulation of nitrate reductase (58) and the response to oxidative stress (17). At least in the case of nitrate reductase, the constitutivity caused by a *spo0A* mutation is reversed by a second mutation at *abrB* (*cpsX*) (36).

A second role of Spo0A seems to be repression or inactivation of *hpr*, a gene that appears to code for another negative regulator of some functions essential for sporulation (71). As indicated in Fig. 2, Spo0A must have at least one more function essential for sporulation since a *spo0A abrB hpr* mutant does not sporulate (71).

Figure 2. A simplified scheme of regulatory events occurring during stationary phase in *B. subtilis*.

ROLES OF SIGMA FACTORS

The DNA-dependent RNA polymerase of *B. subtilis* is a mixed population of molecules in which a common core enzyme associates with a multitude of sigma factors of varying promoter specificity (see C. P. Moran, Jr., this volume). Many genes expressed during stationary phase are transcribed by RNA polymerase containing the major vegetative sigma factor, σ^A (σ^{43}). This factor loses activity within the first 2 h after T_0 (89), causing genes dependent on it to become inactive. (The transcript of the *veg* gene, a σ^A-dependent gene, is still detectable at low levels at later times [33].)

The product of the *spo0H* gene, σ^H (σ^{30}) (19), is necessary for the transcription of many genes that become expressed during the transition from growth to stationary phase (e.g., *spoVG*, *citG*, *com*). It also contributes to continued expression of *rpoD* (the gene for σ^A) during the first 2 h after T_0 (7). σ^H-dependent genes have widely divergent functions, some of which are required for sporulation and some of which are not. In fact, most of the genes known to be expressed in stationary phase depend directly or indirectly on σ^H (see Fig. 2). In nutrient broth medium σ^H is in very low amount during exponential growth and increases substantially within the first hour after growth stops (J. Healy and R. Losick, personal communication; I. Smith and G. Nair, personal communication). One model for regulation of stationary-phase events, including early sporulation gene expression, is that synthesis, activation, or stabilization of σ^H is the key step that unleashes this transcription (18, 100; Healy and Losick, personal communication). For at least some conditions, however, σ^H activity is necessary but not sufficient; some additional signal indicative of stationary phase is required. In a minimal-glucose medium, in which σ^H is present at high level, expression of *spoVG* requires addition of decoyinine (Healy and Losick, personal communica-

tion). Synthesis of σ^H is under *spo0A-abrB* control (18, 100); one or both of these genes may also regulate σ^H activity.

A. Grossman (personal communication) has recently identified a group of genes on the basis of their dependence on σ^H for expression. The genes that are most strongly σ^H dependent are those induced at the end of growth in nutrient broth medium or by addition of decoyinine to minimal-glucose medium. Of these, only a few code for products that are required for sporulation. These results confirm that the σ^H-dependent regulon is not a group of genes of unified function (Fig. 2), reinforcing the notion that stationary phase is a period in the growth cycle during which a number of different phenomena have overlapping regulation.

σ^D (σ^{28}) allows core RNA polymerase to transcribe genes involved in chemotaxis, motility, and autolysis (40; D. Mirel and M. Chamberlin, submitted for publication). The Eσ^D form of RNA polymerase is also responsible for transcription of the *prtR* gene and contributes to transcription of the *sigA* (*rpoD*) and *citG* genes (in both cases from minor promoters) (V. Singer, Ph.D. thesis, University of California, Berkeley, 1988). Utilization of σ^D-dependent promoters occurs during logarithmic phase in nutrient broth medium and is transiently stimulated as cells enter stationary phase, at a time when σ^D accumulation increases (Singer, thesis; Mirel and Chamberlin, submitted; J. D. Helmann and M. Chamberlin, personal communication). Expression of σ^D-dependent genes is blocked in *spo0A*, *spo0B*, *spo0E*, and *spo0F* (but not *spo0H*) mutants (33) and restored in *spo0 abrB* double mutants (Arnosti and Chamberlin, personal communication). Synthesis of σ^D itself is not affected by *spo0A* mutations (M. Chamberlin, personal communication). The σ^D regulon seems to be another example of stationary-phase functions whose regulation overlaps substantially with that of sporulation (Fig. 2).

USEFUL PARADIGMS
FOR SPORULATION

To understand how cells choose the sporulation path, it is important to study in detail the regulation of paradigmatic genes whose function is essential to the process, that is, genes in which mutations cause a Spo⁻ phenotype. The simplifying assumption here is that one can be sure that a particular gene is important if its product is required for sporulation under all conditions. This simplification is too limiting. It turns out that a preponderance of genes in which mutations cause a strong block in sporulation (e.g., *spo0A, spo0B, spo0F, spo0H, spoIIA, spoIID, spoIIE, spoIIG, spoIIJ, spoIIM, spoIIIC, spoIVCB*) code for regulatory proteins or for proteins which allow regulatory proteins to be active; the mutations have pleiotropic effects. While these regulatory proteins are intrinsically interesting, they are not directly responsible for the special morphology and properties of the spore. This simplification, in fact, bars from consideration the genes for small, acid-soluble proteins (which encode major constituents of the spore cytoplasm) and even most spore coat protein genes. In these two cases it is clear that families of genes code for partially or fully redundant products, almost none of which is individually required for sporulation (9, 15). This situation is likely to reflect selection for a mechanism that prevents loss of a necessary function by a single mutation. It is clear that a less restrictive definition of sporulation genes and genes worthy of study is called for.

When is a Spo gene not a *spo* gene? In studying sporulation at the level of phenomena of gene regulation, it is important to choose for study genes whose regulation will be informative both in explaining how the overall process works and in predicting the behavior of other genes. Thus, to analyze the initiation of sporulation as a response to environmental change, the central consideration is how directly a particular gene responds to a particular change in the environment. As long as the gene in question changes its expression soon after the environment becomes altered, it matters little whether the product of that specific gene is absolutely necessary for sporulation; its study is bound to reveal important information that is relevant to sporulation in any case. Expression of genes such as *aprE, amyE, com, fla,* and *sigD* contributes to the overall phenotype of a sporulating cell even though the products of these genes are not required for sporulation. Study of these genes has already provided insight into the mechanisms regulating early sporulation events (for more details, see other chapters of this volume). A second group of genes, including *citB, citG, sdh,* and *citK,* code for products that are required under many conditions of sporulation, but whose requirement can be circumvented by physiological tricks. (Freese et al. [29] showed that the products of these genes are not required for sporulation of a decoyinine-treated culture in glucose-glutamate medium.) Since these genes are expressed early during sporulation, study of their regulation will undoubtedly lead to a clearer sense of how sporulation is controlled. While mutations in these genes do not necessarily lead to inability to sporulate, the products of these genes (as for coat protein genes) are integral components of the sporulating cell; without them the process would be unrecognizable.

Since many changes in the environment lead to sporulation, a gene that responds to most or all of these stimuli is more representative of sporulation than one that only responds to a single stimulus. Inducibility by decoyinine may prove to be the most useful discriminator in choosing the genes most directly related to sporulation from among all those activated during stationary phase. Some care must be given to such an interpretation, however, because inducibility by decoyinine of at least one gene (*amyE*) is masked in the presence of glucose.

GENES WHOSE REGULATION BY METABOLITES OVERLAPS WITH REGULATION OF SPORULATION

One approach to understanding the primary events at the onset of sporulation is to study in detail the regulation of individual genes whose expression responds to one or more of the environmental conditions that lead to sporulation. I summarize below what is known about some of the genes that have been proposed to be useful indicators of early sporulation events. Other reviews in this volume treat in detail the behavior of other stationary-phase genes (*spo0E, spo0F, aprE, com, din, spoVG*) and the regulatory roles of genes such as *spo0A, abrB, sacQ, sacU, hpr, prtR,* and *sin.*

Genes Regulated by the Carbon Source

Genes that are regulated by the carbon source fall into at least four general classes.

Inducible, glucose-repressible genes

A large group of genes that code for degradation of carbon sources other than glucose are subject to induction by substrate and repression by glucose. Examples are genes for metabolism of gluconate, maltose, inositol, sucrose, xylose, acetoin, histidine, and arginine. The best-studied example of induction by substrate is the case of the gluconate utilization (*gnt*) genes. Here it is clear that a three-gene operon is repressed by the product of the first gene (30). The repressor has been shown to bind to a site in the promoter region of the operon and thereby to interfere with binding of RNA polymerase (30). This system is the closest replica of the *Escherichia coli lac* operon that is known to exist in the genus *Bacillus.* Regulation by glucose is substantially different, however. It has been known for some time that *B. subtilis* does not accumulate detectable levels of cyclic AMP during normal aerobic growth (83). Thus, it seemed unreasonable to expect that this organism would use a cyclic AMP–cyclic AMP-binding protein (CAP) system to mediate glucose repression. In fact, when the *gnt* genes were introduced into *E. coli*, they were expressed at high level whether glucose was present or not (62). This suggests that *gnt* genes do not require a positive regulator such as cyclic AMP-CAP complex and that their regulation by glucose in *B. subtilis* is probably by a negative effector. This regulation need not be directed at the *gnt* genes themselves, however. Glucose has a well-known ability to exclude the transport into the cell of other carbon sources, especially those that are transported by the phosphoenolpyruvate-dependent phosphotransferase system (16, 22). Thus, for the moment, there is no need to postulate any special mechanism of glucose repression of the *gnt* operon other than inducer exclusion.

The genes for degradation of xylose and xylose oligomers have been cloned and sequenced (39, 96) and their promoter sites have been identified (31, 39). The genes are in two dicistronic operons, both of which are regulated by the product of *xylR*, the gene that separates the two operons (39). Transcription of these genes in derivatives of strain 168 is dependent on xylose and greatly inhibited by glucose (31). When either promoter region is placed on a high-copy plasmid, expression from the chromosomal promoter becomes partly independent of xylose and partly resistant to glucose repression (31, 39). The most straightforward interpretation of these results is that a negative regulator that is inactivated by xylose regulates both operons; the effect of glucose may only be to prevent xylose uptake.

The histidine utilization genes have been recently cloned and sequenced (66). The first gene of the operon (*hutP*) appears to code for a 16,500-molecular-weight positive regulatory protein without which the downstream genes are not transcribed (66). Since *hutP* is separated from the other genes by a sequence that could form a stem-loop structure, a termination-antitermination mechanism may be responsible for histidine-dependent regulation (66), by analogy with the *trp* (85) and *sacB* (48, 84) genes of *B. subtilis* and the *bgl* operon of *E. coli* (55). The mechanism of glucose repression is completely unknown except to the extent that a cloned fragment (8 kilobases) that contains the promoter region and at least the first two genes of the operon acts as a glucose-regulated promoter (66). There is good reason to think that inducer exclusion is not the mechanism of glucose repression in this case; a mutant that is constitutive with respect to histidine induction is still subject to the glucose effect (8). The regulation by glucose of *hut* gene expression seems to be distinct from that of sporulation-associated genes, however, since the *catA7* mutation in the *hpr* gene (71) relieves glucose repression of extracellular protease and antibiotic production but not of histidase synthesis (46). Moreover, early blocked sporulation mutants (e.g., *spo0A*) are not defective in *hut* expression; in fact, they show an unusually high level of expression when carbon sources other than glucose are supplied (4, 5). Unlike the case for the *hut* genes of *Klebsiella aerogenes* (53), the *B. subtilis hut* genes are not derepressed by nitrogen limitation.

These substrate-inducible, glucose-repressed genes are interesting but not necessarily useful in terms of understanding the induction of sporulation. They do not respond to any signal indicative of stationary phase, including decoyinine addition (51), and do not depend for their expression on any sporulation-associated gene products, such as Spo0A. Detailed study of their regulation is likely to provide important insight into gene regulation in *Bacillus* spp. but is unlikely to explain how cells perceive the signal to sporulate.

Noninducible genes

A second class of genes subject to glucose repression is not subject to induction by substrate. The prime example here is *amyE*, the gene for α-amylase. This gene is expressed only in the absence of glucose and only in cells entering stationary phase. This suggests that there might be two separate regulatory phenomena controlling *amyE* expression. In fact, a *cis*-acting mutation (*gra-10*), which relieves repression by glucose, does not turn on *amyE* expression in exponentially growing cells; cells still have to leave balanced growth in order to express *amyE* (63). (This can be accomplished by allowing cells to exhaust an enriched medium or by adding decoyinine to a minimal-glucose medium [64].) The stationary-phase signal is apparently mediated through the *spo0A* product (G. H. Chambliss, personal communication). The mutation to glucose constitutivity is at position +5 with respect to transcription initiation, within a sequence that shows dyad symmetry and is similar to certain repressor-binding sites (65). These results have led to the hypothesis that glucose repression of *amyE* is a negative regulatory phenomenon. That is, a protein that monitors the availability of glucose is thought to bind to an operator site, thereby interfering with initiation of transcription. If such a protein exists, it should be possible to identify it by mutation. Indeed, from a collection of Tn917 insertions in the *B. subtilis* chromosome, it was possible to isolate several strains in which *amyE* expression was glucose insensitive (Chambliss, personal communication). Some of these mutations were tightly linked to *amyE* and may be similar in effect to the *gra-10* mutation. Others, however, were unlinked and may prove to code for the glucose-sensitive regulator. It will be interesting to know whether these mutations affect *amyE* expression specifically or whether some of them have general effects on catabolite-repressed genes. Among other mutations that interfere with glucose repression, the *cdh-3* mutation described by Fisher and Magasanik (23) is of particular interest. This mutation causes expression of *amyE* and genes for aconitase, histidase, and α-glucosidase to be insensitive to repression by glucose.

The *amyE* gene is a useful paradigm for

early sporulation genes because its expression depends on a stationary-phase signal. Unravelling this phenomenon is likely to provide important insight into how other early sporulation genes are regulated. Interestingly, the *amyE* promoter sequence has homology to that of *citB* (14), but not in the putative glucose repression site of *amyE*. This homology might reflect common regulation determined by growth phase.

Krebs cycle genes

A third class of glucose-repressed genes is typified by the genes of the Krebs cycle.

This cycle provides important biosynthetic building blocks and plays a key role in energy production when fermentation and glycolysis are unable to fill the cell's needs. In many respects it is misleading to think of the Krebs cycle as a single pathway. The tricarboxylic acid component (citrate to 2-ketoglutarate; Fig. 3) has a different role from the dicarboxylic acid component (succinyl coenzyme A [CoA] to oxaloacetate) and seems to be subject to separate, but overlapping, regulation (67, 68). This is necessary for cell economy. If a cell is supplied with glucose and a good source of 2-ketoglutarate (e.g., glutamate or glutamine), it has little need to ex-

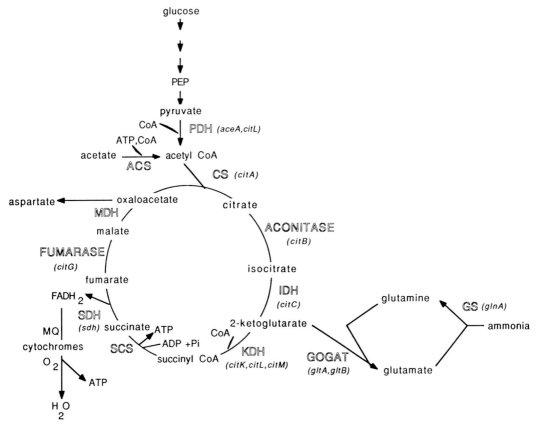

Figure 3. The Krebs cycle and some metabolic pathways with which it interacts. The enzymes responsible for various steps in metabolism are indicated in large, hollow type. For some of the enzymes, the genes that code for them are indicated in italics. Abbreviations: PDH, pyruvate dehydrogenase; ACS, acetyl-CoA synthetase; CS, citrate synthase; IDH, isocitrate dehydrogenase; KDH, ketoglutarate dehydrogenase; SCS, succinyl-CoA synthetase; SDH, succinic dehydrogenase; MDH, malate dehydrogenase; GOGAT, glutamate synthase; GS, glutamine synthetase; MQ, menaquinone.

press the tricarboxylic acid pathway; under these conditions the enzymes that convert citrate to 2-ketoglutarate are greatly reduced in specific activity (38). The cell still needs to have succinic dehydrogenase activity for electron transport and to make succinyl-CoA for biosynthesis, however. It could do so by metabolizing 2-ketoglutarate or by reversing the reactions of the dicarboxylic acid pathway. This is partly analogous to the situation in *E. coli* growing anaerobically. Under these conditions the enzyme ketoglutarate dehydrogenase (KDH) is inactive; the only source of succinyl-CoA is from counterclockwise flow of the dicarboxylic acid pathway. (There is a complication in that succinic dehydrogenase is not used anaerobically to reduce fumarate; *E. coli* codes for a fumarate reductase that is made only under anaerobic conditions.) Thus, the two halves of the Krebs cycle operate under some conditions as independent biochemical pathways.

Genes coding for different components of the Krebs cycle respond differently to the presence of glucose and to the onset of stationary phase. The *citB* gene (coding for aconitase), for instance, is transcribed at low level during exponential growth in a nutrient broth medium and becomes induced as cells enter stationary phase (13). In a minimal medium containing glucose and glutamine, *citB* expression is very low (74); it can be induced rapidly by addition of decoyinine (14). A similar pattern probably holds for citrate synthase (the *citA* product), although this has only been studied at the level of enzyme activity (67, 95). The gene for fumarase (*citG*) is also induced from a very low basal level when cells exhaust a nutrient broth medium; induction under these conditions is repressed by glucose (68). However, *citG* is derepressed during exponential growth in minimal-glucose-Casamino Acids medium (20) and is not further induced by addition of decoyinine (V. Price and A. Moir, personal communication).

The succinic dehydrogenase operon (*sdh*) is induced as cells enter stationary phase in enriched medium, is repressed in the presence of glucose, and is derepressed when the *sdh* promoter is in high copy number (57). The behavior of the *sdh* operon in minimal-glucose medium is not known. Despite the fact that each of the various Krebs cycle genes has its own unique attributes, there are some aspects of common regulation (induction in stationary phase in nutrient broth medium; repression by glucose) (68). It may be significant, therefore, that the *citB* promoter has a 14-base-pair sequence in the -12 to $+2$ region, in common with the same region of the *sdh* promoter and with the $+9$ to $+20$ region of the σ^A promoter (see below) for *citG* (Fig. 4). A similar sequence appears in the -25 region of the *menCD* promoter as well (see below).

Detailed analysis of *citB* gene expression has allowed the construction of a hypothetical scheme that explains its regulation in growing and sporulating cells. The essential facts are: (i) *citB* expression is low in glucose-glutamine minimal medium but high in medium containing citrate as the sole carbon source (74); (ii) *citB* expression is low during exponential growth in nutrient broth medium but becomes induced as cells enter stationary phase (13); (iii) *citB* expression is rapidly induced in minimal-glucose-glutamine medium after addition of decoyinine (13); (iv) induction of *citB* in both nutrient broth and minimal-glucose-glutamine media

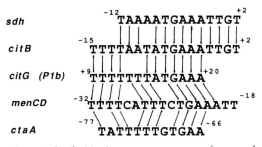

Figure 4. Similarities in promoter sequences for several genes involved in the Krebs cycle and electron transport. The sequences have been aligned as indicated; the superscript numbers refer to distance from the transcription start point. The sequences were obtained from the sources indicated: *sdh* (57); *citB* (14); *citG* (20); *menCD* (61); *ctaA* (Mueller, thesis).

depends on the products of the *citK* (KDH) and *citA* (citrate synthase) genes (13); and (v) multiple copies of the *citB* promoter region lead to derepression of transcription (13).

Taking into account the observations of Uratani-Wong et al. (95) concerning changes in metabolites and enzymes soon after decoyinine addition, we have postulated that the effects of a decrease in GTP concentration caused by exhaustion of nutrient broth or addition of decoyinine stem from the known rapid increase in acetyl-CoA accumulation. Acetyl-CoA is thought to be an activator of KDH (95). The result of such activation is a decrease in the 2-ketoglutarate pool, which leads to induction of *citA*. The consequent production of citrate induces *citB*, presumably by inactivating a repressor (see Fig. 5).

The implications of this model are severalfold. First, it does not rule out a direct effect of glucose on *citB* expression but suggests that the proximate cause of repression of *citB* in minimal-glucose-glutamine medium is the absence of the inducer, citrate. The notion that citrate is an inducer of aconitase was first suggested by Ohne (67). The absence of citrate is due to lack of expression of *citA*, which is thought to be regulated by 2-ketoglutarate, and lack of acetyl-CoA, a precursor of citrate. In that case, one might wonder why *citB* is not induced by addition of citrate to minimal-glucose medium. This can be explained by the known ability of glucose to inhibit citrate transport (97). The second implication is that a change in the pool of acetyl-CoA is the earliest known effect of a decrease in the GTP pool. This raises the possibility that it is pyruvate dehydrogenase activity that responds most directly to a decrease in GTP. Third, this scheme implies that the same mechanism that regulates appearance of *citB* mRNA at the end of growth is responsible for its regulation during growth in different carbon sources (24). This is probably correct, inasmuch as deletion of the bases between positions −67 and −84 with respect to *citB* transcription initiation led to constitutivity of *citB* expression with respect to both glucose repression and inducibility by decoyinine (A. Fouet and A. L. Sonenshein, unpublished data). While this scheme seems to explain some aspects of *citA*, *citB*, and *citK* regulation, there is no evidence at present that it is relevant to regulation of other genes. In fact, a block in KDH activity does not prevent induction by decoyinine of *spoVG* (Zuber, personal communication). This is reinforced by the finding that induction of *citB* is independent of the products of *spo0A*, *spo0B*, and *spo0H* (13). (This conclusion could have been deduced from the observation that *spo0A* mutants are not glutamate auxotrophs [75].) Yet the products of *citA*, *citB*, and *citK* are required for sporulation under most conditions, and their induction is clearly an integral aspect of the events that transpire early during sporulation. The same scheme is likely to apply to *citC* (coding for isocitrate dehydrogenase). The *citC* gene has been cloned and its product has been identified (73), but regulatory studies are incomplete.

The scheme for *citB* induction is clearly

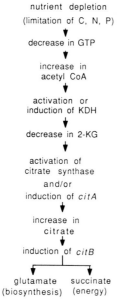

Figure 5. Hypothetical scheme of events leading to induction of the *citB* gene in stationary phase. See text for detailed explanation.

not relevant in a direct sense to the regulation of *citG*. This gene, which may prove to be typical of those coding for enzymes of the dicarboxylic acid pathway, is not inducible by decoyinine (Price and Moir, personal communication). This may have a trivial explanation in the sense that the gene is fully derepressed in minimal-glucose-Casamino Acids medium without addition of decoyinine. In this medium, relatively little carbon can be expected to flow clockwise to succinyl-CoA. This is because most of the pyruvate generated by glycolysis is converted to two- and three-carbon compounds which are excreted pending exhaustion of glucose. Moreover, much of the pyruvate that is converted to 2-ketoglutarate is drawn off for biosynthesis. As a result, the dicarboxylic acid pathway probably flows counterclockwise, and because relatively small amounts of substrates and products are available, the enzymes are expressed at high levels.

By contrast, in nutrient broth medium the *citG* gene is not expressed well until the end of exponential growth phase (20). This can be explained in the following way. Nutrient broth, as manufactured by Difco Laboratories, is a meat product. In meat, the major sources of carbon and nitrogen are fat and protein; in meat protein, the most abundant amino acids are glutamate and aspartate (12; Difco Laboratories, unpublished data). The former can be used directly in biosynthesis, or it can be converted to 2-ketoglutarate (leading to repression of *citA* and *citB*) and subsequently to succinate and fumarate. These are used for biosynthesis and energy production. Aspartate, on the other hand, can be metabolized by the dicarboxylic acid pathway in the counterclockwise direction as far as succinyl-CoA (there is no known reaction to convert succinyl-CoA to 2-ketoglutarate). Under these conditions, the enzymes for both the tricarboxylic and dicarboxylic acid pathways should be at low specific activity because the former would not be needed and the latter would find their substrates and products to be highly abundant.

As the major amino acids in the medium run out, both halves of the Krebs cycle should become more active as the need to synthesize 2-ketoglutarate, succinyl-CoA, and fumarate from other amino acids increases. Whether or not a single regulatory mechanism would control both sets of genes at this stage is unknown.

If glucose is added to nutrient broth medium, the need to use amino acids as carbon and energy sources is spared. As a result, the postexponential induction of both halves of the Krebs cycle does not occur. This rationalization may explain the apparently contradictory effects of glucose on *citG* expression in broth and minimal media.

The differing behavior of *citG* in minimal and nutrient broth media can now be understood at the level of transcriptional control. The *citG* gene is transcribed from two tandem promoters. One promoter, which is utilized by the σ^A form of RNA polymerase, is used for low-level transcription during growth in nutrient broth medium. The second promoter has consensus sequences for recognition by the σ^H-containing form of RNA polymerase (20) and is transcribed by this form of RNA polymerase in vitro (C. P. Moran, Jr., personal communication). At the end of exponential growth phase, this promoter becomes activated and is apparently responsible for the induction of fumarase activity that is observed (20). Addition of glucose to nutrient broth medium reduces this transcription (20). In minimal-glucose-Casamino Acids medium, the apparent σ^H-dependent promoter is used predominantly throughout growth, giving a high level of gene expression (20). As mentioned above, Healy and Losick (personal communication) and Smith and Nair (personal communication) have shown that the intracellular concentration of σ^H is low during growth in nutrient broth and increases in stationary phase; in minimal-glucose medium a high level of σ^H is present all the time. A simple model that derives from these results is that transcription of the *citG* gene depends in a straightforward way on the availability of σ^H and does not necessarily depend on any other factors (Healy and Losick, personal communication). Other σ^H-dependent genes (e.g., spoVG) do not have this property. At least during induction by decoyinine in minimal-glucose medium, they require some signal indicating stationary phase (possibly activation of a positive regulator) in addition to availability of σ^H.

Glucose-stimulated genes

A fourth class of glucose-regulated genes has the unusual property of being induced during stationary phase and stimulated in expression when glucose is added to a rich medium. Examples are the ctc gene (function unknown; 45), the ctaA gene (coding for a regulator of cytochrome oxidase; J. P. Mueller, Ph.D. thesis, Albany Medical College, Albany, N.Y., 1988), the menCD genes (menaquinone biosynthesis genes; 61), and several com genes (coding for competence for genetic transformation; 1; D. Dubnau, this volume). The ctc and menCD genes are particularly stimulated when glutamine is also present. One view of this phenomenon is that these genes are turned on when the tricarboxylic acid pathway is off. In fact, mutations that inactivate citB or citK lead to high levels of ctc and menCD expression (45, 61). This might be explained by a sensitivity of gene expression to internal or external pH; preliminary evidence for such an effect on menCD has been found (K. Hill, J. P. Mueller, and H. Taber, personal communication). Mutations in men or ctaA block sporulation, apparently because of a defect in energy generation from pyruvate derivatives (ctaA mutants have no detectable cytochrome aa_3 oxidase activity and cannot use lactate as the sole carbon and energy source; men mutants do not make menaquinone and therefore have a defect in the electron transport chain [61]). As mentioned above, the menCD promoter has a sequence in common with promoters for Krebs cycle genes (Fig. 4). This would be consistent with coupling of men expression to that of the dicarboxylic acid pathway at a minimum. The ctc gene differs from ctaA and menCD in that its product is not required for sporulation (except at high temperature; 93) and it is expressed in tryptone-yeast extract (L broth) medium but not in nutrient broth medium; the other genes are expressed in both media (at the end of exponential growth phase). ctc is transcribed by a minor form ($E\sigma^B$ [$E\sigma^{37}$]) of vegetative-cell RNA polymerase (37, 44, 47), while ctaA and menCD appear to depend on the major vegetative ($E\sigma^A$) form (61; Mueller, thesis).

Genes Regulated by the Nitrogen Source

In the family Enterobacteriaceae the ability to adapt to growth on a wide range of carbon-energy sources is matched by adaptability to alternative nitrogen sources. These include ammonium salts, many amino acids and their analogs, urea, nitrogen gas (for the N_2 fixers), nitrate, and nitrite. K. aerogenes can use over 50 different sources of nitrogen (94). B. subtilis, by contrast, is much more limited in terms of usable nitrogen sources. The favored source in terms of growth rate supported is glutamine, with ammonium salts a close second. Glutamate is a poor nitrogen source, apparently because its uptake is inefficient (69). Utilization of nitrate, by reduction to ammonia, is also inefficient. Histidine can serve as a sole nitrogen source but only in the absence of glucose (such utilization is repressed by a mixture of amino acids; S. H. Fisher, personal communication). In the presence of glucose, the degradative pathway for this amino acid is so tightly repressed that the absence of other nitrogen sources is insufficient to cause derepression. Induction by arginine of arginase and ornithine aminotransferase is repressed 3-fold by glucose and a further 10-fold by the combination of glucose and glutamine (2). This may reflect regulation akin to that of citB. This is in stark contrast to the situation in the Enterobacteriaceae, in which the absence of a rapidly utilizable nitrogen source leads to expression of many amino acid degradation operons, amino acid transport systems, urease, nitrate reductase, and nitrogen fixation genes (53). This process is mediated by the Ntr system, a complex interplay of genes whose products sense the relative availability of carbon and nitrogen, modify regulatory proteins, and turn transcription of target genes on and off (54). In *B. subtilis*, only glutamate synthase, glutamine synthetase (GS), urease, and asparaginase are regulated by the nature of the nitrogen source or by nitrogen limitation (69; Fisher, personal communication). (In *Bacillus licheniformis*, asparaginase also responds to nitrogen limitation [34]). Proline oxidase activity, like histidase activity, is regulated by a mixture of amino acids (Fisher, personal communication). It is clear that *B. subtilis* does not have the wide-ranging adaptability with respect to the nitrogen source that is typical of the family *Enterobacteriaceae*. On the other hand, *B. subtilis* can sporulate when deprived of a rapidly metabolizable source.

While it was reasonable a priori to think that the regulation of glutamate synthase and GS gene expression during growth in different nitrogen sources might be informative about early sporulation events, it now seems that synthesis of these two enzymes involves operon-specific regulation that is not relevant to spore formation. The glutamate synthase genes, *gltA* and *gltB*, are controlled by the product of *gltC*, the gene immediately upstream. The *gltC* gene is transcribed in the opposite direction from *gltA* from a promoter site that overlaps with the *gltA* promoter (3; D. E. Bohannon and A. L. Sonenshein, manuscript in preparation). Mutations in *gltC* greatly reduce expression from the *gltA* promoter and stimulate transcription from the *gltC* promoter. Thus, it seems that the *gltC* product is a positive regulator of *gltA* (and probably *gltB*) transcription in the absence of glutamate and a negative regulator of its own transcription. The *gltC* product is a 33,000-molecular-weight polypeptide that has sequence similarity to a number of DNA-binding proteins from *E. coli* and other gram-negative bacteria (Bohannon and Sonenshein, manuscript in preparation). These include the products of *nodD* of *Rhizobium meliloti* and *cysB* and *ilvY* of *E. coli*. All of these proteins are positive regulators. In addition, the *gltC* product is very similar in one region to the DNA-binding domain of *B. subtilis* σ^E.

Synthesis of GS is regulated by two proteins encoded in the *glnRA* operon. The *glnA* gene codes for GS; in addition, a product of this gene (presumably GS itself) is necessary for negative regulation of *glnRA* transcription (11, 77, 78). The *glnR* gene codes for a 16,000-molecular-weight polypeptide that has sequence similarity to λ *cro* protein, *trp* repressor, and biotin repressor in a region that has potential to form a helix-turn-helix structure (88; H. J. Schreier, S. W. Brown, K. Hershi, and A. L. Sonenshein, manuscript in preparation). A model for *glnRA* regulation currently being tested is that in nitrogen excess GS protein interacts with or modifies the *glnR* product, causing it to bind to the *glnRA* promoter region and repress transcription. Mutations in *glnA* or *glnR* cause constitutive transcription from the *glnRA* promoter but have no effect on sporulation. They do have some pleiotropic effects, in that certain *glnA* mutations cause several glucose-regulated genes and urease and asparaginase to be partly derepressed in their expression (25; Fisher, personal communication). Whether this reflects a direct role for this regulatory system in control of other genes or an indirect effect caused by altered metabolite pools remains to be determined.

Genes Regulated by Phosphate

In *E. coli*, genes coding for alkaline phosphatase, a periplasmic enzyme, and phosphate transport systems are regulated by the *phoB* product, a positive regulator, whose synthesis or activity or both is modulated by the product of *phoR*. *phoB* and *phoR* are members of the paired classes of regulatory proteins that recognize a partic-ular environmental signal and activate a corresponding global response. *B. subtilis* synthesizes at least two alkaline phospha-tases, one of which is made at the end of exponential growth in a special low-phos-phate medium that does not permit sporu-lation, and the other of which is made during sporulation (60). (It is not known why ex-haustion of phosphate induces sporulation in some media but not in others. In the present case, a very high sugar concentration may override whatever signal phosphate depletion engenders.) The "vegetative" alkaline phos-phatase (encoded by *phoA*) and an unchar-acterized phosphodiesterase are regulated by genes identified genetically as *phoP* and *phoR* (60). Cloning and sequencing of *phoP* have revealed that it is very similar to the *phoB* gene of *E. coli* (81); it presumably codes for a positive regulator of *phoA* and perhaps other genes as well. The *phoR* products of *B. subtilis* and *E. coli*, as deduced from DNA sequencing, are also similar (82).

Expression of *phoA* is also under spor-ulation-associated regulation in that its tran-scription is dependent on the *spo0A* product (43). Moreover, *abrB* mutations suppress the *spo0A* requirement for *phoA* expression, just as they do for many sporulation genes (43). The *spo0A* requirement may be a reflection of the effect of *spo0A* mutations on synthesis or activity of σ^H, since *phoA* transcription is reduced in a *spo0H* mutant and restored in a *spo0A* mutant in which the *spo0H* gene is present on a high-copy plasmid (43). Thus, expression of *phoA*, while not essential for sporulation, has much in common with that of early sporulation genes. In fact, it is a bit misleading to refer to its product as "vege-tative" alkaline phosphatase. It is only made when cells enter stationary phase as a result of phosphate limitation, a condition that in-duces sporulation in other media. *phoA* seems analogous to *amyE* in that both are subject to function-specific (phosphate or glucose) control as well as stationary-phase control.

DIRECT ISOLATION OF EARLY SPORULATION GENES

Defining an early sporulation gene by the time of appearance of its transcript, rather than by the function of its product, circumvents the issue of whether the product is required for sporulation. This was the ra-tionale first used to clone sporulation genes. Successful cloning of the *spoVG* gene (orig-inally the "*0.4 kb*" gene) by hybridization to its purified, low-molecular-weight, stable transcript (80) led to the demonstration that this gene is induced by phosphate limitation, by exhaustion of nutrient broth, and by ad-dition of decoyinine (99). Detailed analysis of this gene has led to the discovery of several alternative sigma factors (see references 6 and 52) and a model for the interaction of the *spo0A* and *abrB* gene products at the onset of sporulation (99).

A general search for genes turned on very early during sporulation is under way in my laboratory (Mathiopoulos and Sonen-shein, in press); the search is based on the assumption that many of the genes induced by decoyinine will prove to be induced under most sporulation conditions. This has been borne out in the first three cases. RNA ex-tracted from early sporulating cells, i.e., cells harvested shortly after exposure to decoyi-nine, was used as a template for synthesis of radioactive cDNA in vitro. After subtractive hybridization to remove cDNAs that were complementary to vegetative RNAs, the cDNA was used to probe libraries of *B. sub-tilis* chromosomal DNA. Clones that hybrid-ized to the cDNA made from sporulation RNA and not to a control cDNA made from vegetative transcripts were isolated. From an initial collection, three cloned segments have been studied in detail. One behaves very much like *spoVG* in that it is transcribed at low level in vegetative cells and is induced within 4 min after addition of decoyinine. Transcripts for two other segments, which have proved to be part of a single large tran-scription unit, are undetectable in vegetative cells but become assayable within 8 min after decoyinine addition. This transcription unit is also induced after exhaustion of nutrient broth medium and after phosphate or carbon limitation. Its expression, like that of *spoVG* and other early sporulation genes, is dependent on SpoOA and σ^H and, surprisingly, on AbrB as well. It is hoped that continued application of this approach will yield a large collection of genes from among which it will be possible to choose those that are expressed at the earliest times of sporulation. Study of the regulation of such genes has a reasonable chance of revealing the regulatory events that most closely follow environmental signals.

UNANSWERED QUESTIONS

This review has sought to raise a number of issues without necessarily resolving them. At the same time, the intention was to give a sense of the interrelatedness of stationary-phase events, the context within which sporulation occurs. It is clear that a signal that coincides with the end of balanced, exponential growth and which can be mimicked by treatments that limit growth by limiting guanine nucleotide accumulation is perceived by the cell as the initiating event of sporulation. It is also clear that there must be more than one stationary-phase signal, since various genes are turned on under some stationary-phase conditions but not others. In some cases (such as the *amyE* gene) this may be due to overriding negative regulation, but in other cases different conditions of growth limitation lead to a different set of responses. For instance, at least nine competence-associated genes in four transcription units are induced at the end of growth, but only if the medium contains glucose (1; D. Dubnau, personal communication). It seems that the signal here is either nitrogen depletion or phosphorus depletion or both. (This can be rationalized by assuming that nucleic acid can be a source of utilizable nitrogen and phosphorus.) Multiple signals imply multiple

signal sensors and transducers. There is now preliminary evidence, based on predicted protein sequence similarities to known regulatory proteins in gram-negative bacteria, to suggest that at least five transducers and three sensors have been identified. Among the transducers, the *spo0A* and *spo0F* gene products (21, 90) regulate both sporulation and other phenomena; a product of the *sacU* locus controls synthesis of a number of extracellular proteins (41, 49); the *phoP* gene codes for a regulator of *phoA* expression (81); and a product of the *comA* locus appears to be the effector for the competence signal (Dubnau, personal communication). The *spoIIJ* product seems to be a sensor involved at a minimum in expression of certain sporulation genes (P. Stragier, personal communication), a second product of the *sacU* locus may be the sensor protein for extracellular enzyme regulation (49), and the *phoR* product is probably the sensor for phosphate limitation (82).

If the signal for sporulation is a metabolite (e.g., GTP), it must be a threshold concentration that matters, since maintenance of cells at a constant slow growth rate leads to sporulation in a significant fraction of the population (10). The signal for sporulation may not be a unique metabolite, however. The fact that the extent of sporulation in a growing culture is inversely proportional to growth rate suggests that the concentration of ppGpp or the rate of ribosome synthesis might serve as a signal. It is also worth noting that the signal may not be a metabolite. P. Demchick and A. L. Koch (personal communication) have offered the interesting hypothesis that the switch to sporulation is a random event that occurs at a constant rate throughout growth under all conditions. Since a cell that chooses the sporulation pathway foregoes growth, its contribution to the total population diminishes rapidly in a fast-growing culture. In a slowly growing culture, the cells that choose sporulation appear to accumulate because they are less rapidly overwhelmed by growing cells. The random switch in this case could be imagined to be a relatively high-frequency DNA rearrangement or a burst of transcription from a key gene.

Whatever the nature of the switch, its target must direct the cell toward the sporulation pathway. It could be that the response of key regulatory genes (e.g., *abrB*, *hpr*) directs the cell immediately toward sporulation. An alternative is that the switch induces a collection of responses as a result of which the cell tests its ability to cope with the environment without sporulating. Perhaps it is only if the flow of events carries the cell as far as formation of the asymmetric septum (and the consequent activation of σ^E; 87) that the cell really becomes committed to the sporulation pathway.

WHY DON'T ALL BACTERIA SPORULATE?

A central question for many sporologists is whether ability to sporulate is the highest manifestation of procaryotic sophistication or only an antediluvian residue of an era in which an ultraconservative life-style was favored. (It should be noted that sophistication is not always beneficial to a species; S. W. Brown [personal communication] points out that the most sophisticated bureaucracies are not always the most efficient or the most adaptable to the needs of their constituencies.) Assuming that ability to sporulate was lost by nonsporulators rather than acquired by sporulators after evolutionary divergence, one might wonder why most bacteria don't form spores. For example, one might ask whether the habitats of nonsporulators select against sporulation. The ability of *E. coli* to adapt to growth at widely ranging rates in highly variable environments might be considered a reflection of its natural habitat in the often rich, but highly variable environment of the intestinal tract (in contrast to the stable, low concentrations of nutrients in soil). But *Clostridium* species grow and sporulate in the intestinal tract, and *K. aero-*

genes inhabits the soil and is even more adaptable than is *E. coli*. Clearly it is not habitat alone that divides the sporulators and the nonsporulators. Perhaps one should ask instead whether sporulators are fundamentally different from their cousins of less interesting morphology. Mechanisms of adaptation to stationary phase in gram-negative bacteria have received little notice, perhaps because of the absence of easily detected morphological changes. As more attention is given to stationary-phase events in gram-negative bacteria (see, for example, reference 79), it may become evident that nonsporulators have retained the essential regulatory program pioneered by *Bacillus* spp. (use of alternative sigma factors to control large regulons, use of multiple *spo0A*-like signal transducers, regulation of multiple stationary-phase pathways by GTP) while sacrificing the genetic security of spore formation.

Acknowledgments. I thank M. Schaechter, S. Fisher, and members of my research group (especially J. P. Mueller and S. W. Brown) for critical comments on the manuscript. I am grateful to colleagues mentioned in the text for helpful discussions and for permission to cite their work before publication.

Unpublished work from my laboratory was supported by Public Health Service research grants (GM30408 and GM36718) from the National Institutes of Health.

LITERATURE CITED

1. Albano, M., J. Hahn, and D. Dubnau. 1987. Expression of competence genes in *Bacillus subtilis*. *J. Bacteriol.* **169**:3110–3117.
2. Baumberg, S., and C. Harwood. 1979. Carbon and nitrogen repression of arginine catabolic enzymes in *Bacillus subtilis*. *J. Bacteriol.* **137**:189–196.
3. Bohannon, D. E., M. S. Rosenkrantz, and A. L. Sonenshein. 1985. Regulation of *Bacillus subtilis* glutamate synthase genes by the nitrogen source. *J. Bacteriol.* **163**:957–964.
4. Boylan, S. A., K. T. Chun, B. A. Edson, and C. W. Price. 1988. Early-blocked sporulation mutations alter expression of enzymes under carbon

control in *Bacillus subtilis. Mol. Gen. Genet.* 212:271–280.

5. Brehm, S. P., S. P. Staal, and J. A. Hoch. 1973. Phenotypes of pleiotropic-negative sporulation mutants of *Bacillus subtilis. J. Bacteriol.* 115:1063–1070.

6. Carter, H. L., and C. P. Moran. 1986. New RNA polymerase σ factor under *spo0* control in *Bacillus subtilis. Proc. Natl. Acad. Sci. USA* 83:9438–9442.

7. Carter, H. L., L.-F. Wang, R. H. Doi, and C. P. Moran. 1988. *rpoD* operon promoter used by σ^H-RNA polymerase in *Bacillus subtilis. J. Bacteriol.* 170:1617–1621.

8. Chasin, L. A., and B. Magasanik. 1968. Induction and repression of the histidine-degrading enzymes of *Bacillus subtilis. J. Biol. Chem.* 243:5165–5178.

9. Connors, M. J., J. M. Mason, and P. Setlow. 1986. Cloning and nucleotide sequencing of genes for three small, acid-soluble proteins from *Bacillus subtilis* spores. *J. Bacteriol.* 166:417–425.

10. Dawes, I. W., and J. Mandelstam. 1970. Sporulation of *Bacillus subtilis* in continuous culture. *J. Bacteriol.* 103:529–535.

11. Dean, D. R., J. A. Hoch, and A. I. Aronson. 1977. Alteration of the *Bacillus subtilis* glutamine synthetase results in overproduction of the enzyme. *J. Bacteriol.* 131:981–987.

12. Diem, K., and C. Lentner. 1970. Scientific tables, p. 511–516. Ciba-Geigy Ltd., Basel.

13. Dingman, D. W., M. S. Rosenkrantz, and A. L. Sonenshein. 1987. Relationship between aconitase gene expression and sporulation in *Bacillus subtilis. J. Bacteriol.* 169:3068–3075.

14. Dingman, D. W., and A. L. Sonenshein. 1987. Purification of aconitase from *Bacillus subtilis* and correlation of its N-terminal amino acid sequence with the sequence of the *citB* gene. *J. Bacteriol.* 169:3062–3067.

15. Donovan, W., L. Zheng, K. Sandman, and R. Losick. 1987. Genes encoding spore coat polypeptides from *Bacillus subtilis. J. Mol. Biol.* 196:1–10.

16. Dowds, B., L. Baxter, and M. McKillen. 1978. Catabolite repression in *Bacillus subtilis. Biochim. Biophys. Acta* 541:18–34.

17. Dowds, B. C. A., P. Murphy, D. J. McConnell, and K. M. Devine. 1987. Relationship among oxidative stress, growth cycle, and sporulation in *Bacillus subtilis. J. Bacteriol.* 169:5771–5775.

18. Dubnau, E., K. Cabane, and I. Smith. 1987. Regulation of *spo0H*, an early sporulation gene, in bacilli. *J. Bacteriol.* 169:1182–1191.

19. Dubnau, E., J. Weir, G. Nair, L. Carter, C. P. Moran, and I. Smith. 1988. *Bacillus* sporulation gene *spo0H* codes for σ^{30} (σ^H). *J. Bacteriol.* 170:1054–1062.

20. Feavers, I. M., V. Price, and A. Moir. 1988. The regulation of the fumarase (*citG*) gene of *Bacillus subtilis* 168. *Mol. Gen. Genet.* 211:465–471.

21. Ferrari, F. A., K. Trach, D. LeCoq, J. Spence, E. Ferrari, and J. A. Hoch. 1985. Characterization of the *spo0A* locus and its deduced product. *Proc. Natl. Acad. Sci. USA* 82:2647–2651.

22. Fisher, S. H. 1987. Catabolite repression in *Bacillus subtilis* and *Streptomyces*, p. 365–385. *In* J. Reizer and A. Peterfsky (ed.), *Sugar Transport and Metabolism in Gram-Positive Bacteria.* Ellis Horwood, Ltd., Chichester, United Kingdom.

23. Fisher, S. H., and B. Magasanik. 1984. Isolation of *Bacillus subtilis* mutants pleiotropically insensitive to glucose catabolite repression. *J. Bacteriol.* 157:942–944.

24. Fisher, S. H., and B. Magasanik. 1984. 2-Ketoglutarate and the regulation of aconitase and histidase formation in *Bacillus subtilis. J. Bacteriol.* 158:379–382.

25. Fisher, S. H., and A. L. Sonenshein. 1984. *Bacillus subtilis* glutamine synthetase mutants pleiotropically altered in glucose catabolite repression. *J. Bacteriol.* 157:612–621.

26. Freese, E. 1981. Initiation of bacterial sporulation, p. 1–12. *In* H. S. Levinson, A. L. Sonenshein, and D. J. Tripper (ed.), *Sporulation and Germination.* American Society for Microbiology, Washington, D.C.

27. Freese, E., and J. Heinze. 1984. Metabolic and genetic control of bacterial sporulation, p. 101–172. *In* A. Hurst, G. Gould, and J. Dring (ed.), *The Bacterial Spore,* vol. 2. Academic Press, Inc. (London), Ltd., London.

28. Freese, E., J. Heinze, and E. M. Galliers. 1979. Partial purine deprivation causes sporulation of *Bacillus subtilis* in the presence of excess ammonium, glucose and phosphate. *J. Gen. Microbiol.* 115:193–205.

29. Freese, E. B., N. Vasantha, and E. Freese. 1979. Induction of sporulation in developmental mutants of *Bacillus subtilis. Mol. Gen. Genet.* 170:67–74.

30. Fujita, Y., and T. Fujita. 1987. The gluconate operon *gnt* of *Bacillus subtilis* encodes its own transcriptional negative regulator. *Proc. Natl. Acad. Sci. USA* 84:4524–4528.

31. Gartner, D., M. Geissendorfer, and W. Hillen. 1988. Expression of the *Bacillus subtilis xyl* operon is repressed at the level of transcription and is induced by xylose. *J. Bacteriol.* 170:3102–3109.

32. Gilman, A. G. 1987. G proteins: transducers of receptor-generated signals. *Annu. Rev. Biochem.* 56:615–649.

33. Gilman, M. Z., and M. J. Chamberlin. 1983. Developmental and genetic regulation of *Bacillus subtilis* genes transcribed by σ^{28} RNA polymerase. *Cell* 35:285–293.

34. Golden, K. J., and R. W. Bernlohr. 1985. Nitrogen catabolite repression of the L-asparaginase of *Bacillus licheniformis. J. Bacteriol.* 164:938–940.

35. Grossman, A. D., and R. Losick. 1988. Extracel-

lular control of spore formation in *Bacillus subtilis*. *Proc. Natl. Acad. Sci. USA* **85**:4369–4373.

36. Guespin-Michel, J. F. 1971. Phenotypic reversion in some early blocked sporulation mutants of *Bacillus subtilis*. Genetic study of polymyxin resistant partial revertants. *Mol. Gen. Genet.* **112**:243–254.

37. Haldenwang, W. G., and R. Losick. 1980. Novel RNA polymerase σ factor from *Bacillus subtilis*. *Proc. Natl. Acad. Sci. USA* **77**:7000–7004.

38. Hanson, R. S., and D. P. Cox. 1967. Effect of different nutritional conditions on the synthesis of tricarboxylic acid cycle enzymes. *J. Bacteriol.* **93**:1777–1787.

39. Hastrup, S. 1988. Analysis of the *Bacillus subtilis* xylose regulon, p. 79–83. *In* A. T. Ganesan and J. A. Hoch (ed.), *Genetics and Biotechnology of Bacilli*, vol. 2. Academic Press, Inc., New York.

40. Helmann, J. D., L. M. Marquez, and M. J. Chamberlin. 1988. Cloning, sequencing, and disruption of the *Bacillus subtilis* σ²⁸ gene. *J. Bacteriol.* **170**:1568–1574.

41. Henner, D. J., M. Yang, and E. Ferrari. 1988. Localization of *Bacillus subtilis* sacU(Hy) mutations to two linked genes with similarities to the conserved procaryotic family of two-component signalling systems. *J. Bacteriol.* **170**:5102–5109.

42. Hoch, J. A., and J. Spizizen. 1969. Genetic control of some early events in sporulation of *Bacillus subtilis* 168, p. 112–120. *In* L. L. Campbell (ed.), *Spores IV*. American Society for Microbiology, Bethesda, Md.

43. Hulett, F. M., and K. Jensen. 1988. Critical roles of *spo0A* and *spo0H* in vegetative alkaline phosphatase production in *Bacillus subtilis*. *J. Bacteriol.* **170**:3765–3768.

44. Igo, M., M. Lampe, C. Ray, W. Schafer, C. P. Moran, Jr., and R. Losick. 1987. Genetic studies of a secondary RNA polymerase sigma factor in *Bacillus subtilis*. *J. Bacteriol.* **169**:3464–3469.

45. Igo, M., and R. Losick. 1986. Regulation of a promoter region that is utilized by minor forms of RNA polymerase holoenzyme in *Bacillus subtilis*. *J. Mol. Biol.* **191**:615–624.

46. Ito, J., and J. Spizizen. 1973. Genetic studies of catabolite repression insensitive sporulation mutants of *Bacillus subtilis*, p. 81–82. *In* J. P. Aubert, P. Schaeffer, and J. Szulmajster (ed.), *Regulation de la Sporulation Microbienne*. Editions du Centre National de la Recherche Scientifique, Paris.

47. Johnson, W. C., C. P. Moran, and R. Losick. 1983. Two RNA polymerase sigma factors from *Bacillus subtilis* discriminate between overlapping promoters for a developmentally regulated gene. *Nature* (London) **302**:800–804.

48. Klier, A., A. Fouet, M. Debarbouille, F. Kunst, and G. Rapoport. 1987. Distinct control sites located upstream from the levansucrase gene of *Bacillus subtilis*. *Mol. Microbiol.* **1**:233–241.

49. Kunst, F., M. Debarbouille, T. Msadek, M. Young, C. Mauel, D. Karamata, A. Klier, G. Rapoport, and R. Dedonder. 1988. Deduced polypeptides encoded by the *Bacillus subtilis* sacU locus share homology with two-component sensor-regulator systems. *J. Bacteriol.* **170**:5093–5101.

50. Lopez, J. M., A. Dromerick, and E. Freese. 1981. Response of guanosine 5′-triphosphate concentration to nutritional changes and its significance for *Bacillus subtilis* sporulation. *J. Bacteriol.* **146**:605–613.

51. Lopez, J. M., B. Uratani-Wong, and E. Freese. 1980. Catabolite repression of enzyme synthesis does not prevent sporulation. *J. Bacteriol.* **141**:1447–1449.

52. Losick, R., and J. Pero. 1981. Cascades of sigma factors. *Cell* **25**:582–584.

53. Magasanik, B. 1982. Genetic control of nitrogen assimilation in bacteria. *Annu. Rev. Genet.* **16**:135–168.

54. Magasanik, B., and F. Neidhardt. 1987. Regulation of carbon and nitrogen utilization, p. 1318–1325. *In* F. C. Neidhardt, J. L. Ingraham, K. B. Low, B. Magasanik, M. Schaechter, and H. E. Umbarger (ed.), *Escherichia coli and Salmonella typhimurium: Cellular and Molecular Biology*. American Society for Microbiology, Washington, D.C.

55. Mahadevan, S., and A. Wright. 1987. A bacterial gene involved in transcriptional antitermination: regulation at a rho-independent terminator in the *bgl* operon of *E. coli*. *Cell* **50**:485–494.

56. Marahiel, M. A., P. Zuber, G. Czekay, and R. Losick. 1987. Identification of the promoter for a peptide antibiotic biosynthesis gene from *Bacillus brevis* and its regulation in *Bacillus subtilis*. *J. Bacteriol.* **169**:2215–2222.

57. Melin, L., K. Magnusson, and L. Rutberg. 1987. Identification of the promoter of the *Bacillus subtilis sdh* operon. *J. Bacteriol.* **169**:3232–3236.

58. Michel, J. F., B. Cami, and P. Schaeffer. 1968. Sélection de mutants de *Bacillus subtilis* bloqués au debut de la sporulation. I. Mutants asporogènes pléotropes selectionnés par croissance en milieu au nitrate. *Ann. Inst. Pasteur* **114**:11–20.

59. Michel, J. F., B. Cami, and P. Schaeffer. 1968. Sélection de mutants de *Bacillus subtilis* bloqués au debut de la sporulation. II. Sélection par adaptation à une nouvelle source de carbone et par viellissement de cultures sporulées. *Ann. Inst. Pasteur* **114**:21–27.

60. Miki, T., Z. Minami, and Y. Ikeda. 1965. The genetics of alkaline phosphatase formation in *Bacillus subtilis*. *Genetics* **52**:1093–1100.

61. Miller, P., J. Mueller, K. Hill, and H. Taber. 1988. Transcriptional regulation of a promoter in the *men* gene cluster of *Bacillus subtilis*. *J. Bacteriol.* **170**:2742–2748.

62. **Miwa, Y., and Y. Fujita.** 1987. Efficient utilization and operation of the gluconate-inducible system of the promoter of the *Bacillus subtilis gnt* operon in *Escherichia coli. J. Bacteriol.* **169**:5333–5335.

63. **Nicholson, W. L., and G. H. Chambliss.** 1985. Isolation and characterization of a *cis*-acting mutation conferring catabolite repression resistance to α-amylase synthesis in *Bacillus subtilis. J. Bacteriol.* **161**:875–881.

64. **Nicholson, W. L., and G. H. Chambliss.** 1987. Effect of decoyinine on the regulation of α-amylase synthesis in *Bacillus subtilis. J. Bacteriol.* **169**:5867–5869.

65. **Nicholson, W. L., Y.-K. Park, T. M. Henkin, M. Won, M. Weickert, J. A. Gaskell, and G. H. Chambliss.** 1987. Catabolite repression-resistant mutations of the *Bacillus subtilis* α-amylase promoter affect transcription levels and are in an operator-like sequence. *J. Mol. Biol.* **198**:609–618.

66. **Oda, M., A. Sugishita, and K. Furukawa.** 1988. Cloning and nucleotide sequences of histidase and regulatory genes in the *Bacillus subtilis hut* operon and positive regulation of the operon. *J. Bacteriol.* **170**:3199–3205.

67. **Ohne, M.** 1974. Regulation of aconitase synthesis in *Bacillus subtilis:* induction, feedback repression, and catabolite repression. *J. Bacteriol.* **117**:1295–1305.

68. **Ohne, M.** 1975. Regulation of the dicarboxylic acid part of the citric acid cycle in *Bacillus subtilis. J. Bacteriol.* **122**:224–234.

69. **Pan, F. L., and J. G. Coote.** 1979. Glutamine synthetase and glutamate synthase activities during growth and sporulation in *Bacillus subtilis. J. Gen. Microbiol.* **112**:373–377.

70. **Perego, M., and J. A. Hoch.** 1987. Isolation and sequence of the *spo0E* gene: its role in initiation of sporulation in *Bacillus subtilis. Mol. Microbiol.* **1**:125–132.

71. **Perego, M., and J. A. Hoch.** 1988. Sequence analysis and regulation of the *hpr* locus, a regulatory gene for protease production and sporulation in *Bacillus subtilis. J. Bacteriol.* **170**:2560–2567.

72. **Perego, M., and J. A. Hoch.** 1988. Molecular cloning of the transcription inhibitor *abrB* of *Bacillus subtilis,* p. 129–134. *In* A. T. Ganesan and J. A. Hoch (ed.), *Genetics and Biotechnology of Bacilli,* vol. 2. Academic Press, Inc., New York.

73. **Phang, C. H., and K. Jeyaseelan.** 1988. Isolation and characterization of *citC* gene of *Bacillus subtilis,* p. 97–100. *In* A. T. Ganesan and J. A. Hoch (ed.), *Genetics and Biotechnology of Bacilli,* vol. 2. Academic Press, Inc., New York.

74. **Rosenkrantz, M. S., D. W. Dingman, and A. L. Sonenshein.** 1985. *Bacillus subtilis citB* gene is regulated synergistically by glucose and glutamine. *J. Bacteriol.* **164**:155–164.

75. **Schaeffer, P.** 1969. Sporulation and the production of antibiotics, exoenzymes, and exotoxins. *Bacteriol. Rev.* **33**:48–71.

76. **Schaeffer, P., J. Millet, and J.-P. Aubert.** 1965. Catabolite repression of bacterial sporulation. *Proc. Natl. Acad. Sci. USA* **54**:704–711.

77. **Schreier, H. J., S. H. Fisher, and A. L. Sonenshein.** 1985. Regulation of expression from the *glnA* promoter of *Bacillus subtilis* requires the *glnA* gene product. *Proc. Natl. Acad. Sci. USA* **82**:3375–3379.

78. **Schreier, H. J., and A. L. Sonenshein.** 1986. Altered regulation of the *glnA* gene in glutamine synthetase mutants of *Bacillus subtilis. J. Bacteriol.* **167**:35–43.

79. **Schultz, J. E., G. I. Latter, and A. Matin.** 1988. Differential regulation by cyclic AMP of starvation protein synthesis in *Escherichia coli. J. Bacteriol.* **170**:3903–3909.

80. **Segall, J., and R. Losick.** 1977. Cloned *Bacillus subtilis* DNA containing a gene that is activated early during sporulation. *Cell* **11**:751–761.

81. **Seki, T., H. Yoshikawa, H. Takahashi, and H. Saito.** 1987. Cloning and nucleotide sequence of *phoP,* the regulatory gene for alkaline phosphatase and phosphodiesterase in *Bacillus subtilis. J. Bacteriol.* **169**:2913–2916.

82. **Seki, T., H. Yoshikawa, H. Takahashi, and H. Saito.** 1988. Nucleotide sequence of the *Bacillus subtilis phoR* gene. *J. Bacteriol.* **170**:5935–5938.

83. **Setlow, P.** 1973. Inability to detect cyclic AMP in vegetative or sporulating cells or dormant spores of *Bacillus megaterium. Biochem. Biophys. Res. Commun.* **52**:365–372.

84. **Shimotsu, H., and D. J. Henner.** 1986. Modulation of *Bacillus subtilis* levansucrase gene expression by sucrose and regulation of the steady-state mRNA level by *sacU* and *sacQ* genes. *J. Bacteriol.* **168**:380–388.

85. **Shimotsu, H., M. I. Kuroda, C. Yanofsky, and D. J. Henner.** 1986. Novel form of transcription attenuation regulates expression of the *Bacillus subtilis* tryptophan operon. *J. Bacteriol.* **166**:461–471.

86. **Sonenshein, A.L.** 1985. Recent progress in metabolic regulation of sporulation, p. 185–193. *In* J. A. Hoch and P. Setlow (ed.), *Molecular Biology of Microbial Differentiation.* American Society for Microbiology, Washington, D.C.

87. **Stragier, P., C. Bonamy, and C. Karmazyn-Campelli.** 1988. Processing of a sporulation sigma factor in *Bacillus subtilis:* how morphological structure could control gene expression. *Cell* **52**:697–704.

88. **Strauch, M. A., A. I. Aronson, S. W. Brown, H. J. Schreier, and A. L. Sonenshein.** 1988. Sequence of the *Bacillus subtilis* glutamine synthetase gene region. *Gene* **71**:257–265.

89. **Tjian, R., and R. Losick.** 1974. An immunological

assay for the sigma subunit of RNA polymerase in extracts of vegetative and sporulating *Bacillus subtilis. Proc. Natl. Acad. Sci. USA* **71**:2872–2876.

90. Trach, K., J. W. Chapman, P. J. Piggot, and J. A. Hoch. 1985. Deduced product of the stage 0 sporulation gene *spo0F* shares homology with the *spo0A*, *ompR*, and *sfrA* proteins. *Proc. Natl. Acad. Sci. USA* **82**:7260–7264.

91. Trach, K., and J. A. Hoch. 1989. The *Bacillus subtilis spo0B* stage 0 sporulation operon encodes an essential GTP-binding protein. *J. Bacteriol.* **171**:1362–1371.

92. Trowsdale, J., S. M. H. Chen, and J. A. Hoch. 1979. Genetic analysis of a class of polymyxin resistant partial revertants of stage 0 sporulation mutants of *Bacillus subtilis:* a map of the chromosomal region near the origin of replication. *Mol. Gen. Genet.* **173**:61–70.

93. Truitt, C. L., E. A. Weaver, and W. G. Haldenwang. 1988. Effects on growth and sporulation of inactivation of a *Bacillus subtilis* gene (*ctc*) transcribed in vitro by minor vegetative cell RNA polymerases (E-σ^{37}, E-σ^{32}). *Mol. Gen. Genet.* **212**:166–171.

94. Tyler, B. 1978. Regulation of the assimilation of nitrogen compounds. *Annu. Rev. Biochem.* **47**:1127–1162.

95. Uratani-Wong, B., J. M. Lopez, and E. Freese. 1981. Induction of citric acid cycle enzymes during initiation of sporulation by guanine nucleotide deprivation. *J. Bacteriol.* **146**:337–344.

96. Wilhelm, M., and C. Hollenberg. 1984. Selective cloning of *Bacillus subtilis* xylose isomerase and xylulokinase in *Escherichia coli* genes by IS5-mediated expression. *EMBO J.* **3**:2555–2560.

97. Willecke, K., and A. B. Pardee. 1971. Inducible transport of citrate in a Gram-positive bacterium, *Bacillus subtilis. J. Biol. Chem.* **246**:1032–1040.

98. Yamashita, S., H. Yoshikawa, F. Kawamura, H. Takahashi, T. Yamamoto, Y. Kobayashi, and H. Saito. 1986. The effect of *spo0* mutations on the expression of *spo0A-* and *spo0F-lacZ* fusions. *Mol. Gen. Genet.* **205**:28–33.

99. Zuber, P., and R. Losick. 1987. Role of *abrB* in *spo0A-* and *spo0B*-dependent utilization of a sporulation promoter in *Bacillus subtilis. J. Bacteriol.* **169**:2223–2230.

100. Zuber, P., M. Marahiel, and J. Robertson, 1988. Influence of *abrB* on the transcription of the sporulation-associated genes *spoVG* and *spo0H* in *Bacillus subtilis*, p. 123–127. *In* A. T. Ganesan and J. A. Hoch (ed.), *Genetics and Biotechnology of Bacilli*, vol. 2. Academic Press, Inc., New York.

Regulation of Procaryotic Development
Edited by Issar Smith, Ralph A. Slepecky, and Peter Setlow
© 1989 American Society for Microbiology, Washington, DC 20006

Chapter 6

Subtilisin: a Redundantly Temporally Regulated Gene?

Fernando Valle and Eugenio Ferrari

Most of the soil bacteria are continually exposed to changes of environmental factors such as nutrients, temperature, pH, and oxygen. To survive they need quick and efficient mechanisms to adapt to the new conditions. These new conditions can induce the expression of certain genes, whose products are necessary to utilize less readily available nutrients or to facilitate the differentiation of the cell into a heat-resistant spore. This, together with the ability to produce enzymes of industrial importance, has encouraged a great number of genetic and physiological studies in *Bacillus subtilis*.

As described in other chapters of this book, numerous genes are involved in the sporulation process of this microorganism, and a very complex regulatory network has been selected to coordinate their expression. It is important to mention that the expression of the genes involved in this process is a response to starvation for carbon, nitrogen, or phosphorus and occurs generally after a period of exponential growth. Obviously under these circumstances of nutrient limitation a new strategy for survival is needed. The sporulation process in *B. subtilis* comprises a series of morphological and physiological changes that occur in a temporally regulated fashion (22, 32, 38). The elucidation of the

mechanism(s) that coordinates the time of appearance and the activity of substances that are required at the different stages of development is one of the major challenges in understanding this process at the molecular level. Sporulation is a very complex phenomenon, and the selection of individual parts of this system for detailed study is a necessary step.

In this review *aprE*, the structural gene for the extracellular protease subtilisin, has served as a model for the study of the regulatory network that allows *B. subtilis* to sporulate. The regulation of the synthesis of this enzyme can also be viewed as a paradigm for the regulation of other enzymes involved in degradative processes. Indeed, several of these other degradative enzymes, whether temporally regulated or not, seem to be controlled by the same regulatory gene(s) that regulates subtilisin. Thus enzymes such as levansucrase, synthesized during vegetative growth, and α-amylase, synthesized during stationary phase but not under sporulation control, are affected by several of the transcriptional activators described in this paper.

THE SUBTILISIN GENE

The *B. subtilis* subtilisin, AprE, belongs to a class of serine alkaline proteases secreted

Fernando Valle • Centro de Investigacion sobre Ingenieria Genetica y Biotecnologia, UNAM, Cuernavaca, Mexico. *Eugenio Ferrari* • Research Department, Genencor Inc., 180 Kimball Way, South San Francisco, California 94080.

into the growth medium by a wide variety of bacilli. Due to the commercial importance of this type of enzyme, a great deal is known about the structural and catalytic properties of *B. subtilis* subtilisin (57). The enzyme is coded for by the *aprE* gene, which has been cloned, mapped, and sequenced (51, 59). DNA sequence analysis, as well as analysis of the product of the *aprE* gene, reveals that the subtilisin gene is translated as a "pre-pro" peptide (40). The 275-amino-acid mature coding sequence is preceded by a 106-amino-acid polypeptide of which the first 29 amino acids have the typical structure of a signal peptide sequence (51, 59). The function of the remaining 77 amino acids is unclear. It has been proposed that the function of the "pro" sequence is either to stabilize the inactive enzyme (40) or to facilitate the correct folding of the enzyme (23, 24). The maturation of the enzyme from pro-subtilisin to mature protein is apparently due to selfprocessing (23, 40). The presence of a "prosequence" seems to be a feature common to at least two other secreted proteases of *B. subtilis*, the neutral protease NprE (64) and a minor extracellular protease, Epr (50).

Despite early speculation about an active involvement of subtilisin in the sporulation process, this suggestion has been demonstrated to be unfounded by showing that a strain carrying a deletion of the *aprE* gene can sporulate normally (51). The cloning and deletion of several other degradative enzymes from *B. subtilis*, i.e., NprE (64), internal serine protease Isp (3), Epr (50), and esterase B (EstB; E. Ferrari and M. Ruppen, unpublished results), seem to point to the conclusion that the major function of these enzymes is a scavenging one. None of these enzymes seems to play an essential role in either the growth or sporulation process, and strains carrying multiple deletions also do not show any obvious sporulation defect (50, 64; E. Ferrari, unpublished results).

REGULATION OF *aprE*

Since the cloning of the *aprE* gene it has been possible to analyze its promoter in detail and study its regulation. As mentioned above, the expression of the subtilisin gene is tightly associated with the onset of sporulation. Owing to the difficulty of analyzing the enzyme activity in a culture supernatant containing several other major and minor proteases expressed concomitantly with subtilisin, almost all the studies described below were done using an *aprE-lacZ* translational fusion (10). Also, to minimize possible artifacts due to plasmid amplification, only the results obtained with a single-copy plasmid integrated into the *B. subtilis* chromosome (48) are being considered in this review.

To locate the promoter of *aprE* gene, two approaches were taken. The first consisted of the determination of the transcriptional start site of the gene by primer extension experiments on mRNA extracted from sporulating cells (9, 18). The start site is located at the adenine labeled +1 in Fig. 1, 58 base pairs upstream of the translational start codon. In a second approach, we generated a series of deletions from the 5' end of the cloned *aprE* promoter fragment (9, 18). The effects of these deletions were tested by analysis of the expression of β-galactosidase by cells carrying *aprE-lacZ* gene fusions. These two approaches locate the promoter in the region between −52 and the start site. The most likely sequence resembling a promoter in this region is outlined in Fig. 1. Although this sequence resembles the consensus sequence for a σ^A promoter (16), it is not clear which of the several sigma factors synthesized in *B. subtilis* is the one responsible for the transcription from the subtilisin promoter. A strain carrying a deletion in the *sigB* (4) gene and also containing the *aprE-lacZ* fusion synthesizes β-galactosidase at the same rate as the wild-type strain (9). This result rules out the possibility that σ^B plays a major role in the transcription of the *aprE* gene, in contrast to a previous proposal

```
                    -400
      GGCGGCCGCA TCTGATGTCT TTGCTTGGCG AATGTTCATC TTATTTCTTC CTCCCTCTCA


                                                              -300
      ATAATTTTTT CATTCTATCC CTTTTCTGTA AAGTTTATTT TTCAGAATAC TTTTATCATC
                                                 ------------>  ------------>

      ATGCTTTGAA AAAATATCAC GATAATATCC ATTGTTCTCA CGGAAGCACA CGCAGGTCAT
                  <------

                                         -200
      TTGAACGAAT TTTTTCGACA GGAATTTGCC GGGACTCAGG AGCATTTAAC CTAAAAAAGC


      ATGACATTTC AGCATAATGA ACATTTACTC ATGTCTATTT TCGTTCTTTT CTGTATGAAA
                                               ------------>    <------

                -100
      ATAGTTATTT CGAGTCTCTA CGGAAATAGC GAGAGATGAT ATACCTAAAT AGAGATAAAA
      --------------->        <------

                      -35                     -10           +1
      TCATCTCAAA AAAATGGGTC TACTAAAATA TTATTCCATC TATTACAATA AATTCACAGA


      ATAGTCTTTT AAGTAAGTCT ACTCTGAATT TTTTTAAAAG GAGAGGGTAA AGAGTGAGAA
                                                                 M   R

      GCAAAAAATT GTGGATC
       S    K    K    L    W    I
```

Figure 1. Nucleotide sequence of the *aprE* regulatory region. The transcription initiation site is denoted by +1. Numbering is with respect to the initiation start site according to Ferrari et al. (9). Inverted repeat sequences are indicated by arrows. The −10 and −35 of the putative promoter are denoted by double overlining.

(59). σ^H, the product of the *sigH* (*spo0H*) gene, can also be excluded from playing a direct role in *aprE* transcription as discussed below.

The plasmid carrying the *aprE-lacZ* fusion was introduced by transformation in several *spo0* mutants (i.e., *spo0A*, *spo0B*, *spo0E*, *spo0F*, *sigH* [*spo0H*], and *spo0J*). The most stringent control was exerted by the *spo0A* mutations, which decreased the rate of synthesis of the fused protein over 10-fold (10; see below). The *spo0J* mutation had no significant effect on the expression of β-galactosidase, while in the other four strains the rate of synthesis was reduced about fivefold (10).

CHARACTERISTICS OF THE *aprE* REGULATORY REGION

In addition to the *spo0* mutations there are at least seven other different genes which are known to affect, either directly or indirectly, the transcription of the subtilisin gene. These genes are *sacU*, *sacQ*, *prtR*, *senN*, *hpr*, *sin*, and *abrB*. It is important to note that in this review we will follow the changes in nomenclature proposed by Henner, Rapoport, and co-workers (20, 28). Therefore the *prtR*, *sacQ*, and *sacU* genes will be referred to as *degR*, *degQ*, and *degS-degU*, respectively. However, because the molecular nature of the different SacU phenotypes [SacU, SacU(Hy), and SacU⁻] is apparently caused by mutations in either the *degS* or *degU* gene, we will keep using the *sacU* symbol throughout this review when referring to the operon, to the phenotypes themselves, or to the strains having, or mutations causing, such phenotypes. The effect of most of these genes on the expression of subtilisin has been analyzed by means of *aprE-lacZ* fusion, using the promoter with an intact upstream region and several constructions using this promoter region preceded by a series of 5′ deletions. The mode of action of these additional regulatory genes and how they affect the synthesis of subtilisin will be analyzed in detail below.

It has been reported that approximately 400 base pairs upstream from the subtilisin promoter is sufficient to allow normal regulation of this gene (18). Inspection of this sequence reveals several interesting features. The promoter is localized in a very A+T-rich region; for example, the 100-base-pair fragment preceding the −35 region is 73% A+T. This abundance of A+T pairs, and their distribution, result in a DNA with a static curvature. This curvature affects the migration properties of this DNA fragment in a polyacrylamide gel (F. Valle, unpublished results). Using site-directed mutagenesis, we have been able to modify this static bending, and the effect of these mutations has been analyzed. We found that the modification of a few bases in the region from −100 to −164 can change the bending drastically, and some of these mutations can alter the expression of the *aprE* promoter (F. Valle and E. Ferrari, manuscript in preparation).

At least three models can be envisaged to explain the result of these changes in the −100 to −164 region, as follows.

1. The introduced changes modify the binding site of one or more of the proteins involved in transcription.
2. The changes fall just outside of the putative sites of interaction of the transcriptional activators. However, the modifications increase or decrease the number of secondary contacts that the

hypothetical regulator(s) can establish in that area.

3. The changes in DNA bending increase or decrease the DNA's ability to form loops and therefore could change the interaction between the regulator(s) and the RNA polymerase.

Obviously more experiments need to be done in order to understand the importance (if any) of the bending of DNA in this region. The presence of bent DNA in regulatory regions has been reported for several bacterial systems (14, 39, 61), and the effect of deletions in such regions has been analyzed (34, 39). Similarly, the analysis of many eucaryotic genes has revealed that sequences upstream of the gene's promoter are important for its regulation. More detailed analysis of these regions reveals that they are complex in their organization and that DNA conformation as well as sequence is important in determining their expression (26). Another interesting feature of the *aprE* promoter is the existence of at least four inverted repeat sequences centered at positions −300, −294, −126, and −98 (see Fig. 1). The presence of these sequences could be fortuitous, or they could be the recognition sites of regulatory proteins.

aprE REGULATION BY *degU-degS*

Several *B. subtilis* mutations capable of causing high levels of secreted enzymes have been isolated. Depending upon the particular type of enzyme analyzed, these mutations were called *sacU* (30), *amyB* (47), or *pap* (2). A more careful analysis revealed that these mutations were able to affect the production of other enzymes as well. These enzymes include subtilisin, neutral protease, α-amylase, glucanase, levansucrase, and intracellular serine protease (5, 19, 31). Strains carrying these mutations also sporulate efficiently in the presence of glucose, which prevents sporulation in wild-type strains of *B. subtilis* (29, 31). These mutated strains also lack flagella and are very poorly transformable (31, 46). Careful mapping showed that all the different mutations mapped to the same locus, referred to initially as *sacU*(Hy) (52).

The mechanism by which the *sacU*(Hy) mutations increase the level of expression of their target genes has recently been investigated. Studies of the levansucrase gene showed that strains carrying a *sacU*(Hy) mutation had an increased level of mRNA, and analysis of the mRNA by nuclease protection showed that the mRNA had the same transcriptional start site in wild-type and *sacU*(Hy) strains (49). A study with the *aprE* gene gave similar results (9, 18), suggesting that the stimulation of expression of these two genes is by increased transcription from their promoters.

Deletion analysis of the *aprE* gene regulatory regions suggests that the target site for this stimulation in *sacU*(Hy) mutant strains is located between nucleotides -36 and -164 (18), and an internal deletion of the DNA region between -100 and -164 destroys the *sacU*(Hy) effect (Valle, unpublished results). As mentioned above, the *aprE* regulatory region has a static bending. A preliminary mapping of the center of the bend suggests that it is around position -140. These data raise the possibility that bent DNA could be involved in the activation process by *sacU*(Hy) mutants. This latter possibility is similar to what has been demonstrated for the recognition site of the T antigen in simian virus 40 (45). This protein is able to bind to DNA specifically, and an unusual DNA conformation is one of the determinants for this binding. Although the exact nature of the structure remains to be determined, it is clear that the DNA in that region is bent (45).

To understand the molecular mechanims by which *sacU*(Hy) mutations increase the transcription of several genes, a DNA fragment containing this locus has been cloned and sequenced (20, 28, 53). The sequenced region revealed four open reading frames (ORFs). The ORFs are numbered here as in the paper by Henner and co-workers (20). Complementation analysis using different portions of the sequenced region suggested that the SacU(Hy) phenotype is due to mutations in either ORF1 or ORF2. The presence of a palindromic sequence similar to transcriptional terminators between ORF2 and ORF3 suggests that ORF3 is in a different transcriptional unit than ORF1 and ORF2. ORF1 and ORF2 have been named *degS* and *degU*, respectively (20, 28), and as has been previously mentioned, *sacU*(Hy) mutations map in either one of these two genes. Sequencing of some of the mutant genes has revealed that the SacU(Hy) phenotype can be due to missense mutations in either of these two genes.

A comparison of the deduced amino acid sequences of *degS* and *degU* with other protein sequences revealed that these two proteins are similar to a conserved procaryotic family of two-component signalling systems (20, 28, 53). The proteins in these groups are able to sense environmental signals, and some of them have been shown to function via a phosphorylation-dephosphorylation mechanism. In general they trigger adaptive behavior involving changes in gene expression or cell movement (these two-component systems are also discussed in this volume by I. Smith). These simple signalling systems are comprised of a "sensor" protein that detects environmental stimuli and a "regulator" protein that controls expression of a group of genes. A model explaining how sensors and regulators work has been discussed in detail by Kofoid and Parkinson (27). Briefly, it is postulated that, after sensing some environmental stimuli, the sensor modifies the regulator. This modified regulator is now able to activate certain specific responses.

Perhaps the best known example of this type of regulatory pathway is the *ntrB-ntrC* system (33). This system is involved in the response of *Escherichia coli* to changes in the availability of ammonia. Lack of ammonia in the growth medium generates an intracellular signal causing the phosphorylation of a specific protein, NRI (the product of *ntrC*). This reaction is catalyzed by a kinase, called

NRII, which is the product of the *ntrB* gene. The phosphorylated NRI is able to bind to the DNA far from the promoter of the regulated genes and interacts with an RNA polymerase containing a special sigma factor. This interaction stimulates transcription. An excess of ammonia generates a signal which results in the dephosphorylation of the DNA-binding protein and therefore in the cessation of the activation of transcription at the regulated promoter (33). Similarity between the product of *degS* and sensor proteins has been reported, but this similarity was confined to small regions of the protein (28). However, an analysis of the *uhp* operon has recently been published. This operon is responsible in the regulation of sugar-phosphate transport in *E. coli* (58). Two genes from this operon, *uhpB* and *uhpA*, have been proposed as the sensor and regulator, respectively. The comparison of *uhpB* with *degS* showed a high degree of similarity throughout the *degS* sequence (Fig. 2). It has been reported that the *uhpB* protein is a bipartite structure with a very nonpolar amino-terminal half, likely to be embedded in the cytoplasmic membrane, and a highly polar carboxyl-terminal region. *degS* does not have similarity with the amino terminus of *uhpB*, and the hydropathic profile of *degS* protein doesn't show any significant transmembrane segment (28), suggesting that the DegS protein could be a cytoplasmic protein.

Comparison of the DegU protein with the protein data bank showed that DegU is similar to several regulatory proteins like NtrC, Spo0A, Spo0F, Dye, OmpR, MalT, PhoB, CheY, and GerE (20, 28). Furthermore, the comparison of UphA, the regulator of the sugar-phosphate transport system, with DegU showed the highest score (Fig. 3). As can be seen, the similarity between these two proteins includes the DNA-binding domain proposed for UphA. A comparison of DNA-binding domains of several transcriptional regulators is presented in Fig. 4. In the putative DNA-binding domain of DegU, instead of the conserved Gly in the turn region

of the helix-turn-helix motif, a Phe is present. The significance of this is unknown. While experiments to prove that DegU is a DNA-binding protein need to be done, it is important to mention that DegU also showed a high degree of similarity with CheY, which is not a DNA-binding protein but acts as a regulator, probably by interacting directly with the flagellar motor (26).

Given the experimental data available on the *degS-degU* system, we cannot exclude the possibility that the function of DegU is to regulate the activity of some other protein, which itself could be the real DNA-binding protein.

Another phenotype of the *sacU* gene has been described in the literature and is referred to as SacU⁻. A characteristic of the SacU⁻ phenotype is the production of lower amounts of secreted enzymes; these cells are also impaired in the transformation process, although not as severely as the *sacU*(Hy) mutants (31, 53). Gene disruption experiments indicate that this phenotype may be caused by lack of either a functional *degS* or *degU* (20, 28). These results seem to indicate that at least an intact *degU* gene product is needed (either activated or not) to obtain what we think is the wild-type phenotype of *B. subtilis*.

OTHER TRANSCRIPTIONAL ACTIVATORS

Although the *degS-degU* genes have received most of the attention, due to the pleiotropic and dramatic effect of the SacU(Hy) phenotype, three other genes have been isolated and characterized that can increase synthesis of some or most of the degradative enzymes. All of these additional genes have strikingly common features such as the size of the encoded peptide, the presence of a helix-turn-helix motif, and the ability to stimulate the expression of degradative enzymes when expressed at higher than wild-type levels. For practical purposes we will consider DegQ, DegR, and SenN to be transcriptional activators, although this needs to be proven.

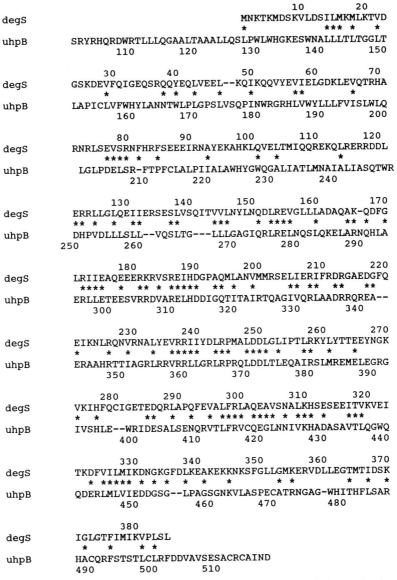

Figure 2. Similarities between *degS* and *uphB* gene products. Only the last 417 amino acids of *uhpB* are shown. * indicates identical residues or conserved substitutions (I, L, V, M; D, E; R, K, H; F, W, Y; Q, N).

The *degQ* gene from three different bacilli has been cloned and characterized (1, 55, 62). The gene isolated from *B. subtilis* encodes a 46-amino-acid polypeptide. The *degQ* gene, when hyperexpressed either because of an up-mutation in its promoter (30, 62) or when present in a multicopy plasmid, causes hyperproduction of certain extracellular enzymes, including subtilisin (62).

The *degR* gene encodes a 60-amino-acid polypeptide. When present in a multicopy plasmid, this gene can transcriptionally activate the expression of degradative enzymes as is the case with *degQ* (35, 54, 63). To date

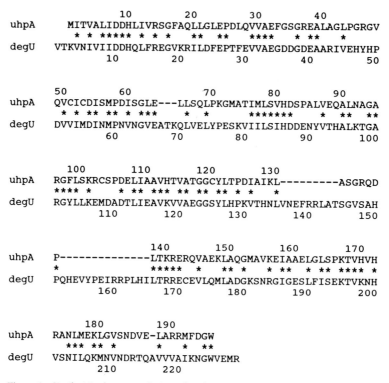

```
                 10        20        30        40
uhpA     MITVALIDDHLIVRSGFAQLLGLEPDLQVVAEFGSGREALAGLPGRGV
          * * ***** * * *   ** *  ** * ****  * **  *
degU     VTKVNIVIIDDHQLFREGVKRILDFEPTFEVVAEGDDGDEAARIVEHYHP
                 10        20        30        40        50


                 50        60        70        80        90
uhpA     QVCICDISMPDISGLE---LLSQLPKGMATIMLSVHDSPALVEQALNAGA
          * * ** ** * ***    * *      ******* *   ** **
degU     DVVIMDINMPNVNGVEATKQLVELYPESKVIILSIHDDENYVTHALKTGA
                 60        70        80        90        100


                 100       110       120       130
uhpA     RGFLSKRCSPDELIAAVHTVATGGCYLTPDIAIKL---------ASGRQD
         ****  *  * ** *** ** ** ** *  *             *
degU     RGYLLKEMDADTLIEAVKVVAEGGSYLHPKVTHNLVNEFRRLATSGVSAH
                 110       120       130       140       150


                            140       150       160       170
uhpA     P--------------LTKRERQVAEKLAQGMAVKEIAAELGLSPKTVHVH
         *               *****  *   ** *   * **  * ** **** *
degU     PQHEVYPEIRRPLHILTRRECEVLQMLADGKSNRGIGESLFISEKTVKNH
                 160       170       180       190       200


                 180       190
uhpA     RANLMEKLGVSNDVE-LARRMFDGW
          *** ** *      *   *  **
degU     VSNILQKMNVNDRTQAVVVAIKNGWVEMR
                 210       220
```

Figure 3. Similarities between *degU* and *uphA* gene products. Symbols and criteria are the same as in Fig. 2.

no chromosomal mutations have been isolated in this gene.

The *senN* gene encodes a 60-amino-acid polypeptide whose activity seems to be similar to that of DegQ and DegR (60). The *senN* gene has been cloned and sequenced from *Bacillus natto*, but its presence in *B. subtilis* has only been demonstrated by Southern blotting.

The function of all these transcriptional activators for degradative enzymes is not clear. The fact that deletion mutations of *degQ* and *degR* do not seem to affect the level at which secreted enzymes are synthesized (19) does not rule out the possibility that they could be transcriptional cofactors. Expression studies done by Henner and co-workers, which indicate that the expression of *degU* and *degR* is negatively controlled by the presence of the *sacU*(Hy) allele (19), do not clarify this issue.

One possibility is that these different ac-tivators respond to different external or internal stimuli and can activate the transcription of a certain number of genes which are necessary for the cell's survival in particular environmental conditions. If this is the case it could be just a coincidence of our experimental approach that these gene products appear to be involved in the transcriptional activation of a common set of genes. This explanation is suggested by the fact that the *degR* gene seems to be preceded by a very strong σ^D promoter, as demonstrated by in vitro transcription experiments (V. Singer and M. Chamberlin, personal communication). σ^D, the product of the *sigD* gene, seems to be responsible for the transcription of genes involved in the synthesis of flagella and the response to chemotactic stimuli (16). It is possible that under this type of external stress, such as the stimuli causing chemotaxis, the cell has the need to synthesize a high amount

Protein	Starting residue	Sequence
degU	181	K S N R G I G E S L F I S E K T V K N H V S
uhpA	153	M A V K E I A A E L G L S P K T V H V H R A
cI	31	L S Q E S V A D K M G M G Q S G V G A L F N
cro	17	F G Q T K T A K D L G V Y Q S A I N L A I H
trpR	66	M S Q R E L K N E L G A G I A T I T R G S N
CRP	167	I L R Q E I G Q I V G C S R E T V G R I L K
deoR	22	L H L K D A A A L L G V S E M T I R R D L N
cro 434	19	M T Q T E L A T K A G V K Q Q S I Q L I E N
asnC	23	T A Y A E L A K Q F G V S P G T I H V R V E
fnr	195	M T R G D I G N Y L G L T V E T L S R L L G
fixJ	156	L P N K S I A Y D L D I S P R T V E V H R A
ompR	88	E V D R I V G L E I G A D D Y I P K P F N P
phoB	199	I R R L R K A L E P G G H D R M V Q T V R G

Figure 4. Comparison of a portion of *degU* gene product with DNA-binding protein regions of other regulatory proteins. Underlined residues in DegU are those in which the same residue is present in other regulatory proteins. Aligment and sequences are from Friedrich and Kadner (11).

of degradative enzymes. The presence of the *degR* gene product would make this possible without turning on or off a variety of other genes as would be the case for *degU*.

REGULATION BY *abrB, spo0A,* AND *spo0H*

B. subtilis mutants blocked in the earliest stages of sporulation by mutations in the *spo0* genes manifest a variety of phenotypes, as discussed elsewhere in this volume (Smith). The mutations with the most severe effect on the expression of subtilisin are also the ones with a stringent effect on the sporulation process itself: *spo0A, spo0B, spo0E, spo0F,* and *sigH* (22, 38). The effect of some of these mutations on the expression of subtilisin has

been studied with the *aprE-lacZ* fusion described earlier and also by determining the amount of *aprE*-specific mRNA (9, 10). The results obtained showed that *spo0A* mutations reduce the rate of β-galactosidase synthesis about 10-fold and almost completely abolish the synthesis of *aprE* mRNA. Interestingly, the same *spo0A* strains are capable of synthesizing the subtilisin as well as a wild-type strain when the entire *aprE* gene is carried in a multicopy plasmid (51). Similarly, a *spo0A* strain carrying a single-copy *aprE-lacZ* fusion synthesizes β-galactosidase at the same rate as a wild-type strain if it carries in a multicopy plasmid a fragment of DNA comprising the regulatory sequences of the *aprE* gene from about −600 to +82 (10). These results strongly indicate that subtilisin expression is controlled, at least in part, by a repressor which is present in the cell in a limited amount. Abundant subtilisin mRNA is present in vegetative cells carrying the *aprE* gene in a multicopy plasmid (Ferrari, unpublished observation), which also suggests that the expression of subtilisin could be controlled through a repressor rather than the lack of a transcriptional factor.

An important further step in the understanding of the relation between *spo0A* mutations and *aprE* expression was the cloning and characterization of the *abrB* gene (37). *abrB* mutants were isolated as capable of partially repressing the Spo0A phenotype (56). Although unable to sporulate, *spo0A abrB* strains can restore the expression of some stationary phase-associated genes such as those encoding proteases and antibiotics (56). With the *aprE-lacZ* fusion, the rate of synthesis of β-galactosidase in a *spo0A abrB* mutant is almost the same as in wild-type strains. Analysis of the expression of an *abrB-lacZ* fusion show that in a *spo*⁺ strain the *abrB* gene is synthesized throughout vegetative growth but is represented at the onset of sporulation. In contrast, in *spo0A* mutants *abrB* synthesis is increased during stationary phase (37). Sequencing of some of the *abrB* mutations has revealed that their phenotype is due either to promoter mutations which reduce the level of transcription of the *abrB* gene or to the lack of an active *abrB* gene product (37).

These results imply that the *spo0A* gene product represses either directly or indirectly the synthesis of *abrB*, which in turn would result in derepression of the synthesis of subtilisin and other sporulation-associated genes (37, 65). This model is further supported by data obtained by Hoch and co-workers which show that the *abrB* gene product can bind to the subtilisin promoter region in gel retardation and footprinting experiments (J. A. Hoch, personal communication). *B. subtilis* strains carrying either point or deletion mutations in the *sigH* gene synthesize β-galactosidase from an *apr-lacZ* fusion at a rate which is fivefold less than that of the wild-

type strain (9). RNA studies also reveal that the synthesis of subtilisin mRNA in *sigH* mutants is reduced to about 30% of the wild-type level (10). This would, at least, rule out the possibility that σ^H, encoded by the *sigH* gene (8), is the only sigma factor responsible for the transcription of the *aprE* gene. However, the possibility always exists that, as is the case for other *B. subtilis* genes, the *aprE* gene is under the control of a dual promoter. This would explain the presence of more than one transcriptional start site (59), although this finding has been attributed to an artifact of the techniques used in the S1 mapping or in the primer extension (18). As mentioned above, a potential σ^A consensus sequence is located in front of the subtilisin gene. This could easily explain the presence of abundant subtilisin mRNA found in vegetative cells carrying the *aprE* gene in a multicopy plasmid (Ferrari, unpublished results). Most likely the function of the *sigH* gene product is a consequence of the sporulation defect rather than a specific lack of a sigma factor necessary for the transcription of *aprE*.

NEGATIVE CONTROL OF *aprE* EXPRESSION

In the previous sections of this chapter we discussed genes such as *degQ*, *degR*, and *senN* that, when expressed at higher than wild-type levels, increase subtilisin expression. There are also the genes *sin* and *hpr*, which have an inhibitory effect on subtilisin production when on a multicopy plasmid (13). Deletions or nonsense mutations in these genes stimulate subtilisin expression (12, 36).

sin

Gaur et al. (13) have reported on the cloning of a DNA fragment of *B. subtilis* which, on a multicopy plasmid, inhibits sporulation and the production of extracellular proteases. Sequence, mutation, and deletion analyses as well as analysis of expression of a *sin-lacZ* fusion and *sin* mRNA revealed that the cloned fragment contained two ORFs and three promoters (12, 13). Both ORFs are translated, but only the second one, ORF2, is responsible for the multicopy Sin phenotype (12). ORF2 encodes a 111-amino-acid protein and has been named Sin. Comparison of its sequence with the protein data bank showed that it has similarity to DNA-binding proteins. These data suggest that Sin might be a repressor of sporulation (12, 13).

A direct interaction of the *sin* gene product with the *aprE* promoter has been suggested on the basis of gel retardation experiments (N. Gaur and I. Smith, personal communication). It was determined that Sin is able to bind to a DNA fragment containing the region between −200 and −414 of the *aprE* regulatory region. Binding of Sin to the region of DNA containing the promoter was also detected (Gaur and Smith, personal communication). The fact that Sin is a tetramer, coupled with the presence of two different Sin-binding sites in the *aprE* promoter region, suggests that the binding of one Sin tetramer to both sites could induce the formation of a DNA loop. Looping has been used in some other regulatory systems to explain repression (for a review see reference 41).

Another interesting feature of Sin is that its inactivation causes a loss of competence and the loss of motility. This phenotype has also been observed in *sacU*(Hy) strains. The presence of a *sacU*(Hy) mutation has no effect on the inhibition of protease production when *sin* is present in multicopy plasmids (13). All these data suggest the possibility that one of the functions of the *degS-degU* system is to alleviate the inhibitory effect of Sin.

hpr

A group of mutants that hyperproduce subtilisin were mapped to one locus called *hpr*. These mutants seem to have a more restricted phenotype than those previously discussed, since they show an effect on the pro-

duction of subtilisin, neutral proteases, and intracellular serine proteases (21, 44) but not on levansucrase or amylase (19, 31). The cloning and sequencing of the *hpr* gene and of some of its mutations (36) demonstrated that the hyperproduction phenotype is due to nonsense mutations or deletions in the coding region of one protein. It has also been determined that the *scoC* (7) and the *catA* (25) genes are identical to *hpr* (36). The notation *hpr* will be used hereafter to refer to any of the three alleles heretofore called *hpr*, *scoC*, or *catA*. The product of the *hpr* gene is a protein of approximately 23 kilodaltons.

These data, together with the fact that when on a multicopy plasmid *hpr* inhibits sporulation and protease production, suggest that Hpr plays a role of a negative regulator of proteases and sporulation as is postulated for the Sin protein. Northern (RNA blot) analysis of *aprE* mRNA levels and initiation points in *hpr* mutants showed that the same transcriptional start site was used as in wild-type strains but that there was an increase in the amount of mRNA produced (9, 18). This suggests that these mutants do not function by activating different promoters but by increasing the steady-state level of mRNA, perhaps at transcriptional level.

Through the use of upstream deletions of the subtilisin promoter it has been shown that the DNA region between −200 and −244 of subtilisin promoter is necessary in order to have stimulation of subtilisin production in an *hpr* mutant (18). It is important to mention that deletions of that region have no effect on the stimulation exerted by *sacU*(Hy) or *degQ*(Hy) (18). Using *hpr-lacZ* fusions, the expression of *hpr* gene has been analyzed (36). It has been found that in *spo0A* mutants the *hpr* gene is constitutively overproduced (36), similar to what has been observed for *abrB* (see above). The fact that a triple *spo0A hpr abrB* mutant is not sporulation proficient indicates that although *abrB* and *hpr* mutation can restore the expression of some temporally regulated genes, at least one more link, controlled by *spo0A*, is missing in the coordination of the sporulation process itself (36).

FINAL CONSIDERATIONS

Plentiful experimental data indicate that, while *aprE* is not an essential gene for sporulation in laboratory conditions, its expression is under the control of a complex regulatory network. Other than the effect of most of the *spo0* genes, which probably affect the synthesis of subtilisin as a "side effect" of the Spo⁻ phenotype, several other genes are involved in the regulation of *aprE* expression.

The AbrB and Sin proteins appear to bind to the *aprE* control region, supporting the notion that these two proteins are DNA-binding proteins which can act as repressors (12, 37). Other proteins have been proposed as positive effectors on *aprE* expression. These proteins include the gene products of *degS*, *degU*, *sacQ*, *prtR*, and *senN*. More complexity is added to the system by the inclusion of Hpr as a negative regulator of subtilisin expression.

This plethora of genes regulating *aprE* expression now allows us to formulate a variety of very complicated models. For example, it is possible to imagine that some of the negative regulators (AbrB, Sin, or Hpr) bind to at least two sites in the subtilisin control region. This binding could induce looping and somehow decrease or block transcription from this promoter. A possible mechanism for this effect could be that the binding of one or more proteins in the regulatory region excludes the binding of some other protein essential for transcription, such as RNA polymerase or perhaps DegU. Several examples of this type of regulation have been reported in procaryotic systems (41). How DegU, DegQ, and the other positive regulators of *aprE* act to overcome this repression and increase gene expression could be envisaged in the same way. Some of these proteins may bind to the subtilisin regulatory region and exclude the binding of the repressors, in this way increasing *aprE* ex-

pression. Another possible role of the transcriptional activators, apart from excluding repressors from the regulatory region, could be that they interact directly with RNA polymerase and help the RNA polymerase to bind and begin transcription. Again, this strategy is utilized in other systems (41, 42).

Utilization of protein phosphorylation as a way to regulate gene expression or enzymatic activities is a common strategy in procaryotic cells (6). The hypothesis that *degS-degU* interaction is mediated by a phosphorylation/dephosphorylation mechanism gives to the system the ability to react to different environmental signals with a graded response, not as a simple on/off switch. The obvious question is what kinds of signals are able to activate the *degS-degU* system? So far this question remains unanswered and open for investigation.

Are all these repressors and activators necessary? One must be particularly careful in interpreting the results obtained in this type of analysis, especially when the regulators are expressed from a multicopy plasmid. The high degree of homology, a possible common ancestry, and a similar mechanism of action of the DNA-binding proteins (17) raise at least two possibilities of incorrect interpretation of the results. One is the danger of "illegitimate cross talk" between components of sensor-regulator molecules of different systems. Indeed, several cases of this anomalous "conversation" have been reported (43). The second possibility of error stems from the high concentration of usually tightly regulated proteins. Some of the phenotypes reported in literature were obtained by analyzing cells carrying the regulator genes in multicopy plasmids (1, 13, 35, 53–55, 60, 62, 63). This could easily create artifacts due to the abnormally high concentration of repressor (or activator) present in the cell. As a result, because of the similarities between the recognition sites of these proteins (15), the hyperproduced regulator could bind to a sequence different from its target.

As a final observation we would like to

Table 1
Genes Affecting Degradative Enzyme Production

Gene	Response by the following gene[a]:						References
	amy	*lvs*	*glu*	*apr*	*npr*	*isp*	
Activators							
sacU(Hy)	H	H	H	H	H	H	19, 29, 30, 44
degQ(Hy)	H	H	NT	H	H	H	19, 29, 30, 44
degR[b]	+	H	NT	H	H	NT	35, 54, 63
senN[b]	H	NT	NT	H	H	NT	60
Repressors							
abrB	+	+	NT	−	−	NT	37, 56
hpr	+	+	NT	−	−	−	21, 36
sin[c]	NT	NT	NT	−	−	NT	12, 13

[a] *amy*, Amylase; *lvs*, levansucrase; *glu*, glucanase; *apr*, subtilisin; *npr*, neutral protease; *isp*, internal serine protease. H, Stimulation; +, no detectable effect (i.e., wild-type phenotype); −, repression as a wild-type gene or when present in a multicopy plasmid; NT, not tested.

[b] In strains carrying the *degR* gene on a multicopy plasmid.

[c] Either on a multicopy plasmid or deleted.

point out that all the transcriptional activators described in this review (DegU, DegQ, DegR, SenN) enhance the expression of all the degradative enzymes analyzed so far, whether temporally regulated or not (Table 1). In contrast, the repressors (*abrB, hpr, sin* gene products) seem to exert an activity only on the temporally regulated enzymes. The real meaning of this finding is not clear at present. One explanation could be that some of the degradative enzymes are needed during stressful situations whether the cell is sporulating or not, while others are needed only during extreme duress, such as conditions triggering sporulation. On the other hand, these results could be biased by our experimental approach and by the overzealousness of studies that focus on the understanding of sporulation.

The presence of all these regulatory elements controlling the expression of the subtilisin, although not fully understood, makes the *aprE* promoter a very interesting tool to investigate the regulation of gene expression in *B. subtilis*. This is particularly appealing not only for the basic knowledge that we can acquire from this type of studies, but also for the possible practical application that can derive from it. Thanks to the existence of well-characterized fermentation processes, we believe that *Bacillus* has the potential for becoming an ideal host for large-scale production of a variety of recombinant-DNA products.

Acknowledgments. We thank all the people mentioned in the text for personal communications which have been helpful in the writing of this review. A special thanks to Dennis Henner who, other than critically reading the manuscript, made a valiant attempt to translate it into English from a rather awkward Mexican-Italian slang. For the same reason, as well for their patience, we would like to thank the editors of this volume. We also thank Rose Thompson and Robb Eby for help in the final setting of the manuscript.

LITERATURE CITED

1. **Amory, A. F., F. Kunst, E. Aubert, A. Klier, and G. Rapoport.** 1987. Characterization of the *sacQ* genes from *Bacillus licheniformis* and *Bacillus subtilis*. *J. Bacteriol.* **169:**324–333.

2. **Ayusaw, D., Y. Yoneda, K. Yamane, and B. Maruo.** 1975. Pleiotropic phenomena in autolytic enzyme(s) content, flagellation, and simultaneous hyperproduction of extracellular α-amylase and protease in a *Bacillus subtilis* mutant. *J. Bacteriol.* **124:**459–469.

3. **Band, L., D. J. Henner, and M. R. Ruppen.** 1987. Construction and characterization of an intracellular serine protease mutant of *Bacillus subtilis*. *J. Bacteriol.* **169:**444–446.

4. Binnie, C., M. Lampe, and R. Losick. 1986. Gene coding the sigma factor from *Bacillus subtilis*. *Proc. Natl. Acad. Sci. USA* 83:5943–5947.

5. Chambert, R., and M.-F. Petit-Glatron. 1984. Hyperproduction of extracellular levansucrase by *Bacillus subtilis*: examination of the phenotype of a *sacU*(Hy) strain. *J. Gen. Microbiol.* 130:3143–3152.

6. Cozzone, A. J. 1988. Protein phosphorylation in prokaryotes. *Annu. Rev. Microbiol.* 42:97–125.

7. Dod, B., and G. Balassa. 1978. Spore control (*Sco*) mutations in *Bacillus subtilis*. III. Regulation of extracellular protease synthesis in the spore control mutations *ScoC*. *Mol. Gen. Genet.* 163:57–63.

8. Dubnau, E., J. Weir, G. Nair, L. Carter III, C. Moran, Jr., and I. Smith. 1988. *Bacillus* sporulation gene *spo0H* codes for σ^{30} (σ^{H}). *J. Bacteriol.* 170:1054–1062.

9. Ferrari, E., D. J. Henner, M. Perego, and J. A. Hoch. 1988. Transcription of *Bacillus subtilis* subtilisin in sporulation mutants. *J. Bacteriol.* 170:289–295.

10. Ferrari, E., S. M. H. Howard, and J. A. Hoch. 1986. Effect of stage 0 sporulation mutations on subtilisin expression. *J. Bacteriol.* 166:173–179.

11. Friedrich, M. J., and R. J. Kadner. 1987. Nucleotide sequence of the *uhp* region of *Escheichia coli*. *J. Bacteriol.* 169:3556–3563.

12. Gaur, N. K., K. Cabane, and I. Smith. 1988. Structure and expression of the *Bacillus subtilis sin* operon. *J. Bacteriol.* 170:1046–1053.

13. Gaur, N. K., E. Dubnau, and I. Smith. 1986. Characterization of a cloned *Bacillus subtilis* gene which inhibits sporulation in multiple copies. *J. Bacteriol.* 168:860–869.

14. Gennaro, M. L., and R. P. Novick. 1988. An enhancer of DNA replication. *J. Bacteriol.* 170:5709–5717.

15. Gicquel-Sanzey, B., and P. Cossart. 1982. Homologies between different prokaryotic DNA-binding regulatory proteins and between their sites of action. *EMBO J.* 5:591–595.

16. Helmann, J. D., and M. J. Chamberlin. 1988. Structure and function of bacterial sigma factors. *Annu. Rev. Biochem.* 57:839–872.

17. Henikoff, S., G. W. Haughn, J. M. Calvo, and J. C. Wallace. 1988. A large family of bacterial activator proteins. *Proc. Natl. Acad. Sci. USA* 85:6602–6606.

18. Henner, D. J., E. Ferrari, M. Perego, and J. A. Hoch. 1988. Location of the targets of the *hpr-97*, *sacU32*(Hy), and *sacQ36*(Hy) mutations in upstream regions of the subtilisin promoter. *J. Bacteriol.* 170:296–300.

19. Henner, D. J., M. Yang, L. Band, H. Shimotsu, M. Ruppen, and E. Ferrari. 1986. Genes of *Bacillus subtilis* which regulate the expression of degradative enzymes, p. 81–90. *In* M. Alacevic, D. Hranueli, and Z. Toman (ed.), *Genetics of Industrial Microorganisms*. Proceedings of the Fifth International Symposium on the Genetics of Industrial Micro-

organisms. Ognjen Prica Printing Works, Karlovac, Yugoslavia.

20. Henner, D. J., M. Yang, and E. Ferrari. 1988. Localization of *Bacillus subtilis sacU*(Hy) mutations to two linked genes with similarities to the conserved procaryotic family of two-component signalling systems. *J. Bacteriol.* 170:5102–5109.

21. Higerd, T. B., J. A. Hoch, and J. Spizizen. 1972. Hyperprotease-producing mutants of *Bacillus subtilis*. *J. Bacteriol.* 122:1026–1028.

22. Hoch, J. A. 1976. Genetics of bacterial sporulation. *Adv. Genet.* 18:69–98.

23. Ikemura, H., and M. Inouye. 1988. In vitro processing of pro-subtilisin produced in *Escherichia coli*. *J. Biol. Chem.* 263:12959–12963.

24. Ikemura, H., H. Takagi, and M. Inouye. 1987. Requirement of prosequence for the production of active subtilisin E in *Escherichia coli*. *J. Biol. Chem.* 262:7859–7864.

25. Ito, J., and J. Spizizen. 1973. Genetic studies of catabolite repression insensitive sporulation mutants of *Bacillus subtilis*. *Colloq. Int. CNRS* 227:81–82.

26. Jeang, K.-T., and G. Khoury. The mechanistic role of enhancer elements in eukaryotic transcription. *Bioassays* 8:104–107.

27. Kofoid, E. C., and J. S. Parkinson. 1988. Transmitter and receiver modules in bacterial signalling proteins. *Proc. Natl. Acad. Sci. USA* 85:4981–4985.

28. Kunst, F., M. Debarbouille, T. Msadek, M. Young, C. Mauel, D. Karamata, A. Klier, G. Rapoport, and R. Dedonder. 1988. Deduced polypeptides encoded by the *Bacillus subtilis sacU* locus share homology with two-component sensor-regulator systems. *J. Bacteriol.* 170:5093–5101.

29. Kunst, F., M. Pascal, J. Lepesant-Kejzlarova, J. A. Lepesant, A. Billault, and R. Dedonder. 1974. Pleiotropic mutations affecting sporulation conditions and the synthesis of extracellular enzymes in *Bacillus subtilis*. *Biochimie* 56:1481–1489.

30. Lepesant, J. A., F. Kunst, J. Lepesant-Kejzlarova, and R. Dedonder. 1972. Chromosomal location of mutations affecting sucrose metabolism in *Bacillus subtilis* Marburg. *Mol. Gen. Genet.* 118:135–160.

31. Lepesant, J. A., F. Kunst, M. Pascal, J. Lepesant-Kejzlarova, M. Steinmetz, and R. Dedonder. 1976. Specific and pleiotropic regulatory mechanisms in the sucrose system of *Bacillus subtilis*, p. 58–69. *In* D. Schlessinger (ed.), *Microbiology—1976*. American Society for Microbiology, Washington, D.C.

32. Losick, R., P. Youngman, and P. J. Piggot. 1986. Genetics of endospore formation in *Bacillus subtilis*. *Annu. Rev. Genet.* 20:625–669.

33. Magasanik, B. 1988. Reversible phosphorylation of an enhancer binding protein regulates the transcription of bacterial nitrogen utilization genes. *Trends Biol. Sci.* 13:475–479.

34. **McAllister, C. F., and E. C. Achberger.** 1988. Effect of polyadenine-containing curved DNA on promoter utilization in *Bacillus subtilis. J. Biol. Chem.* **263:**11743–11749.

35. **Nagami, Y., and T. Tanaka.** 1986. Molecular cloning and nucleotide sequence of a DNA fragment from *Bacillus natto* that enhances production of extracellular proteases and levansucrase in *Bacillus subtilis. J. Bacteriol.* **166:**20–28.

36. **Perego, M., and J. A. Hoch.** 1988. Sequence analysis and regulation of the *hpr* locus, a regulatory gene for protease production and sporulation in *Bacillus subtilis. J. Bacteriol.* **170:**2560–2567.

37. **Perego, M., G. B. Spiegelman, and J. A. Hoch.** 1988. Structure of the gene for the transition state regulator, AbrB: regulator synthesis is controlled by the Spo0A protein. *Mol. Microbiol.* **2:**689–699.

38. **Piggot, P. J., and J. G. Coote.** 1976. Genetics aspects of bacterial endospore formation. *Bacteriol. Rev.* **40:**908–962.

39. **Plaskon, R. R., and R. M. Wartell.** 1987. Sequence distributions associated with DNA curvature are found upstream of strong *E. coli* promoter. *Nucleic Acids Res.* **15:**785–796.

40. **Power, S. D., R. M. Adams, and J. A. Wells.** 1986. Secretion and autoproteolityc maturation of subtilisin. *Proc. Natl. Acad. Sci. USA* **83:**3096–3100.

41. **Ptashne, M.** 1986. Gene regulation by proteins acting nearby and at a distance. *Nature* (London) **322:**697–701.

42. **Ptashne, M.** 1988. How eukaryotic transcriptional activators work. *Nature* (London) **335:**683–689.

43. **Ronson, C. W., B. T. Nixon, and F. M. Ausubel.** 1987. Conserved domains in bacterial regulatory proteins that respond to environmental stimuli. *Cell* **49:**579–581.

44. **Ruppen, M. E., G. L. Van Alstine, and L. Band.** 1988. Control of intracellular serine protease expression in *Bacillus subtilis. J. Bacteriol.* **170:**136–140.

45. **Ryder, K., S. Silver, A. L. DeLucia, E. Fanning, and P. Tegtmeyer.** 1986. An altered DNA conformation in origin region I is a determinant for the binding of SV40 large T antigen. *Cell* **44:**719–725.

46. **Sadaie, Y., and T. Kada.** 1985. *Bacillus subtilis* gene involved in cell division, sporulation, and exoenzyme secretion. *J. Bacteriol.* **163:**648–653.

47. **Sekiguchi, J., N. Takada, and H. Okada.** 1975. Genes affecting the productivity of α-amylase in *Bacillus subtilis* Marburg. *Biochimie* **53:**1059–1066.

48. **Shimotsu, H., and D. J. Henner.** 1986. Construction of a single-copy integration vector and its use in analysis of regulation of the *trp* operon of *Bacillus subtilis. Gene* **43:**85–94.

49. **Shimotsu, H., and D. J. Henner.** 1986. Modulation of *Bacillus subtilis* levansucrase gene expression by sucrose and regulation of the steady-state mRNA level by *sacU* and *sacQ* genes. *J. Bacteriol.* **168:**330–338.

50. **Sloma, A., A. Ally, D. Ally, and J. Pero.** 1988. Gene encoding a minor extracellular protease in *Bacillus subtilis. J. Bacteriol.* **170:**5557–5563.

51. **Stahl, M. L., and E. Ferrari.** 1984. Replacement of the *Bacillus subtilis* subtilisin structural gene with an in vitro-derived deletion mutation. *J. Bacteriol.* **158:**411–418.

52. **Steinmetz, M., F. Kunst, and R. Dedonder.** 1976. Mapping of mutations affecting synthesis of exocellular enzymes in *Bacillus subtilis. Mol. Gen. Genet.* **148:**281–285.

53. **Tanaka, T., and M. Kawata.** 1988. Cloning and characterization of *Bacillus subtilis iep*, which has positive and negative effects on production of extracellular proteases. *J. Bacteriol.* **170:**3593–3600.

54. **Tanaka, T., M. Kawata, Y. Nagami, and H. Uchiyama.** 1987. *prtR* enhances the mRNA level of the *Bacillus subtilis* extracellular proteases. *J. Bacteriol.* **169:**3044–3050.

55. **Tomioka, N., M. Honjo, K. Funakoshi, K. Masnabe, A. Akakoa, I. Mita, and Y. Furutani.** 1985. Cloning, sequencing and some properties of a novel *Bacillus amyloliquefaciens* gene involved in the increase of extracellular protease activities. *J. Biotechnol.* **3:**85–96.

56. **Trowsdale, J., S. M. H. Chen, and J. A. Hoch.** 1979. Genetic analysis of a class of polymyxin resistant partial revertants of stage 0 sporulation mutants of *Bacillus subtilis*: map of the chromosome region near the origin of replication. *Mol. Gen. Genet.* **173:**61–70.

57. **Wells, J. A., and D. A. Estell.** 1988. Subtilisin—an enzyme designed to be engineered. *Trends Biol. Sci.* **13:**291–297.

58. **Weston, L. A., and R. J. Kadner.** 1988. Role of *uhp* genes in expression of the *Escherichia coli* sugar-phosphate transport system. *J. Bacteriol.* **170:**3375–3383.

59. **Wong, S.-L., C. W. Price, D. S. Goldfarb, and R. H. Doi.** 1984. The subtilisin E gene of *Bacillus subtilis* is transcribed from a σ^B promoter in vivo. *Proc. Natl. Acad. Sci. USA* **81:**1184–1188.

60. **Wong, S.-L., L.-F. Wang, and R. H. Doi.** 1988. Cloning and nucleotide sequence of *senN*, a novel *Bacillus natto* (*B. subtilis*) gene that regulates expression of extracellular protein genes. *J. Gen. Microbiol.* **134:**3269–3276.

61. **Wu, H.-M., and D. M. Crothers.** 1984. The locus of sequence-directed and protein induced DNA bending. *Nature* (London) **305:**509–513.

62. **Yang, M., E. Ferrari, E. Chen, and D. J. Henner.** 1986. Identification of the pleiotropic *sacQ* gene of *Bacillus subtilis. J. Bacteriol.* **166:**113–119.

63. **Yang, M., H. Shimotsu, E. Ferrari, and D. J. Henner.** 1987. Characterization and mapping of the *Bacillus subtilis prtR* gene. *J. Bacteriol.* **169**:434–437.

64. **Yang, M. Y., E. Ferrari, and D. J. Henner.** 1984. Cloning of the neutral protease gene of *Bacillus subtilis* and the use of the cloned gene to create an in vitro-derived deletion mutation. *J. Bacteriol.* **160**:15–21.

65. **Zuber, P., and R. Losick.** 1987. Role of AbrB in Spo0A- and Spo0B-dependent utilization of a sporulation promoter in *Bacillus subtilis. J. Bacteriol.* **169**:2222–2230.

Regulation of Procaryotic Development
Edited by Issar Smith, Ralph A. Slepecky, and Peter Setlow
1989 American Society for Microbiology, Washington, DC 20006

Chapter 7

The Competence Regulon of *Bacillus subtilis*

David Dubnau

Competence may be defined as a physiological state which enables a bacterial cell to bind and internalize macromolecular DNA. The end product of this process depends on the genetic endowment of the DNA. If the internalized DNA is homologous to the resident bacterial chromosome or to a resident plasmid, integration into either replicon may occur. If plasmid or bacteriophage DNA is taken up, it can be established as an independently replicating entity. These processes, referred to as transformation (or as transfection in the case of bacteriophage DNA), have proven to be useful for a variety of purposes, including genetic mapping and molecular cloning experiments. We are dealing here with genetically encoded natural competence, not with the artificial competence procedure used, for instance, in *Escherichia coli*.

In *Bacillus subtilis*, competence develops after the transition from steady-state exponential growth to the stationary phase and only in minimal salts-glucose-based media. One focus of interest in the study of competence, therefore, lies in the analysis of its regulation as an example of a late-growth controlled and nutritionally regulated global system of gene expression. A second interest resides in the process of DNA transport. In this, the competence system may be compared with others, including bacterial conjugation and T-strand transfer in *Agrobacterium tumefaciens*. This review will focus on both aspects: on a description of the genes required for DNA transport and on the regulatory interactions governing the expression of these genes.

THE TRANSFORMATION PATHWAY: A DESCRIPTIVE ACCOUNT

Older work utilizing radioactive and density-labeled DNA has revealed the existence of a pathway by means of which transforming DNA is processed and internalized (reviewed in reference 15). High-molecular-weight extracellular DNA is first bound to the surface of the cell. This binding takes place at relatively few sites on each DNA molecule, and the bulk of the DNA is extended into the medium, where it is accessible to exogenous nucleases and susceptible to hydrodynamic shear. There are relatively few binding sites on each competent cell, probably about 40 to 50. The nature of the binding sites is completely obscure. It is not even known whether binding takes place on the cell wall or on the membrane. Unlike the situation in *Haemophilus* sp. (11) or *Neisseria gonorrhoeae* (23), in which binding and uptake exhibit sequence specificity, *B. subtilis* binds and takes up duplex DNA of any origin.

The surface-bound DNA then suffers double-strand cleavage. In *Streptococcus pneumoniae* this is preceded by the intro-

David Dubnau • Department of Microbiology, Public Health Research Institute, New York, New York 10016.

duction of single-strand nicks (34). Although the processing of DNA in these two organisms is generally quite similar, it is not known whether nicking precedes cleavage in *B. subtilis*. The distribution of cleavage sites suggests a pattern of limited random cleavage. If phage T7 DNA (40 kilobases) is bound, each molecule suffers about one cleavage to yield a number average molecular size of about 20 kilobases (14). The distribution of fragment sizes suggests that the cleavage sites are random, but with a bias against cleavage near the molecular termini. When higher-molecular-weight DNA is bound, each molecule is cleaved several times, and the "smaller" end of the distribution resembles that observed with T7 DNA. It is not known what limits the cleavages. Perhaps the determining factors are geometric, depending on the spatial distributions of binding and cleavage sites on the cell surface and on the inherent stiffness of duplex DNA. The cleavage products are double stranded and remain on the cell surface, i.e., accessible to nuclease.

DNA is next converted to single strands and rendered inaccessible to nuclease. These two effects do not seem to be separated in time, leading to the speculation that they are concerted events and reflect the same process, namely, uptake. About one-half of the mass of bound and cleaved DNA is released into the medium as acid-soluble material. Lacks and Greenberg (35) have proposed that a membrane-localized nuclease degrades one strand, while the other intact strand enters the cell. This degradation may even provide energy for the transport event, although there is evidence that uptake requires maintenance of proton motive force (24) (see below, DNA Transport). On the other hand, it is also possible that conversion to single strands, degradation of one strand, and uptake are uncoupled from one another, but occur in quick succession. Single-stranded donor DNA can be recovered from the cell with about the same average length as that of the surface-localized double-stranded fragments. Either donor strand can be internalized, apparently

with equal probability. This single-stranded material has been shown to be a precursor of the DNA that is eventually integrated in the recipient chromosome (12). It is not accessible to external nucleases and is thus considered to be the first intracellular form on the transformation pathway. The first appearance of intracellular DNA occurs after a lag of about 2 min at 37°C. This is taken as the time required for binding, cleavage, and uptake. Strauss (74) has shown that linked marker pairs achieve DNase resistance more slowly than do single markers. This and other data (21) suggest that the DNA is transported across the membrane in a linear fashion, probably through a pore that forms during the development of competence. Integration yields a heteroduplex in which a donor strand and a recipient strand are paired. A discussion of the integration process and of the subsequent resolution of the heteroduplex is not directly relevant to the subject of this chapter and will not be pursued further.

Although little is known about competence proteins, there are clearly many genes involved in the uptake, processing, and integration of transforming DNA. Those involved in the latter event are referred to as *rec* genes. Several have been identified in *B. subtilis*, and the best characterized is *recE*, the bacillus analog of the *E. coli recA* gene (41). The RecE protein is present throughout growth, but increases in concentration in competent cells (see below, The *din* Genes).

PHYSIOLOGY OF THE COMPETENT STATE

The production of high yields of competent cells in *B. subtilis* appears to require growth in minimal salts-glucose media (4). This medium is usually supplemented with an amino acid mixture, e.g., casein hydrolysate, although a defined amino acid mixture that supports the development of competence has been described (83). The substitution of glycerol for glucose has no detectable effect on growth in competence medium, but results in a roughly 10-fold depression in competence (1). In addition, competence begins to develop as the culture passes the transition from exponential growth to the stationary phase (T_0). If the culture is allowed to continue shaking, competence will peak after about 2 h and then slowly decline (1). Most competence protocols, however, involve the dilution of a culture at T_0 into fresh medium, with optimal competence developing 1 or 2 h later, implying that commmitment to competence has occurred by T_0.

A poorly understood feature of competence in B. subtilis is the fact that it develops in only a subfraction (10 to 20%) of the culture. The competent cells can be resolved by isopycnic equilibrium centrifugation on gradients of Renografin, an iodinated polysaccharide (26, 29). Competent cells are not only lighter but also smaller than noncompetent cells and are uninucleate (67).

Competent cells are relatively dormant with respect to DNA and RNA syntheses (13, 50). Whereas the noncompetent fraction exhibits some residual incorporation of radioactive precursors into DNA, the competent cells are relatively inactive. This dormancy is reflected in a longer growth lag for component cells during the "escape" from competence after dilution into fresh medium (50). The cause and nature of this dormancy are not known, although Loveday (42) has suggested that the block is in nucleotide metabolism rather than in replication.

Dooley et al. (13) have reported that cells destined to become competent exhibit a decreased rate of nucleic acid synthesis at least 1 h before transformability appears. Shortly thereafter, these cells become irreversibly committed to competence, even when diluted into fresh medium. The committed cells do not yet exhibit a light buoyant density in Renografin. The latter property appears later and in experiments reported by Dooley et al. precedes the development of competence. It is interesting to consider how cells may begin to alter their pattern of gene expression in this way, decreasing their rates of DNA and stable RNA syntheses at a time when the bulk culture is in a state of apparently steady-state exponential growth. One possibility is that a condition of incipient starvation exists, in which a decreased concentration of some critical nutrient is detected. Another is that accumulation of a factor, possibly extracellular, triggers development along the competence pathway. A precedent for this type of control is provided by the extracellular sporulation factors detected by Grossman and Losick (25), as well as the well-defined competence factor in S. pneumoniae (54, 76). It is possible that an extracellular competence factor also exists in B. subtilis (32), but definitive evidence for this is lacking.

GENETIC NOMENCLATURE

Several genes that affect the development of competence have been characterized on the basis of phenotypes unrelated to competence (e.g., spo0H, spo0A, and degU). These naturally retain their original names, at least for the present. Genes initially identified as affecting transformation receive the designation com together with a letter to identify the locus (e.g., comA and comB). Figure 1 presents a map of the B. subtilis chromosome with loci identified that are known to affect competence. The assignment of com loci is by agreement with G. Venema and M. Polsinelli. Not included in the map are the rec loci. Although required in most cases for successful integrational transformation, these genes will be excluded from this discussion. An exception is recE, which can be regulated as part of the competence system. This case serves as a warning that the exclusion of rec genes is arbitrary and will very likely prove to be unwarranted as more information concerning their regulation is obtained.

REGULATION OF COMPETENCE: INTRODUCTION

The Tn917lac element constructed by Perkins et al. (87) has been used to isolate

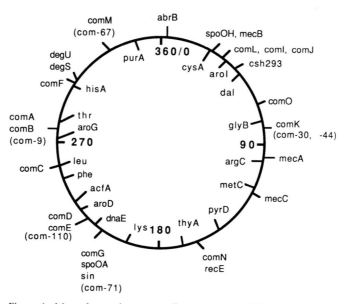

Figure 1. Map of genes known to affect competence. The genes and mutations on the outer rim have been reported to affect competence. Those within parentheses are mutations that have not been assigned to particular loci since their relationships to nearby mutations are uncertain. The sources for the various loci and mutations are given in the text.

com mutants (27). These were recognized on the basis of a plate test for competence. The presence of the promoterless *lacZ* gene permitted an analysis of the transcriptional regulation of the inactivated genes (1). Several *com* genes were defined in this manner. All were deficient in competence but appeared to be normally transducible and mitomycin C resistant (hence, Rec⁺). In addition, based on semiquantitative plate tests, they all seemed to sporulate normally and to produce essentially wild-type levels of the late-growth products amylase, extracellular protease, and antibiotic.

About half of the insertion mutants were Lac⁺, indicating that the Tn*917lac* insertion had occurred in the proper orientation to drive β-galactosidase synthesis from a *com* promoter. Measurement of β-galactosidase levels during growth permitted the classification of these mutants into two groups on the basis of their times of expression.

PHENOTYPE OF "EARLY" MUTANTS

The first group of mutants exhibited levels of transformation about 1,000-fold lower than that of the wild type and expressed β-galactosidase throughout growth in all media tested (1, 27). These mutants expressed rather low levels of β-galactosidase, although well above the level of the endogenous *B. subtilis* enzyme. Cloning, DNA sequencing, and further characterization have revealed that they define two genes, *comA* and *comB*. Although as mentioned above the *comA* and *comB* fusions express β-galactosidase throughout growth, we have recently found that *comA* expression doubles at about T_{-2} and *comB* expression increases severalfold beginning at the same time (Y. Weinrauch and N. Guillen, unpublished data). Finally, the *comA* and *comB* mutants block the development of Renografin separation. The phenotypes of *comA* and *comB* mutants thus suggest that these genes act early in the competence pathway and that they may play a regulatory role.

PHENOTYPE OF "LATE" MUTANTS

A second group of mutants expressed β-galactosidase in competence medium but not in complex media and at a sharply accelerated rate beginning at about T_0, with very little expression evident during exponential growth. All of the mutants in this group were markedly deficient in transformation, depressed by at least 10^5- or 10^6-fold compared with the wild-type strain. These mutants, with the exception of a single insertion (*comG12*), separated in Renografin indistinguishably from the wild type. Remarkably, they expressed β-galactosidase preferentially in the light-density fraction, suggesting that the cognate *com* genes are regulated transcriptionally not only with regard to the growth stage and in response to the nutritional environment, but also in a cell type-specific manner. Cloning and sequencing of the later expressing mutants have revealed that they fall into four transcription units: *comC*, *comD*, *comE*, and *comG*. The organization of these loci will be described further below.

EPISTATIC INTERACTIONS

A survey of an isogenic set of *spo0* mutants has revealed that *spo0A* and *spo0H* mutants are depressed in competence by about 1,000- and 20-fold, respectively. The competence deficiency of *spo0A* mutants has been reported previously (57, 62, 72). Both *spo0A* and *spo0H* are required for separation in Renografin, consistent with the idea that they act early in the development of competence. The *abrB703* mutation restores transformation in the *spo0A* background to about 3% of the wild-type level (1, 77, 78). Interestingly, the *abrB* mutation by itself depresses transformation to this same level (3%). *spo0B*, *spo0E*, *spo0F*, and *spo0J* mutants were essentially normal in transformation, but the *spo0E* and *spo0J* mutants tested did not carry null mutations. Double mutants were constructed carrying *com-lacZ* fusions together with *spo0A* and *spo0H* deletion mutations. These revealed that *spo0A* and *spo0H* are required for the full expression of the late *com* genes, but not for the expression of *comA* and *comB* (1).

Similar experiments, using late *com-lacZ* fusions in *abrB703* backgrounds in the presence and absence of a *spo0A* deletion, showed that the expression of β-galactosidase in these strains parallels the effects on competence described above. Also, *abrB703* did not alter the growth stage-specific control of *com* gene expression. These results suggest that the *abrB* gene product may play a positive role in the development of competence, but is not required to repress the latter during exponential growth. It has been suggested that the *abrB* product may act negatively on certain genes that are normally turned on at the onset of the stationary phase and that the *spo0A* product may act to antagonize the action of the AbrB protein (45). Perego et al. (56) have shown that the transcription of *abrB* is indeed controlled negatively (either directly or indirectly) by the *spo0A* product. They have also inferred from the amino acid sequence of AbrB that the latter is a DNA-binding protein. The dependence of late *com* genes on *spo0H* for full expression implies that at least one essential product in the competence pathway is transcribed by a σ^H RNA polymerase holoenzyme. The *spo0A* gene product has recently been shown to be required for the full expression of *spo0H* (E. Dubnau, K. Cabane, and I. Smith, personal communication). However, the tighter dependence on *spo0A* of both competence and late *com* gene expression suggests that the *spo0A* gene product is required for at least one step in addition to the synthesis of σ^H. One possibility is the suppression of *abrB* overproduction. If so, the observations to date would imply that *abrB* can play both a positive and a negative role during the development of competence and that either overproduction (in a *spo0A* mutant) or underexpression (in an *abrB* mutant) is detrimental to the developmental of competence.

The fact that the dependence on *spo0H* is not absolute, when measured by the effect on either competence or the expression of late competence genes, suggests that some essential *com* gene(s) may be read by alternate holoenzyme forms. A *sigB* knockout mutant (7, 16) was found to express a normal level of competence (1). Triple mutants carrying deletions in *sigB* and *sigH* as well at late *com-lacZ* fusions revealed that the *sigB* mutation had no additional effect on β-galactosidase levels beyond that exerted by the *spo0H* mutation, suggesting that σ^B holoenzyme is not the alternate σ factor suggested above (1).

The *sin* product is believed to be a DNA-binding protein that can function as a positive transcriptional activator of some genes (like *aprE* and motility determinants) and as a negative regulator of sporulation (22). *sin* also appears to be required for competence, since knockouts of this gene are competence deficient (22). An examination of double mutants has revealed that the *sin* product is required for the expression of late *com* fusions (Y. Weinrauch, unpublished data).

We have also found that both *comA* and *comB* are required for the expression of all of the late *com* genes but not of one another (Weinrauch and Guillen, unpublished data). In addition, *comA* is required for the full expression of a gene needed for the production of a growth stage-regulated surfactin antibiotic (P. Zuber, personal communication). Our results confirm the notion that *comA* and *comB* are required early in a dependent competence pathway and increase the number of known gene products required for the full expression of competence and of late *com* genes (*abrB*, *sin*, *spo0A*, *spo0H*, *comA*, and *comB*).

It is probable that this list will grow. Likely additions are *degS* and *degU*. These are two open reading frames (ORFs) that to-

gether constitute what was previously known as the *sacU* locus (29, 33). Mutations in both of these ORFs can increase the synthesis of certain extracellular enzymes [*degS*(Hy) and *degU*(Hy)]. These mutations, which also decrease competence markedly, are now known to be missense mutations (29). Interestingly, knockout mutations, at least in *degU*, also result in competence deficiency (75). Since *degS* and *degU* are possibly cotranscribed, it is not yet possible to test *degS* knockouts for their effect on competence, since polarity effects might complicate the results. In any event it appears that one or both of these genes are required for competence and possibly will prove to be epistatic on the late *com* genes.

ORGANIZATION AND SEQUENCE OF *comA* AND *comB*

comA and *comB* have been cloned, sequenced, and analyzed by S1 nuclease mapping (Weinrauch, unpublished data). They are separated from one another on the chromosome by about 3.5 kilobase pairs. High-resolution transcriptional mapping reveals that *comB* is read from a single major promoter that resembles a consensus σ^A promoter, except that the −35 to −10 spacing is not optimal (Fig. 2). *comA* is read from a similar putative σ^A promoter (Fig. 2) and consists of two cotranscribed ORFs. It has been shown that ORF1 is required for competence, while ORF2 is not (Weinrauch, unpublished data). Just upstream from the *comA* promoter is another ORF which has been partially sequenced. Insertion of a chloramphenicol resistance casette into this ORF results in competence deficiency when this construct replaces the chromosomal wild-type sequence. Thus, a new *com* gene (provisionally called *comP*) has been defined. Weinrauch (unpublished data) has recently found that this gene is epistatic on expression of the late *com* genes. Thus, the list of early genes has now grown to nine, if both *degS* and *degU* are included.

	-35 REGION		-10 REGION
(A)			
comA	TTGGCA	N 16	TATAAA
comB	TAGACA	N 20	TATACT
comG	TTGATT	N 17	TACAAT
consensus	TTGACA	N 17-19	TATAAT
(B)			
comC	GCTATCCAT	N 13	GGGCTTATCCT
comE (P3)	ACAAAACAC	N 13	GGAATTATTTG
spoVG	GCAGGATT	N 13	GGAATTGATAC
rpoD (P3)	GCAGGATT	N 12	AGAATTACTCT
spoOF (P1)	AAAGGAAAT	N 12	AGAATACATAC

Figure 2. Comparison of *com* promoters with σ^A and σ^H promoters. (A) The *comA*, *comB*, and *comG* promoters, identified by high-resolution S1 mapping (Y. Weinrauch, N. Guillen, and M. Albano, unpublished data), are shown together with the consensus σ^A sequence. (B) The *comC* and major *comE* promoter sequences, identified by high-resolution S1 mapping (Mohan, unpublished data) and primer extension (Kozlov and Predich, unpublished data), respectively, are shown together with sequences of the *spoVG* (5), *rpoD* (P3) (10), and *spoOF* (37) promoters. The last three are known to be read by σ^H.

The predicted protein sequence of the *comA* gene product was found to be similar to that of several members of the effector class of procaryotic signal transducers (reviewed in reference 60). Figure 3 shows this relationship. Particularly striking is the relationship to the *degU* protein and to ORF 2 of the *uvrC* locus of *E. coli*. It is noteworthy that the ComA protein appears to contain three highly conserved aspartic acid residues in the N-terminal portion of the protein that are putative targets for phosphorylation (52, 82). Also present within the predicted *comA* sequence is a possible helix-turn-helix determining sequence (Fig. 4). The similarity to the effector proteins, together with the putative helix-turn-helix sequence, suggests that the ComA protein may relay information about the nutritional or growth status to the transcriptional machinery of the cell. The identity of the "sensor" protein that may interact with the *comA* product is unknown. Two candidates come to mind, on the basis of their derived amino acid sequences, which resemble known proteins of the sensor class. The first is the *degS* product (29, 33) discussed above. The second is the *spoIIJ* protein (P. Stragier, personal communication). Weinrauch (personal communication) has tested PY290, a Tn917 insertion in *spoIIJ*, for competence in a BD630 background and found no effect when compared with the isogenic *spoIIJ*⁺ parent. However, PY290 exhibits an extremely leaky Spo⁻ phenotype and possibly does not carry a *spoIIJ* null muta-

Figure 3. Comparison of the *comA* amino acid sequence with those of several other effectors, probable members of the two-component procaryotic regulator family of proteins. *virG* was selected as a well-established member of this family (84). *degU* (29) and ORF2 of *uvrC* (49, 65) were selected because of their close similarity to *comA* and because of their possible relevance to competence. Identical and chemically similar residues are shaded. The latter are defined as follows: (V, I, M, L), (G, A, S, T), (F, W, Y), (E, D), (Q, N), (H, R, K), (P), (C). The sequences were aligned by the FASTP and FASTA programs (38, 55). The figure includes the N-terminal 120 residues of the 214-amino-acid *comA* protein.

```
lambda-cI      QESVADKMGMGQSGVGALFN
ampR           FTHAAIELNVTHSAISQHVK
lysR           LTEAAHLLHTSQPTVSRELA
ilvY           FGRSARAMHVSPSTLSRQIQ
metR           LAAAAAVLHQTQSALSHQFS
tfdO           VGAAARRLHISQPPVTRQIH
comA           NQEIADALHLSKRSIEYSLT
               ****** ********** *
consensus      ppphGp-hGhppppIpph-p
                   A      H       V
```

Figure 4. Possible helix-turn-helix sequence in *comA*. Amino acids 170 to 189 from the C-terminal region of *comA* are compared with a group of probable helix-turn-helix DNA-binding sequences recently compiled (28). The sequences selected for this comparison differ from most of the previously recognized helix-turn-helix motifs, such as that of the λ *cI* repressor (53), most notably in the presence of an H residue in place of the conserved G residue. Also shown is a helix-turn-helix consensus, constructed from published compilations (28, 53, 66). p and h represent polar and hydrophobic residues, respectively. The asterisks denote positions in which the *comA* sequence conforms to the consensus.

tion. Furthermore, this insertion had no inhibitory effect on the expression of a *comG-lacZ* fusion, suggesting that *spoIIJ* is not required for the development of competence.

PROMOTERS OF THE LATE *com* GENES

The *comC*, *comE*, and *comG* loci have been cloned, fully or partially sequenced, and analyzed by high-resolution S1 mapping or primer extension. The promoter of *comC* drives transcription of a single ORF and most closely resembles a σ^H promoter (Fig. 2) (S. Mohan, unpublished data). *comE* appears to be driven by at least three promoters. The strongest of these likewise resembles a σ^H promoter (Fig. 2) (Y. Kozlof and M. Predich, unpublished data). These assignments are plausible, since in vivo transcription of these two loci is *spo0H* dependent. However, they must be regarded as extremely tentative in the absence of further genetic and biochemical evidence. On the other hand, the *comG* promoter appears to be a nearly consensus σ^A promoter (Fig. 2) (M. Albano, unpublished data). This raised the question of the nature of the control that determines the expression of *comG* only after T_0 and in certain media. Clearly this control may be either positive or negative. A fragment bearing the promoter of *comG* has been placed in *trans* on a multicopy plasmid. This plasmid reduces competence as well as the expression of *lacZ* fusions to *comC*, *comG*, *comD*, and *comE* by about 10-fold (Albano, unpublished data). This suggests that these loci are under positive control and that the *comG* promoter fragment is titrating a common factor. If the interpretations of σ specificity from promoter sequences given above are correct, then an interesting possibility (one among many) may be considered: that a common factor may serve to regulate promoters with different σ specificities. The reason for the in vivo σ^H dependence of *comG* espression is unknown; possibly the putative-positive factor requires σ^H for its synthesis.

WHAT TRIGGERS COMPETENCE?

We have explored the nature of the nutritional signal for competence in preliminary fashion (1). The substitution of glucose by other carbon sources in competence medium sharply depresses both the development of competence and the expression of late *com* genes. This is true even in the case of growth on glycerol, an excellent carbon source on which growth in competence medium is indistiguishable from growth on the same medium containing glucose. This latter behavior is reminiscent of the expression of *ctc*, a growth stage-regulated gene which is also dependent on the presence of glucose (31). The expression of *ctc* occurs in the absence of glucose when the *citB* (aconitase) gene is impaired by mutation (31), suggesting that *ctc* expression is normally repressed by an intermediate in the tricarboxylate cycle, which is in turn negatively regulated by glucose. However, this was not true of the late *com* genes, which depended for their expression on the presence of glucose, whether or not *citB* was intact (1).

On the other hand, the expression of the late *com* genes was severely depressed by the addition of glutamine to the competence medium (1). Glutamine is an excellent N source for *B. subtilis*. These effects raise the possibility that the nutritional signal for competence-specific expression involved starvation for an N source and possibly involves the ratio of C to N. In contrast, the expression of *comA* and *comB* was unaffected by the presence or absence of glutamine.

The relationship between the nutritional and growth stage-specific regulations of competence was explored genetically (unpublished data). A strain carrying a *comG* transcriptional fusion was mutagenized with ethyl methanesulfonate and plated on a complex β-galactosidase indicator medium, on which the parent strain fusion does not express. Colonies were selected that expressed β-galactosidase. These carried mutations are called *mec* (for medium-independent expression of competence). They have so far been mapped to three genetic loci (Fig. 1). In the cases of *mecA* and *mecB*, we have found that in liquid cultures the expression of β-galactosidase occurs in complex media. This is unlike the Mec$^+$ strain, in which expression occurs only in competence medium. In fact, the expression of the *comG* fusion in the *mec* backgrounds is identical in the two kinds of media. Also, expression is still subject to growth stage-specific control (*mecC* has not been tested). Thus, the nutritional and stage-specific controls may be distinguished genetically. These effects were reproduced when the remaining late *com* fusions were tested. Thus, the *mec* system mediates the nutritional regulation of all of the known late competence genes. Perhaps the *mec* products are required to process the nutritional signals that trigger the expression of *com* genes.

It was pointed out above that there are multiple factors presumably required for the transcriptional activation of the competence regulon. Several roles for these factors may be envisaged: sigma factors, direct transcriptional activators, sensors of nutritional and growth status, etc. Particularly interesting is the observation that three apparent members of the procaryotic effector family (*spo0A*, *comA*, and *degU*) have been implicated in competence, and only one possible sensor (*degS*) has been identified (although direct evidence that *degS* is required is lacking). What is more, the responding systems overlap extensively in their dependence on the various factors: sporulation, competence, and synthesis of extracellular enzymes and surfactin. Do the effectors act at a common point, at different but simultaneous targets, or sequentially? Do they respond to the same sensors, are they coupled with single sensors in a one-to-one relationship, or can they each respond to a host of sensors so as to funnel a variety of types of information into a general stress response system? The last is a real possibility especially in view of the cross talk detected between other two-component regulatory systems (51). These are some of the fascinating questions that will be addressed by late-growth aficionados in the future.

COMPONENTS OF THE COMPETENCE MACHINERY: INTRODUCTION

The next section of this review will focus on competence genes that are probably directly involved in the binding, uptake, and processing of extracellular DNA. In only a couple of cases is there even a hint as to the functional role of a given protein in these processes. In several cases, however, there is suggestive evidence that a particular gene product is a component of the competence machinery. This evidence may consist of data concerning the pattern of expression of a given product (i.e., expressed beginning at T_0), the nature of the predicted amino acid sequence (i.e., highly hydrophobic, suggesting a membrane localization), and the absence of an epistatic effect on other *com* genes. Although this type of information is suggestive of a role in binding and uptake, it hardly constitutes rigorous evidence. It is certainly possible that a protein satisfying some or all of these criteria will play a regulatory rather than a functional role or may even play both types of roles.

SURVEY OF COMPETENCE MUTANTS POSSIBLY IMPAIRED IN BINDING AND UPTAKE

Several competence-specific proteins have been described, for which functions can be suggested. Eisenstadt et al. (17) have partially purified a single-stranded binding protein from competent cells. This polypeptide was absent from noncompetent cells and was assayed by the protection from nuclease action it conferred on single-stranded DNA. Particularly interesting was the observation that this protein was not expressed in a background carrying an uncharacterized, noncompetent pleiotropic *spo* mutation, presumably an allele of *spo0A*. It is reasonable to postulate a role for a single-strained binding protein in the transformation process: to protect DNA, to facilitate recombination, etc. Unfortunately no genetic evidence exists concerning the essentiality of this protein.

Rosenthal and Lacks (61) have characterized a membrane-associated endonuclease in *S. pneumoniae* that is required for DNA uptake. A possible *B. subtilis* analog was discovered by Smith and co-workers, who have characterized a 75-kilodalton (kDa) competence-specific, membrane-localized protein complex that contains a 17-kDa nuclease as well as an 18-kDa protein (68, 69). It was suggested that the complex is an $\alpha_2\beta_2$-type multimer required for DNA binding and entry. Both polypeptides were required for efficient binding to DNA in vitro, whereas nuclease activity was expressed by both the isolated 17-kDa protein and the 75-kDa complex. The nuclease not only rendered material acid soluble, but also converted DNA to the single-stranded form (70). The 18-kDa protein seemed to limit the activity of the nuclease in in vitro assays. Recently, Vosman et al. have cloned, sequenced, and inactivated

the structural genes for these proteins (79). G. Venema suggests that the nuclease gene be called *comI* and that the 18-kDa gene be called *comJ* (personal communication) (Fig. 1). The two proteins are encoded by overlapping ORFs, with the nuclease cistron upstream from the 18-kDa gene. Inactivation of the nuclease ORF resulted in the loss of both proteins, presumably by a polarity effect, implying that they are cotranscribed and possibly also translationally coupled. This mutant exhibited 5% residual transformation. The total DNA associated with the cell was actually increased, while the amount of DNA taken up was reduced to about one-third of the wild-type level. The augmented apparent binding was presumably due to a reduction of the degradation to acid-soluble material that normally accompanies DNA uptake (see above). These results suggest a role for the nuclease and possibly for the 18-kDa protein as well, but also imply that an alternative uptake pathway exists. Inactivation of the 18-kDa gene resulted in the loss only of this protein. In this case the residual transformation activity was 25% of the wild-type level, and binding was unimpaired. The authors suggest that the 18-kDa protein may function in vivo to limit the action of the nuclease (as it does in vitro), preventing an undue reduction in the molecular weight of transforming DNA.

Fani et al. (20) have isolated four *B. subtilis com* mutants after nitrosoguanidine mutagenesis. The mutations were mapped and their locations are shown in Fig. 1. It is possible that *com-9* and *com-71* are located in *comA* or *comB* and in *comG*, respectively, and new locus numbers have therefore not been assigned. The same is true of *com-30*, which may be located in *comK*, the basis of its map position. The *com-104* mutation appears to define a new gene and is assigned the designation *comM104*. These four mutants were reduced in transformation by 3 to 5 orders of magnitude, but were not reduced in transduction or protoplast transformation. Three of the mutants (*com-9*, *com-71*, and *comM104*) were deficient in binding. The fourth, *com-30*, was deficient in uptake, but bound nearly as much DNA as did the wild type.

Mastromei et al. have utilized Tn*917* as a mutagen to produce and isolate several competence mutants (G. Mastromei, C. Barberio, S. Pistolesi, and M. Polsinelli, personal communication). Although map locations were identified for these insertions, only two of them (*comN114* and *comO18*) define genes that are clearly different from those already discussed. The *comO18* mutant was strongly blocked in DNA binding and was depressed for transformation by about 10,000-fold. Interestingly, *comN114* was depressed by 200-fold in transformation, but bound and took up about two-thirds as much DNA as did the wild type. The series of mutations characterized by Fani et al. and by Mastromei et al. cannot yet be classified as to the time of expression or the nature of their role in competence. However, on the basis of phenotypes, at least *com-30* and *comN114* define functions that are good candidates for components of the competence machinery.

As described above, the use of Tn*917lac* has enabled us to define four loci that are expressed after T_0 and are dependent for their expression on a set of early *com* genes. *comC* and *comG* have been entirely sequenced, and *comE* has been partially sequenced. The last gene specifies a protein of greater than 70 kDa that appears to be quite hydrophobic (19) and probably contains several transmembrane segments (J. Hahn and Y. Kozlof, unpublished data). This protein is of particular interest since *comE* mutants are deficient in uptake, but appear to bind nearly normal amounts of DNA (27). The *comC* ORF specifies a protein of about 19 kDa, which is likewise hydrophobic (Mohan, unpublished data). There is some uncertainty concerning the precise translational start site of the *comC* product. The hydrophobicity profile suggests that *comC* may contain several transmembrane segments. A search of a translated version of the nucleic acid data base failed to turn up

any convincing similarities for either *comC* or *comE*.

A cluster of six Tn*917lac* insertions defines *comG*. These have been used to map the insertion points and to clone and sequence *comG* (Albano, unpublished data). This is a complex locus, comprising seven ORFs. Genetic tests and transcription mapping have proven that *comG* is an operon, with one major promoter driving transcription of all seven ORFs and at least one minor promoter located between ORF1 and ORF2. The predicted protein sequences are of some interest. ORF1 is largely hydrophilic, although it possesses a single domain that is relatively hydrophobic. This ORF shows amino acid sequence similarity to *virB* ORF11 of the *A. tumefaciens* Ti plasmids (58, 81) (Fig. 5). The latter is the only relatively hydrophilic member of a complex locus containing 11 ORFs. The *virB* ORF11 protein has been shown to be an ATPase and to be associated with the cytosolic face of the bacterial membrane (P. Christie and E. Nester, personal communication). The *comG* ORF1 amino acid sequence is particularly similar to that of *virB* ORF11 in a region that contains a putative type A nucleotide-binding sequence (80) (Fig. 5). This similarity to a Ti plasmid protein from an operon thought to be involved in the transfer of single-stranded DNA across the bacterial membrane is suggestive. A further similarity on the level of genetic organization is apparent from the fact that the Ti plasmid virulence system is responsive to a two-component signaling system, of which the effector *virG* is similar in amino acid sequence to *comA* (Fig. 3).

The second ORF of *comG* contains several possible transmembrane segments and shows no similarity to any protein in the data base. ORF3 through ORF7 show hydrophobicity profiles that are similar to one another, being largely hydrophilic except for a single striking N-terminal hydrophobic segment in each polypeptide that exhibits the potential to constitute a transmembrane region. ORF3, ORF4, and ORF5 (and ORF3 in particular)

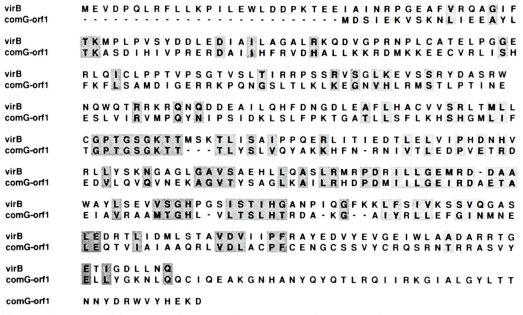

```
virB       M E V D P Q L R F L L K P I L E W L D D P K T E E I A I N R P G E A F V R Q A G I F
comG-orf1  - - - - - - - - - - - - - - - - - - - - - - - - - - - - M D S I E K V S K N L I E E A Y L

virB       T K M P L P V S Y D D L E D I A I L A G A L R K Q D V G P R N P L C A T E L P G G E
comG-orf1  T K A S D I H I V P R E R D A I I H F R V D H A L L K K R D M K K E E C V R L I S H

virB       R L Q I C L P P T V P S G T V S L T I R R P S S R V S G L K E V S S R Y D A S R W
comG-orf1  F K F L S A M D I G E R R K P Q N G S L T L K L K E G N V H L R M S T L P T I N E

virB       N Q W Q T R R K R Q N Q D D E A I L Q H F D N G D L E A F L H A C V V S R L T M L L
comG-orf1  E S L V I R V M P Q Y N I P S I D K L S L F P K T G A T L L S F L K H S H G M L I F

virB       C G P T G S G K T T M S K T L I S A I P P Q E R L I T I E D T L E L V I P H D N H V
comG-orf1  T G P T G S G K T T - - - T L Y S L V Q Y A K K H F N - R N I V T L E D P V E T R D

virB       R L L Y S K N G A G L G A V S A E H L L Q A S L R M R P D R I L L G E M R D - D A A
comG-orf1  E D V L Q V Q V N E K A G V T Y S A G L K A I L R H D P D M I I L G E I R D A E T A

virB       W A Y L S E V V S G H P G S I S T I H G A N P I Q G F K K L F S I V K S S V Q G A S
comG-orf1  E I A V R A A M T G H L - V L T S L H T R D A - K G - - A I Y R L L E F G I N M N E

virB       L E D R T L I D M L S T A V D V I I P F R A Y E D V Y E V G E I W L A A D A R R T G
comG-orf1  L E Q T V I A I A A Q R L V D L A C P F C E N G C S S V Y C R Q S R N T R R A S V Y

virB       E T I G D L L N Q
comG-orf1  E L L Y G K N L Q Q C I Q E A K G N H A N Y Q Y Q T L R Q I I R K G I A L G Y L T T

comG-orf1  N N Y D R W V Y H E K D
```

Figure 5. Comparison of the amino acid sequences of *comG* ORF1 and *virB* ORF11 from *A. tumefaciens* Ti plasmid pTiA6NC (81). The sequence of *virB* ORF11 from Ti plasmid pTiC58 (58) is also similar. Identical and chemically similar residues (defined as in the legend to Fig. 3) are shaded. The completely conserved GPTGSGKTT sequence corresponds to an A-type nucleotide-binding consensus (80). The R(M)II(L)G sequence, followed closely by a conserved D residue, corresponds to a possible B-type consensus (80). The sequences were aligned using the FASTP and FASTA programs (38, 55).

are similar to the N-terminal regions of a class of bacterial pilins found in *Pseudomonas aeruginosa*, *Bacteroides nodosus*, *N. gonorrhoeae*, and *Moraxella bovis*. These similarities are shown in Fig. 6. The conserved phenylalanine residue near the N terminus of these proteins is the N-terminal residue of the mature forms of these pilins and is always N methylated. This residue is conserved in the *comG* pilinlike proteins. It is interesting that the *Neisseria* and *Moraxella* pilins appear to be required for the full expression of competence in these organisms (8, 71; M. So, personal communication). The *Pseudomonas* pilin appears to assemble into a structure with fivefold symmetry and with a 1.2-nm pore in the center, wide enough to admit a single strand of DNA. It is possible that the pilinlike proteins may cooperate to form a pore capable of admitting single-stranded DNA.

The single *lacZ* fusion insertion in *comG*

ORF1 (*comG12*) prevents Renografin separation, although it expresses β-galactosidase at T_0. The remaining *comG* insertion mutants separate as does the wild type in Renografin, including one with an insertion in *comG* ORF2 (Albano, unpublished data). Thus, ORF1 appears to be uniquely required for Renografin separation, among the *comG* ORFs. This was proven by an experiment in which a single copy of the *comG* promoter, together with ORF1, was shown to complement *comG12* for Renografin separation, but not for competence. The latter was presumably because of polarity on downstream *comG* ORFs. ORF1 is also uniquely required among the *comG* ORFs for the full expression of two Tn917lac fusions to *comE*. This putative ATPase is therefore implicated in achieving the full expression of an unlinked gene and in promoting separation in Renografin. It is possible that the epistatic effect on *comE* is a secondary consequence of a

```
B.nodosus       M K S L Q K G F T L I E L M I V V A I I G I L A A I A I P Q Y Q N Y I A R S Q V
P.aeruginosa    - M K A Q K G F T L I E L M I V V A I I G I L A A I A I P Q Y Q N Y V A R S E G
N.gonorrheae    M N T L Q K G F T L I E L M I V I A I V G I L A A I A L P A Y Q D Y T A R A Q V
M.bovis         - M N A Q K G F T L I E L M I V I A I I G I L A A I A L P A Y Q D Y I S K S Q T
comG-orf3       - - M N E K G F T L V E M L I V L F I I S I L L L I T I P N V T K H N Q T I Q K
comG-orf4       K L N E E K G F T L L E S L L V L S L A S I L L V A V F T T L P P A T D N T A V
comG-orf5       M W R E N K G F S T I E T M S A L S L W L F V L L T V V P L W D K L M A D E K M
```

Figure 6. Comparison of pilins with the predicted amino acid sequences of *comG* ORF3, ORF4, and ORF5. Included are the N-terminal 38 to 40 amino acids of pilins from *P. aeruginosa* PAK (63), *B. nodosus* (18), *N. gonorrhoeae* (47), and *M. bovis* (46). Identical and chemically similar amino acids (defined as in the legend to Fig. 3) are shaded. The remaining portions of these proteins show little similarity (not shown). The sequences were aligned using the FASTP and FASTA programs (38, 55). The figure includes the N-terminal 38, 40, and 40 residues of the ORF3, ORF4, and ORF5 proteins, respectively. The entire proteins consist of 98, 143, and 115 amino acid residues, respectively.

dependence of *comE* expression on the prior development of the light-buoyant-density cells destined to become competent. This would then constitute a further example of morphological coupling (73). It is also noteworthy that *comG123* is not epistatic on the expression of *comG* or *comD*, implying that the step leading to Renografin separation is not required for the expression of the latter genes.

The *comG* data are subject to major uncertainty because of the polar nature of Tn*917* insertions. We can only be certain that two of the seven *comG* ORFs are required for competence: ORF1, because it is required uniquely for Renografin separation and the expression of *comE*, and ORF7, because it is the final ORF of the operon. It will be necessary to test each of the remaining five ORFs by introducing nonpolar mutations. R. Breitling (unpublished data) has shown in preliminary experiments that a nonpolar mutation in ORF3 renders a strain noncompetent.

DNA TRANSPORT

It seems certain that the elaborate competence apparatus includes hydrophobic protein components that assemble in the membrane, presumably to form a DNA transport machine. These components may include the products of *comC*, *comE*, *comG*, and the 75-kDa nuclease complex, as well as other undiscovered proteins. Understanding the topology of this machine and its function will be a formidable task. This is clear when one

considers that the mechanism of lactose transport, requiring a single membrane protein, is still not understood. Not only does the proposed competence machine probably include several components, but the known binding and processing steps that precede entry introduce a complexity that will render kinetic analysis difficult. It may be useful, however, to speculate about the kinds of processes that may be responsible for DNA transport. Two proposals have been published.

The Rosenthal and Lacks model posits a central role for a membrane nuclease (61). It is suggested that the nuclease forms an aqueous pore or channel, through which single-stranded DNA enters. An asymmetry in the position of the active site of the nuclease in the channel would ensure that only one strand is hydrolyzed as the DNA enters the cell. It is implicit in this model that the uptake process is driven by the action of the nuclease. As stated above, Lacks and co-workers have demonstrated that a membrane-localized nuclease is required for DNA entry in *S. pneumoniae* and that DNA enters the cell in a single-stranded form. Rosenthal and Lacks have also shown that the nuclease can be isolated as a rather large complex, presumably containing several copies of the nuclease monomer (61). The description of *B. subtilis* transformation already given is also consistent with the Rosenthal and Lacks model (DNA enters in a single-stranded form), and Venema and co-workers have identified a membrane-localized nuclease required for

DNA uptake in that organism. The model does not propose a role for the proton motive gradient or for other protein components of the competence machine and does not address the sensitivity of uptake to inhibitors of energy metabolism.

Grinius and co-workers (reviewed in reference 24) have proposed a different model. Using uncoupling agents in a variety of experiments, these investigators have concluded that DNA uptake in *B. subtilis* requires maintenance of both components of the proton motive force. Grinius proposed that the transforming DNA binds to proteins on the outside of the cell; this complex binds protons acquiring a positive charge; and the entire complex electrophoreses through an aqueous channel in the membrane. Once in the relatively alkaline cytoplasm, protons are released. This electrogenic transport model explains the requirement for both components of the proton motive force. Nieuwenhoven et al. (50a) have criticized aspects of the Grinius work, suggesting that the ΔpH alone provides the driving force. Mechanistically theirs is an electroneutral proton symport proposal for DNA entry, and its continued operation would require pumps to prevent the intracellular accumulation of protons. Neither model explicitly addresses the conversion of transforming DNA to the single-stranded form, but both are certainly compatible with a role for a membrane-localized nuclease in converting transforming DNA to single strands. An important criticism must be leveled at both the Grinius and Nieuwenhoven studies. In both cases, measurements of ΔpH and of $\Delta \psi$ were carried out on bulk competent cultures. However, in *B. subtilis* only 10 to 20% of the bulk culture consists of competent cells, and these are physiologically quite distinct from the noncompetent subpopulation (13). Measurements carried out directly on the competent fraction are therefore needed.

A further model may be worth considering. Sugar phosphates (e.g., glucose 6-phosphate [G-6-P]) are taken up by *Streptococ-*

cus lactis and *E. coli* in an anion-exchange reaction (2, 3, 44). It is suggested that, at a pH above the pK_2 for the sugar phosphate, this is a process in which two molecules of monovalent G-6-P enter and lose protons in the cell interior and a single divalent G-6-P molecule is expelled. The result is an electroneutral exchange, with the net uptake of one molecule of G-6-P. Of course, the continued operation of this process also depends on the expulsion of protons and may thus be (indirectly) driven by proton circulation. DNA is a sugar phosphate polymer. Perhaps duplex DNA enters an aqueous channel and encounters a nuclease (or perhaps a nuclease-helicase complex) and a nucleotide residue from one strand is cleaved and expelled. The process may thus be driven by anion exchange, but with continual coupling to proton circulation, in a manner analogous to the transport of G-6-P. For every two residues (base pair) entering the cell, one nucleotide would be expelled, presumably by reorientation of a membrane exchange center. This model is consistent with the experimental data marshaled in support of the Rosenthal and Lacks and Grinius proposals. It assigns a role for a membrane-associated nuclease and for the dependence of entry on proton motive force. An interesting difference from the Rosenthal and Lacks model lies in the implied location of the nuclease-active site with respect to the membrane. The fact that in *B. subtilis* and *S. pneumoniae* acid-soluble products from the degraded strand, including nucleotides, are found in the medium has been accepted as evidence that the entry nuclease is on the external face of the membrane. However, the anion-exchange model suggests that the nuclease-active site may be on the cytosolic face. This is a testable prediction.

THE *din* GENES

Love and co-workers have used Tn*917lac* to isolate insertions that express β-galactosidase after treatment with DNA-damaging agents such as mitomycin C (*din* mutants) (39). These insertions serve to identify genes that comprise members of the B. *subtilis* SOS regulon (or the SOB system [40, 86]). Also induced by DNA-damaging agents is the product of the *recE* gene, the B. *subtilis* equivalent of the E. *coli recA* gene (41, 43). Remarkably, it has been found that the expression of β-galactosidase from the *din* fusions occurs not only after treatment of the strains with mitomycin C but also when the mutant strains are grown to the competent state (39). What is more, this expression occurs preferentially in the light-buoyant-density fraction after separation in Renografin, yet the *din* mutants are not competence deficient. These results demonstrate that the SOB and competence systems of B. *subtilis* are closely related, at least as far as their regulation is concerned. Like the expression of the *din* genes in response to DNA damage, competence-induced expression requires the presence of a functional *recE* gene. This result was interpreted to mean that the competence- and DNA damage-dependent inductions, although triggered by different events, were in fact both dependent on some component(s) of the SOB system. In this respect the B. *subtilis* SOB system resembles the E. *coli* SOS system, whose expression depends on the *recA* gene product. The expression of the *recE* gene was studied using anti-RecA protein antibody, with which the B. *subtilis* protein cross-reacts (85). It was found that after entry into the competent state, the expression of the RecE protein is induced. The expression of the *recE* product induced by DNA damage requires that an intact *recE* gene product be present. This, however, is not true of the competence-induced expression of the mutant *recE4* product. The mutant protein was induced by growth to the competent state, but not by DNA damage, suggesting that the *recE* gene is under dual control and that its induction in the competent state is not due simply to competence-associated DNA damage.

Yasbin et al. (85) also determined the effect of deletion mutations in *spo0A* and *spo0H* on competence-induced expression of the RecE protein. First, they reported that residual transformation in the *spo0A* strain was 20% of the wild-type level, while that in the *spo0H* strain was was 2% of the wild-type level. In view of the well-documented severe competence deficiency of *spo0A* mutants (1, 57, 62, 72, 77), it is likely that this result may have been due to the inadvertent accumulation of *abr* mutations or possibly to the use of the YB886 background, since the latter is known to contain several adventitious mutations (e.g., *amy* [J. Hahn, unpublished data] and *sigB* [30]). Given this, it was reported that competence-specific induction of the RecE protein occurred in the two *spo0* mutants (85). However, when the *recE4* mutation was also introduced, this induction did not take place. The authors concluded from these data that *recE* is under two forms of regulation, either of which can suffice to turn on expression during the development of competence: one requiring the *recE* product itself and the other requiring *spo0A* and *spo0H*.

The striking interrelationships shown between competence and SOB induction are germane to our understanding of the biological role of competence. Also intriguing in this regard is the close similarity between the predicted protein sequence of *comA* and that of ORF2 of the E. *coli uvrC* locus (Fig. 3). The latter is an ORF of unknown function that is cotranscribed (at least part of the time) with *uvrC*, a DNA repair gene (65). Perhaps *comA* serves to link expressions of the SOB and competence systems. This similarity may indicate that also in E. *coli* the DNA repair system is modulated by signals that are transduced through a two-component relay system and that this system also may interact with other global stress response pathways.

RELATIONSHIP OF COMPETENCE AND SPORULATION

The process of sporulation is classically reported to be repressed by excess glucose.

Competence, on the other hand, appears to be favored by this C source. When *comA* is overproduced on a multicopy plasmid, sporulation is inhibited, but competence is not (Weinrauch, unpublished data). The Sin protein acts as an inhibitor of sporulation when overproduced, but appears to be an activator of competence (22). These fragmentary observations are at least consistent with the notion that the two processes are mutually exclusive, perhaps alternative solutions to the problems of life during the stationary phase. The most extreme view would hold that a cell committed to competence is irrevocably prevented from sporulating, without an intervening cycle of growth. The common dependence of the two processes on *spo0A* and *spo0H* can be rationalized by the assertion that these genes are part of an apparatus that collects information concerning nutritional and growth status and relays this information to systems that determine a variety of possible outcomes during the stationary phase, some concomitant and some mutually exclusive. The branch points in these processes would occur after the action of the two *spo0* genes. Perhaps slightly more difficult to rationalize is the *csh-293* mutant of A. Grossman (personal communication). This mutant carries a Tn*917lac* fusion that is dependent for expression on *spo0H*, exhibits an increased synthesis of β-galactosidase at T_0, is several hundred-fold deficient in competence, and is also somewhat oligosporogenic. It is possible that this mutation occurs in a regulatory gene that acts before the competence-sporulation branch point and is itself turned on in a *spo0H*-dependent manner at T_0. In more direct contradiction to the idea that competence and sporulation are mutually exclusive events, however, are the experiments of Lencastre and Piggot (36), in which it was shown that competent cells after transformation can go on to sporulate. This suggests that these processes can occur sequentially and in the absence of an intervening cycle of growth. Perhaps competence develops under conditions that repress sporulation, possibly

in response to N starvation in the presence of a good C source. Later, as the C source is depleted, the population may begin sporulation. This sequence of events would bear an intriguing similarity to the situation in other organisms, e.g., yeasts (and humans?), in which a sexual phase precedes dormancy. It is clear that the state of competence is not an obligatory step for successful sporulation, since *com* mutations as a class do not confer sporulation deficiency. Nevertheless, it is possible that transient competence accompanies the development of spores. It would be of interest to determine whether and under what conditions transient expression of *com* genes can be detected during sporulation.

Viewed in this way, competence and sporulation may not be mutually exclusive events, but may comprise stages in a common stress response pathway. It is interesting that in other organisms extracellular competence factors have been defined and that such a factor may even exist in *B. subtilis* (32, 54, 76). This makes biological sense since a system of sexual exchange might reasonably be regulated in response to population density. Why then does sporulation in *B. subtilis* appear to be regulated by extracellular factors (25)? It is not obvious why it would be advantageous for sporulation to be regulated by population density. Perhaps there are extracellular stress response pheromones that signal a common competence-sporulation pathway, the initial (sexual) component of which has evolved to respond to population density.

WHY COMPETENCE?

There are at least three hypotheses to explain the evolution of competence. These are not necessarily mutually exclusive. First, competence may be a stress mechanism that eats DNA in response to starvation. In fact, the competent cell can take up massive amounts of DNA, and nonhomologous DNA is rapidly degraded, so this is not a farfetched idea. On the other hand, *B. subtilis* generally scavenges macromolecules, using extracellular hydrolytic enzymes, and possesses a perfectly functional extracellular nonspecific nuclease produced during the stationary phase. This would seem to be a more expedient evolutionary adaptation than the development of a rather elaborate competence system. Second, competence may have evolved as a mechanism of DNA repair, as a way to import templates to repair accumulated damage (6, 48). Of course, the exogenous DNA might also carry its load of mutational damage, but Redfield (59), in an interesting computer simulation published with a wonderful title, has suggested that competence could nevertheless confer a selective advantage in this way. The close relationship between the development of competence and SOB induction is consistent with the idea that these processes evolved to repair damage to DNA. Third, competence may have evolved as a means of introducing genetic variation. It appears that the last two theories in particular are complementary and that both may have contributed to the development of competence. It is also worth keeping in mind that competence may have evolved in response to pressure to adapt to some special circumstance. In *N. gonorrhoeae*, for instance, it has been suggested that transformation is needed for antigenic switching of pilus serotypes (64).

Acknowledgments. The work described in this review was supported by Public Health Service grant AI10311 from the National Institutes of Health. Computer facilities at the Public Health Research Institute are supported by Public Health Service grant RR-02990 from the National Institutes of Health and by grant DBM-8502189 from the National Science Foundation.

I thank the investigators quoted for personal communications and gratefully acknowledge valuable discussions with J. Aghion, M. Albano, F. Breidt, R. Breitling, E. Dubnau, N. Guillen, J. Hahn, L. Mindlich, S. Mohan, I. Smith, L. Sonenshein, S. Su, and Y. Weinrauch.

LITERATURE CITED

1. **Albano, M., J. Hahn, and D. Dubnau.** 1987. Expression of competence genes in *Bacillus subtilis*. *J. Bacteriol.* **169**:3110–3117.

2. **Ambudkar, S. V., T. J. Larson, and P. C. Maloney.** 1986. Reconstitution of sugar phosphate transport systems of *Escherichia coli*. *J. Biol. Chem.* **261**: 9083–9086.

3. **Ambudkar, S. V., L. A. Sonna, and P. C. Maloney.** 1986. Variable stoichiometry of phosphate-linked anion exchange in *Streptococcus lactis*: implications for the mechanism of sugar phosphate transport by bacteria. *Proc. Natl. Acad. Sci. USA* **83**:280–284.

4. **Anagnostopoulos, C., and J. Spizizen.** 1961. Requirements for transformation in *Bacillus subtilis*. *J. Bacteriol.* **81**:741–746.

5. **Banner, C. D. B., C. P. Moran, Jr., and R. Losick.** 1983. Deletion analysis of a complex promoter for a developmentally regulated gene from *Bacillus subtilis*. *J. Mol. Biol.* **168**:351–365.

6. **Bernstein, H., H. C. Byerly, F. A. Hopf, and R. E. Michod.** 1984. Origin of sex. *J. Theor. Biol.* **110**: 323–351.

7. **Binnie, C., M. Lampe, and R. Losick.** 1986. Gene encoding the σ^{37} species of RNA polymerase σ factor from *Bacillus subtilis*. *Proc. Natl. Acad. Sci. USA* **83**:5943–5947.

8. **Bovre, K., and L. O. Froholm.** 1972. Competence in genetic transformation related to colony type and fimbriation in three species of *Moraxella*. *Acta Pathol. Microbiol. Scand.* **80**:649–659.

9. **Cahn, F. H., and M. S. Fox.** 1968. Fractionation of transformable bacteria from competent cultures of *Bacillus subtilis* on Renografin gradients. *J. Bacteriol.* **95**:867–875.

10. **Carter, H. L., III, L.-F. Wang, R. H. Doi, and C. P. Moran, Jr.** 1988. *rpoD* operon promoter used by σ^H-RNA polymerase in *Bacillus subtilis*. *J. Bacteriol.* **170**:1617–1621.

11. **Danner, D. B., R. A. Deitch, K. Sisco, and H. O. Smith.** 1980. An eleven-base-pair sequence determines the specificity of DNA uptake in *Haemophilus*. *Gene* **11**:311–318.

12. **Davidoff-Abelson, R., and D. Dubnau.** 1973. Kinetic analysis of the products of donor deoxyribonuclease in transformed cells of *Bacillus subtilis*. *J. Bacteriol.* **116**:154–162.

13. **Dooley, D. C., C. T. Hadden, and E. W. Nester.** 1971. Macromolecular synthesis in *Bacillus subtilis* during development of the competent state. *J. Bacteriol.* **108**:668–679.

14. **Dubnau, D.** 1976. Genetic transformation of *Bacillus subtilis*: a review with emphasis on the

recombination mechanism, p. 14–27. *In* D. Schlessinger (ed.), *Microbiology—1976*. American Society for Microbiology, Washington D.C.

15. Dubnau, D. 1982. Genetic transformation in *Bacillus subtilis*, p. 147–178. *In* D. Dubnau (ed.), *The Molecular Biology of the Bacilli*, vol. 1. *Bacillus subtilis*. Academic Press, Inc., New York.

16. Duncan, M. L., S. S. Kalman, S. M. Thomas, and C. W. Price. 1987. Gene encoding the 37,000-dalton minor sigma factor of *Bacillus subtilis* RNA polymerase: isolation, nucleotide sequence, chromosomal locus, and cryptic function. *J. Bacteriol.* 169:771–778.

17. Eisenstadt, E., R. Lange, and K. Willecke. 1975. Competent *Bacillus subtilis* cultures synthesize a denatured DNA binding activity. *Proc. Natl. Acad. Sci. USA* 72:323–327.

18. Elleman, T. C., P. A. Hoyne, N. M. McKern, and D. J. Stewart. 1986. Nucleotide sequence of the gene encoding the two-subunit pilin of *Bacteroides nonosus* 265. *J. Bacteriol.* 167:243–250.

19. Engelman, D. M., T. A. Steitz, and A. Goldman. 1986. Identifying nonpolar transbilayer helices in amino acid sequences of membrane proteins. *Annu. Rev. Biophys. Biophys. Chem.* 15:321–353.

20. Fani, R., G. Mastromei, M. Polsinelli, and G. Venema. 1984. Isolation and characterization of *Bacillus subtilis* mutants altered in competence. *J. Bacteriol.* 157:153–157.

21. Gabor, M., and R. Hotchkiss. 1966. Manifestation of linear organization in molecules of pneumococcal transforming DNA. *Proc. Natl. Acad. Sci. USA* 56:1441–1448.

22. Gaur, N. K., E. Dubnau, and I. Smith. 1986. Characterization of a cloned *Bacillus subtilis* gene which inhibits sporulation in multiple copies. *J. Bacteriol.* 168:860–869.

23. Goodman, S. D., and J. J. Scocca. 1988. Identification and arrangement of the DNA sequence recognized in specific transformation of *Neisseria gonorrhoeae*. *Proc. Natl. Acad. Sci. USA* 85:6982–6986.

24. Grinius, L. 1982. Energetics of gene transfer into bacteria. *Sov. Sci. Rev. Sect. Biol. Rev. D* 3:115–165.

25. Grossman, A. D., and R. Losick. 1988. Extracellular control of spore formation in *Bacillus subtilis*. *Proc. Natl. Acad. Sci. USA* 85:4369–4373.

26. Hadden, C., and E. W. Nester. 1968. Purification of competent cells in the *Bacillus subtilis* transformation system. *J. Bacteriol.* 95:876–885.

27. Hahn, J., M. Albano, and D. Dubnau. 1987. Isolation and characterization of competence mutants in *Bacillus subtilis*. *J. Bacteriol.* 169:3104–3109.

28. Henikoff, S., G. W. Haughn, J. M. Calvo, and J. C. Wallace. 1988. A large family of bacterial activator proteins. *Proc. Natl. Acad. Sci. USA* 85:6602–6606.

29. Henner, D. J., M. Yang, and E. Ferrari. 1988. Localization of *Bacillus subtilis sacU*(Hy) mutations to two linked genes with similarities to the conserved procaryotic family of two-component signalling systems. *J. Bacteriol.* 170:5102–5109.

30. Igo, M., M. Lampe, C. Ray, W. Schafer, C. P. Moran, Jr., and R. Losick. 1987. Genetic studies of a secondary RNA polymerase sigma factor in *Bacillus subtilis*. *J. Bacteriol.* 169:3464–3469.

31. Igo, M., and R. Losick. 1986. Regulation of a promoter region that is utilized by minor forms of RNA polymerase holoenzyme in *Bacillus subtilis*. *J. Mol. Biol.* 191:615–624.

32. Joenje, H., M. Gruber, and G. Venema. 1972. Stimulation of the development of competence by culture fluids in *Bacillus subtilis* transformation. *Biochim. Biophys. Acta* 262:189–199.

33. Kunst, F., M. Debarbouille, T. Msadek, M. Young, C. Mauel, D. Karamata, A. Klier, G. Rapaport, and R. Dedonder. 1988. Deduced polypeptides encoded by the *Bacillus subtilis sacU* locus share homology with two-component sensor-regulator systems. *J. Bacteriol.* 170:5093–5101.

34. Lacks, S. 1979. Uptake of circular deoxyribonucleic acid and mechanism of deoxyribonucleic acid transport in genetic transformation of *Streptococcus pneumoniae*. *J. Bacteriol.* 138:404–409.

35. Lacks, S., and B. Greenberg. 1976. Single-strand breakage on binding of DNA to cells in the genetic transformation of *Diplococcus pneumoniae*. *J. Mol. Biol.* 101:255–275.

36. Lencastre, H., and P. J. Piggot. 1979. Identification of different sites of expression for *spo* loci by transformation of *Bacillus subtilis*. *J. Gen. Microbiol.* 114:377–389.

37. Lewandowski, M., E. Dubnau, and I. Smith. 1986. Transcriptional regulation of the *spo0F* gene of *Bacillus subtilis*. *J. Bacteriol.* 168:870–877.

38. Lipman, D. J., and W. R. Pearson. Rapid and sensitive protein similarity searches. *Science* 227:1435–1441.

39. Love, P. E., M. J. Lyle, and R. E. Yasbin. DNA-damage-inducible (*din*) loci are transcriptionally activated in competent *Bacillus subtilis*. *Proc. Natl. Acad. Sci. USA* 82:6201–6205.

40. Love, P. E., and R. E. Yasbin. 1984. Genetic characterization of the inducible SOS-like system of *Bacillus subtilis*. *J. Bacteriol.* 160:910–920.

41. Love, P. E., and R. E. Yasbin. 1986. Induction of the *Bacillus subtilis* SOS-like response by *Escherichia coli* RecA protein. *Proc. Natl. Acad. Sci. USA* 83:5204–5208.

42. Loveday, K. S. 1978. DNA synthesis in competent *Bacillus subtilis* cells. *J. Bacteriol.* 135:1158–1161.

43. Lovett, C. M., Jr., P. E. Love, R. E. Yasbin, and J. W. Roberts. 1988. SOS-like induction in *Bacillus*

subtilis: induction of the RecA protein analog and a damage-inducible operon by DNA damage in Rec⁺ and DNA repair-deficient strains. *J. Bacteriol.* 170:1467–1474.

44. **Maloney, P. C.** 1987. Coupling to an energized membrane: role of ion-motive gradients in the transduction of metabolic energy, p. 222–243. *In* F. C. Neidhardt, J. L. Ingraham, K. B. Low, B. Magasanik, M. Schaechter, and H. E. Umbarger (ed.), *Escherichia coli and Salmonella typhimurium: Cellular and Molecular Biology*, vol. 1. American Society for Microbiology, Washington, D.C.

45. **Marahiel, M. A., P. Zuber, G. Czekay, and R. Losick.** 1987. Identification of the promoter for a peptide antibiotic biosynthesis gene from *Bacillus brevis* and studies on its regulation in *Bacillus subtilis. J. Bacteriol.* 169:2215–2222.

46. **Marrs, C. F., G. Schoolnik, J. M. Koomey, J. Hardy, J. Rothbard, and S. Falkow.** 1985. Cloning and sequencing of a *Moraxella bovis* pilin gene. *J. Bacteriol.* 163:129–132.

47. **Meyer, T. F., E. Billyard, R. Haas, S. Storzbach, and M. So.** 1984. Pilus genes of *Neisseria gonorrhoeae:* chromosomal organization and DNA sequence. *Proc. Natl. Acad. Sci. USA* 81:6110–6114.

48. **Michod, R. E., M. F. Wojciechowski, and M. A. Hoelzer.** 1988. DNA repair and the evolution of transformation in the bacterium *Bacillus subtilis. Genetics* 118:31–39.

49. **Moolenaar, G. F., C. A. Sluis, C. Backendorf, and P. Putte.** 1987. Regulation of the *Escherichia coli* excision repair gene *uvrC.* Overlap between the *uvrC* structural gene and the region coding for a 24 kD protein. *Nucleic Acids Res.* 15:4273–4289.

50. **Nester, E. W., and B. A. D. Stocker.** 1963. Biosynthetic latency in early stages of deoxyribonucleic acid transformation in *Bacillus subtilis. J. Bacteriol.* 86:785–796.

50a. **Nieuwenhoven, M. H., K. J. Hellingworth, G. Venema, and W. N. Konings.** 1982. Role of proton motive force in genetic transformation of *Bacillus subtilis. J. Bacteriol.* 151:771–776.

51. **Ninfa, A. J., E. G. Ninfa, A. N. Lupas, A. Stock, B. Magasanik, and J. Stock.** 1988. Crosstalk between bacterial chemotaxis signal transduction proteins and regulators of transcription of the Ntr regulon: evidence that nitrogen assimilation and chemotaxis are controlled by a common phosphotransfer mechanism. *Proc. Natl. Acad. Sci. USA* 85:5492–5496.

52. **Nixon, B. T., C. W. Ronson, and F. M. Ausubel.** 1986. Two component regulatory systems responsive to environmental stimuli share strongly conserved domains with the nitrogen assimilation regulatory genes *ntrB* and *ntrC. Proc. Natl. Acad. Sci. USA* 83:7850–7854.

53. **Pabo, C. O., and R. T. Sauer.** 1984. Protein-DNA recognition. *Annu. Rev. Biochem.* 53:293–321.

54. **Pakula, R., and W. Walczak.** 1963. On the nature of competence of transformable *Streptococci. J. Gen. Microbiol.* 31:125–133.

55. **Pearson, W. R., and D. J. Lipman.** 1988. Improved tools for biological sequence comparison. *Proc. Natl. Acad. Sci. USA* 85:2444–2448.

56. **Perego, M., G. B. Spiegelman, and J. A. Hoch.** 1988. Structure of the gene for the transition state regulator *abrB:* regulator synthesis is controlled by the *spoOA* sporulation gene in *Bacillus subtilis. Mol. Microbiol.* 2:689–699.

57. **Piggot, P., and J. G. Coote.** 1976. Genetic aspects of bacterial endospore formation. *Bacteriol. Rev.* 40:908–962.

58. **Powell, B. S., G. K. Powell, R. O. Morris, P. M. Rogowsky, and C. I. Kado.** 1987. Nucleotide sequence of the virG locus of the *Agrobacterium tumefaciens* plasmid pTiC58. *Mol. Microbiol.* 1:309–316.

59. **Redfield, R. J.** 1988. Evolution of bacterial transformation: is sex with dead cells ever better than no sex at all? *Genetics* 119:213–221.

60. **Ronson, C. W., B. T. Nixon, and F. Ausubel.** 1987. Conserved domains in bacterial regulatory proteins that respond to environmental signals. *Cell* 49:579–581.

61. **Rosenthal, A. L., and S. D. Lacks.** 1980. Complex structure of the membrane nuclease of *Streptococcus pneumoniae* revealed by two dimensional electrophoresis. *J. Mol. Biol.* 141:133–146.

62. **Sadaie, Y., and T. Kada.** 1983. Formation of competent *Bacillus subtilis* cells. *J. Bacteriol.* 153:813–821.

63. **Sastry, P. A., B. L. Pasloske, W. Paranchych, J. R. Pearlstone, and L. B. Smillie.** 1985. Comparative studies of the amino acid and nucleotide sequences of pilin derived from *Pseudomonas aeruginosa* PAK and PAO. *J. Bacteriol.* 164:571–577.

64. **Seifert, H. S., R. S. Ajioka, C. Marchal, P. F. Sparling, and M. So.** 1988. DNA transformation leads to pilin antigenic variation in *Neisseria gonorrhoeae. Nature* (London) 336:392–395.

65. **Sharma, S., T. F. Stark, W. G. Beattie, and R. E. Moses.** 1986. Multiple control elements for the *uvrC* gene unit of *Escherichia coli. Nucleic Acids Res.* 14:2301–2318.

66. **Siegele, D. A., J. C. Hu, and C. A. Gross.** 1988. Mutations in *rpoD,* the gene encoding the σ⁷⁰ subunit of *Escherichia coli* RNA polymerase, that increase expression of the *lac* operon in the absence of CAP-cAMP. *J. Mol. Biol.* 203:29–37.

67. **Singh, R. N., and M. P. Pitale.** 1967. Enrichment of *Bacillus subtilis* transformants by zonal centrifugation. *Nature* (London) 213:1262–1263.

68. **Smith, H., K. Wiersma, S. Bron, and G. Venema.** 1983. Transformation in *Bacillus subtilis:* purifi-

cation and partial characterization of a membrane-bound DNA-binding protein. *J. Bacteriol.* **156:** 101–108.

69. **Smith, H., K. Wiersma, S. Bron, and G. Venema.** 1984. Transformation in *Bacillus subtilis:* a 75,000-dalton protein complex is involved in binding and entry of donor DNA. *J. Bacteriol.* **157:**733–738.

70. **Smith, H., K. Wiersma, G. Venema, and S. Bron.** 1985. Transformation in *Bacillus subtilis:* further characterization of a 75,000-dalton protein complex involved in binding and entry of donor DNA. *J. Bacteriol.* **164:**201–206.

71. **Sparling, P. F.** 1966. Genetic transformation of *Neisseria gonorrhoeae* to streptomycin resistance. *J. Bacteriol.* **92:**1364–1371.

72. **Spizizen, J.** 1965. Analysis of asporogenic mutants in *Bacillus subtilis* by genetic transformation, p. 125–137. *In* L. L. Campbell and H. O. Halvorsen (ed.), *Spores III.* American Society for Microbiology, Washington, D.C.

73. **Stragier, P., C. Bonamy, and C. Karmazyn-Campelli.** 1988. Processing of a sporulation sigma factor in *Bacillus subtilis:* how morphological structure could control gene expression. *Cell* **52:**697–704.

74. **Strauss, N.** 1966. Further evidence concerning the configuration of transforming deoxyribonucleic acid during entry into *Bacillus subtilis. J. Bacteriol.* **91:**702–708.

75. **Tanaka, T., and M. Kawata.** 1988. Cloning and characterization of *Bacillus subtilis iep,* which has positive and negative effects on production of extracellular proteases. *J. Bacteriol.* **170:**3593–3600.

76. **Tomasz, A., and J. L. Mosser.** 1966. On the nature of the pneumococcal activator substance. *Proc. Natl. Acad. Sci. USA* **55:**58–66.

77. **Trowsdale, J., S. M. H. Chen, and J. A. Hoch.** 1978. Genetic analysis of phenotypic revertants of *spo0A* mutants in *Bacillus subtilis:* a new cluster of ribosomal genes, p. 131–135. *In* G. Chambliss and J. C. Vary (ed.), *Spores VII.* American Society for Microbiology, Washington D.C.

78. **Trowsdale, J., S. M. H. Chen, and J. A. Hoch.** 1979. Genetic analysis of a class of polymixin resistant partial revertants of stage 0 sporulation mutants of *Bacillus subtilis:* map of the chromosome region near the origin of replication. *Mol. Gen. Genet.* **173:**61–70.

79. **Vosman, B., G. Kuiken, and G. Venema.** 1988. Transformation in *Bacillus subtilis:* involvement of the 17-kilodalton DNA-entry nuclease and the competence-specific 18-kilodalton protein. *J. Bacteriol.* **170:**3703–3710.

80. **Walker, J. E., M. Saraste, M. J. Runswick, and N. J. Gay.** 1982. Distantly related sequences in the α- and β-subunits of ATP synthase, myosin, kinases and other ATP-requiring enzymes and a common nucleotide binding fold. *EMBO J.* **1:**945–951.

81. **Ward, J. E., D. E. Akiyoshi, D. Regier, A. Datta, M. P. Gordon, and E. W. Nester.** 1988. Characterization of the *virB* operon from an *Agrobacterium tumefaciens* Ti plasmid. *J. Biol. Chem.* **263:**5804–5814.

82. **Weiss, V., and B. Magasanik.** 1988. Phosphorylation of nitrogen regulator I (NR_I) of *Escherichia coli. Proc. Natl. Acad. Sci. USA* **85:**8919–8923.

83. **Wilson, G. A., and K. Bott.** 1968. Nutritional factors influencing the development of competence in the *Bacillus subtilis* transformation system. *J. Bacteriol.* **95:**1439–1449.

84. **Winans, S. C., P. R. Ebert, S. E. Stachel, M. P. Gordon, and E. W. Nester.** 1986. A gene essential for *Agrobacterium tumefaciens* virulence is homologous to a family of positive regulatory loci. *Proc. Natl. Acad. Sci. USA* **83:**8278–8282.

85. **Yasbin, R., J. Jackson, P. Love, and R. Marrero.** 1988. Dual regulation of the *recE* gene, p. 109–113. *In* A. T. Ganesan and J. A. Hoch (ed.), *Genetics and Biotechnology of Bacilli,* vol. 2. Academic Press, Inc., New York.

86. **Yasbin, R. E.** 1977. DNA repair in *Bacillus subtilis.* II. Activation of the inducible system in competent cells. *Mol. Gen. Genet.* **153:**219–225.

87. **Youngman, P., P. Zuber, J. B. Perkins, K. Sandman, M. Igo, and R. Losick.** 1985. New ways to study developmental genes in spore-forming bacteria. *Science* **228:**285–291.

Regulation of Procaryotic Development
Edited by Issar Smith, Ralph A. Slepecky, and Peter Setlow
© 1989 American Society for Microbiology, Washington, DC 20006

Chapter 8

Sigma Factors and the Regulation of Transcription

Charles P. Moran, Jr.

OVERVIEW

Many, probably most, species of eubacteria contain multiple forms of RNA polymerase. The "core" polymerase, which is shared by each type of RNA polymerase, is composed of several protein subunits (α^2, β, β'). A different specificity protein, called a sigma factor, is found associated with each different type of RNA polymerase. The association of a sigma factor with the core RNA polymerase forms a holoenzyme that can bind to specific sites on DNA, known as promoters, from where transcription is initiated. These sigma factors govern the specificity of the polymerase-promoter interaction; therefore, each type of sigma factor directs the holoenzyme to a different type of promoter. This specificity is due to interactions between the sigma factor and a sequence of nucleotides within a promoter that signal recognition by its cognate form of RNA polymerase holoenzyme.

Most bacteria appear to contain a holoenzyme form that has a specificity for promoter utilization similar to holoenzymes that cointain the primary sigma factors from *Escherichia coli* (σ^{70}) and *Bacillus subtilis* (σ^A; Table 1) (71). Secondary bacterial sigma factors, first discovered in *B. subtilis*, also have been found in enteric bacteria (18, 26, 28, 38) and *Streptomyces coelicolor* (4, 70).

In most cases the secondary sigma factors enable RNA polymerase to utilize numerous promoters that are dispersed around the bacterial chromosome. Some of the secondary sigma factors have been shown to be essential for the transcription of genes that are involved in complex responses to environmental signals (e.g., endospore formation by *B. subtilis*, the development of motility in *B. subtilis*, and the heat shock response and the response to nitrogen limitation, both in enteric bacteria). Obviously, additional sigma factors make a more versatile transcription apparatus. Some sigma factors are not only necessary for transcription of certain genes but also are exploited to regulate gene transcription. However, in most cases the relative roles of the sigma factor and ancillary factors in regulating promoter activity have not been completely resolved.

The primary purpose of this review is to describe the roles that sigma factors can play in bacterial gene expression. These roles are illustrated by focusing on some of the sigma factors that have been isolated from *B. subtilis*—the greatest number from any one organism. Before focusing on the biological roles of the sigma factors in *B. subtilis*, I briefly describe the structure of sigma factors and a model for the mechanism by which sigma factors direct the use of promoters by

Charles P. Moran, Jr. • Department of Microbiology and Immunology, Emory University School of Medicine, Atlanta, Georgia 30322.

Table 1
Sigma Factors in *Bacillus subtilis*

Sigma factor	Previous designations	Gene	Functions	Cognate promoter consensus sequence	
				−35	−10
σ^A	σ^{55}, σ^{43}	*rpoD*	Housekeeping	TTGACA	TATAAT[a]
σ^B	σ^{37}	*sigB*	Unknown and *ctc* transcription	AGGNTT	GGNATTGNT[b]
σ^C	σ^{32}	*sigC*[c]	Unknown	AAATC	TANTGNTTNTA[d]
σ^D	σ^{28}	*sigD*	Flagellar synthesis	CTAAA	CCGATAT[e]
σ^E	σ^{29}	*sigE*, *spoIIGB*	Sporulation	GAANAANT	CATATT[f]
σ^F	$\sigma^{spoIIAC}$	*sigF*, *spoIIAC*	Sporulation	Unknown but similar to σ^G [g]	
σ^G		*sigG*, *spoIIIG*	Sporulation: forespore specific	YGHATR	CAHWHTAH[h]
σ^H	σ^{30}	*sigH*, *spo0H*	Sporulation	CAGGA	GAATTWWT[i]
σ^K	σ^{27}	*spoIVCB*, *spoIIIC*[j]	Sporulation: mother cell specific	Unknown[k]	

[a] Base pairs at positions −14 to −18 also contribute to promoter activity (25, 45).
[b] Single-base substitutions in this sequence have been shown to prevent utilization of the *ctc* promoter by Eσ^B (46, 65).
[c] This gene has not been identified.
[d] This consensus sequence is based on the comparison of only two promoters that are used in vitro by Eσ^C and is therefore of little predictive value.
[e] This sequence is based on comparison of promoters, but no mutational analysis of sequence has been reported (15).
[f] Base substitutions at three positions of this sequence have been shown to reduce in vitro utilization of the G4 promoter by Eσ^E (50).
[g] Recognition in vitro is similar to σ^G (Setlow, personal communication).
[h] Based on comparison of 12 promoters (M. Nicholson, D. Sun, B. Setlow, and P. Setlow, personal communication). H = A, C, or T; R = A or G; W = A or T; Y = C or T.
[i] Sequence based on comparison of six promoters (Carter and Moran, unpublished data). Base substitutions in four positions have been found to reduce utilization of the *spoVG* promoter by Eσ^H (74a). Sequences upstream from the −35 region are essential for activity of several promoters. This enhancement is not σ specific (44).
[j] These two genes must be fused during sporulation to encode σ^K (see text).
[k] The sequences of the *cotA* and *cotD* promoters are similar at the −10 and −35 regions but also near −20. It is not known, however, which of these sequences signal recognition by Eσ^K and which signal repression by the *spoIIID* product.

RNA polymerase. For a more detailed account of the structure and function of sigma factors, the reader should see the review by Helman and Chamberlin (23). There are other recent reviews on sigma factors and on gene expression during sporulation (7, 39).

SIGMA FACTORS: A BRIDGE BETWEEN PROMOTERS AND RNA POLYMERASE

When a sigma factor binds to the core RNA polymerase it may direct the RNA polymerase to a specific promoter by acting as an adaptor or bridge. The core RNA polymerase can bind nonspecifically to DNA. The sigma factor, which binds first to the core RNA polymerase, may lie between RNA polymerase and the promoter, where the sigma factor makes sequence-specific contacts with the promoter. These additional contacts increase the affinity of the holoenzyme for the promoter; hence, the sigma bridges the RNA polymerase to the promoter. Some of the evidence for this model is summarized below.

In most cases the nucleotide sequences of promoters that are recognized by the same holoenzyme are most similar to one another at two regions of the promoters. These are five- to eight-nucleotide stretches centered about 10 base pairs (bp) and often about 35 bp upstream from the start point of transcription: the -10 and -35 regions, respectively. The consensus sequences for promoters that are used by each type of holoenzyme from *B. subtilis* are shown in Table 1. In several cases some of the nucleotides in these sequences have been shown by mutagenesis to be necessary for utilitization of a promoter by its cognate form of RNA polymerase. In other cases, however, the sequences have not been rigorously tested, and therefore, the predictive value of these consensus sequences is less reliable (see Table 1 footnotes). Since unique sequences, usually at two regions of promoters, govern the utilization of these promoters by forms of RNA polymerase that differ only by their sigma factors, Losick and Pero (39) proposed that the sigma factor makes sequence-specific contacts at these two regions of the promoters. The best evidence for this model comes from recent observations (described below) that specific amino acid substitutions in a sigma factor can suppress the detrimental effects of specific base substitutions in a promoter.

The nucleotide sequences of the structural genes for 12 sigma factors have been determined. Comparisons of the inferred amino acid sequences of the sigma factors reveal four regions that are conserved among these proteins. These regions are numbered 1 through 4, beginning at the N-terminal end of the protein (Fig. 1) (17, 23, 64). Region 1 includes a stretch of 120 amino acids of which 60% are identical or chemically similar between the primary sigma factors from *E. coli* (σ^{70}) and *B. subtilis* (σ^A) (16, 17). This region is not evident in other sigma factors, and its function is unknown.

Region 2 is the most highly conserved region of the sigma factors. It is present to some degree in all the known sigma factors. This region has been further subdivided by Helman and Chamberlin (23). They designated these subdivisions as 2.1, at the N terminus of region 2, through 2.4 at the C

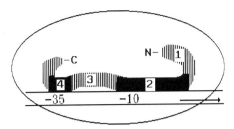

Figure 1. A model for sigma factor function. The sigma factor may lie between RNA polymerase and the promoter, where the sigma factor makes sequence-specific contacts with the promoter. Sigma factors have four conserved domains, indicated as regions 1 through 4. Mutational analyses described in the text indicate that amino acids in region 2 contact the -10 region of the promoter and region 4 of the sigma contacts the -35 region of the promoter. The figure is not intended to imply that the sigma factor contacts the -10 and -35 regions simultaneously. The contacts could be made sequentially.

terminus of region 2. The stretch of 15 amino acids that makes up the core of region 2.2 is the most highly conserved among sigma factors. Since sigma factors from one organism can function in vitro with core RNA polymerase from other bacteria (57), it is thought that the domain of the sigma factor that interacts with the core must be very similar in all sigma factors. Since region 2.2 is the most highly conserved region among sigma factors, it has been proposed to be a core binding domain (23, 38). At this time, however, there is no compelling evidence for this model.

Region 2.4 may interact directly with the -10 regions of its cognate promoters. A substitution of an isoleucine for the threonine at position 100 (designated T-100-I) in this region of σ^H from *B. subtilis* has been shown to change the specificity of this sigma factor (74a). A single-base transition at position -13 of a promoter that is used by Eσ^H (*spoVG*) was found to greatly reduce utilization of this promoter in vivo and in vitro by Eσ^H. A strain with the mutant allele of the σ^H structural gene (*spo0H81*, T-100-I) does not utilize the wild-type *spoVG* promoter, but the T-100-I substitution does enable utilization of the *spoVG* promoter that has the transition at position −13. This effect also was observed in vitro with Eσ^H that was isolated from the *spo0H81* strain. Other mutations in *spo0H* did not suppress the effect of this mutation in the *spoVG* promoter, and the *spo0H81* allele did not suppress the deleterious effects of mutations at other positions in the *spoVG* promoter. The remarkable specificity of this suppression of one promoter mutation by the T-100-I substitution was further demonstrated by the finding that the effects of the other two possible base substitutions at position −13 of the *spoVG* promoter were not suppressed by the T-100-I substitution. The simplest interpretation of these results is that the threonine at position 100 of σ^H makes a specific contact with the G:C base pair at position −13 of the *spoVG* promoter. This threonine is unable to make a useful contact with an A:T base pair at position −13 of the mutant *spoVG* promoter. When isoleucine is substituted at position 100 of σ^H, it can make a specific contact with an A:T base pair at position −13 but not with other base pairs in this position. Additional evidence for this model comes from the recent observation that removal of the side chains from threonine and isoleucine that may contact the −13 position of the *spoVG* promoter, by substitution of an alanine at position 100 of σ^H, results in a loss of specificity. This mutant σ^H (T-100-A) is unable to discriminate between *spoVG* promoters that have different base pairs at position −13 (D. Daniels and R. Losick, personal communication). Similarly, a mutation (T-440-I) in region 2.4 of σ^{70} of *E. coli* has been found to suppress the effect of a base substitution in the −10 region of two cognate promoters (58a). Taken together, these data strongly support the model that some of the amino acids in region 2.4 of sigma factors make base-pair-specific contacts in the −10 region of their cognate promoters.

Recently, Helman and Chamberlin (23)

made the intriguing proposal that region 2.3 and possibly 2.1 of sigma factors may be directly involved in melting the DNA at the promoter. Their proposal was based on the observations that these regions contain a remarkably large number of aromatic amino acids, which are also found in single-stranded DNA-binding proteins. Furthermore, the sequence in region 2.3 is similar to a consensus sequence that is considered to be a signature of single-stranded DNA-binding proteins from eucaryotes. Approximately 17 bp of DNA is unwound when RNA polymerase binds to a promoter to form the open complex (58). Sigma factors have been chemically cross-linked to the unwound region of promoters; therefore, when the holoenzyme is bound to a promoter, the sigma factor is in close proximity to the unwound region (59). There is, however, no direct evidence at this time that the sigma factor directly participates in melting DNA. Helman and Chamberlin speculated that the amino acids in region 2.3 and possibly 2.1 may facilitate the unwinding of DNA, possibly by interacting with the nontranscribed strand. It will be interesting to see the effects of mutations in regions 2.3 and 2.1 on promoter utilization.

Regions 3 and 4 of sigma factors have been predicted to form helix-turn-helix structures similar to those found in several sequence-specific DNA-binding proteins from procaryotes. Region 3 is the least conserved of these regions, and currently there is no evidence of a role for this region. However, region 4 probably interacts with the −35 regions of promoters. This conclusion is based on mutational analyses similar to those discussed above for region 2.4 (14a, 58a). These analyses showed that mutations in the −35 regions of promoters were suppressed by specific amino acid substitutions in region 4 of σ^{70} from *E. coli* (14a, 58a). Although most sigma factors evidently have a region that is homologous to region 4, the sigma factor encoded by gene *55* of phage T4 does not (17, 23). In this exceptional case, use of promoters by RNA polymerase containing this sigma factor was found to be governed only by sequences near the -10 region (10). These results are consistent with the model that region 4 interacts specifically with sequences near the -35 region of promoters.

The above results are summarized in a model (Fig. 1). While associated with RNA polymerase, the sigma factor makes sequence-specific contacts, usually with two regions of the promoter. Amino acids in region 4 contact specific base pairs that are usually located near position -35 of the promoter, whereas amino acids in region 2.4 contact base pairs in the -10 region of the promoter. It is not known whether the sigma factor contacts the -10 and -35 regions sequentially or simultaneously. The amino acids in regions 2.3 and 2.1 of the sigma factor may facilitate unwinding of the DNA at the promoter. After several ribonucleotides are incorporated into the transcript the sigma factor is released, reducing the affinity of the polymerase for the promoter. Hence, the polymerase is able to leave the promoter and continue its transcription of the DNA.

This model can be used to make several interesting predictions that can soon be tested. It may be possible to use mutational analysis to establish a colinear correlation between the amino acids in regions 2.4 and 4 of sigma factors and the base pairs that they interact with in the -10 and -35 regions, respectively, of a cognate promoter. Amino acid substitutions in regions 2.1 or 2.3 may prevent or slow unwinding of the DNA at promoters, and some substitutions in region 2.2 may cause such tight binding of the sigma factor to the core subunits that the sigma factor cannot be released from the polymerase, thus preventing the polymerase from leaving the promoter. These latter mutations may be dominant to the wild-type allele since a polymerase that contains this type of mutant sigma factor could effectively repress promoter activity.

THE SIGMA FACTORS IN *B. SUBTILIS*

Nine sigma factors have been isolated from *B. subtilis* (Table 1). Each sigma factor confers on the RNA polymerase a different specificity for promoter utilization, which has been demonstrated by in vitro transcription assays. The most abundant sigma factor in growing cells of *B. subtilis* is σ^A. This primary sigma probably directs transcription of most of the genes needed during growth in rich media but also probably plays a role in the early stages of sporulation. It is most homologous to the primary sigma factor in *E. coli*, σ^{70}. The secondary sigma factors can be divided into two groups, those that are found in growing cells and those that are produced only in sporulating cells. Of the four characterized secondary sigma factors in growing cells, only one, σ^H, has been shown to be necessary for sporulation (8). σ^H may also be involved in the development of competence for DNA uptake (1). σ^D is necessary for the development of motility (24), and the roles of σ^B and σ^C are unknown. Those sigma factors that are produced only during sporulation may activate transcription of genes that must be transcribed during specific periods of endospore development or in a specific compartment of the cell. I discuss below the current models for the role of each sigma factor in *B. subtilis*, emphasizing their role in endospore formation.

σ^H IS NECESSARY FOR THE INITIATION OF SPORULATION

In response to nutrient depletion, *B. subtilis* can undergo a remarkably complex cellular differentiation that culminates in the production of an endospore. Endospore formation requires the expression of more than 50 loci, each of which may include several genes. Expression of many of these genes is regulated at the level of transcription; i.e., transcription of these genes is activated at a specific time after the onset of sporulation

(40). Transcription of the sporulation-specific gene *spoVG* served as the first paradigm for the study of gene transcription during the earliest times after the onset of sporulation. The rate of transcription of *spoVG* dramatically increases within minutes after the initiation of spore formation (47, 76). The increased transcription presumably reflects an increase in initiation of transcription since a fusion of the *spoVG* promoter to a promoterless derivative of *lacZ* from *E. coli* results in a rapid accumulation of β-galactosidase after the onset of sporulation (76). Carter and Moran (5) used transcription from the *spoVG* promoter in vitro as an assay to isolate an RNA polymerase sigma factor from sporulating *B. subtilis* that directed efficient use of the *spoVG* promoter (σ^H, originally called σ^{30}).

σ^H-RNA polymerase ($E\sigma^H$) uses the *spoVG* promoter efficiently in vitro, but several criteria must be satisfied to clearly establish that a specific form of RNA polymerase utilizes a promoter in vivo. The polymerase should be shown to utilize the promoter accurately in vitro, and transcription from the promoter in vivo should occur during that part of the life cycle of the bacterium when the sigma is present. Inactivation of the gene encoding the sigma factor can be used to demonstrate that the sigma factor is essential for utilization of the promoter, and mutations in the promoter that prevent its utilization in vitro by the purified RNA polymerase should reduce its utilization in vivo. The most compelling evidence for the interaction of a sigma factor with a promoter in vivo would be the demonstration that an amino acid substitution in the sigma factor suppresses the effect of a base substitution at one position in the promoter. All of these criteria have been satisfied in the case of a σ^H and the *spoVG* promoter.

spoOH, which is essential for endospore formation, is the structural gene for σ^H. Several lines of evidence support this conclusion. The nucleotide sequence of *spoOH* showed that the gene could encode a 25,447-dalton

protein with an amino acid sequence that is 25% identical to those of other sigma factors (8). I. Smith and his co-workers constructed a translational fusion between *spoOH* from *Bacillus licheniformis* and *lacZ* from *E. coli*. The *spoOH-lacZ* fusion protein was isolated and used to elicit antiserum, which was subsequently enriched for *spoOH*-specific antibodies. The *spoOH*-specific antibodies were shown to react specifically with the sigma factor that was called σ^{30} by Carter and Moran (5), but was henceforth known as σ^H (8). As expected, $E\sigma^H$ was not detected in strains that contained a deletion allele of *spoOH*. Recently R. Losick (personal communication) and his co-workers used in vitro recombinant DNA methods to isolate an *E. coli* strain that produced the *spoOH* gene product. This protein was purified from this strain and used with core RNA polymerase from *B. subtilis* to reconstruct transcription of *spoVG*. Taken together, these results demonstrate that *spoOH* encodes a sigma factor, σ^H, that will direct transcription of *spoVG* in an in vitro assay.

There is compelling evidence that σ^H also directs transcription of *spoVG* in vivo. Several mutations in *spoOH*, including two different deletion alleles of *spoOH*, prevent utilization of the *spoVG* promoter in vivo (74, 74a, 76). However, the strongest evidence that $E\sigma^H$ directly uses the *spoVG* promoter in vivo is the observation, discussed earlier, that a mutation in *spoOH* (the *spoOH81* allele) is able to suppress the deleterious effect of one specific base substitution in the *spoVG* promoter both in vitro and in vivo (74a).

Although it is clear that σ^H-RNA polymerase uses the *spoVG* promoter, it is not known how the activity of this promoter is regulated. Evidently, increased accumulation of σ^H at the end of exponential growth probably cannot account for all of the induction of *spoVG* promoter. Recently, two groups have used σ^H-specific antibodies to measure the amount of σ^H in vegetative and sporulating cells (Losick, personal communication;

I. Smith, personal communication). They found that in complex sporulation medium (DS) σ^H is present in growing cells but that the amount of σ^H increases about fivefold by 2 h after the end of exponential growth. In contrast to the results seen in DS medium, the amount of σ^H does not increase in minimal glucose medium after sporulation and *spoVG* transcription are induced by the addition of decoyinine. It is not likely that the σ^H in vegetative cells is completely inactive since σ^H from vegetative cells is active in vitro and appears to direct transcription from some promoters in vegetative cells (e.g., *citGP2*, see below). Evidently, in minimal medium, and probably DS medium, activation of *spoVG* promoter activity requires a factor in addition to σ^H. In one model the *abrB* gene product has been proposed to be a negative regulator of *spoVG* transcription (74, 76–78).

Recently several other promoters that may be used by RNA polymerase containing σ^H have been identified. The *rpoD* operon of *B. subtilis*, which includes the structural gene for σ^A, is transcribed from at least three promoters. Promoters P1 and P2 are used during the exponential growth phase, whereas P3 is used only during stationary phase (6). Evidently, $E\sigma^H$ uses the P3 promoter to transcribe the *rpoD* operon during stationary phase. Transcription from P3 was prevented by a mutation in *spo0H* (6). Moreover, $E\sigma^H$ efficiently and accurately used the P3 promoter in vitro (6). Although it seems likely that $E\sigma^H$ transcribes the *rpoD* operon during the early stages of sporulation, it is not known whether this transcription has any significant consequence for the cell.

In contrast to the promoters *rpoDP3* and *spoVG*, which are used by $E\sigma^H$ during the early stages of sporulation, the P2 promoter of *citG* evidently is used by $E\sigma^H$ during the exponential growth phase. *citG*, the structural gene for fumarase, is transcribed from two promoters, P1 and P2 (13). The P2 promoter is not active in a *spo0H* mutant (13). Moreover, $E\sigma^H$ uses the P2 promoter

efficiently and accurately in vitro (H. L. Carter, K. M. Tatti, and C. P. Moran, unpublished data). Taking these findings together, it is likely that $E\sigma^H$ uses this promoter in vivo. S1 mapping experiments (13), primer extension analyses (K. M. Tatti and C. P. Moran, unpublished), and promoter-*lacZ* fusions (A. Moir, personal communication) indicate that P2 is used predominantly during the exponential growth phase; therefore, it is likely that σ^H is active in vegetative cells.

Two other approaches have been used to identify additional promoters that are used by $E\sigma^H$. H. L. Carter and C. P. Moran (unpublished data) screened a plasmid library that contained random fragments of the *B. subtilis* chromosome for promoters that could be used in vitro by purified $E\sigma^H$. One such promoter, PH-1, was used in a promoter probe plasmid to direct transcription of *xylE* in *B. subtilis*. As expected, the PH-1 promoter was not active in a *spo0H* mutant, whereas in wild-type cells the promoter was activated at about the onset of sporulation. It appears, therefore, that promoters can be used by $E\sigma^H$ during growth as well as after the onset of sporulation.

In another approach, A. Grossman (personal communication) isolated strains of *B. subtilis* that carried insertions of a Tn*917* derivative that contained a promoterless version of *lacZ*. He screened the strains for those that produced β-galactosidase from the Tn*917-lacZ* insertions when σ^H was expressed at high levels from an inducible promoter. Many of these insertions probably lie within genes that are transcribed by $E\sigma^H$ since β-galactosidase expression was reduced in a *spo0H* deletion mutant. As anticipated by the preceding discussion, the *spo0H*-dependent transcription of some of these genes was observed during growth, some during the early stages of sporulation, and some during growth and sporulation.

Since the σ^H regulon includes genes that are transcribed at different phases of the growth cycle, it seems unlikely that σ^H plays a primary role in the temporal regulation of transcription. Evidently σ^H is a necessary transcription factor, but the regulation of many of these promoters probably depends upon additional factors. This model is analogous to that proposed for σ^{60} of *E. coli*, which is necessary for transcription of nitrogen-regulated genes (26, 28). Transcription of these genes also requires the *ntrC* product, which binds to DNA upstream from at least one nitrogen-regulated promoter, *glnA* (26, 28). In this case, the activity of the *ntrC* gene product, not the sigma factor, is modulated by nitrogen levels (26, 28). Therefore, in this example the *ntrA*-encoded sigma factor, σ^{60}, is an essential transcription factor, but it does not appear to play a regulatory role. σ^H is essential for gene expression that enables *B. subtilis* to begin differentiation into an endospore, but σ^H may play this type of non-regulatory role.

The results described above apparently present a paradox. *spo0H* is essential for cells to proceed past stage 0 of sporulation, and *spo0H* encodes an RNA polymerase sigma factor. However, inactivation of any one of the genes that have been identified, thus far, as being transcribed by σ^H does not block sporulation at stage 0. There are at least three explanations for this observation. The Tn*917* insertion library used by Grossman and other searches may have been incomplete, or only inactivation of more than one gene that is transcribed by $E\sigma^H$ will result in a *spo0H* phenotype. Another possibility is that at the beginning of sporulation $E\sigma^H$ transcribes a gene that is essential for sporulation but that is also essential for growth. During exponential growth this gene may be transcribed by another form of RNA polymerase. Most of the identified genes that are transcribed by σ^H have been shown to be transcribed from more than one promoter; therefore, it would not be surprising to find an essential sporulation gene that is transcribed by $E\sigma^H$ and another type of polymerase.

$E\sigma^H$ also may be involved in the development of competence for DNA uptake. The efficiency of DNA uptake and transformation of *spo0H* mutants is lowered and the transcription of several genes that are necessary for competence is impaired by mutations in *spo0H* (1).

THE OTHER SECONDARY SIGMA FACTORS IN GROWING *B. SUBTILIS*

Exponentially growing *B. subtilis* contain at least three secondary sigma factors in addition to σ^H. σ^B, originally called σ^{37}, was identified by in vitro transcription from the *spoVG* promoter (21, 22). Subsequent experiments that included the inactivation of the structural gene for σ^B, *sigB*, demonstrated that σ^B was not necessary for transcription of *spoVG* (3). Evidently, there is enough overlap in the specificity of σ^B and σ^H that $E\sigma^B$ can use the *spoVG* promoter in the very sensitive in vitro transcription assay, albeit less efficiently than does $E\sigma^H$. The role of σ^B is unknown. Inactivation of *sigB* does not prevent endospore formation (3, 9) but does prevent transcription from the *ctc* promoter (3). It is likely that the *ctc* promoter is used directly in vivo by $E\sigma^B$, since examination of a series of mutations in the *ctc* promoter revealed that those mutations that prevent use of the promoter in vitro by $E\sigma^B$ also reduce its utilization in vivo (52). The *ctc* promoter is activated most efficiently in vivo when cells enter stationary phase under conditions that repress the tricarboxylic acid cycle (29); however, the function of *ctc* in stationary phase is unknown. *ctc* may also function in growing cells since growth of a *ctc* mutant was found to be transiently inhibited after a temperature shift (69). The increased transcription of *ctc* in stationary phase coincides with increased transcription of *sigB* (C. Price, personal communication); however, it is not known whether this results in increased accumulation of σ^B. The study of σ^B and *ctc* has been important for understanding the interactions between promoters and their cognate form of RNA polymerase (65, 66), but the role of σ^B in the life cycle of *B. subtilis* remains a mystery. It is not known whether homologs of *sigB* are present in other species of *Bacillus*, but if so, then *sigB* may be useful in natural habitats. The discovery of the role of σ^B could uncover an heretofore unrecognized feature of the *Bacillus* life cycle.

σ^C, formerly σ^{32}, was discovered by its ability to direct transcription from the *spoVG* promoter in vitro. However, $E\sigma^C$ initiates transcription 10 bp downstream from the site used by $E\sigma^H$ (30). This downstream promoter is referred to as the P2 promoter of *spoVG*. It seems unlikely that $E\sigma^C$ accounts for a significant fraction of the *spoVG* transcription in vivo. Mutations in *spoOH* that inactivate σ^H completely block *spoVG* transcription (76), although $E\sigma^C$ can be isolated from these mutants (Carter and Moran, unpublished results). Nuclease protection experiments with RNA isolated from sporulating cells have identified a *spoVG* transcript with a 5' terminus at the P2 start point (44, 75, 76), but this transcript could be the result of processing of the longer transcript generated by $E\sigma^H$ from P1. Obviously, the structural gene for σ^C must be identified and inactivated before the role of σ^C in *spoVG* transcription and sporulation can be assessed. However, there is indirect evidence that $E\sigma^C$ is active in early sporulating cells. When the *spoVG* promoter is present on a multicopy plasmid, a large amount of the P2 transcript is produced from the plsamid (75). In these cells very little, if any, P2 transcript is generated from a chromosomal copy of the *spoVG* promoter (75). Possibly the plasmid-borne copies of *spoVG* titrate the $E\sigma^C$, preventing its use on the chromosomal *spoVG* promoter. The multiple copies of the *spoVG* promoter also prevent sporulation. Of course σ^C may not be the only factor that is titrated by the *spoVG* promoter. Therefore the role of σ^C remains unknown.

σ^D, also referred to as σ^{28}, evidently is not involved in sporulation. Helman et al. disrupted the structural gene for σ^D and found that $E\sigma^D$ is essential for transcription of genes that are needed for flagellar synthesis and motility (24).

σ^E AND OTHER FACTORS PRODUCED DURING ENDOSPORE FORMATION

Four sigma factors that are produced exclusively in sporulating cells have been identified. σ^E, formerly σ^{29}, appears about 2 h after the onset of sporulation and disappears about 3 h later (20). It is clear that σ^E is essential for endospore formation since mutations in the structural gene for σ^E, *sigE*, block sporulation. Furthermore, this phenotype can be complemented to sporulation proficiency by lysogenization of the mutant strain with a specialized transducing phage that carries only *sigE* (33). It is likely that the production of σ^E directly leads to transcription of a specific set of genes during sporulation, but only a few of these genes have been tentatively identified. One gene that evidently is transcribed by $E\sigma^E$ is *spoIID* (55). The *spoIID* promoter is usually activated between 1 and 2 h after the onset of sporulation. It is not activated in mutants that fail to produce σ^E, and the promoter is used in vitro by purified $E\sigma^E$ (55).

Although it seems likely that $E\sigma^E$ uses the *spoIID* promoter in vivo, a direct interaction in vivo between σ^E and the *spoIID* promoter (e.g., allele-specific suppression of a promoter mutation) has not been demonstrated. However, Stragier et al. (64) found that when production of σ^E was placed under the control of an inducible promoter, induction of σ^E in vegetative cells resulted in activation of the *spoIID* promoter. This result strengthens the argument that $E\sigma^E$ uses the *spoIID* promoter and is consistent with the possibility that the production of σ^E is sufficient for activation of the *spoIID* promoter, at least in vegetative cells. In this model σ^E acts as a regulatory factor, the production of which activates a specific class of promoters. However, there probably are other regulatory factors interacting with this class of promoters, since for example *spoIID* transcription appears to end after only 1 h of active transcription while $E\sigma^E$ is still present at high levels.

W. G. Haldenwang and his colleagues have considered at least three possible relationships among genes transcribed by $E\sigma^E$. The σ^E regulon may consist of genes that (i) encoded similar structural components, or (ii) must be transcribed during a specific temporal period of development, or (iii) must be transcribed in one specific compartment of the differentiating cell. Ray and Haldenwang (54) transcribed chromosomal DNA from *B. subtilis* in vitro with $E\sigma^E$ to produce radiolabeled RNA. This RNA was used to probe a library of *B. subtilis* DNA fragments cloned in phage lambda. They found and characterized several promoters that were used in vitro by $E\sigma^E$. The activity of these promoters in vivo was examined by S1 nuclease mapping experiments. The 5' termini of transcripts were mapped to the start of five of the promoters that were used in vitro by $E\sigma^E$. These transcripts were found only in sporulating cells and were not found during sporulation of mutants that failed to produce σ^E. Although it seems likely, there is no direct evidence for use of these promoters in vivo by $E\sigma^E$. Mutations in one of these promoters (G4) that reduce its use in vitro by $E\sigma^E$ also prevent its activity in vivo, as would be expected if the promoter is used directly by $E\sigma^E$ (53). At least three of the five genes isolated by Ray and Haldenwang were found to be transcribed in vegetative cells from adjacent promoters (54). This observation led them to suggest that the role of σ^E is to transcribe genes during a specific temporal period in development. They thought that the products of these genes were not likely to encode similar functions since some of the genes that they isolated were transcribed during growth whereas others were expressed only during sporulation. Moreover, this group (R. Jonas, unpublished results) has used a σ^E-specific antibody to detect σ^E in both the forespore and mother cell compartment of the developing cell; therefore, it is unlikely that the role of σ^E is to direct compartment-specific transcription.

REGULATION OF σ^E PRODUCTION

Since the role of σ^E appears to be the activation of gene transcription during a specific temporal period of development, it is not surprising that the production of σ^E is tightly regulated. The production of σ^E is controlled at two levels. The structural gene for σ^E, *sigE*, is not transcribed until about 30 min after the onset of sporulation (32, 33). Once translated, the 29 N-terminal amino acids of the primary product of *sigE*, P31, must be proteolytically removed to produce the active form of σ^E (37).

Haldenwang and his co-workers have used a monoclonal antibody that reacts specifically with σ^E and P31 to study the accumulation of these products during sporulation (37, 68). P31 appears by 1 h after the onset of sporulation, whereas σ^E is found to be the predominant form at 2 h after the onset of sporulation. Pulse-chase experiments were used to demonstrate the precursor-product relationship between P31 and σ^E (37). The time at which P31 is processed to form σ^E closely corresponds to completion of the asymmetric cell division that separates the forespore precursor from the mother cell. This observation led Haldenwang and co-workers (37) to suggest that the processing of P31 may be coupled to this morphological event. They further suggested that ". . . the incorporation of the activating enzyme as a component of the division septum might be an effective means of coupling these events" (37). The intriguing implication of their suggestion is that the temporal expression of the *sigE* regulon may be modulated by the completion of a specific morphological landmark.

Mutations in several genes prevent the processing of P31. The structural gene for σ^E, *sigE* (67), is part of a two-gene operon known as *spoIIG*. Kenney and Moran (33) used mutation and complementation analysis to show that a second gene that is essential for sporulation, *spoIIGA*, lies between *sigE* and the promoter for the *spoIIG* operon (Fig. 2). Disruption of this gene prevents the processing of P31 (31). Stragier et al. (62) found that some processing of P31 can occur in vegetative cells if *spoIIGA* is expressed from a strong inducible promoter. They reported that the *spoIIGA* nucleotide sequence indicates that it could encode a protein with a sequence that is similar to that of aspartic proteases. Therefore, they speculated that the *spoIIGA* product directly processes P31. Furthermore, they pointed out that the N-terminal end of the *spoIIGA* product is similar to that of membrane-spanning proteins. Therefore the protease may be associated with the membrane, possibly the sporulation-specific septum. This is an important and interesting result, but this model must be considered highly speculative until more direct evidence is presented. If true, it will be interesting to learn how the protease is activated upon the completion of septation. It should be noted that mutations in several other genes that block normal sporulation-specific septation also prevent P31 processing as anticipated by this model. However, other mutations that block septation prevent transcription of several sporulation genes (see below). Therefore, the mutations that prevent processing could do so indirectly by preventing transcription of genes that are necessary for processing of P31.

Transcription of the *spoIIG* operon, which includes *spoIIGA* and *sigE*, is activated about 30 min after the onset of spor-

ulation (32). Kenney and Moran used integrational plasmids to introduce a series of polar mutations into the *spoIIG* operon (33). The effects of these mutations on *sigE* expression and the results they obtained by cloning various fragments of the *spoIIG* operon into a promoter probe plasmid identified a 217-bp DNA fragment that contained the *spoIIG* promoter. The activity of the *spoIIG* promoter and mutant derivatives also has been studied after lysogenization of *B. subtilis* with specialized transducing phage (SPβ derivatives) that carry *spoIIG-lacZ* transcriptional fusions. In wild-type lysogens, β-galactosidase begins to accumulate about 30 min after the onset of sporulation (32, 33). This sporulation-induced activity of the *spoIIG* promoter is reminiscent of the *spoVG* promoter, discussed earlier, which is activated within minutes after the onset of sporulation. The evidence is compelling that *spoVG* is transcribed by $E\sigma^H$. Moreover, σ^H also is essential for *spoIIG* transcription since *spo0H* mutants fail to activate the *spoIIG* promoter (33). However, recent evidence indicates that $E\sigma^H$ does not utilize the *spoIIG* promoter. For example, $E\sigma^H$ did not use the promoter in vitro (32). Surprisingly, $E\sigma^A$ was the only form of polymerase that could be shown to use the *spoIIG* promoter in vitro (32).

The sequence of the *spoIIG* promoter is shown in Fig. 3. This promoter contains sequences that are similar to those found at the −10 and −35 regions of promoters that are used by $E\sigma^A$. The unusual feature of this promoter is that these putative σ^A contact sites are separated by 22 bp, rather than the typical 17 or 18 bp. The promoter of the *spoIIE* operon, which is expressed about the same time during sporulation as the *spoIIG* promoter, has a similar structure with −10 and −35 σ^A contact sequences separated by 21 bp (20). This type of promoter structure, therefore, may be important for gene expression during one period of spore formation. Four single-base substitutions in the −10 region (at positions −6 and −11) and two in

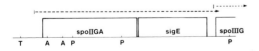

Figure 2. Organization of the *spoIIG* operon. Shown is a partial restriction map of the *spoIIG* operon. Restriction endonuclease cleavage sites indicated are: T, *Taq*I; A, *Aha*III; and P, *Pst*I. The open blocks represent the coding regions of *spoIIGA* and *sigE* and part of the coding region of an adjacent gene, *spoIIIG*. Transcripts are indicated by the broken arrows.

```
TCGACAAATTAAGCAGATTTCCCTGAAAAATTGTATTTTCCTCTCAACATTAA
           -80                    -60                    -40


  -35                           -10
TTGACA                          TATAAT

TTGACAGACTTTCCCACAGAGCTTGCTTTATACTTATGAAGCAAGAAGGGGA
   ▼    *                  -20        ▼      ▼    +1 ***
   C                                 G      G
   A                                 C      C
```

Figure 3. Nucleotide sequence of the *spoIIG* promoter. Shown is the nucleotide sequence of the nontranscribed strand. Transcription occurs from left to right and is initiated at the position indicated as + 1. The canonical sequences found at the − 10 and − 35 regions of promoters used by Eσ^A are indicated above the promoter sequence. Six single-base substitutions that reduce use of the promoter are indicated below the arrows. The asterisks indicate the positions of base substitutions that have little or no effect on promoter activity. Deletion of the base pairs between − 88 and · − 62 has been shown to abolish promoter activity (32).

the − 35-like sequence (position − 37) have been found to prevent *spoIIG* promoter function in vivo (Fig. 3) (32; unpublished data). Therefore, these sequences are critical for promoter utilization. Recently, merodiploids that carry a mutant allele of *rpoD*, the structural gene for σ^A, have been used to examine interaction of σ^A and the *spoIIG* promoter. One amino acid substitution in σ^A appears to suppress the effect of one base substitution in the − 10 region of the *spoIIG* promoter (T. Kenney and C. P. Moran, unpublished results). This allele-specific interaction supports the model that the *spoIIG* promoter is used in vivo by Eσ^A.

If Eσ^A uses the *spoIIG* promoter, additional factors must regulate promoter activity since the promoter is not active until after the start of sporulation. Mutations in several genes prevent *spoIIG* transcription. These include several *spo0* genes (*spo0A, spo0B, spo0E, spo0F,* and *spo0H*) as well as several *spoII* genes (*spoIIF, spoIIL, spoIIN*) (32). It is likely that most of these gene products affect *spoIIG* promoter activity only indirectly. For example, mutations in *spo0H* prevent *spoIIG* transcription but, as discussed above, it is unlikely that σ^H interacts with the *spoIIG* promoter. Rather, Eσ^H may transcribe a gene that is directly or indirectly needed for

spoIIG transcription. It is not known whether the products of *spo0A, spo0B, spo0E,* or *spo0F* interact with promoters. In one model the products of *spo0B, spo0E,* and *spo0F* are necessary to activate *spo0A* (27; see discussion of *spo0* genes by I. Smith in this volume). It has been proposed that the *spo0A* product prevents expression of negative regulators such as *abrB* (48, 76, 77). However, a mutation in *abrB* does not restore *spoIIG* promoter activity in a *spo0A* background (33); therefore, *spo0A* affects *spoIIG* promoter activity through a different regulator or by a different mechanism. Moreover, the effect of *spo0A* is not simply due to the decreased levels of σ^H in *spo0A* since the *abrB* mutation restores wild-type σ^H levels but not *spoIIG* transcription (5). It will be interesting to identify the factor(s) that is affected by the action of *spo0H* and *spo0A* to activate *spoIIG* transcription.

The products of the *spoII* genes are candidates for this factor. *spoIIN* was identified by a temperature-sensitive allele, *spo-279*, which at the nonpermissive temperature results in a block of sporulation at stage II (73) and prevents transcription from the *spoIIG* promoter (32). *spoIIN* has been cloned and sequenced by T. Leighton and P. Stragier (personal communication) and was found to be an allele of *ftsA*. The *ftsA* product is necessary for septation during growth (2). Lutkenhaus and co-workers found that the *spo-279* mutation caused filamentous growth in addition to preventing completion of the spore septum at the nonpermissive temperature (2). It is interesting that a mutation that prevents normal septation can prevent transcription of *spoIIG*, but it seems unlikely that the *ftsA* product directly regulates the *spoIIG* promoter. The *spoIIL* product is necessary for *spoIIG* transcription, but since this mutation maps very close to *spo0A* it is uncertain that it identifies a new gene. Mutations in *spoIIF* also prevent transcription of *spoIIG*, but the product of this gene is not yet well characterized.

Regardless of the action of *spoIIF*, another factor may be involved in regulation of the *spoIIG* promoter. A mutation at position −38 of the *spoIIG* promoter causes expression from the promoter in vegetative cells. This expression is independent of the *spoII* genes such as *spoIIF* (Kenney and Moran, unpublished data). This expression, even in vegetative cells, however, is still dependent on *spo0A* and *spo0H*. Therefore *spo0A* and *spo0H* may lead to inactivation of a repressor in vegetative cells or produce an activator that is needed in addition to that produced by *spoIIF*.

σ^F AND A PARALLEL PATHWAY OF GENE EXPRESSION

The *spoIIA* operon consists of three open reading frames designated *spoIIAA*, *spoIIAB*, and *spoIIAC* (49). Mutations in *spoIIAA* and *spoIIAC* block sporulation at stage II, but thus far no mutations in *spoIIAB* have been identified. The *spoIIAC* gene encodes a sigma factor designated as σ^F. This conclusion is based on two observations. First, the nucleotide sequence of *spoIIAC* indicates that the amino acid sequence is similar to those of other sigma factors (11, 61). Second, P. Setlow and his co-workers (personal communication) isolated an RNA polymerase sigma factor from sporulating *B. subtilis* by virtue of its ability to direct transcription in vitro from the *sspE* promoter. This sigma probably is the product of *spoIIAC* since the sequence of the N-terminal 15 amino acids of σ^F is identical to that predicted by the nucleotide sequence of the *spoIIAC* open reading frame. It is not likely that $E\sigma^F$ transcribes *sspE* in vivo since the *sspE* promoter is used much more efficiently in vitro by another form of RNA polymerase, $E\sigma^G$ (discussed below), and mutants that lack σ^G but contain σ^F do not transcribe *sspE*. Currently it is not known which promoters are used in vivo by the sporulation-essential sigma factor σ^F.

The *spoIIA* operon, which encodes σ^F, and the *spoIIG* operon, which encodes σ^E, appear to be transcribed at about the same time during sporulation. It is not known why the cell appears to use two new sigma factors at about the same time during sporulation. Also, if these sigma factors play similar roles, then it is not clear why σ^E is proteolytically processed from an inactive precursor whereas σ^F is not.

The *spoIIA* and *spoIIG* operons are transcribed about the same time during sporulation, and their transcription evidently depends upon the activity of the same *spo0* and *spoII* gene products (P. Stragier, personal communication). However, the nucleotide sequences of the promoter regions of *spoIIA* and *spoIIG* differ substantially. As discussed earlier, the *spoIIG* promoter, like that of *spoIIE*, appears to be used by $E\sigma^A$. The nucleotide sequence upstream from the start point of transcription in the *spoIIA* operon is not similar to promoters used by $E\sigma^A$; rather, it is similar to the sequence that is believed to signal recognition of promoters by $E\sigma^H$ (72). Furthermore, Piggot and his co-workers (74) have found that multiple copies of the *spoIIA* promoter cannot overcome the dependence on *spo0H* for promoter activity, whereas multiple copies of the *spoIIA* promoter can partially overcome the requirements for other *spo0* gene products. These results led Piggot and co-workers to suggest that $E\sigma^H$ may transcribe the *spoIIA* operon. Recently, $E\sigma^H$ was shown to use the *spoIIA* promoter in vitro (J.-J. Wu, K. M. Tatti, P. Piggot, and C. P. Moran, unpublished data). It seems likely that $E\sigma^H$ transcribes the *spoIIA* operon, but genetic analyses will be necessary to confirm this conclusion.

It may seem disconcertingly complicated that the gene for one sporulation-essential sigma factor, σ^E, is transcribed by $E\sigma^A$, while the gene for another essential sigma factor, σ^F, is being transcribed by $E\sigma^H$. If this is true, it raises the possibility that under some circumstances the cell could respond to environmental signals by activating only one of two parallel sporulation pathways. Alternatively, one could view the necessity of activating two parallel pathways during the sporulation, the σ^A-σ^E path and the σ^H-σ^F path, as a fail-safe mechanism to insure that the cell did not prematurely commit itself to differentiation.

σ^G AND THE COMPARTMENTALIZATION OF GENE EXPRESSION

Endospore formation in *B. subtilis* is preceded by an asymmetric cell division that produces two cells within a single cell wall, a mother cell and forespore. The forespore and mother cell each contain a copy of the genome, but different sets of genes are expressed in these two cells. The mother cell donates several products to the developing forespore, for example proteins that form a coat around the spore, as well as smaller molecules, such as dipicolinic acid, which are taken into the developing forespore. In contrast, some proteins, such as the small acid-soluble proteins that protect spore DNA from damage by UV light (42), and glucose dehydrogenase are produced exclusively in the forespore (14, 60). (See the chapters by R. Losick and L. Kroos and by P. Setlow in this volume for additional discussions of this process.) A fundamental problem for the control of gene expression is posed by the necessity to activate the correct set of genes in the two cellular compartments. Since several genes (e.g., *gdh* [51], *spoVA* [12]) have been shown to be transcribed exclusively in the forespore, the problem can be rephrased to ask how the transcription of specific genes is limited to one compartment.

Recently, Setlow and his co-workers found a sigma factor, σ^G, that may be present exclusively in the forespore compartment (64a; Setlow, this volume). *sspE* encodes one of the small acid-soluble proteins found in the spore (41). σ^G was isolated from sporulating cells and identified by its ability to direct RNA polymerase to use the *sspE* promoter in vitro. The sequence of the 15 N-terminal amino acids of σ^G is identical to

that predicted by the nucleotide sequence of *spoIIIG*. Stragier and his colleagues (31a) and Kobayashi and his co-workers (34, 43) had recently determined the nucleotide sequence of *spoIIIG*, which lies immediately downstream from the *spoIIG* operon. This sequence is similar to other sigma factors. In addition, both groups found that disruption of *spoIIIG* blocked sporulation at stage III. Furthermore, inactivation of *spoIIIG* prevents transcription of several genes that are transcribed exclusively in the forespore (e.g., *sspE*, *gdh*) (P. Setlow, personal communication). Additional evidence that Eσ^G transcribes genes such as *sspE* and *gdh* in vivo comes from the recent observation that transcription from the *sspE* promoter can be activated in vegetative cells when σ^G is produced from a strong inducible promoter (P. Setlow and P. Stragier, personal communication). This result, however, does not exclude the possibility that additional factors regulate promoter activity during sporulation.

A transcriptional fusion between the *spoIIIG* promoter and *lacZ* from *E. coli* resulted in accumulation of β-galactosidase exclusively in the forespore compartment (Stragier, personal communication). This result raises the possibility that σ^G is produced predominantly in the forespore, where it directs transcription of the forespore-specific genes (Stragier, personal communication). It is not known how expression of σ^G is limited to the forespore compartment. Stragier has suggested that a low level of *spoIIIG* transcription results from readthrough from the *spoIIG* operon (personal communication). The σ^G that is produced then directs transcription from the *spoIIIG* promoter. However, Y. Kobayashi (personal communication) found that *spoIIIG* promoter activity is not blocked by a disruption of *spoIIIG* as predicted by this autocatalytic model. This disagreement has not yet been resolved. However, in either case the autocatalytic model does not directly address how σ^G expression is limited to the forespore.

σ^K AND COAT PROTEIN GENE TRANSCRIPTION

In contrast to the genes discussed in the previous section, the spore coat protein genes and *spoIVC* are transcribed exclusively in the mother cell compartment (56). Transcription of two structural genes for coat proteins *cotA* and *cotD* begins between 4 and 5 h after the onset of sporulation. Recently, Kroos et al. discovered a 27-kilodalton protein that, when added to core RNA polymerase, directed transcription in vitro from the *cotA* and *cotD* promoters (35). This 27-kilodalton sigma factor, σ^K, has also been shown to be necessary for transcription in vitro from the *spoIVC* promoter. However, transcription of *spoIVC* in vitro requires a 14-kilodalton protein in addition to σ^K. This 14-kilodalton factor, the product of *spoIIID* (35), inhibits transcription by $E\sigma^K$ from the *cotA* and *cotD* promoters. During sporulation *spoIVC* is transcribed about 1 h before *cotA* and *cotD* (36); therefore, Kroos et al. have proposed that *spoIVC* transcription may be activated by the production of σ^K and the *spoIIID* product, 3.5 h after the onset of sporulation. About 1 h later the *spoIIID* product becomes inactive, resulting in derepression of *cotA* and *cotD* transcription by $E\sigma^K$.

σ^K may play a critical role in the temporal control of transcription in the mother cell between 3 and 7 h after the start of sporulation. However, the temporal control during this period is subject to additional factors (e.g., the *spoIIID* product) that fine-tune expression of a subset of genes which are expressed during this time. On the basis of genetic evidence, Losick and his colleagues also propose that the *gerE* product may activate expression from the *cotC* promoter at the end of this temporal period while σ^K is directing transcription (see Losick and Kroos, this volume).

The production of σ^K 3.5 h after the onset of sporulation requires a DNA rearrangement (63; see Losick and Kroos, this volume). The N-terminal half of σ^K is encoded by *spoIVCB*, and the C-terminal half is encoded by *spoIIIC*. These genes are separated by at least 10 kilobases on the chromosome in vegetative cells. A site-specific recombination event during sporulation fuses these two genes, presumably only in the mother cell. This rearrangement may account for the presence of σ^K exclusively in the mother cell; therefore, it will be interesting to learn how this rearrangement is limited to the mother cell. As suggested for σ^G, σ^K appears to be autoregulatory, by directing transcription of its structural gene from the *spoIVC* promoter. Moreover, comparison of the N-terminal amino acid sequence of σ^K with the DNA sequence of *spoIVCB* revealed that σ^K may be proteolytically processed from a preprotein as is the case for σ^E.

CONCLUSIONS

Nine sigma factors have been isolated from *B. subtilis* and more are likely to be found. The sigma factors that have been found appear to play several roles during endospore formation. σ^H is necessary for the initiation of sporulation, but some genes transcribed at the earliest stage of sporulation are transcribed by $E\sigma^A$. $E\sigma^H$ also transcribes nonsporulation genes. It is likely that additional factors regulate the activity of promoters that are used by $E\sigma^H$.

The sigma factors that are produced exclusively during sporulation probably play critical roles in regulating the temporal transcription of genes, but almost certainly other factors interact with these sigma factors to fine-tune the temporal expression of specific genes. The production of some sigma factors exclusively in one cellular compartment such as the forespore (e.g., σ^G) probably insures that certain genes are transcribed in the proper compartment. It is not known, however, what mechanisms control the compartment-specific production of a sigma factor. At the current rapid rate at which the specific role of each sigma factor is being determined, it may soon be possible to determine which general problems of gene regulation during cell differentiation have been solved by the appearance of sigma factors and which have been solved by additional types of regulatory factors.

Acknowledgments. The work in my laboratory has been supported by Public Health Service grants AI20319 and GM39917 and a Research Career Development Award, AI00760, from the National Institutes of Health.

LITERATURE CITED

1. **Albano, M., J. Hahn, and D. Dubnau.** 1987. Expression of competence genes in *Bacillus subtilis*. *J. Bacteriol.* **169:**3110–3117.

2. **Beall, B., M. Lowe, and J. Lutkenhaus.** 1988. Cloning and characterization of *Bacillus subtilis* homologs of *Escherichia coli* cell division genes *ftsZ* and *ftsA*. *J. Bacteriol.* **170:**4855–4864.

3. **Binnie, C., M. Lampe, and R. Losick.** 1986. Gene encoding the σ^{37} species of RNA polymerase σ factor from *Bacillus subtilis*. *Proc. Natl. Acad. Sci. USA* **83:**5943–5947.

4. **Buttner, M. J., A. M. Smith, and M. J. Bibb.** 1988. At least three different RNA polymerase holoenzymes direct transcription of the agarose gene (*dagA*) of *Streptomyces coelicolor* A3(2). *Cell* **52:**599–607.

5. **Carter, H. L., III, and C. P. Moran, Jr.** 1986. New RNA polymerase sigma factor under *spo0* control in *Bacillus subtilis*. *Proc. Natl. Acad. Sci. USA* **83:**9438–9442.

6. **Carter, H. L., III, L.-F. Wang, R. H. Doi, and C. P. Moran, Jr.** 1988. *rpoD* operon promoter used by σ^H-RNA polymerase in *Bacillus subtilis*. *J. Bacteriol.* **170:**1617–1621.

7. **Doi, R. H., and L.-F. Wang.** 1986. Multiple procaryotic ribonucleic polymerase sigma factors. *Microbiol. Rev.* **50:**227–243.

8. **Dubnau, E., J. Weir, G. Nair, H. L. Carter III, C. P. Moran, Jr., and I. Smith.** 1988. *Bacillus* sporulation gene *spo0H* codes for σ^{30} (σ^H). *J. Bacteriol.* **170:**1054–1062.

9. **Duncan, M. L., S. S. Kalman, S. M. Thomas, and C. W. Price.** 1987. Gene encoding the 37,000-dalton minor sigma factor of *Bacillus subtilis* RNA polymerase: isolation, nucleotide sequence, chromosomal locus, and cryptic function. *J. Bacteriol.* **169:**771–778.

10. **Elliott, T., and E. P. Geiduschek.** 1984. Defining a bacteriophage T4 late promoter: absence of a "−35" region. *Cell* **36:**211–219.

11. **Errington, J., P. Fort, and J. Mandelstam.** 1985. Duplicated sporulation genes in bacteria. *FEBS Lett.* **188:**184–188.

12. **Errington, J., and J. Mandelstam.** 1986. Use of a *lacZ* gene fusion to determine the dependence pattern and the spore compartment expression of sporulation operon *spoVA* in *spo* mutants of *Bacillus subtilis*. *J. Gen. Microbiol.* **132:**2977–2985.

13. **Feavers, I. M., V. Price, and A. Moir.** 1988. The regulation of the fumarase (*citG*) gene of *Bacillus subtilis* 168. *Mol. Gen. Genet.* **211:**465–471.

14. **Fujita, V., A. Ramaley, and E. Freese.** 1977. Location and properties of glucose dehydrogenase in sporulating cells and spores of *Bacillus subtilis*. *J. Bacteriol.* **132:**282–293.

14a. **Gardella, T., H. Moyle, and M. M. Susskind.** 1989. A mutant *Escherichia coli* σ^{70} subunit of RNA polymerase with altered promoter specifity. *J. Mol. Biol.* **206:**579–590.

15. **Gilman, M. Z., J. L. Wiggs, and M. J. Chamberlin.** 1981. Nucleotide sequences of two *Bacillus subtilis* promoters used by *Bacillus subtilis* sigma-28 RNA polymerase. *Nucleic Acids Res.* **9:**5991–6000.

16. **Gitt, M. A., L.-F. Wang, and R. H. Doi.** 1985. A strong sequence homology exists between the major RNA polymerase sigma factors of *Bacillus subtilis* and *Escherichia coli*. *J. Biol. Chem.* **260:**7178–7185.

17. **Gribskov, M., and R. R. Burgess.** 1986. Sigma factors for *E. coli*, *B. subtilis*, phage SPO1, and phage T4 are homologous proteins. *Nucleic Acids Res.* **14:**6745–6763.

18. **Grossman, A. D., J. W. Erickson, and C. A. Gross.** 1984. The *htpR* gene product of *E. coli* is a sigma factor for heat shock promoters. *Cell* **38:**383–390.

19. **Guzman, P., J. Westpheling, and P. Youngman.** 1988. Characterization of the promoter region of the *Bacillus subtilis spoIIE* operon. *J. Bacteriol.* **170:**1598–1609.

20. **Haldenwang, W. G., N. Lang, and R. Losick.** 1981. A sporulation-induced sigma-like regulatory protein from *B. subtilis*. *Cell* **23:**615–624.

21. **Haldenwang, W. G., and R. Losick.** 1979. A modified RNA polymerase transcribes a cloned gene under sporulation control in *Bacillus subtilis*. *Nature* (London) **282:**256–260.

22. **Haldenwang, W. G., and R. Losick.** 1980. A novel RNA polymerase sigma factor from *Bacillus subtilis*. *Proc. Natl. Acad. Sci. USA* **77:**7000–7004.

23. **Helman, J. D., and M. J. Chamberlin.** 1988. Structure and function of bacterial sigma factors. *Annu. Rev. Biochem.* **57:**839–879.

24. **Helman, J. D., L. M. Marquez, and J. J. Chamberlin.** 1988. Cloning, sequencing and disruption of the *Bacillus subtilis* σ^{28} gene. *J. Bacteriol.* **170:**1568–1574.

25. **Henkin, T. M., and A. L. Sonenshein.** 1987. Mutations of the *Escherichia coli lacUVS* promoter resulting in increased expression in *Bacillus subtilis*. *Mol. Gen. Genet.* **209:**467–474.

26. **Hirschman, J., P. K. Wong, K. Sei, J. Keener, and**

S. Kustu. 1985. Products of nitrogen regulatory genes *ntrA* and *ntrC* of enteric bacteria activate *glnA* transcription in vitro: evidence that the *ntrA* product is a sigma factor. *Proc. Natl. Acad. Sci. USA* 82:7525–7529.

27. Hoch, J. A., K. Trach, F. Kawamura, and H. Saito. 1985. Identification of the transcriptional suppressor *sof-1* as an alteration in the *spo0A* protein. *J. Bacteriol.* 161:552–555.

28. Hunt, T. P., and B. Magasanik. 1985. Transcription of *glnA* by purified *Escherichia coli* components: core RNA polymerase and the products of *glnF*, *glnG*, and *glnL*. *Proc. Natl. Acad. Sci. USA* 82:8453–8457.

29. Igo, M. M., and R. Losick. 1986. Regulation of a promoter that is utilized by minor forms of RNA polymerase holoenzyme in *Bacillus subtilis*. *J. Mol. Biol.* 191:615–624.

30. Johnson, W. C., C. P. Moran, and R. Losick. 1983. Two RNA polymerase σ factors from *Bacillus subtilis* discriminate between overlapping promoters for a developmentally regulated gene. *Nature* (London) 302:800–804.

31. Jonas, R. M., E. A. Weaver, T. J. Kenney, C. P. Moran, Jr., and W. G. Haldenwang. 1988. The *Bacillus subtilis spoIIG* operon encodes both σ^E and a gene necessary for σ^E activation. *J. Bacteriol.* 170:507–511.

31a. Karmazyn-Campelli, C., C. Bonamy, B. Savelli, and P. Stragier. 1989. Tandem genes encoding sigma factors for consecutive steps of development in *Bacillus subtilis*. *Genes Dev.* 3:150–157.

32. Kenney, T. J., P. A. Kirchman, and C. P. Moran, Jr. 1988. Gene encoding σ^E is transcribed from a σ^A-like promoter in *Bacillus subtilis*. *J. Bacteriol.* 170:3058–3064.

33. Kenney, T. J., and C. P. Moran, Jr. 1987. Organization and regulation of an operon that encodes a sporulation-essential sigma factor in *Bacillus subtilis*. *J. Bacteriol.* 169:3329–3339.

34. Kobayashi, Y. 1987. The role of sigma factors in transcription regulation. *Cell. Technol.* 6:593–604. (In Japanese.)

35. Kroos, L., B. Kunkel, and R. Losick. 1989. Developmental regulatory protein from *Bacillus subtilis* switching RNA polymerase recognition specificity. *Science* 243:526–528.

36. Kunkel, B., K. Sandman, S. Panzer, P. Youngman, and R. Losick. 1988. Identification of the promoter for the *Bacillus subtilis* sporulation locus *spoIVC* and its use in studies of temporal and spatial control of gene expression. *J. Bacteriol.* 170:3515–3522.

37. LaBell, T. L., J. E. Trempy, and W. G. Haldenwang. 1987. Sporulation-specific σ factor σ^{29} of *Bacillus subtilis* is synthesized from a precursor protein, P31. *Proc. Natl. Acad. Sci. USA* 84:1784–1788.

38. Landick, R., V. Vaughn, E. T. Lau, R. A. VanBogelen, J. W. Erickson, and F. C. Neidhart. 1984. Nucleotide sequence of the heat shock regulatory gene of *E. coli* suggests its protein product may be a transcriptional factor. *Cell* 38:175–182.

39. Losick, R., and J. Pero. 1981. Cascades of sigma factors. *Cell* 25:582–584.

40. Losick, R., P. Youngman, and P. J. Piggot. 1986. Genetics of endospore formation. *Annu. Rev. Genet.* 20:625–669.

41. Mason, J. M., R. H. Hackett, and P. Setlow. 1988. Regulation of expression of genes coding for small, acid-soluble proteins of *Bacillus subtilis* spores: studies using *lacZ* gene fusions. *J. Bacteriol.* 170:239–244.

42. Mason, J. M., and P. Setlow. 1986. Essential role of small, acid-soluble spore proteins in resistance of *Bacillus subtilis* spores to UV light. *J. Bacteriol.* 167:174–178.

43. Masuda, E. S., H. Anaguehi, K. Yamada, and Y. Kobayashi. 1988. Two developmental genes encoding σ factor homologs are arranged in tandem in *Bacillus subtilis*. *Proc. Natl. Acad. Sci. USA* 85:7637–7641.

44. Moran, C. P., Jr., N. Lang, C. D. B. Banner, W. G. Haldenwang, and R. Losick. 1981. Promoter for a developmentally regulated gene in *Bacillus subtilis*. *Cell* 25:783–791.

45. Moran, C. P., Jr., N. Lang, S. F. J. LeGrice, G. Lee, M. Stephens, A. L. Sonenshein, J. Pero, and R. Losick. 1982. Nucleotide sequences that signal the initiation of transcription and translation in *Bacillus subtilis*. *Mol. Gen. Genet.* 186:339–346.

46. Moran, C. P., Jr., N. Lang, and R. Losick. 1981. Nucleotide sequence of a *Bacillus subtilis* promoter recognized by *Bacillus subtilis* RNA polymerase containing σ^{37}. *Nucleic Acids Res.* 9:5979–5990.

47. Ollington, J. F., W. G. Haldenwang, T. V. Huynh, and R. Losick. 1981. Developmentally regulated transcription in a cloned segment of the *Bacillus subtilis* chromosome. *J. Bacteriol.* 147:432–442.

48. Perego, M., G. B. Spiegelman, and J. A. Hoch. 1988. Structure of the gene for the transition state regulator *abrB*: regulator synthesis is controlled by the spo0A sporulation gene in *Bacillus subtilis*. *Mol. Microbiol.* 2:689–699.

49. Piggot, P. J., C. A. M. Curtis, and H. DeLencastre. 1984. Use of integrational plasmid vectors to demonstrate the polycistronic nature of a transcriptional unti (*spoIIA*) required for sporulation of *Bacillus subtilis*. *J. Gen. Microbiol.* 130:2123–2136.

50. Rather, P. N., R. E. Hay, C. L. Ray, W. G. Haldenwang, and C. P. Moran, Jr. 1986. Nucleotide sequences that define promoter that are used by *Bacillus subtilis*. Sigma-29 RNA polymerase. *J. Mol. Biol.* 192:557–565.

51. Rather, P. N., and C. P. Moran, Jr. 1988. Compartment-specific transcription in *Bacillus subtilis*: identification of the promoter for *gdh*. *J. Bacteriol.* 170:5086–5092.

52. Ray, C., R. E. Hay, H. L. Carter, and C. P. Moran, Jr. 1985. Mutations that affect utilization of a pro-

moter in stationary-phase *Bacillus subtilis. J. Bacteriol.* **163**:610–614.

53. **Ray, C., K. M. Tatti, C. H. Jones, and C. P. Moran, Jr.** 1987. Genetic analysis of RNA polymerase-promoter interaction during sporulation in *Bacillus subtilis. J. Bacteriol.* **169**:1807–1811.

54. **Ray, G. L., and W. G. Haldenwang.** 1986. Isolation of *Bacillus subtilis* genes transcribed in vitro and in vivo by a major sporulation-induced, DNA-dependent RNA polymerase. *J. Bacteriol.* **166**:472–478.

55. **Rong, S., M. S. Rosenkrantz, and A. L. Sonenshein.** 1986. Transcriptional control of the *Bacillus subtilis spoIID* gene. *J. Bacteriol.* **165**:771–779.

56. **Sandman, K., L. Kroos, S. Cutting, P. Youngman, and R. Losick.** 1988. Identification of the promoter for a spore coat protein gene in *Bacillus subtilis* and studies on the regulation of its induction at a late stage of sporulation. *J. Mol. Biol.* **200**:461–473.

57. **Shorenstein, R. G., and R. Losick.** 1973. Comparative size and properties of the sigma subunits of ribonucleic acid polymerase from *Bacillus subtilis* and *Escherichia coli. J. Biol. Chem.* **248**:6170–6173.

58. **Siebenlist, U., R. B. Simpson, and W. Gilbert.** 1980. *E. coli* RNA polymerase interacts homologously with two different promoters. *Cell* **20**:269–281.

58a.**Siegle, D. A., J. C. Hu, W. A. Walter, and C. A. Gross.** 1989. Altered promoter recognition by mutant forms of the σ^{70} subunit of *Escherichia coli* RNA polymerase. *J. Mol. Biol.* **206**:591–603.

59. **Simpson, R. B.** 1979. The molecular topography of RNA polymerase promoter interaction. *Cell* **18**:277–285.

60. **Singh, R. P., B. Setlow, and P. Setlow.** 1977. Levels of small molecules and enzymes in the mother cell compartment and the forespore of sporulating *Bacillus megaterium. J. Bacteriol.* **130**:1130–1138.

61. **Stragier, P.** 1986. Comment on "Duplicated Sporulation Genes in Bacteria" by J. Errington, P. Fort and J. Mandelstam. *FEBS Lett.* **195**:9–11.

62. **Stragier, P., C. Bonamy, and C. Kalmazyn-Campelli.** 1988. Processing of a sporulation sigma factor in *Bacillus subtilis*: how morphological structure could control gene expression. *Cell* **52**:697–704.

63. **Stragier, P., B. Kunkel, L. Kroos, and R. Losick.** 1989. Chromosomal rearrangement generating a composite gene for a developmental transcription factor. *Science* **243**:507–512.

64. **Stragier, P., C. Parsot, and J. Bouvier.** 1985. Two functional domains conserved in major and alternate sigma factors. *FEBS Lett.* **187**:11–15.

64a.**Sun, D., P. Stragier, and P. Setlow.** 1989. Identification of a new σ-factor involved in compartmentalized gene expression during sporulation of *Bacillus subtilis. Genes Dev.* **3**:141–149.

65. **Tatti, K. M., and C. P. Moran.** 1984. Promoter recognition by σ^{37} RNA polymerase from *Bacillus subtilis. J. Mol. Biol.* **175**:285–297.

66. **Tatti, K. M., and C. P. Moran.** 1985. Utilization of one promoter by two forms of RNA polymerase from *Bacillus subtilis. Nature* (London) **314**:190–192.

67. **Trempy, J. E., C. Bonamy, J. Szulmajster, and W. G. Haldenwang.** 1985. *Bacillus subtilis* sigma factor σ^{29} is the product of the sporulation-essential gene *spoIIG. Proc. Natl. Acad. Sci. USA* **82**:4189–4192.

68. **Trempy, J. E., J. Morrison-Plummer, and W. G. Haldenwang.** 1985. Synthesis of σ^{29}, an RNA polymerase specificity determinant, is a developmentally regulated event in *Bacillus subtilis. J. Bacteriol.* **161**:340–346.

69. **Truitt, C. L., E. A. Weaver, and W. G. Haldenwang.** 1988. Effects on growth and sporulation of inactivation of a *Bacillus subtilis* gene (*ctc*) transcribed in vitro by minor vegetative cell RNA polymerases ($E\sigma^{37}$, $E\sigma^{32}$). *Mol. Gen. Genet.* **212**:166–171.

70. **Westpheling, J., M. Ranes, and R. Losick.** 1985. RNA polymerase heterogeneity in *Streptomyces coelicolor. Nature* (London) **313**:22–27.

71. **Wiggs, J. L., J. W. Bush, and M. J. Chamberlin.** 1979. Utilization of promoter and terminator sites on bacteriophage T7 DNA by RNA polymerases from a variety of bacterial orders. *Cell* **16**:97–109.

72. **Wu, J.-J., M. G. Howard, and P. J. Piggot.** 1989. Regulation of transcription of the *Bacillus subtilis spoIIA* locus. *J. Bacteriol.* **171**:692–698.

73. **Young, M.** 1976. Use of temperature-sensitive mutants to study gene expression during sporulation in *Bacillus subtilis. J. Bacteriol.* **126**:928–936.

74. **Zuber, P.** 1985. Localizing the site of *spo0*-dependent regulation in the *spoVG* promoter of *Bacillus subtilis*, p. 149–156. *In* J. A. Hoch and P. Setlow (ed.), *Molecular Biology of Microbial Differentiation*. American Society for Microbiology, Washington, D.C.

74a.**Zuber, P., J. Healy, H. L. Carter III, S. Cutting, C. P. Moran, Jr., and R. Losick.** 1989. Mutation changing the specificity of an RNA polymerase sigma factor. *J. Mol. Biol.* **206**:605–614.

75. **Zuber, P., J. M. Healy, and R. Losick.** 1987. Effects of plasmid propagation of a sporulation promoter on promoter utilization and sporulation in *Bacillus subtilis. J. Bacteriol.* **169**:461–469.

76. **Zuber, P., and R. Losick.** 1983. Use of a *lacZ* fusion to study the role of the *spo0* genes of *Bacillus subtilis* in developmental regulation. *Cell* **35**:275–283.

77. **Zuber, P., and R. Losick.** 1987. Role of AbrB in Spo0A- and Spo0B-dependent utilization of a sporulation promoter in *Bacillus subtilis. J. Bacteriol.* **169**:2223–2230.

78. **Zuber, P., M. Marahiel, and J. Robertson.** 1988. Influence of *abrB* on the transcription of the sporulation-associated genes *spoVG* and *spo0H* in *Bacillus subtilis*, p. 123–127. *In* A. T. Ganesan and J. A. Hoch (ed.), *Genetics and Biotechnology of Bacilli*. Academic Press, Inc., San Diego, Calif.

Regulation of Procaryotic Development
Edited by Issar Smith, Ralph A. Slepecky, and Peter Setlow
© 1989 American Society for Microbiology, Washington, DC 20006

Chapter 9

Initiation of Sporulation

Issar Smith

Commitment, for sporulating microorganisms as well as for contemporary human relationships, is not a decision to be taken lightly. The choice between entering a developmental cycle as opposed to the continuation of simple cell division is a momentous one for these life forms. It may be advantageous for cells to be able to respond to environmental stress by forming dormant, resistant spores, but this process should only occur when all other possibilities for dealing with a hostile environment have been exhausted. Since spores exhibit very long delays in the resumption of vegetative growth, early sporulation can be lethal in a population sense as other, more cautious sporulators will continue growth or will resume growth more rapidly than the prematurely differentiating cell.

Nutrient stress can elicit global cellular responses in sporulating as well as nonsporulating microorganisms. Examples are mating, meiosis, and sporulation in fungi, amino acid and carbon starvation in enteric bacteria, the production of secondary metabolites and spores in *Streptomyces* spp., and sporulation in the genera *Bacillus* and *Clostridium*. Sporulation and other aspects of late-growth regulation have been studied most extensively in *Bacillus subtilis*, for reasons presented in the introduction to this volume. When these normally soil-inhabiting organisms face depletion of external carbon, nitrogen, or phosphorus sources, they begin a series of processes geared to meeting the environmental challenge. They become motile, secrete hydrolytic enzymes and antibiotics, show induction of the SOS repair system, and become competent (able to accept DNA). These alternative responses allow bacilli to swim away from starvation, to modify the external milieu by degrading biopolymers so that the permeable end products of digestion can replenish depleted internal pools, to kill competing bacteria in the same ecological niche, and finally, to possibly repair damaged chromosomal DNA. This last process may be important for the successful completion of the last cycle of chromosomal replication, as it appears that only cells with terminated chromosomes can form spores, the ultimate defense against the adverse environment.

The purpose of this chapter is to discuss the initiation of sporulation. Recent advances in our understanding of spore initiation, coupled with an explosion of knowledge concerning environmentally interactive systems in bacteria, allow a more mechanistic approach to the topic than was previously possible. This review will be organized in the context of the following questions. When a *B. subtilis* cell is faced with nutrient stress, what are the specific environmental signals the cell will respond to; i.e., what is the flow of information from the environment to the cell? How is the information processed so

Issar Smith • Department of Microbiology, The Public Health Research Institute, New York, New York 10016.

that a decision to continue vegetative growth or to differentiate is made? How is the transduced information relayed to the biosynthetic machinery? In other words, what are the elements and the circuitry involved in the change from an exponentially growing cell to one which is committed to sporulation? Clearly, a sporulation-competent cell must have an elaborate system of environmentally responsive receptors which receive nutrient stress signals. There should be transducers which transfer this information to effectors that cause the transcriptional and translational machinery of the cell to synthesize the proteins which reflect the differentiated state. In addition, there should be a series of fail-safe controls to prevent unwarranted entry into sporulation. There should also be common pathways linking the control of sporulation to the other late-growth phenomena discussed above, as well as branched circuits which separate these different events.

Given the subjective approach, this chapter is not meant to be an exhaustive review of the topic. For example, the relationship between the termination of DNA replication, cell septation, and the initiation of sporulation will not be covered (see P. Stragier, this volume, for a brief discussion of this area). It will try to highlight areas in which questions can be asked and, it is hoped, can be answered (or ways of answering questions can be presented).

ENVIRONMENTAL SIGNALS AND SIGNAL TRANSDUCTION

Signals and Small Molecules

In recent years there has been an exponential increase in our understanding of the control of sporulation (the evidence for this statement is displayed in this volume). Ironically, we have learned very little about the role of environmental signals in this process since the germinal studies of Pierre Schaeffer and Joel Mandelstam (15, 90). These workers and their groups demonstrated that star-

vation for rapidly metabolizable carbon sources or easily available nitrogen (and, to a significantly lesser extent, exogenous phosphate) induced sporulation. They also showed that, even in continuous culture, cells could sporulate at a certain frequency and that this parameter could be changed by varying the levels of carbon and/or nitrogen sources or by altering growth rate. This information led to the concept, in 1965, that "the probability of a cell in growth medium to become committed to sporulate must be determined by the intracellular concentration of at least one nitrogen-containing catabolite repressing, directly or indirectly, the expression of all the sporulation genes" (90). The same idea was expressed as recently as 1981: "It seems likely (and is experimentally more productive to assume) that there is ultimately one compound of low molecular weight which, in combination with a protein, suppresses sporulation; . . . this compound must contain carbon, nitrogen and phosphorus atoms" (74). Freese and co-workers, in a comprehensive series of experiments, have shown that GTP has the properties of the hypothetical corepressor (74). Unfortunately, there is no knowledge of how nutrient stress causes reduction in GTP levels or what role is played by this nucleotide in the repression of sporulation. It is intriguing to think of a specific role for a GTP-binding protein, like the *Saccharomyces cerevisiae* sporulation regulator Ras2 (109), in the regulation of *Bacillus* development or perhaps the involvement of a GTP-binding component of the translation apparatus (EF-Tu, EF-G, etc.) in this process. Until more data are obtained, however, one can only speculate on the function of GTP in sporulation. There is also an indication that methylation, mediated by *S*-adenosylmethionine-utilizing methylases, may play a repressive role in the initiation of sporulation (2, 113). The signals for this postulated downregulation and the mechanism of its action, as for GTP levels, are unknown. It is even more surprising that the mechanism of carbon repression of sporulation remains undefined. It was reported by Schreiber over 90 years ago that glucose inhibited or delayed this process (47). However, there is absolutely no information on cell receptors for glucose which could relay information to the sporulation control system (see A. L. Sonenshein, this volume, for a more complete discussion of GTP levels, intermediary metabolism, and glucose effects on other genes).

Elements of the Signal Transduction Mechanism

Mutations causing insensitivity of the sporulation process to glucose have been isolated. Some of them are in genes for components of the transcriptional apparatus, *rpoB*, *rpoC* (107), and *rpoD* (*sigA*) (45). A similar mutation, *catA7*, has been mapped to the *hpr* locus, which codes for a repressor of sporulation and protease formation (77). These will be discussed in later sections. Certain other mutations are in genes coding for the regulatory proteins SpoOA (G. Olmedo and P. Youngman, personal communication) and DegU (32, 50). These latter two molecules, on the basis of amino acid sequence, are related to a group of proteins, variously defined as effectors, regulators, or receivers, which are elements of procaryotic two-component sensory transduction systems (Fig. 1). However, in no case are these proteins in direct contact with the external milieu. Mutations in *degS* also cause a glucose-insensitive phenotype (32). The *degS* open reading frame (ORF), which is very hydrophilic and has no potential membrane-spanning domains, shows some relatedness in amino acid sequence to a class of proteins called sensors, modulators, or transmitters which comprise the other part of the two-component system (Fig. 2). In this review the terminology of Kofoid and Parkinson (48), i.e., "transmitters" and "receivers," will be utilized. These terms are less specific as to function or mechanism of information transfer than some of the others, yet evoke an imagery of circuits. Although it has not yet been demonstrated that a similar sensory transduction pathway (or pathways) is involved in sporulation, it is of value to discuss the two-component systems of other bacterial systems and to relate them to the functions of similar (at least in terms of amino acid sequence) sporulation proteins. (These systems are reviewed in reference 86.) In its now-classic form, the two-component system is a motif consisting of two proteins, one of which, the transmitter, responds to an external signal. The transmitter then transfers this information to a receiver protein, which in turn modifies some aspect of cell function (usually, but not exclusively, gene expression) to cause adjustment to a changed environment (86). The transmitters are frequently integral membrane proteins in direct contact with the environment, and all have extensive homology at their carboxy-terminal ends. Each transmitter interacts, or as in the case of sporulation, is predicted to interact, with its cognate receiver, and proteins of this latter group possess extensive relatedness at the amino-terminal terminus. In some cases, the transmitter is not associated with the cytoplasmic membrane (e.g., CheA and NtrB) but can interact with another protein which is in direct contact with the environment (CheA, via CheW, interacts with integral membrane proteins, the methyl-accepting receptors [6]). Table 1 lists some of these environmentally responsive systems, and Fig. 1 and 2 illustrate some of the more striking similarities in the amino acid sequences of the receiver and transmitter classes of proteins, respectively.

In three cases, the NtrB-NtrC, EnvZ-OmpR, and CheA-CheB, CheY systems, the mechanism of information transfer has been shown to be an ATP-utilizing autophosphorylation of the transmitter, followed by a transfer of the phosphate to the receiver (35, 40, 46, 71, 121). The CheA autophosphorylation site, a histidine residue in a highly conserved sequence found in most receivers (Fig. 2), has been identified (34). NtrB is also autophosphorylated at a histidine residue, probably in the conserved sequence (118).

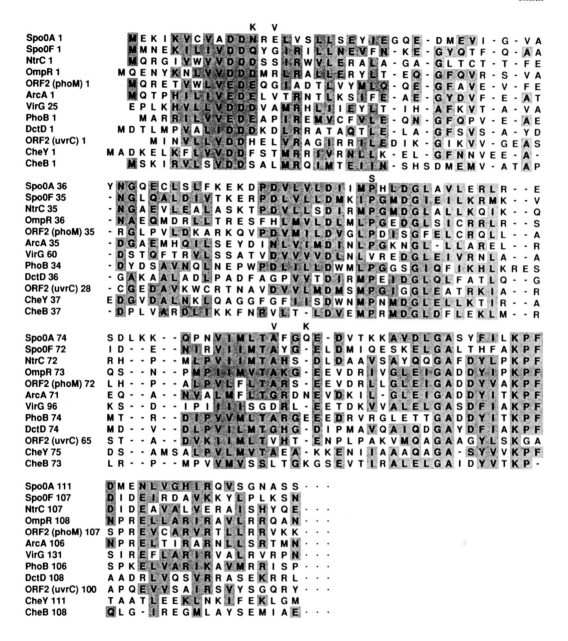

Figure 1. Amino acid sequence similarities in bacterial receiver proteins. Identical and chemically similar amino acids (F, W, Y), (I, L, M, V), (A, G), (S, T), (E, D, N, Q), (H, K, R), (P), and (C) are shaded when six or more proteins have identical or similar sets of amino acids at the same position. Sequences are from the following references: SpoOF (112); NtrC (*E. coli*) (67); OmpR (*E. coli*) (13); ORF2 (*phoM*) (*E. coli*) (3); ArcA (*E. coli*) (18); VirG (*Agrobacterium tumefaciens*) (119); PhoB (*E. coli*) (61); DctD (*Rhizobium leguminosarum*) (85); ORF2 (*uvrC*) (*E. coli*) (69); Spo0A (26); and CheY and CheB (*Salmonella typhimurium*) (102). Above the Spo0A sequence are depicted mutations which cause insensitivity to glucose repression of sporulation (see text for further description). The changes at positions 12 and 14, *sof-1* and *surOB20*, respectively, are published (36, 94). The changes at positions 60, 87, and 90 are defined as *coi-1*, *coi-12*, and *coi-15*, respectively (Olmedo and Youngman, personal communication).

Table 1
Two-Component Sensing Systems and Their Functions[a]

Receiver	Transmitter	Organism	Environmental stimulus	Response
NtrC	NtrB	*Escherichia coli* *Klebsiella pneumoniae* *Rhizobium meliloti*	Nitrogen limitation	Transcription of nitrogen regulon
OmpR	EnvZ	*E. coli*	Osmolarity changes	Transcription of *ompF* and *ompC*
ArcA	CpxA?	*E. coli*	Lack of O_2	Repression of aerobic respiration genes
PhoB	PhoR	*E. coli*	Phosphate limitation	Transcription of phosphate regulon
ORF2 (*phoM*)	PhoM (ORF3)	*E. coli*	Phosphate limitation	Catabolite control of phosphate regulon transcription
VirG	VirA	*Agrobacterium tumefaciens*	Plant exudate	Transcription of T1 plasmid *vir* genes
DctD	DctB	*Rhizobium leguminosarum*	4C-dicarboxylic acids	Transcription of *dctA*
PgtA	?	*Salmonella typhimurium*	Phosophoglycerates	Induction of phosphoglycerates transport system
ORF2 (*uvrC*)	?	*E. coli*	DNA damage?	?
CheY	CheA	*E. coli* *S. typhimurium*	Attractants/repellents	Chemotaxis (increased tumbling)
CheB	CheA	*S. typhimurium*	Attractants/repellents	Chemotaxis (adaptation)
Spo0A	?	*Bacillus subtilis*	Starvation	Sporulation
Spo0F	?	*B. subtilis*	Starvation	Sporulation

[a]In response to nitrogen limitation, NtrC/NtrB regulates transcription of the nitrogen regulon in *E. coli* as well as the *nif* system of the nitrogen-fixing bacteria (8, 17, 59, 73, 84). In response to changing osmolarity, OmpR/EnvZ regulates the transcription of the structural genes of the outer membrane porin proteins OmpF and OmpC (30, 63). Genes encoding enzymes associated with aerobic metabolism are transcriptionally repressed during anaerobic growth; the ArcA protein is necessary for this repression (43). ArcA (as SfrA), along with CpxA, is also necessary for F plasmid conjugation; however, it is not clear whether CpxA plays a role in aerobic regulation (97). The four Pho proteins regulate the transcription of the genes necessary for phosphate assimilation in response to phosphate limitation; regulation via the *phoM* operon is subject to catabolite control (115, 116). During crown gall tumor formation, the VirA/VirG proteins mediate the transcription of the virulence genes of the Ti plasmid in *A. tumefaciens* in response to plant exudate (52, 119). In response to succinate, fumarate, or malate, the DctD/DctB proteins regulate the transcription of *dctA*, which encodes the 4C-dicarboxylic acid transport protein (85). Similarly, in response to phosphoglycerates the *pgtA* gene induces PgtP, the phosphoglycerate transport protein (123). The second ORF of the *uvrC* operon is included in the receiver set on the basis of sequence similarity with Spo0F; inactivation of the ORF indicates that this protein, unlike UvrC, has no role in resistance to UV radiation (69). During chemotaxis, CheY interacts with the flagellar motor, causing clockwise rotation (and thus tumbly behavior), and the CheB protein is a methylesterase which, by controlling the degree of methylation of the membrane-bound receptor, affects adaptation to changing concentrations of attractant or repellant (58, 82, 117). Both proteins are regulated by CheA (101, 104).

The transmitter transfers this phosphate group to an aspartate residue on the receiver (35, 46, 71, 121). Three aspartates, two at the extreme amino-terminal end, usually at or near the 10th and 11th positions, and one at an approximate location at residues 50 to 60 (Fig. 1), are found in receiver proteins (in a few cases, the first of the two adjacent aspartates in the N-terminal region is replaced by a glutamic residue). The recent elucidation of the CheY crystal structure has shown that these acidic residues form a pocket that is presumably the site of phosphorylation (103). Upon phosphorylation, the receivers are activated as transcriptional enhancers, in the case of NtrC (71, 79) and OmpR (A. Ninfa and T. Silhavy, personal communication), or as modulators of the chemotactic response, i.e., CheY and CheB (6). Activated NtrC functions by catalyzing the isomerization of a closed recognition complex between $E\sigma^N$ and the *glnA* P2 to an open complex (79). Phospho-CheY interacts with the flagellar motor to induce clockwise rotation, while CheB, a methylesterase, is activated for the removal of methyl groups from the methyl-accepting receptors (58). Though the phosphorylation activation of receivers has only been shown for the three above-mentioned systems, the high degree of amino acid sequence similarity in the other members of the transmitter-receiver system suggests that a similar information transfer mechanism is also utilized.

Spo0A and another essential sporulation protein, Spo0F, are similar in sequence to the receiver class, but a corresponding transmitter has not been identified as yet. Several receiverlike proteins have been recently described in *B. subtilis*: PhoR (92) and SpoIIJ (P. Stragier, personal communication) as well as the above-mentioned DegS (32, 50). Since mutations in *spoIIJ* cause a sporulation-defective phenotype (88), and it is expressed during vegetative growth (Stragier, personal communication), its gene product could be the postulated sporulation transmitter or one of several such proteins involved in the process (SpoIIJ is further discussed below). Hydropathy profiles indicate that SpoIIJ is not a membrane protein (Stragier, personal communication).

Role of the *spo0* Proteins in Signal Transduction

The transition from vegetative growth to the first morphologically distinct stage of sporulation, the formation of the spore septum, requires seven genes, *spo0A*, *-B*, *-E*, *-F*, *-H*, *-J*, and *-K*, though it was recently reported that *spo0J* may actually be a later-acting element (105). All have been cloned and sequenced (57, 76), but the data on the

CheA 46	A	A	H	S	I	K	G	G	A G
NtrB 138	L	A	H	E	I	K	N	P	L S
EnvZ 242	V	S	H	D	L	R	T	P	L T
CpxA 248	I	S	H	E	L	R	T	P	L T
PhoM 263	R	T	H	E	L	K	S	P	L A
E.PhoR 211	V	S	H	E	L	R	T	P	L T
B.PhoR 359	V	S	H	E	L	K	T	P	I T
VirA 483	I	A	H	E	F	N	N	I	L G
DctD 413	V	A	H	E	T	N	Q	P	V A
SpoIIJ 423	I	A	H	E	I	R	N	P	L T

CheA 487	S	G	R	G	V	G	M	D	V V
NtrB 312	G	G	T	G	L	G	L	S	I A
EnvZ 381	S	G	T	G	L	G	L	A	I V
CpxA 417	G	G	T	G	L	G	L	A	I V
PhoM 433	K	S	S	G	L	G	L	A	F V
E.PhoR 385	G	G	S	G	L	G	L	A	I V
B.PhoR 535	G	G	T	G	L	G	L	A	I V
VirA 666	G	G	T	G	L	G	L	A	S V
DctD 584	S	G	L	G	L	G	L	V	I S
SpoIIJ 566	K	G	T	G	L	G	L	M	V T

Figure 2. Regions of high similarity in bacterial transmitter proteins. Two highly conserved regions in transmitter proteins are displayed. The shading has the same meaning as in Fig. 1, but when five or more related amino acids are at the same position. The origin of the proteins can be found in the footnote to Table 1 and the legend of Fig. 1. The leftmost sequences possess the histidine residue believed to be the site of autophosphorylation (34, 118). The references to sequences can be found, for the most part, in reference 86. Others are: CheA (*E. coli* and *S. typhimurium*) (34, 101); B-PhoP (*B. subtilis*) (92) (E-PhoR denotes the homologous protein from E. coli); and SpoIIJ (Stragier, personal communication).

latter two have not been published (J. Hoch, personal communication). Of these genes, only *spo0H* has a clearly defined function, coding for an alternative σ factor. Since it is not part of the signal transduction mechanism, it will be discussed in a later part of this review. There is very little information concerning the role of *spo0J* and *spo0K* in sporulation, except for the fact that mutations in these genes affect the in vitro methylation of an uncharacterized protein (K. Golden and R. Bernlohr, personal communication).

spo0A

As discussed above, Spo0A, on the basis of its amino acid sequence, is part of the sporulation sensory mechanism. Functional *spo0B*, *spo0E*, and *spo0F* genes are also essential for differentiation, but mutations in these genes can be suppressed by the *sof-1* mutation, an asparagine-to-lysine substitution in codon 12 of the *spo0A* ORF (36). A similar mutation, *sur0B20*, is a valine-for-glutamate substitution at amino acid residue 14 of Spo0A (94). Interestingly, these residues are very close to the postulated phosphorylation site of the receiver proteins, which in the case of Spo0A is at amino acid positions 10 and 11 (Fig. 1). It is therefore conceivable that the function of Spo0B, Spo0E, and Spo0F is to modify Spo0A or to prepare it for phosphorylation (it has been reported, as unpublished results, that Spo0A is a substrate for the phosphotransferase reaction mediated by NtrB [46] or by an uncharacterized kinase [Hoch, personal communication]).

The end result of sporulation signal transduction can be viewed as the activation of Spo0A, which has been postulated to be a key step in the initiation of sporulation (76). This protein is the master regulator of most late growth processes, including antibiotic formation, extracellular protease synthesis, competence, the SOS response (see the chapters on these subjects in this volume), $E\sigma^D$-dependent transcription (see C. P. Moran, Jr., this volume), and the expression of several

early sporulation genes. Many of these processes require SpoB, SpoE, and SpoF as well, and the *sof-1* mutation eliminates this requirement. Spo0A is a DNA-binding protein (42), binding to the promoter of at least one gene, *abrB* (Hoch, personal communication), which codes for AbrB, a major regulator of sporulation and other late-growth-regulated genes (the function of AbrB is discussed in a subsequent section). The use of a conditional mutation in the *spo0A* gene has shown that for sporulation to occur, the protein must be functional until stage II (F. Kawamura, personal communication). As with many other early sporulation genes, *spo0A* expression increases at T_0 and requires functional *spo0B*, -E, -F, and -H genes as well as *spo0A* itself, and the *sof-1* mutation bypasses the requirement for the first three genes (122). The rise in *spo0A* expression at T_0 is repressed by glucose, and decoyinine, which lowers GTP levels (74), suppresses the glucose effect (S. Yamashita, F. Kawamura, H. Yashikawa, H. Takahashi, Y. Kobayashi, and H. Saito, *J. Gen. Microbiol.*, in press). As discussed above, the *sof-1* mutation renders the sporulation process insensitive to glucose, as do several newly isolated mutations in the *spo0A* ORF amino-terminal region (Olmedo and Youngman, personal communication). One of these, *coi-1*, a change in the highly conserved proline at position 60 to a serine residue, is very close to aspartate 56 (Fig. 1), part of the postulated phosphorylation acceptor site (103). It is tempting to speculate that Spo0A is sensing glucose levels and that Spo0A phosphorylation is somehow involved in this process. Possibly Spo0B, -E, and -F are acting at this step, and *sof-1* and *coi-1* could be acting to make Spo0A phosphorylation unnecessary (or easier).

spo0B

The *spo0B* gene is expressed at maximal levels during vegetative growth, and its expression declines as T_0 is approached (7, 26), which is very different from other early sporulation genes (see below). Another difference between *spo0B* and *spo0A*, *spo0F*, and, most likely, *spo0E* is that the latter three require functional *spo0* genes for expression, while *spo0B* does not. The window of Spo0B function has been determined, using a temperature-sensitive *spo0B* gene. The protein must be present in an active form from the beginning of sporulation through stage II (51). The gene is found downstream from the ribosomal protein L27 genetic determinant (111). In addition, it is in a polycistronic unit which also codes for an essential gene with an ORF resembling that of the *Escherichia coli era* (*ras*-like) gene (1), which, like its *E. coli* counterpart, binds GTP (111). It has been noted that the amino-terminal region of Spo0B shows some similarity to the carboxy-terminal region of Spo0A: 19% identity plus 24% conservative changes (41). Little else is known about *spo0B*, except for the fact that mutations in the *abrB* gene partially suppress *spo0B* lesions (see below). Mutations in *spo0B* also affect the in vitro methylation/demethylation of a 40-kilodalton protein (Golden and Bernlohr, personal communication). The relationship between this in vitro reaction and sporulation is unknown but is being investigated (5).

spo0E

The *spo0E* gene is induced at the end of logarithmic growth, and although complete epistatic requirements have not yet been published, *spo0E* expression is dependent on *spo0A*. There is no obvious similarity between the derived Spo0E amino acid sequence and any other protein (76). The presence of multiple copies of the *spo0E* promoter inhibits sporulation (76), but it is not known whether this is caused by sequestration of a protein essential for sporulation, as has been postulated for a similar phenomenon observed with the *spoVG* promoter (124).

spo0F

By virtue of its amino acid sequence, Spo0F is a receiver in the signal transduction pathway (Fig. 1), but its actual effector role in sporulation, outside of possibly being involved in Spo0A activation, is unknown. One of the more intriguing aspects of Spo0F function is the observation that when present in multiple copies, *spo0F* inhibits sporulation (11, 12, 54). This inhibition is caused by the intact protein, not the promoter, as is the case with *spo0E* and *spoVG*. Mutations in the ORF which cause a loss of the inhibition phenotype when *spo0F* is present in multicopy also causes a Spo$^-$ phenotype when the mutated gene is reintroduced into the chromosome. The reverse is also true; i.e., mutations in the chromosomal *spo0F* ORF which cause null sporulation abolish the inhibition when the mutations are transferred to multicopy plasmids containing *spo0F*. This indicates that the positive and negative effects on sporulation are due to the *spo0F* protein (54). Competence and *aprE* (alkaline protease structural gene) expression (M. Lewandoski and I. Smith, unpublished observations), as well as the rise in *spo0A* transcription at T_0 (12), are also inhibited when Spo0F is overproduced. The expression of *spo0B*, *spo0H*, and *spoVG* is not inhibited by Spo0F overproduction (Lewandoski and Smith, unpublished results). Other genes have not been tested.

It is possible that the inhibition of sporulation caused by overproduction of Spo0F is due to the sequestering of "transmitter" activity (possibly phosphorylation). This could prevent the modification of another sporulation protein like Spo0A, which might be an essential step for sporulation. One would predict, given this hypothesis, that overproduction of a transmitter protein that can interact with Spo0F and the putative essential sporulation protein would restore wild-type sporulation levels to a culture that was unable to produce spores because of excess Spo0F. This prediction has been borne out, using SpoIIJ as the putative transmitter (I. Smith and P. Stragier, unpublished data). When a high-copy-number plasmid with a functional *spoIIJ* gene is introduced into a strain that cannot sporulate (less than 1% of wild type) due to elevated levels of Spo0F, normal levels of sporulation are obtained (30 to 60%). Sporulation deficiency, due to null mutations in various early sporulation genes, and the sporulation inhibition phenotype caused by overproduction of the Sin protein (I. Smith, unpublished experiments) or by multiple copies of *comA* (I. Smith and Y. Weinrauch, unpublished experiments) are not suppressed by overexpression of SpoIIJ. These results strongly suggest that SpoIIJ is a transmitter that interacts with an essential sporulation gene (possibly Spo0A) and Spo0F during the initiation of sporulation. Spo0A (46) (Hoch, personal communication) and Spo0F can be phosphorylated in vitro (Hoch, personal communication; U. Bai, A. Ninfa, and I. Smith, unpublished data), so the interaction may be similar to that described for the better-characterized transmitter-receiver systems (see above). This theory remains at a highly speculative level in the absence of biochemical evidence for interaction between SpoIIJ and Spo0F and/or Spo0A. Such evidence will not be definitive, however, since "cross talk" between transmitters and receivers from different systems can occur in vitro and under special conditions in vivo but may not be physiological (72). Genetic evidence like allele-specific suppression between *spoIIJ* and *spo0F* and/or *spo0A* mutations, which has been obtained in the *ompR-envZ* system (63, 98), would be more convincing. In addition, there may be more than one sporulation transmitter, and it should be possible to use a cloning strategy based on the suppression of Spo0F spore inhibition to isolate genetic determinants for such proteins.

Related Sensory Transduction Systems in Other Growth-Regulated Processes

Bacilli

As discussed above, the DegS/DegU system, which is involved in the synthesis of several extracellular enzymes (see F. Valle and

E. Ferrari, this volume), is related to other two-factor sensory systems. *comA*, one of the early genes in the competence regulon, codes for a protein with strong resemblance to the receiver class (Fig. 3 in the chapter by D. Dubnau, this volume). The *pho* regulon in *B. subtilis*, as in gram-negative bacteria, is controlled by a two-component sensory transduction system (91, 92) (Fig. 1 and 2). Starvation for P_i induces a vegetative alkaline phosphatase, and this process requires *spo0A* (38). Interactions between these two-component regulatory systems and sporulation are discussed in the section on fail-safe controls and conditional loops.

Other bacteria

Two-component sensory systems are also involved in the control of various late-growth-regulated processes in several bacterial genera. The plasmid-linked microcin B17 operon, which is induced at the end of exponential growth in *E. coli*, is under OmpR/EnvZ transcriptional control (33). The *Staphylococcus aureus agr* locus controls the synthesis of several enyzmes, including virulence factors, which appear at the end of exponential growth (83). The *agrA* gene has recently been sequenced (75), and analysis of its derived amino acid sequence indicates the protein is of the receiver class (A. Ninfa, personal communication). Upstream of *agrA* is an ORF, ORF2, which shows strong resemblance to the *B. subtilis* DegS protein and other transmitters (B. Kreiswirth and R. Novick, personal communication). The virulence factors of a number of other pathogenic bacteria, many of which also appear at the end of stationary growth, are controlled by proteins of the transmitter and receiver type (16, 66, 100).

EFFECTOR MECHANISMS IN EARLY STAGES OF SPORULATION

It is difficult to neatly separate a discussion of the signal transduction pathways for the initiation of sporulation from the effector mechanisms which they activate. Most of the components of the signalling steps are themselves up-regulated at the beginning of the developmental process. Thus, components of the transcriptional machinery are made in higher amounts and more Spo0A, Spo0E, and Spo0F are made, in a kind of autogenous cascade. In this section, the interactions between the signal transduction system (largely Spo0A) and the effectors ($E\sigma^A$ and $E\sigma^H$) are discussed. Since negative regulation seems to play a key role in these interrelationships (via the late growth regulatory protein AbrB) as well as others (controlled by the *sin* protein), there will be a separate section on sporulation repressors and divergence pathways.

In recent years, much information has accumulated on the effector components involved in the earliest stages of sporulation. These data are consistent with the belief that most regulation occurs at the level of transcription (see Moran, this volume, for a complete discussion of transcriptional regulation and alternative σ factors). The isolation of alternative σ factors which conferred different promoter specificity on the same core polymerases bearing these transcriptional factors led to the hypothesis of a σ cascade during *B. subtilis* sporulation (56). This hypothesis was based on the role of different σ factors in the temporal control of early, middle, and late gene expression in the virulent *B. subtilis* bacteriophage SPO1 and the purification of a unique holoenzyme during sporulation. It was postulated that under proper signalling, an RNA polymerase holoenzyme would transcribe a block of genes, one of which coded for a σ factor different from the one present in the transcribing RNA polymerase. This new σ factor would form a holoenzyme with unique promoter recognition, and a new set of genes could then be transcribed. One of these would be another σ factor, etc. The concept of a temporally ordered selection of promoter subsets by newly functioning holoenzymes with matching promoter specificities during sporulation has now been largely substantiated. However, the circuitry of synthesis and processing of the alternative σ factors, and of regulation of activity of the newly appearing holoenzymes containing these proteins, is more complex than originally envisaged and cannot be explained by a simple linear pathway.

Role of $E\sigma^A$ in Sporulation

The key holoenzymes during the transition from vegetative growth to stage II are $E\sigma^A$ and $E\sigma^H$. The former transcribes *spo0H* (*sigH*) (20; I. Smith and E. Dubnau, unpublished results), the genetic determinant for σ^H (21; P. Zuber, J. Healy, H. L. Carter III, S. Cutting, C. P. Moran, Jr., and R. Losick, *J. Mol. Biol.*, in press), but also transcribes the later expressed genes *spoIIE* (P. Youngman, personal communication) and *spoIIG* (C. Moran, Jr., personal communication). Other genes turned on at the end of stationary growth, e.g., *aprE*, are also transcribed by $E\sigma^A$ (R. Doi, personal communication). How is activity of this holoenzyme regulated so that housekeeping genes and stage-specific sporulation genes can be expressed at different times? As in *E. coli*, the activity of the major vegetative RNA polymerase seems to be regulated by ancillary factors, i.e., repressors and activators, so that an orderly program of gene expression is followed. There is genetic and physiological evidence that there are interactions between $E\sigma^A$, the catabolite-sensing system, and some of the *spo* proteins during the transition from vegetative growth to stationary phase and sporulation. As discussed above, *crs* mutations, which allow sporulation in the presence of normally inhibitory levels of glucose, have been mapped to the *rpoB* locus (106) and *rpoD* (*sigA*) (45). This latter mutation, *crsA47*, also suppresses the sporulation-deficient phenotype of *spo0B*, -*F*, -*J*, and -*K* mutants (45, 53). Mutations which suppress some of the poor growth characteristics of *crsA47* map in the *spoA*, -*B*, -*F*, and -*K* loci (45). In addition, an *rpoB* mutation, *rfm-11*, suppresses the glucose resistance phenotype of several *crs* mutations

(106). These results strongly suggest that $E\sigma^A$ must interact, at some level, with *spo0* proteins and that glucose sensing is involved in this relationship.

rpoD mutations phenotypically similar to *crsA47* have been isolated in *E. coli*. These lesions allow increased expression from the catabolite-sensitive *ara* and *mal* regulons and the *lac* operon in the absence of cyclic AMP (95). None of these mutations, however, which are predominantly localized in the σ^{70} region 4 helix-turn-helix domain (a presumptive DNA-binding domain) and to a lesser degree in region 2 (see Moran, this volume, for a further explanation of the σ domains), is very close to the equivalent residue of *crsA47*, which is approximately 40 amino acids upstream from the helix-turn-helix region (45).

One of the major controlling elements for postexponential $E\sigma^A$ activity (mediated by Spo0A) is AbrB. As described below, AbrB is a repressor of several genes which are transcribed by this holoenzyme, including *spo0H*, and a discussion of *spo0H* regulation, which follows, will be used to describe some of the interactions between Spo0A, AbrB, and the transcriptional machinery.

Regulation of *spo0H* Expression and $E\sigma^H$ Activity

One of the earliest steps in the initiation of sporulation is the transcription of specific genes by $E\sigma^H$. For a complete understanding of how this effector mechanism is temporally controlled, it is important to know how *spo0H*, which encodes σ^H, is regulated.

Regulation of the *spo0H* gene

High-resolution mapping studies and in vitro assays have shown that $E\sigma^A$ transcribes *spo0H*. In rich media, *spo0H* mRNA is found early in vegetative growth at low levels which increase 5- to 10-fold by T_0, a result also observed with translational *spo0H-lacZ* translational fusions (19, 81; E. Dubnau and I. Smith, unpublished observations; P. Zuber, personal communication). Quantitative immunoassays of the *spo0H* gene product, σ^H, during growth give similar results, but with a delay of approximately 60 min so that a maximum level of σ^H is observed at T_1 to T_2 (Smith and Dubnau, unpublished data; J. Healey and R. Losick, personal communication). Spo0A, activated by Spo0B, Spo0E, and Spo0F interaction or by the *sof-1* mutation, positively regulates the synthesis of *spo0H* (J. Weir, E. Dubnau, and I. Smith, unpublished results; Zuber, personal communication). There is no evidence that Spo0A, a DNA-binding protein (42; Hoch, personal communication), interacts with the *spo0H* promoter. Rather, it is becoming clear that Spo0A acts, to a large extent, by inhibiting the expression of *abrB* via binding to its promoter (78; Hoch, personal communication), so that genes which are repressed by AbrB can be transcribed. This confirms a hypothesis, originally proposed to explain the Spo0A/AbrB relationship, which stated that Spo0A, in an unknown way, down-regulated *abrB* (126). Some genes seem to be exclusively regulated by AbrB so that in *spo0A abrB* double or simple *abrB* mutants, their expression is constitutive, i.e., is observed during vegetative growth, instead of being growth regulated. In this group of genes are *spo0H* (19; Weir et al., unpublished results; Zuber, unpublished results), *spo0E* (76), and *tycA* (62). Other genes, such as *spoVG* (126) and *aprE* (25), are regulated at another level, in addition to AbrB, since temporal regulation is still present in *abrB* mutants even though the *spo0A* requirement is lost. The complex regulation of *aprE* is discussed elsewhere in this volume (Valle and Ferrari), as is *spoVG* transcription (see below). AbrB is a DNA-binding protein interacting with the promoters of genes it regulates (Hoch, personal communication; Zuber, personal communication) (see below).

Studies with *spo0H-lacZ* fusions on multicopy plasmids have shown that *spo0H* expression is not dependent on a functional *spo0A* gene in this setting (J. Weir and I. Smith, unpublished results). This phenomenon has also been observed with *spoVG* (125) and *aprE* (24). It is possible that AbrB does not bind efficiently to these promoters when they are plasmid borne.

Glucose has no effect on *spo0H* expression, and while extensive studies have not been performed, the gene appears to be expressed constitutively in minimal media (Dubnau and Smith, unpublished results).

Role of Eσ^H in sporulation

Eσ^H transcribes genes which are expressed at the beginning of the sporulation process. One of these is the *rpoD* (*sigA*) gene, encoding σ^A. This gene has three promoters of which two are transcribed only during vegetative growth and the third, P3, is read only during stationary phase (114). P3 has a consensus Eσ^H promoter sequence (Fig. 3), its transcription in vivo is dependent on *spo0H*, and it serves as an in vitro template for Eσ^H (10). This situation is analogous to the synthesis of *E. coli* σ^70 and σ^32 after a heat shock. There is a σ^32 consensus promoter sequence for *rpoD* which is transcribed after heat shock (14, 110). The *rpoH* gene, encoding σ^32, has three promoters, two which are recognized by Eσ^70 and one which is transcribed only at elevated temperatures (22) by a different holoenzyme with a newly described σ factor, σ^27 (C. Gross, personal communication).

It is not known whether *B. subtilis rpoD* P3 transcription is essential for sporulation, but the other genes which are transcribed by Eσ^H at the end of exponential growth (*citG* and *spoVG*) are not necessary for the early stage of this process. Since mutations in *spo0H* block sporulation before stage II, there must be one or more essential early sporulation genes which are transcribed by this holoenzyme. *spo0F* is a likely candidate, since its expression in vivo shows the same temporal pattern, nutritional responses, and *spo0H* requirement (122; Lewandoski and Smith, unpublished data) as *spoVG*, which is transcribed by Eσ^H (see Moran, this volume, and R. Losick and L. Kroos, this volume). In addition, the *spo0F* promoter closely resembles the consensus promoter recognized by this holoenzyme (Fig. 3). Eσ^H is also involved in the transition to stage II. It probably transcribes *spoIIA*, as in vivo expression of this latter gene is dependent on *spo0H* (120). The *spoIIA* promoter is similar to the Eσ^H consensus sequence (120) (Fig. 3), and it is transcribed, in vitro, by this holoenzyme (Moran, personal communication). *spoIIAC* encodes σ^F, a later-acting transcription factor (P. Setlow, personal communication). Since Eσ^A transcribes *spoIIG*, which codes for σ^E, it is clear that even at the earliest stages of sporulation, prior to the separation of the forespore and its transcriptional machinery from the spore mother cell, there are at least two

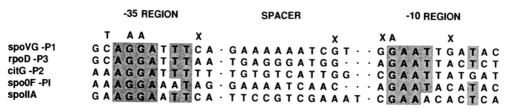

Figure 3. Promoters recognized by Eσ^H. Sequences are from the following references: *citG* (P2) (23); *rpoD* (P3) (10); *spoVG* (P1) (4); *spoIIA* (120); and *spo0F* (54). Expression of all promoters has been shown to be dependent on an intact *sigH* gene (*spo0H*) in vivo (10, 23, 120, 122, 125). Transcription from the *citG* (P2), *rpoD* (P3), and *spoVG* (P2) promoters by Eσ^H has been demonstrated in vitro (9, 10; Moran, personal communication). The base changes over the *spoVG* sequence are those which give severe down effects on transcription. The crosses indicate positions in which mutations have little or no efect. The mutational analysis is from the Losick laboratory (Zuber et al., in press). Bases present at the same position in at least four of the five promoters are shaded.

divergent pathways of σ factor-controlled gene expression.

As discussed in this chapter and others in this book, most late-growth processes require functional *spo0A* and *spo0H* genes. Since *spo0A* is required to down-regulate AbrB, a repressor of *spo0H* synthesis, it is difficult to decide, in some cases, whether a need for Spo0A is direct or indirect (is actually a specific *spo0H*, i.e., σ^H requirement). While *abrB* mutations can be used to study this question, this approach is only applicable to genes which are under AbrB control. There are genes essential for sporulation which are not subject to AbrB repression, since *spo0A abrB* double mutants cannot sporulate. It has been noted that expression of the *spo0H* gene is independent of *spo0A* when on a multicopy plasmid (see above). This fact was utilized to determine *spo0A* versus *spo0H* requirements for the expression of certain late-growth-regulated genes (Weir et al., unpublished data). As expected, *spoVG* expression was restored to wild-type levels in a *spo0A* mutant carrying the *spo0H* gene on a plasmid. Temporal control was normal, indicating other factors besides expression of *spo0H* were involved in the regulation of *spoVG* expression (see below). Identical results were obtained with *spo0F*, providing further circumstantial evidence that this gene is transcribed by $E\sigma^H$. The same experimental protocol has been used to show that *phoA* expression, which is under *spo0A*/*spo0H* control, only requires the σ^H genetic determinant (38). In all of the above experiments, the suppressing effects of plasmid-borne *spo0H* were due to the intact gene, presumably reflecting synthesis of σ^H, and not due to the sequestering of a repressor by the *spo0H* promoter. This method is being used to study the regulation of other *spo0A*/*spo0H*-dependent genes, but it is clear that *spo0A* has functions in sporulation other than the dialectical one of negating a negator, i.e., freeing *spo0H* from AbrB control. *spo0A* mutants carrying a plasmid-borne *spo0H* gene are still unable to sporulate.

Regulation of $E\sigma^H$ activity

While the regulation of $E\sigma^H$ activity is extensively covered elsewhere in this book (Moran, this volume; Losick and Kroos, this volume), there are a few points which are of special relevance to the areas covered in this chapter. Studies, discussed above, in which the amounts of σ^H during growth in rich medium are measured show a direct correlation between the absolute amount of this protein and *spoVG* expression (Smith and Dubnau, unpublished data; Healy and Losick, personal communication). This suggests that under these growth conditions, the level of $E\sigma^H$ is the determining factor in the transcription of promoters utilized by this holoenzyme. There are some unanswered questions, however, in this relatively simple scenario. Under certain conditions when the *spo0H* gene is transcribed in vegetative growth, by virtue of being placed behind an inducible promoter (Healy and Losick, personal communication) or on a multicopy plasmid (Weir and Smith, unpublished observations), *spoVG* shows normal temporal induction (see above). This could be explained by the fact that AbrB, a direct-acting repressor of *spoVG*, is present. A *cis*-acting mutation in the *spoVG* promoter (Fig. 4) has the same phenotype (independence of a *spo0A* requirement for *spoVG* expression) as an *abrB* mutation (126), and it has recently been demonstrated that AbrB binds to this region of *spoVG* (Zuber, personal communication). One would expect, therefore, that in an *abrB* mutant, *spoVG* expression should be constitutive, since this is true for *spo0H*. However, this is not the case, and it is possible that even though *spo0H* can be transcribed in vegetative growth, there is posttranscriptional regulation, either at the level of mRNA or protein stability. This could result in higher amounts of σ^H only at the end of stationary growth, even when the gene is transcribed during vegetative growth. There could also be ancillary factors necessary for the temporally regulated transcription of $E\sigma^H$-dependent genes.

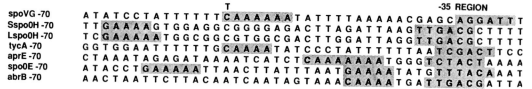

Figure 4. AbrB-binding sites on temporally regulated promoters. This figure displays temporally regulated promoters which are affected by AbrB in vivo (all of the promoters) and which bind to this protein in vitro: *spoVG* (Zuber, personal communication) and *aprE*, *spoOE*, and *abrB* (Hoch, personal communication). Similar sequences, those showing a G(C)AAAA sequence at or near a putative AbrB-binding site, are underlined, as are −35 hexamer sequences. References to in vivo results are in the text. The C-to-T mutation in *spoVG* (−57) renders this promoter independent of *abrB* (126). L*spo0H* and S*spoH* refer to the *spo0H* genes from *B. licheniformis* and *B. subtilis*, respectively.

FAIL-SAFE MECHANISMS AND CONDITIONAL LOOPS IN LATE-GROWTH DEVELOPMENT

Formal models could be elucidated to explain sporulation initiation by using only positive control (all the *spo0* genes are defined as positive factors since mutations in these elements stop the process at the earliest stage). However, it is apparent that negative regulation plays a very important role in the cell's decision to sporulate. The pleiotropic nature of mutations coding for some of these negative elements suggests that these proteins also act as switches, channeling stationary-phase cells away from sporulation to other developmentally regulated processes. It is the purpose of this section to analyze the functions of some of the better-characterized sporulation repressors in light of this possible role. This approach will then lead to a discussion of the interactions between the different global processes which are initiated at the commencement of the stationary phase of growth.

Sporulation Repressors

There are three proteins which repress sporulation or certain aspects of the developmental program, AbrB, Hpr, and Sin (see Valle and Ferrari, this volume, and Dubnau, this volume, for a related discussion of these proteins). The genes coding for these proteins are expressed during exponential growth (27, 77, 78). The first two are under the control of Spo0A, as this protein represses *abrB* expression by binding to its promoter (Hoch, personal communication) and AbrB is a positive activator of *hpr* expression, binding to the promoter as well (Hoch, personal communication). Expression of *sin* is independent of direct Spo0A control (27), but Sin, its gene product, is extremely unstable in crude extracts of wild-type cells and has only been purified from *spo0A* mutants (N. Gaur and I. Smith, unpublished results). This suggests that Spo0A may be involved in the down-regulation of Sin, at the level of proteolysis. This would be analogous to RecA action on LexA and certain bacteriophage repressors after induction of the SOS response in *E. coli* (55). Spo0A does not necessarily have to be involved directly in the degradation of Sin but might stimulate the synthesis of a specific protease. Interestingly, the same possibility was originally entertained for the role of Spo0A in the regulation of AbrB (126).

Hpr

Hpr inhibits sporulation when it is present in abnormally high quantities (77). It is not known which essential sporulation processes are inhibited by Hpr, but loss-of-function mutations in the *hpr* gene cause insensitivity to glucose repression of sporulation (77). This suggests some link between Hpr and nutrient-sensing systems. It is also not known whether Hpr is a DNA-binding protein, as is the case with both Sin and AbrB.

Sin

Sin, like Hpr, inhibits sporulation when overexpressed (28). Until recently, sporulation targets for Sin inhibition had not been identified. In experiments using integrated *lacZ* gene fusions, the expression of all early sporulation genes tested (*spo0B*, *spo0F*, *spo0H*, and *spoVG*), and that of some other late-growth-regulated genes as well (*citB* and *isp-1*), was unaffected by Sin overexpression (Gaur and Smith, unpublished data). Under the same experimental conditions, more than 90% repression of the *aprE* gene was observed (28). Since Sin overexpression blocks sporulation prior to the appearance of alkaline phosphatase (28), a biochemical marker for successful completion of stage II, the Sin effect on several *spoII* genes was then examined (I. Mandic-Mulec, N. Gaur, and I. Smith, unpublished data). It was observed that *spoIIA* and *spoIIE* expression was severely repressed by overabundance of Sin, while *spoIIG* was unaffected. When the *sin* gene was disrupted, the level of *spoIIE* expression was higher than in the wild type (the other *spoII* genes have not been tested with a *sin* disruption, as yet). It is not known, yet, whether Sin binds directly to the *spoIIE* and *spoIIA* promoters, as it does to the upstream region of *aprE* (see Valle and Ferrari, this volume), or whether the inhibitory effect is indirect, via the repression of a positive regulatory gene. In any case, the Sin effect seems specific, since not all of the stage II genes tested are affected by the Sin protein. It is too early to speculate on the mechanism of Sin repression of gene function, since definitive footprinting has only been obtained for the Sin/*aprE* interaction. In this case, there are two binding sites for Sin, one at position -260 to -240 and the second near the transcriptional initiation site (Gaur and Smith, unpublished results). This suggests that DNA looping, caused by Sin binding, may be involved in repression of *aprE*, as has been demonstrated for other systems (*gal* [60], *lac* [89], *ara* [39], R6K replication [70], and λ p_{RM} regulation [37]). Sin is also

an autogenous repressor of its own synthesis, but there is probably one binding site for the protein, close to the transcription initiation site, and looping is not likely. In any case, gene repression caused by Sin is not due to interaction between it and a specific holoenzyme. *spoIIE* is transcribed by Eσ^A, while Eσ^H transcribes *spoIIA*.

AbrB

It has not been demonstrated that abnormally high levels of AbrB will inhibit sporulation. Since this protein does inhibit expression of the essential sporulation genes *spo0E* and *spo0H*, it would be expected to block the developmental process, if overexpressed. The mechanism of AbrB repression may be similar to that of Sin. Genes which are regulated by this protein have binding sites in their promoters (Fig. 4). There are certain similarities in these DNA sequences which are shared by other genes whose expression is affected, in vivo, by AbrB, but which have not yet been shown to bind AbrB (*tycA* and *spo0H*). Prominent among these sequences is the C(G)AAAA sequence found in the -40 to -70 region of all AbrB target genes. A mutation in this motif, *spoVG42*, which is a C-to-T transition at the -57 position of *spoVG*, renders this gene independent of *spo0A* control and gives a phenotype (with regard to *spoVG* expression) identical to that observed with a mutation in *abrB* (126). In addition, the *B. subtilis spo0H* promoter has several upstream GAAAA sequences. The equivalent *Bacillus licheniformis* sequence, as originally cloned, contains only the promoter-proximal GAAAA at position -68 (Fig. 4). Both *spo0H* genes are under *spo0A abrB* control (19; Weir et al., unpublished data). Since the G(C)AAAA sequence in *aprE*, *spo0E*, and *abrB* (the latter gene is controlled autogenously by AbrB [78]) is protected by AbrB binding (Hoch, personal communication), it seems clear that this short sequence is important for AbrB-DNA interaction. The mechanism of AbrB repression, e.g., by rendering the promoter inaccessible to RNA polymerase or by preventing formation of an open transcription complex, etc., is presently unknown. However, since this protein binds near the -35-base-pair region of all AbrB-sensitive promoters, it should interfere with an early stage of transcription initiation. As observed with Sin, AbrB repression is not polymerase specific. This protein represses $spoVG$, transcribed by $E\sigma^H$, as well as inhibiting $aprE$, $spo0H$, $spo0E$, and $tycA$. $aprE$ and $spo0H$ are transcribed by $E\sigma^A$, and this is probably true for $spo0E$, and $tycA$, judging from their promoter sequences (62, 76). AbrB is also a repressor of flagellin synthesis (D. Arnosti and M. Chamberlin, personal communication).

Alternate Pathways in Late-Growth Development

Even though our knowledge of the regulation of the various late-growth phenomena discussed in this chapter and others in this volume has increased dramatically in the last few years, our understanding is still too fragmentary to allow a definitive description of their control and interaction with each other. However, the sporulation repressors described in the previous section, Sin and AbrB, function in two capacities, as positive and negative regulators of gene expression. Sin represses specific sporulation genes and $aprE$, but is required for motility and competence, while AbrB represses certain sporulation and other late-growth-regulated genes, yet is necessary for competence and hpr expression. This duality of function suggests that these two proteins act as regulatory switches, inhibiting sporulation, but serving to positively regulate other late-growth processes, which can be viewed as alternative pathways for differentiation. As previously discussed, sporulation is costly and potentially dangerous to the cell facing a changing environment. Consequently, the existence of such alternative responses to an adverse environment, before terminal differentiation, is not surprising.

The concept of developmental switches, i.e., dual functional regulators of gene function which control pathways of development, is not a new one, and there are many examples in both procaryotes and eucaryotes. SpoIIID serves this role in regulating $E\sigma^K$ transcription in the spore mother cell (49; Losick and Kroos, this volume). The λ cI repressor also has this capability (65). In yeasts, the RAN1 and REM1 proteins function to repress meiosis and sporulation and are essential for certain vegetative functions (64, 68). The lin-14 gene of $Caenorhabditis$ $elegans$ encodes a protein which is essential in determining the temporal regulation of developmental fate. When overexpressed (gain-of-function mutations), cells which would normally differentiate at later stages maintain an earlier phenotype. Null mutations in lin-14 (loss-of-function lesions) give the opposite phenotype; i.e., cell lineages which would ordinarily arise at later developmental periods make a precocious appearance (87). This latter example is remarkably reminiscent of Sin function in late-growth regulation.

At this point in our discussion, it will be useful to describe some of the general features of the alternative developmental pathways, with special attention to their integration with the sporulation response. All late-growth responses require Spo0A, so the branch points in development must diverge after the functioning of this key component. Glucose inhibits many of the late-growth processes, and mutations in some of regulatory genes for these pathways allow sporulation in the presence of normally inhibitory levels of this carbon source. Among these are hpr mutations, which cause higher levels of $aprE$ expression, thought to be loss-of-function lesions (77). Missense mutations in both DegS or DegU [$sacU$(Hy) mutations] allow sporulation in the presence of normally inhibiting levels of glucose, as well as cause overexpression of $aprE$ and other growth-regulated genes (32). As discussed above, these proteins form a transmitter-receiver sensory transducing dyad. One of these lesions, $degU$(Hy)32, is at residue 12, adjacent to the conserved aspartate at position 11 in all receivers, and is the same residue changed in the Spo0A *sof-1* mutation which also causes glucose-insensitive sporulation (see above). The immediate result of the *sacU*(Hy) mutations, e.g., increased protein stability or function, etc., is not known, but overexpression of the wild-type DegU causes high level of alkaline protease production (108) (the effects of this condition on glucose inhibition of sporulation were not reported). Null mutations in these genes (*sacU*-negative mutations) do not cause a sporulation-negative phenotype. This indicates that DegS and DegU are not essential for sporulation, but that there may be some cross talk between sporulation signal transducers (or effectors) and these regulators, and perhaps a common transmitter interacts with both a sporulation receiver and DegU.

The competence regulon (reviewed by Dubnau, this volume) is extremely interesting in regard to sporulation. Late *com* genes are dependent on *spo0A* and *spo0H*, are growth regulated (show increases in gene expression at T_0), but require glucose for expression, which is consistent with the fact that the development of competence is dependent on the presence of this carbon source. Thus a divergence between sporulation and competence pathways must involve glucose sensing and response. The existence of mutations which alter glucose susceptibility of sporulation and competence in opposing ways supports this concept. *sacU*(Hy) mutations have very low competence (99) (overproduction of wild-type DegU also inhibits competence [108]). Inactivation of *sin* prevents competence by blocking expression of late *com* genes (see Dubnau, this volume), and Sin represses certain stage II sporulation genes (Mandic-Mulec et al., unpublished results). Thus, Sin seems to play the role of a switch, possibly acting in conjuction with another protein which is controlled by glucose levels. The isolation of *mec* mutations, which allow expression of late *com* genes in the absence of

glucose, suggests the presence of Sin-like repressors of the *com* regulon. Since some of the *mec* mutations cause a Spo$^-$ phenotype, there seems to be a complementary control network regulating sporulation initiation and competence. One can speculate that glucose plus Sin and low levels of functional Mec protein(s) result in high competence and low sporulation, while the absence of both glucose and functional Sin, plus the presence of Mec protein(s), gives the reverse phenotype.

Both sporulation and competence require Eσ^H. A transposon-induced mutation defining a new gene, *csh-293*, which is dependent on *spo0H*, causes defects in both sporulation and competence. This mutation, similar in phenotype to *spo0A* lesions, is linked to *aroI* and *dal*, far from *spo0A* (K. J. Jaacks, J. Healy, R. Losick, and A. D. Grossman, submitted for publication). There are other significant relationships between sporulation and competence. One of the early (constitutive) competence genes, *comA*, has an ORF with a conserved receiver amino-terminal domain (see Fig. 3 in Dubnau, this volume), and this gene in multicopy inhibits sporulation. This phenomenon is different from the Spo0F inhibition phenotype, since overexpression of SpoIIJ doesn't suppress the *comA* effect. Similarly, overexpression of Spo0F inhibits competence (Lewandoski and Smith, unpublished data), and the mechanism is not known. The role of AbrB in the decision to become competent or to sporulate is intriguing, but it is not known where in the competence pathway the AbrB requirement is manifested.

There are other pathways which show relationships to sporulation. For example, flagellar formation is inhibited in *sacU*(Hy) strains (99). *sin* deletions cause nonmotility (Gaur and Smith, unpublished results), and the recent sequencing of the *flaD* gene, which is essential for flagellar formation and motility (93), shows it is identical to *sin* (J. Sekiguchi, personal communication). Flagellar synthesis is under the positive control of Eσ^D (31). Not all of the genes which are tran-

scribed by this holoenzyme have been characterized as to epistatic requirements, but at
least two of these are dependent on Spo0A
(29). The *hag* gene, coding for flagellin, is
transcribed by Eσ^D and is repressed by AbrB
(Arnosti and Chamberlin, personal communication). Little else is known about the relationship between motility and late growth
regulation, but on the basis of the Sin role in
the *com* regulon, it is possible that Sin acts
as a positive regulator for motility genes,
working in conjunction with Eσ^D, and that
AbrB is a general repressor for this regulon.

We leave this discussion of interacting
networks with some interesting relationships
and several unanswered questions. These will
be addressed in the final summary and per-
spectives section below.

SUMMARY AND PERSPECTIVES

This article has viewed the initiation of
sporulation in the light of recent developments in the field of procaryotic physiology
and responses to environmental stress. The
Bacillus cell faced with an environmental
stress, usually starvation, must process environmental information so that it can make
the best choice as to entering one of the possible alternative pathways of development.
The signals are not yet known, even though
low levels of efficiently utilizable carbon, nitrogen, or phosphate can induce sporulation.
Lowering internal GTP pools can also induce
this process. The signal transduction system
for sporulation utilizes motifs similar to those
found in other environmentally responsive
systems. In these highly conserved signalling
systems, an environmental signal is relayed
to a transmitter protein, which in several cases
has been shown to be an ATP-utilizing, au-
tophosphorylating kinase. The phosphorylated transmitter transfers its phosphate to a
receiver protein which activates the latter for
its role in the response to the environmental
signal. The essential early sporulation proteins Spo0A and Spo0F are of the receiver
type, and both are phosphorylated in vitro.

While a sporulation-specific transmitter protein has not been conclusively identified,
physiological evidence suggests that SpoIIJ,
which has high sequence similarity to all
transmitters, can interact with the two *spo0*
receivers.

Activated Spo0A (by interaction with
Spo0B, Spo0E, and Spo0F or by the *sof-1*
mutation) is essential for all late-growth processes. A major role for this key protein is
the down-regulation of the *abrB* gene which
codes for AbrB, a negative regulator for these
temporally regulated phenomena. One of the
important genes regulated by AbrB is *spo0H*,
an essential early sporulation gene coding for
σ^H. RNA polymerase holoenzyme containing
this σ factor transcribes genes which are expressed, in vivo, at the beginning of the developmental process. The major cellular RNA
polymerase, containing σ^A, transcribes *spo0H*
and also the gene for a later-acting σ factor,
σ^E. Eσ^H transcribes both a minor promoter
of the σ^A gene and also *spoIIA*, which codes
for the later-acting σ^F. This indicates that a
cascade for sigma factors does exist, but it
proceeds in a branched pathway.

Repressors of sporulation have been isolated, and it is believed that these proteins,
AbrB and Sin, both promoter-binding proteins, may be involved in developmental
switching. This hypothesis is based on their
dual regulatory nature, i.e., they are repressors of sporulation and also activators of other
late-growth processes such as competence
(AbrB and Sin) and motility (Sin). Since sporulation is a serious decision for a cell, it is
believed essential to have several layers of
fail-safe control and alternative pathways so
that it does not sporulate prematurely and
can respond to an environmental stress in a
less than terminal fashion, unless absolutely
necessary. (Figure 5 presents a cartoon view
of the summary).

There are several areas of ignorance and
uncertainty in this oversimplified description
of the early stages of sporulation. In the area
of signal transduction, it is essential to learn
about the specific signal(s) which is sensed

by cellular components to alert the cell of impending starvation. Very little is known about the receptors, as well. For example, how are external glucose levels sensed? It would be surprising if the phosphotransferase system pathway were not involved, since this membrane-localized regulatory system is also involved in catabolite repression in enteric bacteria (80). How is GTP binding involved in the repression of sporulation? Is the B. subtilis Ras-like *era* gene involved in sporulation? How is information of this type relayed to a hypothetical transmitter? Even if it is demonstrated that SpoIIJ is an autokinase and that it interacts, both in vitro and in vivo, with SpoOF and SpoOA, will other transmitters also function in sporulation? An analogous situation exists in the chemotaxis system, where there are dedicated membrane-bound sensory molecules, Tsr, Tar, Trg, Tap, etc., which interact with the soluble transmitter, CheA. In addition, information from aerotaxis receptors, the phosphotransferase system, and several periplasmic sugar-binding proteins is also relayed to the chemotaxis effector machinery (58). There is also in vitro and in vivo evidence for cross talk between heterologous transmitter-receiver pairs in the nitrogen assimilation and chemotaxis pathways (72). Since starvation for carbon, nitrogen, and phosphate can induce sporulation, it is logical to assume that several sensing systems will interact with the sporulation pathway. Many of the late-growth-regulated processes such as competence and extracellular enzyme secretion utilize transmitter-receiver dyad sensing systems and also require SpoOA. What is their relation to sporulation and vice versa? Since many repressors of sporulation are essential for other late-growth processes and vice versa (in the case of the competence regulon, at least one *mec* gene functions in this manner), what is the network of interaction?

While the pathways of effector action during early stages of sporulation are becoming more comprehensible, there are still some unresolved questions. What determines the

ability of $E\sigma^A$ to transcribe first vegetative genes and then *spo0H* at the end of growth, and then to transcribe *spoIIG* and *spoIIE* 1 to 2 h later, during stage II? The same question could be raised for the transcription of *spoIIA* by $E\sigma^H$, which occurs 1 to 2 h after the rise in *spo0F* and *spoVG* transcription at T_0. As discussed in this review, inactivation of AbrB does not change the temporal regulation of certain genes repressed by this protein. This suggests that other repressors, like Sin which inhibits *spoIIE* and *spoIIA*, or positive activators play a role in the transition from stage 0 to stage II. Relatively modern techniques, such as "Southwestern" protein blotting (96), in vivo footprinting (89), and gel retardation with crude extracts (44), have made it possible to identify and purify procaryotic and eucaryotic proteins which bind to specific DNA probes. It should be relatively simple to isolate proteins which interact with temporally regulated promoters, like *spoIIA*, *spoIIE*, and *spoIIG*, etc., in a sporulation-stage-specific manner. Perhaps it will be found that SpoOA is one of these hypothetical binding proteins!

The last sentence leads to the major unanswered query in the initiation of sporulation. What is the role of SpoOA in this process and how does activation/phosphorylation of this protein alter its function? Also, what does the *sof-1* mutant really do? Will it suppress *spo0B*, *spo0E*, and *spo0F* double or triple mutations? If so, this would indicate that all of their gene products are involved in the same step and that *sof-1* is a true bypass mutation. Is *sof-1*-altered SpoOA a better substrate for phosphorylation, or is it so changed that it is no longer able to accept phosphotransfer; i.e., is it constitutively altered? SpoOA is essential for all late-growth processes, but its only known task is to repress the synthesis of AbrB. However, absence of AbrB, a repressor of sporulation, is not sufficient to allow sporulation if SpoOA is not present. The same result was observed in an *hpr abrB spo0A* triple mutant (77). Inactivation of Sin is also insufficient to suppress the sporulation defect

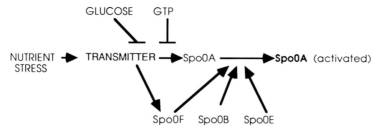

A Stress sensing and activation of Spo0A

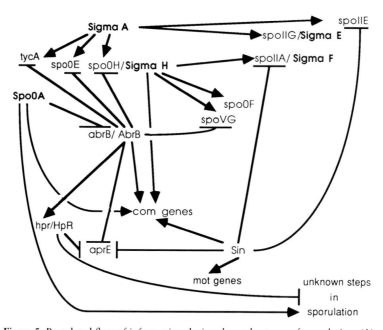

B Early changes in gene expression leading to sporulation

Figure 5. Postulated flow of information during the early stages of sporulation. (A) Sensing of nutrient stress and activation of Spo0A. A nutrient stress signal is received by one or more sensors, and this protein(s) interacts with Spo0F and/or Spo0A. Spo0A becomes activated by the action of Spo0B, Spo0E, and Spo0F. High levels of GTP and glucose block the process in an unknown manner. The *sof-1* mutation in Spo0A eliminates the need for SpoB, SpoE, and SpoF in sporulation and also makes this developmental process insensitive to glucose repression, as if glucose inhibits the Spo0A activation step. The arrows indicate an activation or positive information flow, while the T-shaped symbol indicates a down-regulatory event. (B) Early changes in gene expression leading to sporulation. Spo0A functions as a repressor of transcription of the negative regulator, AbrB, and as a positive regulator of the competence regulon and still-undefined steps in sporulation. The activation of Spo0A illustrated in panel A is essential for all of these functions, except for the competence requirement (D. Dubnau, personal communication). A major known function for activated Spo0A is to down-regulate *abrB* expression. AbrB is a negative regulator of the late-growth-regulated genes *tycA* and *aprE*, as well as sporulation genes *spo0E*, *spo0H*, and *spoVG*. AbrB is also a repressor of some genes of the *mot* regulon (Arnosti and Chamberlin, personal communication), but the interaction is

of *spo0A* mutations. It would be interesting to create a quadruple mutant, with all of the three sporulation repressor genes and *spo0A* inactivated. Such a strain might never grow, as it may sporulate prematurely. This possibility could be controlled experimentally by putting one of the repressor genes behind a controllable promoter, or making a temperature-sensitive Sin or AbrB protein. Assuming a plausible result, i.e., inactivating all three repressors still does not bypass the Spo0A requirement for sporulation, how does one look for the essential Spo0A function? Hopefully, all of the workers interested in the initiation of sporulation are designing biochemical, physiological, and genetic experiments to answer the key question of Spo0A function.

Acknowledgments. I would like to thank past and present members of my group, Jeanie Dubnau, Kettly Cabane, Uma Bai, Nand Gaur, Mark Lewandoski, Ines Mandic-Mulec, Mima Predich, Gopal Nair, Narayanarao Ramakrishna, and Joyce Weir, for their hard work and intellectual contributions. I also thank my colleagues David Dubnau, Yvette Weinrauch, Richard Losick, Judy Healy, Peter Zuber, Patrick Stragier, Patrick Piggot, Alex Ninfa, Jeffrey Stock, James Hoch, Eugenio Ferrari, Dennis Henner, Georges Rapoport, Roy Doi, Peter Setlow, Philip Youngman, Gabriela Olmedo, Charles Moran, Jr., Alan Grossman, Abraham Sonenshein, Fujio Kawamura, Richard Novick, Barry Kreiswirth, and Robin Smith for helpful discussions and unpublished information. My deepest appreciation is expressed to Annabel Howard for invaluable secretarial assistance.

The work performed in my laboratory was supported by Public Health Service grants GM19693 and GM32651 from the National Institutes of Health and by a grant from the National Science Foundation (PCM-8313516, for computer facilities).

LITERATURE CITED

1. **Ahnn, J., P. E. March, H. E. Takiff, and M. Inouye.** 1986. A GTP-binding protein of *Escherichia coli* has homology to yeast RAS proteins. *Proc. Natl. Acad. Sci. USA* **83**:8849–8853.

2. **Allen, E. R., C. Orrego, H. Wabiko, and E. Freese.** 1986. An *ethA* mutation in *Bacillus subtilis* 168 permits induction of sporulation by ethionine and increases DNA modification of bacteriophage φ105. *J. Bacteriol.* **166**:1–8.

3. **Amemura, M., K. Makino, H. Shinagawa, and A. Nakata.** 1986. Nucleotide sequence of the *phoM* region of *Escherichia coli*: four open reading frames may constitute an operon. *J. Bacteriol.* **168**:294–302.

4. **Banner, C. D., C. P. Moran, Jr., and R. Losick.** 1983. Deletion analysis of a complex promoter for a developmentally regulated gene from *Bacillus subtilis*. *J. Mol. Biol.* **168**:351–365.

5. **Bernlohr, R. W., A. L. Saha, C. C. Young, B. R. Toth, and K. J. Golden.** 1988. Nutrient-stimulated methylation of a membrane protein in *Bacillus licheniformis*. *J. Bacteriol.* **170**:4113–4118.

6. **Borkovich, K. A., N. Kaplan, J. F. Hess, and M. I. Simon.** 1989. Transmembrane signal transduction in bacterial chemotaxis involves ligand-dependent activation of phosphate group transfer. *Proc. Natl. Acad. Sci. USA* **86**:1208–1212.

7. **Bouvier, J., P. Stragier, C. Bonamy, and J. Szulmajster.** 1984. Nucleotide sequence of the *spo0B*

---

not depicted in the figure, for reasons of clarity. It positively regulates Hpr and the competence (*com*) regulon. Sin is a repressor of *aprE* and some stage II sporulation genes. It is an activator for the *com* regulon and motility (*mot*) genes, which are transcribed by $E\sigma^D$. RNA polymerase containing the major vegetative σ factor, σ^A, transcribes the *spo0H* gene, which codes for σ^H. RNA polymerase containing σ^H transcribes *spoVG* and possibly *spo0F*. RNA polymerases containing σ^A and σ^H also function at stage II as the former enzyme transcribes *spoIIE* and *spoIIG*, which is the genetic determinant for σ^E. RNA polymerase $E\sigma^H$ transcribes *spoIIA*, which codes for σ^E. In this figure, all events, except for those on the extreme upper right, take place during the transition from vegetative to stationary growth (T_0). References for this figure are in the text.

gene of *Bacillus subtilis* and regulation of its expression. *Proc. Natl. Acad. Sci. USA* 81:7012–7016.

8. Buikema, W. B., W. W. Szeto, P. V. Lemley, W. H. Orme-Johnson, and F. M. Ausubel. 1985. Nitrogen fixation specific regulatory genes of *Klebsiella pneumoniae* and *Rhizobium meliloti* share homology with the general nitrogen regulatory gene *ntrC* of *K. pneumoniae. Nucleic Acids Res.* 13:4539–4555.

9. Carter, H. L., III, and C. P. Moran, Jr. 1986. New RNA polymerase σ factor under *spo0* control in *Bacillus subtilis. Proc. Natl. Acad. Sci. USA* 83:9438–9442.

10. Carter, H. L., III, L. Wang, R. H. Doi, and C. P. Moran, Jr. 1988. *rpoD* operon promoter used by σ^H-RNA polymerase in *Bacillus subtilis. J. Bacteriol.* 170:1617–1621.

11. Chapman, J. W., and P. J. Piggot. 1987. Analysis of the inhibition of sporulation of *B. subtilis* caused by increasing the number of copies of the *spo0F* gene. *J. Gen. Microbiol.* 133:2079–2088.

12. Chibazakura, T., S. Yamashita, H. Yoshikawa, F. Kawamura, H. Takahashi, and H. Saito. 1988. The multicopy *spo0F* gene inhibits an enhancement of the *spo0A* transcription at the early stage of sporulation in *Bacillus subtilis. J. Gen. Appl. Microbiol.* 34:451–455.

13. Comeau, D. E., K. Ikenaka, K. Tsung, and M. Inouye. 1985. Primary characterization of the protein products of the *Escherichia coli ompB* locus: structure and regulation of synthesis of the OmpR and EnvZ proteins. *J. Bacteriol.* 164:578–584.

14. Cowing, D. W., J. C. A. Bardwell, E. A. Craig, C. Woolford, R. W. Hendrix, and C. A. Gross. 1985. Consensus sequence for *Escherichia coli* heat shock gene promoters. *Proc. Natl. Acad. Sci. USA* 82:2679–2683.

15. Dawes, I. W., and J. Mandelstam. 1970. Sporulation of *Bacillus subtilis* in continuous culture. *J. Bacteriol.* 103:529–535.

16. Deretic, V., R. Dikshit, W. M. Konyecsni, A. M. Chakrabarty, and T. K. Misra. 1989. The *algR* gene, which regulates mucoidy in *Pseudomonas aeruginosa*, belongs to a class of environmentally responsive genes. *J. Bacteriol.* 171:1278–1283.

17. Dixon, R. A. 1984. The genetic complexity of nitrogen fixation. *J. Gen. Microbiol.* 130:2745–2755.

18. Drury, L. S., and R. S. Buxton. 1985. DNA sequence analysis of the *dye* gene of *Escherichia coli* reveals amino acid homology between the Dye and OmpR proteins. *J. Biol. Chem.* 260:4236–4242.

19. Dubnau, E., K. Cabane, and I. Smith. 1987. Regulation of *spo0H*, an early sporulation gene in bacilli. *J. Bacteriol.* 169:1182–1191.

20. Dubnau, E., N. Ramakrishna, K. Cabane, and I. Smith. 1981. Cloning of an early sporulation gene in *Bacillus subtilis. J. Bacteriol.* 147:622–632.

21. Dubnau, E., J. Weir, G. Nair, L. Carter III, C. Moran, Jr., and I. Smith. 1988. *Bacillus* sporulation gene *spo0H* codes for σ^{30} (σ^H). *J. Bacteriol.* 170:1054–1062.

22. Erickson, J. W., V. Vaughn, W. A. Walter, F. C. Neidhardt, and C. A. Gross. 1987. Regulation of the promoters and transcripts of *rpoH*, the *Escherichia coli* heat shock regulatory gene. *Genes Dev.* 1:419–432.

23. Feavers, I. M., V. Price, and A. Moir. 1988. The regulation of the fumarase (*citG*) gene of *Bacillus subtilis* 168. *Mol. Gen. Genet.* 211:465–471.

24. Ferrari, E., S. M. H. Howard, and J. A. Hoch. 1986. Effect of stage 0 sporulation mutations on subtilisin expression. *J. Bacteriol.* 166:173–179.

25. Ferrari, E., D. J. Henner, M. Perego, and J. A. Hoch. 1988. Transcription of *Bacillus subtilis* subtilisin and expression of subtilisin in sporulation mutants. *J. Bacteriol.* 170:289–295.

26. Ferrari, F. A., K. Trach, D. LeCoq, J. Spence, E. Ferrari, and J. A. Hoch. 1985. Characterization of the *spo0A* locus and its deduced product. *Proc. Natl. Acad. Sci. USA* 82:2647–2651.

27. Gaur, N. K., K. Cabane, and I. Smith. 1988. Structure and expression of the *Bacillus subtilis sin* operon. *J. Bacteriol.* 170:1046–1053.

28. Gaur, N. K., E. Dubnau, and I. Smith. 1986. Characterization of a cloned *Bacillus subtilis* gene which inhibits sporulation in multiple copies. *J. Bacteriol.* 168:860–869.

29. Gilman, M., and M. Chamberlin. 1983. Developmental and genetic regulation of *Bacillus subtilis* genes transcribed by σ^{28}-RNA polymerase. *Cell* 35:285–293.

30. Hall, M. H., and T. J. Silhavy. 1981. Genetic analysis of the *ompB* locus in *Escherichia coli. J. Mol. Biol.* 151:1–15.

31. Helmann, J. D., L. M. Marquez, and M. J. Chamberlin. 1988. Cloning, sequencing, and disruption of the *Bacillus subtilis* σ^{28} gene. *J. Bacteriol.* 170:1568–1574.

32. Henner, D. J., M. Yang, and E. Ferrari. 1988. Localization of *Bacillus subtilis sacU*(Hy) mutations to two linked genes with similarities to the conserved procaryotic family of two-component signalling systems. *J. Bacteriol.* 170:5102–5109.

33. Hernandez-Chico, C., J. L. San Millan, R. Kolter, and F. Moreno. 1986. Growth phase and OmpR regulation of transcription of microcin B17 genes. *J. Bacteriol.* 167:1058–1065.

34. Hess, J. F., R. B. Bourret, and M. I. Simon. 1988. Histidine phosphorylation and phosphoryl group transfer in bacterial chemotaxis. *Nature* (London) 336:139–143.

35. Hess, J. F., K. Oosawa, N. Kaplan, and M. I. Simon. 1988. Phosphorylation of three proteins in the signaling pathway of bacterial chemotaxis. *Cell* 53:79–87.

36. Hoch, J. A., K. Trach, F. Kawamura, and H. Saito. 1985. Identification of the transcriptional suppressor *sof-1* as an alteration in the *spo0A* protein. *J. Bacteriol.* **161**:552–555.

37. Hochschild, A., and M. Ptashne. 1988. Interaction at a distance between λ repressors disrupts gene activation. *Nature* (London) **336**:353–357.

38. Hulett, F. M., and K. Jensen. 1988. Critical roles of *spo0A* and *spo0H* in vegetative alkaline phosphatase production in *Bacillus subtilis*. *J. Bacteriol.* **170**:3765–3768.

39. Huo, L., K. J. Martin, and R. Schleif. 1988. Alternative DNA loops regulate the arabinose operon in *Escherichia coli*. *Proc. Natl. Acad. Sci. USA* **85**:5444–5448.

40. Igo, M. M., and T. J. Silhavy. 1988. EnvZ, a transmembrane environmental sensor of *Escherichia coli* K-12, is phosphorylated in vitro. *J. Bacteriol.* **170**:5971–5973.

41. Ikeuchi, T., J. Kudoh, and S. Tsunasawa. 1986. Amino-terminal structure of *spo0A* protein and sequence homology with *spo0F* and *spo0B* proteins. *Mol. Gen. Genet.* **203**:371–376.

42. Ikeuchi, T., S. Tsunasawa, and F. Sakiyama. 1987. Purification and characterization of the Spo0A protein of *Bacillus subtilis* from an overproducing strain of *Escherichia coli*. *Eur. J. Biochem.* **167**:233–238.

43. Iuchi, S., and E. C. C. Lin. 1988. *arcA* (*dye*), a global regulatory gene in *Escherichia coli* mediating repression of enzymes in aerobic pathways. *Proc. Natl. Acad. Sci. USA* **85**:1888–1892.

44. Johnson, P. F., W. H. Landschulz, B. J. Graves, and S. L. McKnight. 1987. Identification of a rat liver nuclear protein that binds to the enhancer core element of three animal viruses. *Genes Dev.* **1**:133–146.

45. Kawamura, F., L. Wang, and R. H. Doi. 1985. Catabolite-resistant sporulation (*crsA*) mutations in the *Bacillus subtilis* RNA polymerase σ⁴³ gene (*rpoD*) can suppress and be suppressed by mutations in *spo0* genes. *Proc. Natl. Acad. Sci. USA* **82**:8124–8128.

46. Keener, J., and S. Kustu. 1988. Protein kinase and phosphoprotein phosphatase activities of nitrogen regulatory proteins NTRB and NTRC of enteric bacteria: roles of the conserved amino-terminal domain of NTRC. *Proc. Natl. Acad. Sci. USA* **85**:4976–4980.

47. Keynan, A., and N. Sandler. 1984. Spore research in historical perspective, p. 1–48. *In* A. Hurst and G. W. Gould (ed.), *The Bacterial Spore*. Academic Press Inc. (London), Ltd., London.

48. Kofoid, E. C., and J. S. Parkinson. 1988. Transmitter and receiver modules in bacterial signaling proteins. *Proc. Natl. Acad. Sci. USA* **85**:4981–4985.

49. Kroos, L., B. Kunkel, and R. Losick. 1989. Switch protein alters specificity of RNA polymerase containing a compartment-specific sigma factor. *Science* **243**:526–528.

50. Kunst, F., M. Debarbouille, T. Msadek, M. Young, C. Mauel, D. Karamata, A. Klier, G. Rapoport, and R. Dedonder. 1988. Deduced polypeptides encoded by the *Bacillus subtilis* sacU locus share homology with two-component sensor-regulator systems. *J. Bacteriol.* **170**:5093–5101.

51. Lamont, I. A., and J. Mandelstam. 1984. Use of temperature-sensitive mutants to study gene expression of two closely linked sporulation loci in *Bacillus subtilis*. *J. Gen. Microbiol.* **130**:1263–1269.

52. Leroux, B., M. F. Yanofsky, S. C. Winans, J. E. Ward, S. F. Ziegler, and E. W. Nester. 1987. Characterization of the *virA* locus of *Agrobacterium tumefaciens*: a transcriptional regulator and host range determinant. *EMBO J.* **6**:849–856.

53. Leung, A., S. Rubinstein, C. Yang, L. Jing-Wei, and T. Leighton. 1985. Suppression of defective-sporulation phenotypes by mutations in the major sigma factor gene (*rpoD*) of *Bacillus subtilis*. *Mol. Gen. Genet.* **201**:88–95.

54. Lewandoski, M., E. Dubnau, and I. Smith. 1986. Transcriptional regulation of the *spo0F* gene of *Bacillus subtilis*. *J. Bacteriol.* **168**:870–877.

55. Little, J. W., and D. W. Mount. 1982. The SOS regulatory system of *Escherichia coli*. *Cell* **29**:11–22.

56. Losick, R., and J. Pero. 1981. Cascades of sigma factors. *Cell* **25**:582–584.

57. Losick, R., P. Youngman, and P. J. Piggot. 1986. Genetics of endospore formation in *Bacillus subtilis*. *Annu. Rev. Genet.* **20**:625–669.

58. Macnab, R. M. 1987. Motility and chemotaxis, p. 732–760. *In* F. C. Neidhardt, J. L. Ingraham, K. B. Low, B. Magasanik, M. Schaechter, and H. E. Umbarger (ed.), *Escherichia coli and Salmonella typhimurium: Cellular and Molecular Biology*. American Society for Microbiology, Washington, D.C.

59. Magasanik, B., and F. C. Neidhardt. 1987. Regulation of carbon and nitrogen utilization, p. 1318–1326. *In* F. C. Neidhardt, J. L. Ingraham, K. B. Low, B. Magasanik, M. Schaechter, and H. E. Umbarger (ed.), *Escherichia coli and Salmonella typhimurium: Cellular and Molecular Biology*. American Society for Microbiology, Washington, D.C.

60. Majumdar, A., and S. Adhya. 1984. Demonstration of two operator elements in *gal*: in vitro repressor binding studies. *Proc. Natl. Acad. Sci. USA* **81**:6100–6104.

61. Makino, K., H. Shinagawa, M. Amemura, and A. Nakata. 1986. Nucleotide sequence of the *phoB* gene, the positive regulatory gene for the phos-

phate regulon of *Escherichia coli* K-12. *J. Mol. Biol.* **190**:37–44.

62. Marahiel, M. A., P. Zuber, G. Czekay, and R. Losick. 1987. Identification of the promoter for a peptide antibiotic biosynthesis gene from *Bacillus brevis* and its regulation in *Bacillus subtilis*. *J. Bacteriol.* **169**:2215–2222.

63. Matsuyama, S., T. Mizuno, and S. Mizushima. 1986. Interaction between two regulatory proteins in osmoregulatory expression of *ompF* and *ompC* genes in *Escherichia coli*: a novel *ompR* mutation suppresses pleiotropic defects caused by an *envZ* mutation. *J. Bacteriol.* **168**:1309–1314.

64. McLeod, M., and D. Beach. 1988. A specific inhibitor of the *ran1*⁺ protein kinase regulates entry into meiosis in *Schizosaccharomyces pombe*. *Nature* (London) **332**:509–514.

65. Meyer, B. J., and M. Ptashne. 1980. Gene regulation at the right operator (O$_R$) of bacteriophage λ. III. λ repressor directly activates gene transcription. *J. Mol. Biol.* **139**:195–205.

66. Miller, J. F., J. J. Mekalanos, and S. Falkow. 1989. Coordinate regulation and sensory transduction in the control of bacterial virulence. *Science* **243**:916–922.

67. Miranda-Rios, J., R. Sanchez-Pescador, M. Urdea, and A. A. Corvarrubias. 1987. The complete nucleotide sequence of the *glnALG* operon of *Escherichia coli* K12. *Nucleic Acids Res.* **15**:2725–2770.

68. Mitchell, A. P., and I. Herskowitz. 1986. Activation of meiosis and sporulation by repression of the RME1 product in yeast. *Nature* (London) **319**:738–742.

69. Moolenaar, G. F., C. A. Sluis, C. Backendorf, and P. Putte. 1987. Regulation of the *Escherichia coli* excision repair protein *uvrC*. Overlap between the *uvrC* structural gene and the region coding for a 24-kD protein. *Nucleic Acids Res.* **15**:4273–4289.

70. Mukherjee, S., H. Erickson, and D. Bastia. 1988. Detection of DNA looping due to simultaneous interaction of a DNA-binding protein with two spatially separated binding sites on DNA. *Proc. Natl. Acad. Sci. USA* **85**:6287–6291.

71. Ninfa, A. J., and B. Magasanik. 1986. Covalent modification of the *glnG* product, NR$_I$, by the *glnL* product, NR$_{II}$, regulates the transcription of the *glnALG* operon in *Escherichia coli*. *Proc. Natl. Acad. Sci. USA* **83**:5909–5913.

72. Ninfa, A. J., E. G. Ninfa, A. N. Lupas, A. Stock, B. Magasanik, and J. Stock. 1988. Crosstalk between bacterial chemotaxis signal transduction proteins and regulators of transcription of the Ntr regulon: evidence that nitrogen assimilation and chemotaxis are controlled by a common phosphotransfer mechanism. *Proc. Natl. Acad. Sci. USA* **85**:5492–5496.

73. Nixon, T., C. W. Ronson, and F. M. Ausubel. 1986. Two-component regulatory systems responsive to environmental stimuli share strongly conserved domains with the nitrogen assimilation regulatory genes *ntrB* and *ntrC*. *Proc. Natl. Acad. Sci. USA* **83**:7850–7854.

74. Ochi, K., J. C. Kandala, and E. Freese. 1982. Evidence that *Bacillus subtilis* sporulation induced by the stringent response is caused by a decrease in GTP or GDP. *J. Bacteriol.* **151**:1062–1065.

75. Peng, H., R. P. Novick, B. Kreiswirth, J. Kornblum, and P. Schlievert. 1988. Cloning, characterization, and sequencing of an accessory gene regulator (*agr*) in *Staphylococcus aureus*. *J. Bacteriol.* **170**:4365–4372.

76. Perego, M., and J. A. Hoch. 1987. Isolation and sequence of the *spo0E* gene: its role in initiation of sporulation in *Bacillus subtilis*. *Mol. Microbiol.* **1**:125–132.

77. Perego, M., and J. A. Hoch. 1988. Sequence analysis and regulation of the *hpr* locus, a regulatory gene for protease production and sporulation in *Bacillus subtilis*. *J. Bacteriol.* **170**:2560–2567.

78. Perego, M., G. B. Spiegelman, and J. A. Hoch. 1988. Structure of the gene for the transition state regulator, *abrB*: regulator synthesis is controlled by the *spo0A* sporulation gene in *Bacillus subtilis*. *Mol. Microbiol.* **2**:689–699.

79. Popham, D. L., D. Szeto, J. Keener, and S. Kustu. 1989. Function of a bacterial activator protein that binds to transcriptional enhancers. *Science* **243**:629–635.

80. Postma, P. W., and J. W. Lengeler. 1985. Phosphoenolpyruvate:carbohydrate phosphotransferase system of bacteria. *Microbiol Rev.* **49**:232–269.

81. Ramakrishna, N., E. Dubnau, and I. Smith. 1984. The complete DNA sequence and regulatory regions of the *Bacillus licheniformis spo0H* gen. *Nucleic Acids Res.* **12**:1779–1790.

82. Ravid, S., P. Matsumura, and M. Eisenbach. 1986. Restoration for flagellar clockwise rotation in bacterial envelopes by insertion of the chemotaxis protein CheY. *Proc. Natl. Acad. Sci USA* **83**:7157–7161.

83. Recsei, P. B. Kreiswirth, M. O'Reilly, P. Schlievert, A. Gruss, and R. P. Novick. 1985. Regulation of exoprotein gene expression by *agr*, p. 700–706. *In* J. Jeljaszewicz (ed.), *The Staphylococci: Proceedings of V International Symposium on Staphylococci and Staphylococcal Infections*. Gustav Fischer Verlag, Stuttgart.

84. Reitzer, L. J., and B. Magasanik. 1987. Ammonia assimilation and the biosynthesis of glutamine, glutamate, aspartate, asparagine, L-alanine, and D-alanine, p. 302–320. *In* F. C. Neidhardt, J. L. Ingraham, K. B. Low, B. Magasanik, M. Schaechter, and H. Umbarger (ed.), *Escherichia coli and Salmonella typhimurium: Cellular and Molecular Bi-*

ology. American Society for Microbiology, Washington, D.C.

85. Ronson, C. W., P. M. Astwood, B. T. Nixon, and F. M. Ausubel. 1987. Deduced products of C4-dicarboxylate transport regulatory genes of *Rhizobium leguminosarum* are homologous to nitrogen regulatory gene products. *Nucleic Acids Res.* 15:7921–7934.

86. Ronson, C. W., B. T. Nixon, and F. M. Ausubel. 1987. Conserved domains in bacterial regulatory proteins that respond to environmental stimuli. *Cell* 49:579–581.

87. Ruvkun, G., and J. Giusto. 1989. The *Caenorhabditis elegans* heterochronic gene *lin-14* encodes a nuclear protein that forms a temporal developmental switch. *Nature* (London) 338:313–319.

88. Sandman, K., R. Losick, and P. Youngman. 1987. Genetic analysis of *Bacillus subtilis spo* mutations generated by Tn917-mediated insertional mutagenesis. *Genetics* 117:603–617.

89. Sasse-Dwight, S., and J. D. Gralla. 1988. Probing co-operative DNA-binding *in vivo*. The *lac* O_1:O_3 interaction. *J. Mol. Biol.* 202:107–119.

90. Schaeffer, P., J. Millet, and J. Aubert. 1965. Catabolic repression of bacterial sporulation. *Proc. Natl. Acad. Sci. USA* 54:704–711.

91. Seki, T., H. Yoshikawa, H. Takahashi, and H. Saito. 1987. Cloning and nucleotide sequence of *phoP*, the regulatory gene for alkaline phosphatase and phosphodiesterase in *Bacillus subtilis*. *J. Bacteriol.* 169:2913–2916.

92. Seki, T., H. Yoshikawa, H. Takahashi, and H. Saito. 1988. Nucleotide sequence of the *Bacillus subtilis phoR* gene. *J. Bacteriol.* 170:5935–5938.

93. Sekiguchi, J., B. Ezaki, K. Kodama, and T. Akamatsu. 1988. Molecular cloning of a gene affecting the autolysin level and flagellation in *Bacillus subtilis*. *J. Gen. Microbiol.* 134:1611–1621.

94. Shoji, K., S. Hiratsuka, F. Kawamura, and Y. Kobayashi. 1988. New suppressor mutation *surOB* of *spoOB* and *spoOF* mutations in *Bacillus subtilis*. *J. Gen. Microbiol.* 134:3249–3257.

95. Siegele, D. A., J. C. Hu, and C. A. Gross. 1988. Mutations in *rpoD*, the gene encoding the σ^{70} subunit of *Escherichia coli* RNA polymerase, that increase expression of the *lac* operon in the absence of CAP-cAMP. *J. Mol. Biol.* 203:29–37.

96. Silva, C. M., D. B. Tully, L. A. Petch, C. M. Jewell, and J. A. Cidlowski. 1987. Application of a protein-blotting procedure to the study of human glucocorticoid receptor interactions with DNA. *Proc. Natl. Acad. Sci. USA* 84:1744–1748.

97. Silverman, P. M. 1985. Host cell-plasmid interactions in the expression of DNA donor activity by F$^+$ strains of *Escherichia coli* K-12. *BioEssays* 2:254–260.

98. Slauch, J. M., S. Garret, D. E. Jackson, and T. J. Silhavy. 1988. EnvZ functions through OmpR to control porin gene expression in *Escherichia coli* K-12. *J. Bacteriol.* 170:439–441.

99. Steinmetz, M., F. Kunst, and R. Dedonder. 1976. Mapping of mutations affecting synthesis of exocellular enzymes in *Bacillus subtilis*. *Mol. Gen. Genet.* 148:281–285.

100. Stibitz, S., W. Aaronson, D. Monack, and S. Falkow. 1989. Phase variation in *Bordetella pertussis* by frameshift mutation in a gene for a novel two-component system. *Nature* (London) 338:266–269.

101. Stock, A., T. Chen, D. Welsh, and J. Stock. 1988. CheA protein, a central regulator of bacterial chemotaxis, belongs to a family of proteins that control gene expression in response to changing environmental conditions. *Proc. Natl. Acad. Sci. USA* 85:1403–1407.

102. Stock, A., D. E. Koshland, Jr., and J. Stock. 1985. Homologies between the *Salmonella typhimurium* CheY protein and proteins involved in the regulation of chemotaxis, membrane protein synthesis and sporulation. *Proc. Natl. Acad. Sci. USA* 82:7989–7993.

103. Stock, A. M., J. M. Mottonen, J. B. Stock, and C. E. Schutt. 1989. Three-dimensional structure of CheY, the response regulator of bacterial chemotaxis. *Nature* (London) 337:745–752.

104. Stock, J. 1987. Mechanisms of receptor function and the molecular biology of information processing in bacteria. *BioEssays* 6:199–203.

105. Stragier, P., C. Bonamy, and C. Karmazyn-Campelli. 1988. Processing of a sporulation sigma factor in *Bacillus subtilis*: how morphological structure could control gene expression. *Cell* 52:697–704.

106. Sun, D., and I. Takahashi. 1984. A catabolite resistance mutation is localized in the *rpo* operon of *Bacillus subtilis*. *Can. J. Microbiol.* 30:423–429.

107. Sun, D., and I. Takahashi. 1985. A catabolite resistance mutation is localized in the *rpo* operon of *Bacillus subtilis*. *Can. J. Microbiol.* 31:429–435.

108. Tanaka, T., and M. Kawata. 1988. Cloning and characterization of *Bacillus subtilis iep*, which has positive and negative effects on production of extracellular proteases. *J. Bacteriol.* 170:3593–3600.

109. Tatchell, K. 1986. *RAS* genes and growth control in *Saccharomyces cerevisiae*. *J. Bacteriol.* 166:364–367.

110. Taylor, W. E., D. B. Straus, A. D. Grossman, Z. F. Burton, C. A. Gross, and R. R. Burgess. 1984. Transcription from a heat-inducible promoter causes heat shock regulation of the sigma subunit of *E. coli* RNA polymerase. *Cell* 38:371–381.

111. Trach, K., and J. A. Hoch. 1989. The *Bacillus subtilis spoOB* stage 0 sporulation operon encodes an essential GTP-binding protein. *J. Bacteriol.* 171:1362–1371.

112. Trach, K. A., J. W. Chapman, P. J. Piggot, and J. R. Hoch. 1985. Deduced product of the stage 0

sporulation gene *spoOF* shares homology with the *spoOA*, *ompR*, and *sfrA* proteins. *Proc. Natl. Acad. Sci. USA* 82:7260–7264.

113. Wabiko, H., K. Ochi, D. M. Nguyen, E. R. Allen, and E. Freese. 1988. Genetic mapping and physiological consequences of *metE* mutations of *Bacillus subtilis*. *J. Bacteriol.* 170:2705–2710.

114. Wang, L., and R. H. Doi. 1987. Promoter switching during development and the termination site of the σ⁴³ operon of *Bacillus subtilis*. *Mol. Gen. Genet.* 207:114–119.

115. Wanner, B. L. 1987. Phosphate regulation of gene expression in *Escherichia coli*, p. 1318–1326. *In* F. C. Neidhardt, J. L. Ingraham, K. B. Low, B. Magasanik, M. Schaechter, and H. E. Umbarger (ed.), *Escherichia coli and Salmonella typhimurium: Cellular and Molecular Biology*. American Society for Microbiology, Washington, D.C.

116. Wanner, B. L., M. R. Wilmes, and D. C. Young. 1988. Control of bacterial alkaline phosphatase synthesis and variation in an *Escherichia coli* K-12 mutant by adenyl cyclase, the cyclic AMP receptor protein, and the *phoM* operon. *J. Bacteriol.* 170:1092–1102.

117. Weis, R. M., and D. E. Koshland, Jr. 1988. Reversible receptor methylation is essential for normal chemotaxis of *Escherichia coli* in gradients of aspartic acid. *Proc. Natl. Acad. Sci. USA* 85:83–87.

118. Weiss, V., and B. Magasanik. 1988. Phosphorylation of nitrogen regulator I (NR$_I$) of *Escherichia coli*. *Proc. Natl. Acad. Sci. USA* 85:8919–8923.

119. Winans, S. C., P. R. Ebert, S. E. Stachel, M. P. Gordon, and E. W. Nester. 1986. A gene essential for *Agrobacterium* virulence is homologous to a family of positive regulatory loci. *Proc. Natl. Acad. Sci. USA* 83:8278–8282.

120. Wu, J., M. G. Howard, and P. J. Piggot. 1989. Regulation of transcription of the *Bacillus subtilis* *spoIIA* locus. *J. Bacteriol.* 171:692–698.

121. Wylie, D., A. Stock, C. Wong, and J. Stock. 1988. Sensory transduction in bacterial chemotaxis involves phosphotransfer between Che proteins. *Biochem. Biophys. Res. Commun.* 151:891–896.

122. Yamashita, S., H. Yoshikawa, F. Kawamura, H. Takahashi, T. Yamamoto, Y. Kobayashi, and H. Saito. 1986. The effect of *spo0* mutations on the expression of *spo0A*- and *spoOF-lacZ* fusions. *Mol. Gen. Genet.* 205:28–33.

123. Yu, G., and J. Hong. 1986. Identification and nucleotide sequence of the activator gene of the externally induced phosphoglycerate transport system of *Salmonella typhimurium*. *Gene* 45:51–57.

124. Zuber, P., J. M. Healy, and R. Losick. 1987. Effects of plasmid propagation of a sporulation promoter on promoter utilization and sporulation in *Bacillus subtilis*. *J. Bacteriol.* 169:461–469.

125. Zuber, P., and R. Losick. 1983. Use of a *lacZ* fusion to study the role of the *spo0* genes of *Bacillus subtilis* in developmental regulation. *Cell* 35:275–283.

126. Zuber, P., and R. Losick. 1987. Role of *abrB* in *spoOA*- and *spoOB*-dependent utilization of a sporulation promoter in *Bacillus subtilis*. *J. Bacteriol.* 169:2223–2230.

Regulation of Procaryotic Development
Edited by Issar Smith, Ralph A. Slepecky, and Peter Setlow
© 1989 American Society for Microbiology, Washington, DC 20006

Chapter 10

Forespore-Specific Genes of *Bacillus subtilis*: Function and Regulation of Expression

Peter Setlow

One of the key morphological events in the sporulation of *Bacillus subtilis* is the formation of an asymmetrically positioned septum approximately 1.5 h into the developmental program. This asymmetric septum divides the sporulating cell into two compartments of unequal size, the larger mother cell and the smaller forespore compartment, each of which contains a chromosome but with different developmental fates. The mother cell initially engulfs the forespore compartment, thus bounding the latter by a double membrane; enclosed within the mother cell, the forespore develops into the dormant spore. Subsequently the mother cell lyses, releasing the spore into the surrounding medium. As is evident from the different fates of the two cell compartments noted above, there must be different patterns of gene expression in the two compartments, with many genes expressed only in one of the two compartments (see R. Losick and L. Kroos, this volume). In this chapter I will focus on genes expressed only in the forespore compartment, concentrating on both the regulation of their expression and the function(s) of the products of these genes.

A variety of techniques have been used to determine the cellular compartment or compartments in which specific genes are expressed during sporulation. In the genetic method of De Lencastre and Piggot (5), sporulating cells of strains carrying mutations in various *spo* genes were transformed by wild-type DNA at or near the beginning of sporulation, but before asymmetric septation. The number of spores resulting from this transformation was scored after sporulation was completed, as was the percentage of these spores which themselves were genetically asporogenous. Presumably, the latter would arise only if the original *spo* mutation was in a gene whose expression was needed only in the mother cell. In contrast, for genes whose expression was required in the forespore, all spores produced from the transformation should be *spo⁺*. Application of this technique to a variety of *spo* mutations indicated that only the *spoVA* locus could be classified as forespore specific (5).

An analogous approach tested the complementation of various *spo* mutations by fusion of protoplasts of these mutants with protoplasts of wild-type cells, when protoplasts of both strains were prepared at ca. t_3 of sporulation (4). In this technique the only *spo* mutations complemented should be those in genes expressed only in the mother cell. The *spoVA* locus was also classified as forespore specific in this test. However, genes expressed prior to asymmetric septation, but whose products are not needed until later in spor-

Peter Setlow • Department of Biochemistry, University of Connecticut Health Center, Farmington, Connecticut 06032.

ulation (even if only in the forespore), will not be complemented in this assay. A second drawback to this approach, as well as the transformation method described above, is that they can only be easily applied to genes in which mutations result in a clear phenotype (i.e., *spo*). Unfortunately, as noted below, for many of the known genes expressed only in the forespore, their destruction results in either no obvious phenotype or only a very subtle phenotype.

A third general approach to determining a gene's locus of expression is to delineate the dependence of the expression of that gene on other *spo* genes, since mother-cell- and forespore-specific genes show different dependence patterns (6, 17; Losick and Kroos, this volume). Thus, if the expression of a particular gene shows a pattern of dependence on other *spo* loci which is identical to that for other forespore genes, this strongly suggests that the gene in question is itself a forespore-specific gene. However, care must be taken not to overinterpret such dependence relationships. For example, stage III and IV mother-cell-specific genes are not directly dependent on the *spoIIIA* and *spoIIIE* loci, whereas forespore-specific genes are (Losick and Kroos, this volume). However, the expression of neither the *spoIIIA* or *spoIIIE* loci themselves appears to be forespore specific, as the *spoIIIE* locus is expressed in vegetative cells (D. Foulger and J. Errington, *Abstr. X Internat. Spores Conf.*, Woods Hole, Mass., 1988 [*Spores X*], p. 97), while the *spoIIIA* locus may well be mother cell specific (5; Losick and Kroos, this volume).

The fourth method for determining the site of expression of a specific gene is the direct demonstration that a gene's product is located in one or both cellular compartments. In cases where there is an assay for the intact gene product, this has been done by assay of the level of gene product in forespore and mother cell compartments after cell fractionation (11, 15, 31). In one case this has also been achieved by immunoelectron microscopy of whole sporulating cells (10). While

it could well be argued that some of these gene products are actually synthesized in the mother cell and then rapidly transported into the forespore, in a few cases a gene's mRNA has also been localized in the forespore (31).

In cases where there is no direct assay for a gene's product, various *lacZ* fusions have been constructed and the site of expression of the β-galactosidase has been determined. The most direct determination of this parameter is by assay of β-galactosidase in mother cell and forespore extracts after cell fractionation. While these analyses have generally been unambiguous, it has been reported that *lacZ* fusions may turn over much faster in the mother cell compartment than in the forespore (Foulger and Errington, *Spores X*, p. 97). Consequently, it is possible that mother cell expression of a *lacZ* fusion could be underestimated or missed. Rather than carry out cell fractionation experiments, some workers have also shown that essentially all of the β-galactosidase produced from a *lacZ* fusion to a forespore-specific gene is incorporated into the mature spore, in which it can be assayed after either germination or removal of spore coats and lysozyme treatment (7, 18).

In practice, for all genes which to date have been unambiguously assigned as forespore specific (Table 1), this assignment has been based on application of at least two of the approaches described above. For all of such genes, the location of the gene product, either the intact gene's product or the β-galactosidase produced from a *lacZ* fusion, has been shown to be predominantly if not exclusively in the forespore (Table 1). Analysis of those forespore genes which have been mapped indicates that such genes are not confined to any one region of the chromosome (Table 1). Similarly, analysis of the codon usage in forespore-specific genes shows that these genes generally exhibit codon usage patterns similar to those for vegetative genes and other sporulation, genes which are also similar to each other (33, 34; unpublished results).

Table 1
Evidence for Forespore-Specific Expression of B. subtilis Genes[a]

Gene or operon	Map position (degrees)	Forespore-specific gene expression[b]				
		Transformation	Protoplast fusion	Dependence on spo genes	Localization of gene product	Location of β-galactosidase from lacZ fusion
gdh	34	ND	ND	+	+	+
gerA	290	ND	ND	+	?	+
gpr	ND	ND	ND	ND	+	ND
0.3 kb	3	ND	ND	+	?	+
spoIIIG	135	ND	ND	+	ND	+
spoVA	211	+	+	+	?	+
ssp	65, 70, 115, 180, 260	ND	ND	+	+	+

[a] Data taken from references 4, 5, 7, 10, 11, 13a, 15, 16, 18, 23, 26, 41, and 36; Moir et al., *Spores X*, p. 12; and Setlow et al., unpublished results.
[b] Symbols: ND, not determined; +, results consistent with forespore-specific expression; ?, intact gene product not identified.

Genes expressed only in the forespore compartment have generally been initially identified as potential genes of interest in two different ways. For one group of genes (those coding for glucose dehydrogenase [gdh], small acid-soluble spore proteins [SASPs, coded for by ssp genes], and the SASP-specific protease [gpr]), their gene products were identified first and were found only in the dormant spore and developing forespore (11, 15, 31). Subsequently, these genes were cloned and mutated. Surprisingly, analysis of knockout mutations in this group of genes has shown that such mutations have either no phenotype or only a rather subtle phenotype (24, 31; R. F. Ramaley, K. A. Lampel, and P. Nakatani, *Spores X*, p. 100). A second group of forespore-specific genes was initially identified only as genes in which mutations blocked either sporulation (spoIIIG, spoVA) or germination (gerA). Only after subsequent analysis of the site of expression of genes identified by these mutations was their expression in the forespore confirmed.

FORESPORE-SPECIFIC GENES

spoIIIG

The spoIIIG gene is located immediately downstream of the spoIIIG operon (13a, 19).

Mutations in spoIIIG block sporulation in stage III and also block expression of all forespore-specific genes which have been analyzed to date (13a, 19, 37a). Determination of the nucleotide sequence of the spoIIIG locus indicated that this gene coded for an RNA polymerase sigma factor (13a, 19). This was confirmed by isolation of RNA polymerase containing the spoIIIG gene product (now called σ^G) from sporulating cells (37a). Analysis of the localization of the β-galactosidase produced from a spoIIIG-lacZ fusion indicated that spoIIIG is expressed only in the forespore between t_2 and t_3 of sporulation (13a).

The function of σ^G appears to be to direct transcription of forespore-specific genes, and for a number of these genes (ssp, gdh, spoVA) it appears that σ^G synthesis alone is sufficient to explain the regulation of their expression, since induction of σ^G synthesis in vegetative cells of B. subtilis immediately turns on the group of genes (37a). Genes in this group have highly conserved sequences spaced 17 to 18 base pairs apart which are located around 10 and 35 bases preceding their transcription start sites (20). These sequences define a consensus sequence for σ^G-dependent promoter recognition which is TGAATA in the −35 region and CATACTA in the −10 region (20). These sequences are distinct from those recognized by other σ factors in vivo (20; C. P. Moran, Jr., this volume). In addition to their transcription in vivo by Eσ^G, this group of genes is also transcribed in vitro by this form of RNA polymerase (37a). Where tested, the σ^G-dependent in vitro transcription start sites are identical to those found in vivo (20). Surprisingly, σ^G-dependent promoters are also recognized in vitro by σ^F, although σ^F apparently does not recognize such promoters in vivo (20, 37a).

σ^G appears to be synthesized between t_2 and t_3 of sporulation and before the turn on of other known forespore-specific genes, i.e., gdh, spoVA, and ssp (13a). Save for the removal of the amino-terminal methionine, σ^G undergoes no processing at its amino terminus (37a), in contrast to several other sporulation-specific σ factors (σ^E and σ^K [Losick and Kroos, this volume]). Processing at its carboxy terminus has not been ruled out, but seems unlikely, since removal of more than a few residues would destroy its -35 binding domain (13a).

As noted above, expression of a transcriptional spoIIIG-lacZ fusion integrated at the amy locus is confined to the forespore. Expression from this lacZ fusion is abolished or greatly reduced in all tight asporogenous mutations in stage 0 or II genes which have been tested (13a). The spoIIIG-lacZ fusion is also shut off by a spoIIIA mutation and a spoIIIG deletion mutation, but not by spoIIIC, spoIIID, and spoIIIE mutations (13a). These results mirror those found for expression of forespore-specific genes known to be σ^G dependent, with the exception of the effect of the spoIIIE mutation. It is known that expression of gdh, spoVA, and ssp genes is abolished by the same spoIIIE mutation which permits spoIIIG-lacZ expression (7, 18). This suggests that the spoIIIE gene product is somehow required for σ^G function in vivo. However, the spoIIIE product (which is synthesized in both growth and sporulation [Foulger and Errington, Spores X, p. 97]) is not required in vitro for σ^G function (37a).

Similarly, induction of synthesis of σ^G in vegetative cells of the spoIIIE mutant results in immediate turn on of σ^G-dependent genes with kinetics indistinguishable from those in wild-type vegetative cells (37a). Consequently, the role of the spoIIIE gene in σ^G function and expression is at present unclear. However, it should be noted that recent results (D. Sun and P. Setlow, unpublished data) have indicated (i) that Eσ^G is not present (<10% of wild-type levels) in spoIIIE cells at t_3 and t_4 of sporulation and (ii) that a translational spoIIIG-lacZ fusion integrated at the spoIIIG locus is not expressed in a spoIIIE background. Consequently, the regulation of the spoIIIG gene when placed in the amy locus may be anomalous with regard to its dependence on spoIIIE.

Analysis of the transcripts encoding σ^G which are present at various times in sporulation indicates that the spoIIIG gene is initially transcribed as a large transcript including at least some, and possibly all, of the spoIIG operon (19). Subsequently, this larger transcript disappears and is replaced by a shorter one whose transcription starts just prior to the spoIIIG gene (19). The sequences located 10 and 35 nucleotides upstream of this latter transcription start site conform well to the consensus sequence for σ^G promoter recognition (19, 20), suggesting that much σ^G synthesis is autocatalytic. Indeed, expression of the spoIIIG-lacZ fusion integrated at the amy locus is almost completely abolished in a spoIIIG deletion mutation (13a). These results suggest that spoIIIG transcription is initially driven by an upstream promoter, possibly that for spoIIG, and that this transcription "primes the pump" with σ^G, which now allows more rapid transcription from a σ^G-dependent promoter. However, it has been reported that a spoIIIG deletion mutation blocks neither the synthesis of the short spoIIIG transcript (Y. Kobayashi, X Internat. Spore Conf., Woods Hole, Mass., 1988) or the expression of a spoIIIG-lacZ fusion integrated in the spoIIIG locus (Sun and Setlow, unpublished data). Consequently, the identity of the RNA polymerase synthesizing the short transcript is not clear. While $E\sigma^G$ can initiate transcription in vitro at the promoter immediately upstream from the *spoIIIG* gene, $E\sigma^F$ is actually more active on this promoter in vitro than is $E\sigma^G$ (Sun and Setlow, unpublished data). Clearly, many unanswered questions remain about the regulation of σ^G synthesis, and answers to these questions are crucial to an understanding of the reason for compartmentation of σ^G synthesis and, thus, forespore-specific gene expression. However, even if the *spoIIG* promoter is initially used to direct *spoIIIG* transcription, one would predict that σ^G would be present in both mother cell and forespore, since σ^E is itself in both compartments (2). Furthermore, this simple model does not explain the fact that *spoIIIA* and *spoIIE* mutations block σ^G synthesis, since these genes are not needed for synthesis of σ^E (Moran, this volume).

While the understanding of the regulation of σ^G synthesis is by no means complete, it has been noted that a *spoIIIG* transcript originating prior to the σ^G-dependent promoter could form a strong stem-loop structure which might well block access to the *spoIIIG* gene's ribosome binding site (19). In contrast, with the shorter transcript, the gene's ribosome binding site would be accessible (19). This suggests the possibility of forespore-specific translational control of initial σ^G synthesis, although other models, such as antitermination of initial *spoIIIG* transcription, are also possible.

A Forespore-Specific Regulon: *gdh*, *spoVA*, and *ssp*

The *gdh* and *spoVA* operons and the *ssp* genes are all turned on in parallel at ca. t_3 of sporulation, shortly after induction of σ^G synthesis (13a, 18). These genes are not expressed in *spoIIIA*, *spoIIIE*, and *spoIIIG* mutants (7, 18). Expression of these genes has been shown to be forespore specific by a combination of genetic methods (4, 5), lo-

calization in sporulating cells of β-galactosidase produced from *lacZ* fusions (7, 18, 26), and localization in sporulating cells of the intact gene product (11, 13, 31) (Table 1). For the *gdh* and *ssp* genes it has been shown that essentially all of their gene products accumulated in sporulation are found in the mature spore (11, 31).

The major determinant for the regulation of this group of genes appears to be their σ^G-dependent promoter sequences. These genes are turned on in vegetative *B. subtilis* immediately after induction of σ^G synthesis (37a) and are also accurately transcribed in vitro by σ^G-containing RNA polymerase (20). For both the *gdh* operon and several of the *ssp* genes, analysis of the in vivo expression and σ^G-dependent in vitro transcription of genes carrying deletion mutations up to and into the -35 and -10 regions, as well as point mutations in conserved residues in these regions, has indicated that the primary determinant of the transcription of these genes is the nature of their -35 and -10 sequences (19a, 26; P. Fajardo-Cavazos, F. Tovar-Rojo, and P. Setlow, unpublished results). The expression of the *ssp* genes is also subject to some type of feedback regulation exerted at the posttranscriptional level (*sspE* gene) or on the level of *ssp* mRNA (*sspA* and -*B* genes) (31). While the mechanism of this feedback control is not understood, it appears to involve the intact *ssp* gene products or mRNAs themselves and is not exerted at the promoters for these genes (31).

gdh

The enzyme glucose dehydrogenase was identified as a forespore-specific gene product over 10 years ago (11). More recently, the gene coding for this enzyme (*gdh*) was cloned and sequenced and found to map between *aroI* and *mtlB* (14, 25). *gdh* mRNA, which appears only at ca. t_3 of sporulation, is ca. 1.6 kilobases in length, more than twice as long as is needed to code for glucose dehydrogenase (14, 39). Surprisingly, the only promoter for *gdh* mRNA is located ca. 900 base pairs upstream of the glucose dehydrogenase coding sequence (19a, 26). DNA sequence analysis of this region has shown the presence of an open reading frame of 855 base pairs preceding the *gdh* gene (26, 39), suggesting that the *gdh* gene is the second gene of a bicistronic operon. However, the identity of the first gene in this operon has not been established, and its protein product has not been identified.

Early work involving isolation of *gdh* mutants by classical means led to the suggestion that glucose dehydrogenase was essential for spore germination on glucose (37). However, these results may have been compromised by the presence of other mutations in these strains. In more recent work in which *gdh* mutants were constructed by molecular genetic methods, spores lacking glucose dehydrogenase germinated on glucose as well as did wild-type spores (26; Ramaley et al., *Spores X*, p. 100). It was also reported that insertional mutants in the first open reading frame in the *gdh* operon were asporogenous (39). However, a recent report noted that analogous mutants sporulated well (26).

While an essential role for glucose dehydrogenase in spore germination now seems unlikely, it is clear that this enzyme does function during germination of wild-type spores on glucose. Under such conditions a large fraction of the glucose is converted to gluconate by glucose dehydrogenase, with the gluconate subsequently being phosphorylated to 6-phosphogluconate and further metabolized by the hexose-monophosphate pathway (22). While operation of this pathway is not essential for spore germination or outgrowth, it does ascribe a function to at least one gene of the *gdh* operon.

spoVA

The *spoVA* operon was initially identified as the site of mutations which blocked sporulation in stage V. (The *spoVH* locus,

which was also suggested to be forespore specific [3], is now known to be identical to the *spoVA* locus [3, 9; J. Errington, personal communication].) Subsequently this locus was cloned and sequenced and found to map immediately downstream of the *spoIIA* locus (9). The *spoVA* operon directs synthesis of a large (>2-kilobase) transcript whose synthesis is turned on at ca. t_3 of sporulation (27). This transcript is polycistronic, containing five distinct open reading frames which do not overlap (9). Mutations in the *spoVA* locus have been divided into at least four complementation groups (9), with a number of *spoVA* mutations tentatively assigned to individual open reading frames of the operon. Unfortunately, the protein products of the *spoVA* operon have not yet been identified, and analysis of the codon usage of these genes suggests that they are not highly expressed (34). Examination of the five proteins predicted from the nucleotide sequence of the *spoVA* operon indicates that three have quite high isoelectric points, with one of the latter also being quite hydrophobic. While the function of the *spoVA* gene products is unknown, they are required for further forespore development which results in some morphogenic signal detected in the mother cell, since *spoVA* mutants block mother cell gene expression in stage V (Losick and Kroos, this volume).

ssp

The *ssp* genes encode a large group of SASPs whose properties, function, and synthesis have recently been reviewed in detail (31). In *B. subtilis*, SASPs comprise ca. 10% of total spore protein, but are rapidly degraded early in spore germination. The amino acids produced in this process support much of the protein synthesis during spore germination and outgrowth. Three proteins comprise the majority of the SASP fraction in *B. subtilis*; these are termed SASP-α, -β, and -γ. SASP-α and -β are almost identical in primary sequence and are coded for by the *sspA* and *sspB* genes. In addition to these genes, two additional genes (*sspC* and *sspD*) which code for proteins extremely homologous to SASP-α and -β have been cloned and sequenced. The *sspC* and *sspD* genes have been shown to code for proteins present in spores at much lower levels than SASP-α and -β. In addition to these four genes coding for α/β-type SASPs, *B. subtilis* undoubtedly contains at least three more homologous genes. The various members of this multigene family are scattered about on the chromosome, and all appear to be expressed in parallel during sporulation as monocistronic mRNAs.

While one function of these proteins is to serve as amino acid storage proteins, the α/β-type SASPs are also essential for the resistance of spores to UV light. However, to demonstrate this involvement, spores lacking both SASP-α and -β must be analyzed. In strains lacking either the *sspA* or the *sspB* gene, there is sufficient SASP synthesis directed by the remaining gene to provide significant or full UV resistance to the resulting spores (30, 31). While the mechanism whereby α/β-type SASPs provide spore UV resistance is not known, α/β-type SASPs are associated with spore DNA in vivo (10). Somehow this interaction alters the photochemistry of the spore DNA such that UV irradiation of spores produces the less lethal thyminyl-thymidine adduct (originally called spore photoproduct), rather than the more lethal cyclobutane-type thymine dimer (30). However, it is not known how α/β-type SASPs bring about this change in DNA photochemistry. While it has been suggested that α/β-type SASPs alter the conformation, hydration, or both (30) of spore DNA, the details of this process are not understood. The α/β-type SASPs also appear to play a minor role in spore heat resistance.

In contrast to the multigene nature of the α/β-type *ssp* genes, there is only a single gene (termed *sspE*) coding for γ-type SASP in *B. subtilis*. Removal of this gene by mutation results in no change in the heat or radiation resistance of the spores lacking SASP-γ. Indeed, in contrast to α/β-type SASPs, SASP-γ is not associated with DNA in vivo (10).

In addition to the α/β- and γ-type SASPs, *B. subtilis* spores also contain a large number of other unrelated SASPs present at levels well below those of major SASPs (13, 29). The exact number of these minor species is not clear, but it is certainly more than 10 (29). These proteins are spore specific, and work in *Bacillus megaterium* suggests that most are synthesized in the forespore in parallel with major SASPs (29). These minor proteins represent a large number of potential forespore-specific genes, all with unknown functions, and thus may be a rich field for study. One possible gene coding for such a minor SASP is the *0.3 kb* gene, whose protein coded for could well be an SASP (21, 36; see below).

gerA Operon

The *gerA* operon was initially identified by isolation of mutants whose spores were defective in germination on alanine (28). This locus was found to contain three complementation groups (40), and the operon was subsequently cloned and sequenced (8, 41). The operon appears to encode three different proteins, the coding sequences for which overlap slightly (8, 41). Analysis of the *gerA* region with integrational plasmids indicated that the three open reading frames found in the *gerA* locus probably comprise a single transcriptional unit and that these open reading frames probably represent the three complementation groups into which *gerA* mutants can be divided (41). The *gerA* locus maps very close to *citG*. Indeed, *gerA* and *citG* are transcribed in opposite directions with their translational start sites separated by only ca. 370 base pairs (8, 41; A. Moir, I. M. Feavers, A. R. Zuberi, and J. McCarvil, *Spores X*, p. 12).

Analysis of the expression of a *gerA-lacZ* fusion has indicated that *gerA* is only expressed in the forespore somewhere around

t_3 (41; Moir et al., *Spores X*, p. 12) and in parallel with glucose dehydrogenase (B. Setlow, P. Setlow, and A. Moir, unpublished results). Expression of this *gerA-lacZ* fusion during sporulation is blocked in a *spoIIIG* deletion mutant, and the fusion can be turned on in vegetative cells by induction of synthesis of σ^G, in parallel with induction of glucose dehydrogenase synthesis (Setlow et al., unpublished results). These findings indicate that, like the *ssp* genes and *gdh* and *spoVA* operons, the *gerA* operon may also be transcribed directly by Eσ^G. However, neither in vitro transcription of this operon by Eσ^G nor refined analysis of the in vivo transcription start site of the *gerA* operon has yet been achieved. Consequently, comparison of the -10 and -35 regions of the *gerA* promoter with the consensus σ^G sequences is not possible. While the intercistronic region between *citG* and *gerA* does contain at least one potential σ^G promoter with the correct spacing between the -10 and -35 regions (8, 20), further work is needed on the precise localization of the *gerA* transcription start site in vivo.

The phenotype of *gerA* mutants is that they are defective in germination on alanine but not other germinants (28). Consequently, it seems unlikely that the *gerA* gene products are involved in the central pathway of spore germination. They may instead be involved solely in the recognition of the germinant alanine (28; S. J. Foster and K. Johnstone, this volume). Analysis of the proteins encoded by the three *gerA* open reading frames indicates that they contain a number of hydrophobic regions, and it has been suggested that these proteins might be membrane associated and together comprise the alanine receptor involved in alanine-dependent spore germination (8, 41). However, the *gerA* gene products have not yet been localized in spores; indeed, the very low levels of expression of the *gerA-lacZ* fusion (41) suggest that these proteins may be present at very low levels in spores. This is further suggested by the rather unbiased (for *B. subtilis* genes) codon usage

in the three cistrons of the *gerA* operon (34, 41).

0.3 kb Gene

The *0.3 kb* gene was originally identified as a gene encoding a relatively stable RNA species which was synthesized only at a late stage of sporulation (21, 36). Upon cloning and sequencing of the *0.3 kb* gene, it was found to encode a 61-residue protein which is rather basic, suggesting it might encode one of the SASP family (36). However, the protein product of this gene has not been identified in spores, and its function is unknown. More recently, analysis of *0.3 kb* expression, using a single-copy *0.3 kb-lacZ* fusion integrated in the chromosome, has indicated that expression of this gene (i) is forespore specific; (ii) depends on σ^G synthesis during sporulation; (iii) also requires the *spoIIIA* and *spoIIIE* loci; and (iv) is switched on in vegetative cells upon induction of σ^G synthesis (23). However, these studies have further indicated that the regulation of *0.3 kb* expression is somewhat different from that of other σ^G-dependent genes such as *gerA*, *gdh*, *spoVA*, and *ssp*, all of which are expressed during sporulation, significantly before expression of a *0.3 kb-lacZ* fusion (23, 37a). Similarly, expression of the *0.3 kb-lacZ* fusion in vegetative cells in which σ^G synthesis has been induced is significantly slower than that of the *gerA*, *gdh*, *spoVA*, or *ssp* genes (23, 37a). These data suggest that σ^G alone is not sufficient for *0.3 kb* transcription, and this is consistent with a lack of success in obtaining *0.3 kb* transcription in vitro with Eσ^G (W. L. Nicholson and P. Setlow, unpublished results). While the -10 and -35 sequences of the *0.3 kb* gene do exhibit significant homology to the consensus of σ^G promoter sequence, they do so only if a 15-nucleotide spacing is assigned (20, 36). Consequently, some other protein factor may be needed to facilitate transcription of this gene by Eσ^G, possibly even a new σ factor.

Other Forespore-Specific Genes

In addition to the forespore-specific genes described above, there are a number of proteins found exclusively or preferentially in the forespore, but whose genes have not yet been cloned or characterized. These include the SASP-specific spore protease, aspartase, and glutamate decarboxylase (1, 15, 35). Work in *B. megaterium* has shown that the SASP-specific protease is synthesized only during sporulation, approximately in parallel with its SASP substrates, and is found only in the forespore (15). The function of this serine protease is to cleave SASP endoproteolytically within one or two specific sequences, thereby allowing complete degradation of the resulting oligopeptides (31, 32). The gene for this protease (termed *gpr*) has been cloned and sequenced from *B. megaterium*, and this has now led to the cloning of the *B. subtilis* gene (38; M. D. Sussman and P. Setlow, unpublished results). Future work should indicate whether the *B. subtilis gpr* gene is another gene directly dependent on σ^G for its transcription.

Both aspartase and glutamate decarboxylase are present in vegetative cells of *Bacillus* species (1, 12, 35). However, during sporulation of *Bacillus cereus* or *B. megaterium*, forespore levels of these enzymes are 10 to 50 times higher than in the mother cell (1, 35). The reason for the elevated levels of these two enzymes is not clear, but they may be involved in forespore metabolism, possibly using amino acids derived from the mother cell (35). Unfortunately, the genes for these enzymes have not been cloned in *B. subtilis*. Analysis of the function and regulation of these genes might give insight into not only forespore-specific gene expression, but also forespore metabolism.

In addition to the enzymes described above, there are undoubtedly other forespore-specific gene products yet to be discovered. Possibilities include (i) various enzymes of germ-cell wall biosynthesis; (ii) receptors for germinants other than alanine; (iii) proteins of central germination pathways; (iv) proteins involved in the processing of the zymogen of the SASP-specific spore protease (15); (v) proteins involved in the uptake of dipicolinic acid and/or divalent cations by the developing spore; and (vi) the enzyme (or enzymes) involved in the repair during spore germination of "spore photoproducts" formed in spore DNA by UV irradiation (30).

CONCLUSIONS

Examining what we know to date about forespore-specific gene expression, one is struck by how simple much of it seems. Clearly there is much we don't understand about the mechanism ensuring σ^G synthesis only in the developing forespore. However, with this major uncertainty aside, regulation of most forespore genes can be explained in whole or in part simply by the forespore-specific synthesis of σ^G. This in turn allows coordinate transcription of the σ^G-dependent genes of the *gerA*, *gdh*, and *spoVA* operons, as well as *ssp* genes. Only the delayed expression of the *0.3 kb* gene relative to these others is an anomaly. Is this apparent simplicity of regulation of gene expression in the forespore real, or only an artifact of our lack of knowledge? I believe it is the latter and that this reflects the dearth of knowledge about forespore-specific genes. This paucity of knowledge may well stem from the methods used to date to identify developmentally specific genes in *B. subtilis*: i.e., by looking for loss of differentiation (sporulation or germination). While this has resulted in identification of many such genes, it is striking that mutations in many forespore-specific gene products (SASP, glucose dehydrogenase, spore protease, spore photoproduct repair) have very little major effect on subsequent differentiation, although close examination of such mutants does reveal subtle phenotypes. In some of these cases it is clear that only multiple mutations will reveal striking phenotypes because of the redundancy of some of these systems. If this finding based on a rather limited sample is more generally true, it may be difficult to identify forespore-specific genes and functions by classical genetic methods.

Acknowledgments. The work in my laboratory was supported by a Public Health Service grant from the National Institutes of Health (GM-19698) and by a grant from the Army Research Office.

LITERATURE CITED

1. Andreoli, A. J., J. Saranto, N. Caliri, E. Escamilla, and E. Pina. 1978. Comparative study of proteins from forespore and mother cell compartments of *Bacillus cereus*, p. 260–264. *In* G. Chambliss and J. C. Vary (ed.), *Spores VII*. American Society for Microbiology, Washington, D.C.
2. Carlson, H. C., and W. G. Haldenwang. 1989. The σ^E subunit of *Bacillus subtilis* RNA polymerase is present in both forespore and mother cell compartments. *J. Bacteriol.* 171:2216–2218.
3. Cutting, S. M., and J. Mandelstam. 1988. Cloning and dependence pattern of the sporulation operon *spoVH*. *J. Bacteriol.* 170:802–809.
4. Dancer, B. N., and J. Mandelstam. 1981. Complementation of sporulation mutations in fused protoplasts of *Bacillus subtilis*. *J. Gen. Microbiol.* 123:17–26.
5. de Lencastre, H., and P. J. Piggot. 1979. Identification of different sites of expression for *spo* loci by transformation of *Bacillus subtilis*. *J. Gen. Microbiol.* 114:377–389.
6. Errington, J., S. M. Cutting, and J. Mandelstam. 1988. Branched pattern of regulatory interactions between late sporulation genes in *Bacillus subtilis*. *J. Bacteriol.* 170:796–801.
7. Errington, J., and J. Mandelstam. 1986. Use of a *lacZ* gene fusion to determine the dependence pattern and the spore compartment expression of sporulation operon *spoVA* in *spo* mutants of *Bacillus subtilis*. *J. Gen. Microbiol.* 132:2977–2985.
8. Feavers, I. M., J. S. Miles, and A. Moir. 1985. The nucleotide sequence of a spore germination gene (*gerA*) of *Bacillus subtilis* 168. *Gene* 38:95–102.
9. Fort, P., and J. Errington. 1985. Nucleotide sequence and complementation analysis of a polycistronic sporulation operon, *spoVA*, in *Bacillus subtilis*. *J. Gen. Microbiol.* 131:1091–1105.
10. Francesconi, S. C., T. J. MacAlister, B. Setlow, and P. Setlow. 1988. Immunoelectron microscopic localization of small, acid-soluble spore proteins in sporulating cells of *Bacillus subtilis*. *J. Bacteriol.* 170:5963–5967.
11. Fujita, V., R. Ramaley, and E. Freese. 1977. Location and properties of glucose dehydrogenase in sporulating cells and spores of *Bacillus subtilis*. *J. Bacteriol.* 132:282–293.
12. Iijima, T., M. D. Diesterhaft, and E. Freese. 1977. Sodium effect of growth on aspartate and genetic analysis of a *Bacillus subtilis* mutant with high aspartase activity. *J. Bacteriol.* 129:1440–1447.
13. Johnson, W. C., and D. J. Tipper. 1981. Acid-soluble spore proteins of *Bacillus subtilis*. *J. Bacteriol.* 146:972–987.
13a. Karmazyn-Campelli, C., C. Bonamy, B. Savelli, and P. Stragier. 1981. Tandem genes encoding sigma factors for consecutive steps of development in *Bacillus subtilis*. *Genes Dev.* 13:150–157.
14. Lampel, K. A., B. Uratani, G. Rasul Chaudhry, R. F. Ramaley, and S. Rudikoff. 1986. Characterization of the developmentally regulated *Bacillus subtilis* glucose dehydrogenase gene. *J. Bacteriol.* 166:238–243.
15. Loshon, C. A., B. M. Swerdlow, and P. Setlow. 1982. *Bacillus megaterium* spore protease: synthesis and processing of precursor forms during sporulation and germination. *J. Biol. Chem.* 257:10838–10845.
16. Losick, R., P. Youngman, and P. J. Piggot. 1986. Genetics of endospore formation in *Bacillus subtilis*. *Annu. Rev. Genet.* 20:625–669.
17. Mandelstam, J., and J. Errington. 1987. Dependent sequences of gene expression controlling spore formation in *Bacillus subtilis*. *Microbiol. Sci.* 4:238–244.
18. Mason, J. M., R. H. Hackett, and P. Setlow. 1988. Studies on the regulation of genes coding for small, acid-soluble proteins of *Bacillus subtilis* spores using *lacZ* gene fusions. *J. Bacteriol.* 170:239–244.
19. Masuda, E. S., H. Anaguchi, K. Yamada, and Y. Kobayashi. 1988. Two developmental genes encoding σ factor homologs are arranged in tandem in *Bacillus subtilis*. *Proc. Natl. Acad. Sci. USA* 85:7637–7641.
19a. Nakatani, Y., W. L. Nicholson, K.-D. Nietzke, P. Setlow, and E. Freese. 1989. Sigma-G RNA polymerase controls forespore specific expression of the glucose dehydrogenase operon in *Bacillus subtilis*. *Nucleic Acids Res.* 17:999–1017.
20. Nicholson, W. L., D. Sun, B. Setlow, and P. Setlow. 1989. Promoter specificity of σ^G-containing RNA polymerase from sporulating cells of *Bacillus subtilis*: identification of a group of forespore-specific promoters. *J. Bacteriol.* 171:2708–2718.
21. Ollington, J. F., and R. Losick. 1981. A cloned gene that is turned on at an intermediate stage of spore formation in *Bacillus subtilis*. *J. Bacteriol.* 147:443–451.
22. Otani, M., N. Ikara, C. Umezawa, and K. Sano. 1986. Predominance of gluconate formation from glucose during germination of *Bacillus megaterium* QM B1551 spores. *J. Bacteriol.* 167:148–151.
23. Panzer, S., D. Sun, P. Setlow, and R. Losick. 1989.

Evidence for an additional temporal class of gene expression in the forespore compartment of sporulating *Bacillus subtilis*. *J. Bacteriol.* **171**:561–564.

24. Postemsky, C. J., S. S. Dignam, and P. Setlow. 1978. Isolation and characterization of *Bacillus megaterium* mutants containing decreased levels of spore protease. *J. Bacteriol.* **135**:841–850.

25. Rasul Chaudhry, G., Y. S. Halpern, C. Saunders, N. Vasantha, B. J. Schmidt, and E. Freese. 1984. Mapping of the glucose dehydrogenase gene in *Bacillus subtilis*. *J. Bacteriol.* **160**:607–611.

26. Rather, P. N., and C. P. Moran, Jr. 1988. Compartment-specific transcription in *Bacillus subtilis*: identification of the promoter for *gdh*. *J. Bacteriol.* **170**:5086–5092.

27. Saava, D., and J. Mandelstam. 1986. Synthesis of *spoIIA* and *spoVA* mRNA in *Bacillus subtilis*. *J. Gen. Microbiol.* **132**:3005–3011.

28. Sammons, R. L., A. Moir, and D. A. Smith. 1981. Isolation and properties of spore germination mutants of *Bacillus subtilis* 168 deficient in the initiation of germination. *J. Gen. Microbiol.* **124**:229–241.

29. Setlow, P. 1978. Purification and characterization of additional low-molecular-weight basic proteins degraded during germination of *Bacillus megaterium* spores. *J. Bacteriol.* **136**:331–340.

30. Setlow, P. 1988. Resistance of bacterial spores to ultraviolet light. *Comments Mol. Cell. Biophys.* **5**:253–264.

31. Setlow, P. 1988. Small, acid-soluble spore proteins of *Bacillus* species: structure, synthesis, genetics, function and degradation. *Annu. Rev. Microbiol.* **42**:319–338.

32. Setlow, P., C. Gerard, and J. Ozols. 1980. The amino acid sequence specificity of a protease from spores of *Bacillus megaterium*. *J. Biol. Chem.* **255**:3624–3628.

33. Sharp, P. M., E. Cowe, D. G. Higgins, D. C. Shields, K. H. Wolfe, and F. Wright. 1988. Codon usage patterns in *Escherichia coli*, *Bacillus subtilis*, *Saccharomyces cerevisiae*, *Schizosaccharomyces pombe*, *Drosophila melanogaster* and *Homo sapiens*: a review of the considerable within-species diversity. *Nucleic Acids Res.* **16**:8207–8211.

34. Shields, D. C., and P. M. Sharp. 1987. Synonymous codon usage in *Bacillus subtilis* reflects both translational constraints and mutational biases. *Nucleic Acids Res.* **15**:8023–8040.

35. Singh, R. P., B. Setlow, and P. Setlow. 1977. Levels of small molecules and enzymes in the mother cell and the forespore of sporulating *Bacillus megaterium*. *J. Bacteriol.* **130**:1130–1138.

36. Stephens, M. A., N. Lang, K. Sandman, and R. Losick. 1984. A promoter whose utilization is temporarily regulated during sporulation in *Bacillus subtilis*. *J. Mol. Biol.* **176**:333–348.

37. Strauss, N. 1983. Role of glucose dehydrogenase in germination of *Bacillus subtilis* spores. *FEMS Microbiol. Lett.* **20**:379–384.

37a.Sun, D., P. Stragier, and P. Setlow. 1989. Identification of a new sigma factor involved in compartmentalized gene expression during sporulation of *Bacillus subtilis*. *Genes Dev.* **3**:141–149.

38. Sussman, M. D., P. S. Vary, C. Hartman, and P. Setlow. 1988. Integration and mapping of *Bacillus megaterium* genes which code for small, acid-soluble spore proteins and their protease. *J. Bacteriol.* **170**:4942–4945.

39. Uratani, B., K. A. Lampel, R. H. Lipsky, and E. Freese. 1985. Characterization of the gene for glucose dehydrogenase and flanking genes of *Bacillus subtilis*, p. 71–76. *In* J. Hoch and P. Setlow (ed.), *Molecular Biology of Microbial Differentiation*. American Society for Microbiology, Washington, D.C.

40. Zuberi, A. R., I. M. Feavers, and A. Moir. 1985. Identification of three complementation units in the *gerA* spore germination locus of *Bacillus subtilis*. *J. Bacteriol.* **162**:756–762.

41. Zuberi, A. R., A. Moir, and I. M. Feavers. 1987. The nucleotide sequence and gene organization of the *gerA* spore germination operon of *Bacillus subtilis* 168. *Gene* **51**:1–11.

Regulation of Procaryotic Development
Edited by Issar Smith, Ralph A. Slepecky, and Peter Setlow
© 1989 American Society for Microbiology, Washington, DC 20006

Chapter 11

Dependence Pathways for the Expression of Genes Involved in Endospore Formation in *Bacillus subtilis*

Richard Losick and Lee Kroos

Endospore formation in *Bacillus subtilis* is a process of cellular differentiation that involves sequential changes in cell physiology and ultrastructure occurring as a response to conditions of nutrient depletion. The sporulating cell proceeds through a series of well-defined, morphological stages that culminate in the formation within the sporangium of an alternate cell type, the endospore. The hallmark of this process and the earliest morphological manifestation of sporulation are the formation of an asymmetrically positioned septum, which partitions the sporangium into mother cell and forespore compartments. The two compartments each receive a chromosome generated during the last vegetative round of DNA replication, but they then undergo divergent developmental fates. The forespore ultimately becomes the core of the spore, is responsible for the production of development-specific proteins that are packaged within the spore core, and is the source of the spore genome. The mother cell, on the other hand, governs the production and assembly of the outer protective layers of the spore, but it and its chromosome are ultimately discarded when maturation of the spore is complete.

The conventionally defined stages of sporulation are the following: 0, the stage representing vegetative or growing cells which have not entered the sporulation pathway; I, the stage of chromosome condensation into an axial filament (because no mutations that block at this stage are known, stage I is frequently not considered to be a true stage of sporulation); II, the stage of asymmetric septum formation; III, the stage of engulfment, in which the mother cell membrane migrates around the forespore and engulfs it in a second membrane layer, thereby pinching off the forespore as a protoplast within the mother cell; IV, the stage of cortex formation, in which cell wall-like material is deposited between the inner and outer membranes that surround the spore protoplast; V, the stage of coat formation, in which the tough protein shell that encases the spore is deposited around the exterior of the cortex outer membrane; VI, the stage during which the spore matures to acquire its full set of characteristic resistance properties; and, finally, VII, the stage at which the fully ripened spore is released by lysis of the mother cell. The entire process lasts for 6 to 8 h at 37°C. When exposed to nutrients, the spore ger-

Richard Losick • Department of Cellular and Developmental Biology, Biological Laboratories, Harvard University, Cambridge, Massachusetts 02138. *Lee Kroos* • Department of Biochemistry, Michigan State University, East Lansing, Michigan 48824.

minates and produces a growing cell, thus completing the cycle of development.

A tenet of research on sporulation is that progression through these morphological stages is governed by an underlying temporal program of developmentally regulated gene expression (37). Indeed, much progress has been made in identifying, isolating, and sequencing genes that govern spore formation, but only recently have the expression and control of genes involved in sporulation been studied on a sufficiently large scale to permit the organization of information on the regulation of individual genes into a rudimentary network that describes the overall program of sporulation gene expression. Here we summarize recent results from many laboratories that allow one to determine the times at which individual developmental genes are switched on, to assign the differential expression of genes to the mother cell or forespore chambers of the sporangium, and to understand how the activation of developmental genes depends on the products of other such genes. Extending and updating an earlier effort (36), we propose an integrated scheme that organizes the results of regulatory studies into a network of dependent pathways. An emerging feature of the network is the existence of apparent coupling points that link the program of sporulation gene expression to landmark events in spore morphogenesis.

GENES INVOLVED IN SPORULATION AND GERMINATION

Traditionally, genes involved in the differentiation cycle have been identified by the isolation of mutants blocked in sporulation and germination (44, 45). Such mutants become arrested at a particular step in the cycle and are classified accordingly. Thus, *spo* mutants are blocked in sporulation, generally at a characteristic stage; *ger* mutants are defective in germination; and *out* mutants are impaired in the metamorphosis of a germinated (phase-dark) spore into a growing cell. The

distinction between *ger* and *spo* genes is rather arbitrary, however. Many (probably all) *ger* genes are expressed during sporulation, as is expected since their products participate in the formation of the maturing spore to confer on it the capacity to germinate. Furthermore, certain *ger* mutants, such as *gerJ*, exhibit defects in late sporulation and hence could justifiably be designated *spo* mutants (61). Conversely, certain mutants (e.g., *spoVIA*, *spoVIB*, and *spoVIC*) identified on the basis of late sporulation defects turn out to exhibit germination abnormalities (24–26). Thus, in the context of this review, *ger* and *spo* genes are treated as indistinguishable components of the program of sporulation gene expression. In contrast, *out* genes are thought to be expressed during the outgrowth period of germination, and as such these genes are not relevant to our discussion of the program of sporulation gene expression.

Recently, the application of "reverse genetic" approaches (i.e., the use of partial amino acid sequences or specific antibodies to identify and clone structural genes for particular proteins) has revealed several new genes whose products are involved in spore formation but which are not among genes uncovered by the traditional approach of gathering mutants with Spo⁻ or Ger⁻ phenotypes. These newly identified genes fall into two classes: the *cot* genes (8, 67), which encode the polypeptide components of the spore coat, and the *ssp* genes (5, 17, 20; P. Setlow, this volume), which encode the partially homologous family of small, acid-soluble polypeptides located in the spore core. So far seven *cot* and five *ssp* genes have been identified. Coat proteins and small, acid-soluble polypeptides make up 50 to 60% and 8 to 10%, respectively, of the total spore protein, yet mutants of these genes (made by replacing the corresponding chromosomal genes with in vitro-constructed mutant genes) are found to be altered only in subtle properties of the spore (e.g., lysozyme resistance in the case of certain *cot* genes [67] and

UV resistance in the case of certain *ssp* genes [39]), but not in spore formation itself. These results explain why these genes were not found in traditional mutant searches and indicate that individual species of the principal spore structural proteins are not needed for the formation of a normal-looking (optically refractile) spore.

A third approach to finding genes involved in the program of sporulation gene expression is the identification of developmentally regulated genes on the basis of their time of induction. For example, *0.3 kb* is a gene whose transcription is turned on at a late stage of sporulation and which was identified on the basis of the appearance of its transcript, a stable RNA, in cells at late stages of sporulation (41). It is not known whether the protein product of *0.3 kb* is involved in the formation of the spore or in the acquisition of spore resistance properties, but its transcription is nonetheless rigidly entrained in the sporulation developmental program (42, 50). A general approach to the identification of genes whose expression is associated with sporulation is the use of the fusion-generating transposon Tn*917-lac* to create transcriptional fusions on a large scale at many sites in the chromosome as a direct consequence of the insertion of the element into genes (43, 66). Methods have been described for screening large numbers of bacteria containing Tn*917-lac* at different sites for tagged genes whose expression occurs during sporulation. This methodology can also be used in conjunction with fusions of various sporulation regulatory genes (see below) to the isopropyl-β-D-thiogalacto-pyranoside (IPTG)-inducible *spac* promoter (63) to identify genes whose expression is under the control of specific sporulation transcription factors (A. Grossman, unpublished results).

Thus, a full description of the range of genes that are involved in sporulation and/or whose expression is under developmental control requires the use of a variety of approaches, and the known list of genes that constitute the sporulation developmental program is likely to be incomplete.

REGULATORY GENES AND THEIR PRODUCTS

In almost all examples so far known, sporulation genes have been found to be regulated primarily or exclusively at the level of transcription and, when analyzed, at the level of transcription initiation (e.g., reference 33). An exception is autoregulation of the small, acid-soluble polypeptide structural gene *sspE*, which is exerted at a posttranscriptional level (39). Undoubtedly, other examples of posttranscriptional regulation will emerge as the regulation of additional genes involved in sporulation is studied, but for present purposes we will assume that developmental regulation operates primarily at the level of promoter activity.

In considering regulatory genes and their products, we distinguish between true transcription factors (e.g., activators, repressors, and sigma factors), which directly govern gene expression, and other kinds of proteins (e.g., membrane proteins and structural proteins) that are indirectly required for gene expression. This issue arises from the observation that many *spo* mutations have highly pleiotropic effects on gene expression, blocking the transcription of several (many) developmental genes. Often such mutations are located in genes (e.g., *spo0H* [9] and *spoIIGB* [53, 57]) that turn out upon further analysis to be direct transcriptional determinants that directly govern the transcription of other genes, and their pleiotropic effects on gene expression reflect their involvement in a dependent regulatory pathway. However, in other cases the mutation is located in the gene for a membrane or structural protein. In the case of the *spoIIE* operon, at least one of whose products is likely to be a membrane protein (P. Guzman, J. Westpheling, and P. Youngman, personal communication), the pleiotropic effect of mutations on gene expression is understood to be indirect, reflecting the requirement for the products of all three *spoIIE* cistrons in the processing of pro-σ^E to its mature form (34, 52). To avoid confusion, we reserve the designation "regulatory" for genes that are known or inferred to encode direct transcriptional determinants.

Three broad categories of genes that are known or inferred to be regulatory can be distinguished. The first category, which is considered more fully in a separate chapter by I. Smith (this volume), includes genes whose predicted products are homologous to members of a set of two-component sensory systems that are widespread among eubacteria. The two-component sensory systems consist of pairs of sensor and regulator proteins, known in at least two cases to modulate their activity by phosphorylation. Included in this category are the *spo0* gene products Spo0A and Spo0F (16, 56, 64), which are homologous to each other in an amino-terminal region that is highly conserved among the regulator members of the two-component sensory systems. Because *spo0A* and *spo0F* mutations (like mutations in other kinds of *spo0* genes) cause pleiotropic effects on gene transcription and because in other systems regulator genes (e.g., *ntrC*; see, for example, reference 22) have been shown to be transcriptional determinants, it is presumed that Spo0A and Spo0F are themselves direct regulators of promoter activity. This has not been established, however, and at least some members (*cheY* and *cheB* [51]) of the regulator family are clearly not transcriptional regulators. Whether or not these stage 0 genes encode direct transcriptional determinants, one function of Spo0A protein seems to be to block the action of *abrB* (68; M. Perego, G. Spiegelman, and J. A. Hoch, *Mol. Microbiol.*, in press), a gene whose product is inferred to be a negative regulator of the transcription of certain sporulation and sporulation-associated genes.

If *spo0A* and *spo0F* function as regulator genes, then it should be expected that

their activity is governed by the products of one or more sensor genes, members of which are homologous to each other in amino acid sequences at their carboxy terminus. A candidate for such a sensor gene on the basis of sequence homology is *spoIIJ* (P. Stragier, personal communication), a gene in which an insertional mutation (49) causes an oligosporogenous, stage 0 to II defect. It is, of course, not known whether SpoIIJ functions as a sensor protein and, if so, whether it responds to nutritional signals by modulating the activity (by phosphorylation?) of Spo0A or Spo0F or both.

The second category of regulatory genes and the topic of the chapter by C. P. Moran, Jr. (this volume), encompasses those that encode RNA polymerase sigma factors that are required for sporulation but not for growth. So far, the following five developmental sigma factors have been identified: σ^E, the product of *spoIIGB* or *sigE* (53, 57); σ^F, the product of *spoIIAC* or *sigF* (11, 18, 55); σ^H, the product of *spo0H* or *sigH* (3, 9); σ^G, the product of *spoIIIG* or *sigG* (55); and σ^K or σ^{27} (32), which is encoded by a composite gene composed of cistron *B* of the *spoIVC* locus and *spoIIIC* (see below; 32, 54). The proposed relationship of these sigma factors to the temporal and spatial program of sporulation gene expression is diagrammed in Fig. 1.

The third category of developmental regulatory genes represents those encoding proteins that are believed to act as direct activators or repressors of transcription but are not homologous to members of the two-component sensory systems and are not RNA polymerase sigma factors. An example of a gene in this category is *gerE*, whose product, an 8.5-kilodalton (kDa) polypeptide (6), is inferred to be a regulatory protein on the following bases: (i) the existence of a possible α-helix–β-turn–α-helix structure near its carboxyl terminus (23), a structural motif characteristic of the DNA-binding domain of many repressors and activators; (ii) its high overall homology to the carboxyl terminus

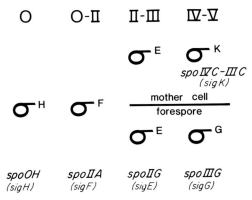

Figure 1. Temporal and spatial relationship of RNA polymerase sigma factors to sporulation. Stages of sporulation are indicated across the top. The compartmentalization of sigma factors after stage II is indicated by the mother cell/forespore line. The presence of σ^E in both compartments is based on the work of Carlson and Haldenwang (2). Structural genes are listed below the corresponding sigma factor.

of the *sacU* gene product, a regulatory protein of *B. subtilis* involved in controlling the production of extracellular proteins (21); and (iii) regulatory studies that suggest an intimate role for GerE in the activation and repression of certain *cot* genes (48; L. Zheng, unpublished results). An additional example is *spoIIID*, whose product, a polypeptide of about 14 kDa, is a repressor and an activator of the transcription in vitro of certain mother-cell-specific genes whose promoters are utilized by σ^K-containing RNA polymerase (32; see below).

PATTERNS OF DEVELOPMENTALLY REGULATED GENE EXPRESSION

The regulation of *spo*, *ger*, *cot*, and *ssp* genes can be considered at two levels. At the first level it is sufficient to determine the time of transcriptional activation of individual developmental genes and the quantitative dependence of this expression on the products of other developmental genes. The gathering of this kind of information on a large scale provides the basis for organizing dependence patterns into a regulatory network that de-

scribes the overall program of sporulation gene expression. At the second level it is necessary through the application of appropriate genetic and biochemical approaches to understand the regulation of individual genes in terms of specific *cis*- and *trans*-acting elements. This kind of information provides a basis for describing the developmental program in terms of specific transcriptional proteins and their target DNA sequences. In this section we summarize regulatory information on examples of genes that have been the subject of these kinds of investigations. We consider, in order, genes activated at progressively later stages of sporulation, beginning with a gene activated at the onset of sporulation. For simplicity, we do not include the regulation of *spo0* genes, a topic which is considered at length in the chapter by Smith (this volume).

spoVG

Although a mutation in *spoVG* impairs spore formation at a relatively late stage, this gene is induced at the very onset of sporulation (reviewed in reference 37). Under conditions in which sporulation is induced by the treatment of cells with decoyinine, stimulation of *spoVG* transcription can be detected within 15 min after the addition of drug. The *cis*-acting regulatory region of *spoVG* consists of an upstream, A+T-rich element, which strongly enhances the level of *spoVG* transcription, and a promoter region which gives rise to an upstream, primary transcript called P1 and a downstream, secondary transcript called P2. It is not known whether P2 arises by transcription initiation or as a result of nucleolytic cleavage of P1 RNA, but for present purposes we will assume that both transcripts are the products of a single promoter.

The transcription of *spoVG* is controlled, in part, by the products of *spo0A* and *spo0H*, which exert their effects through partially independent pathways of negative and positive control. Evidence that *spo0A* acts through a pathway of negative control comes from the finding that the dependence of *spoVG* expression on *spo0A* can be bypassed by mutations at *abrB* (mutations that also suppress some other aspects of the phenotype associated with *spo0A* mutants [59]) or by a *cis*-acting mutation within the *spoVG* promoter (68). It is believed that the *abrB* gene product acts, directly or indirectly, to block the transcription of *spoVG* and that the *spo0A* gene product somehow causes the inactivation of the AbrB protein or blocks the expression of the *abrB* gene (68; Perego et al., in press). The product of the *spo0H* gene, on the other hand, is a positively acting regulatory protein, the RNA polymerase sigma factor σ^H, which interacts directly with the *spoVG* promoter (3, 9, 68). *spo0A*-dependent loss of AbrB at the onset of sporulation could contribute to the induction of *spoVG*, but *spoVG* expression is not fully constitutive in a *spo0A abrB* double mutant; that is, some induction of *spoVG* is still observed in the double mutant. Thus, a *spo0A*-independent mechanism for the stimulation of *spoVG* transcription must also exist. This inducing mechanism could be an increase in the level or activity of σ^H or transcriptional stimulation of *spoVG* through the action of an additional, as yet unidentified, regulatory protein.

spoIIA, *spoIIE*, and *spoIIG*

The stage II loci *spoIIA*, *spoIIE*, and *spoIIG* are polycistronic operons that are activated at approximately 30 to 90 min after the start of sporulation (13, 19, 30, 62). The *spoIIA* operon comprises three genes, one of which encodes σ^F (18, 55). The *spoIIG* operon contains the structural gene (*spoIIGB* or *sigE*) for σ^E and a gene (*spoIIGA*) required for the conversion of pro-σ^E to its mature form (27, 52). The *spoIIE* operon comprises three cistrons, all of which are somehow required in pro-σ^E processing (58; P. Guzman, J. Westpheling, W. Haldenwang, and P. Youngman, personal communication), a

topic that is discussed further below and in the chapter by Moran (this volume).

Mutations in *spo0A*, *spo0H*, and all other stage 0 genes tested except *spo0J* abolish the transcription of *spoIIA*, *spoIIE*, and *spoIIG* (13, 19, 30, 62). In contrast, few *spoII* mutations affect the expression of these operons, and in particular, the transcription of all three stage II operons is not blocked by mutations in *spoIIAC* (encoding σ^F) or *spoIIGB* (encoding σ^E). Thus, among known sporulation sigma factors, only the product of *spo0H*, σ^H, is required for the transcription of *spoIIA*, *spoIIE*, and *spoIIG*. This is puzzling because the RNA polymerase-binding regions of the *spoIIE* and *spoIIG* operons do not appear to be recognized by σ^H-RNA polymerase. Rather, they contain sequences that conform perfectly in the case of *spoIIE* (19) and almost perfectly (29) in the case of *spoIIG* to " − 10" and " − 35" canonical sequences recognized by the primary, vegetative holoenzyme form $E\sigma^A$. The spacing between these $E\sigma^A$-like consensus sequences is, however, significantly larger than the 17 to 19 base pairs usually required for efficient promoter recognition. This raises the possibility that the *spoIIE* and *spoIIG* operons are transcribed by a modified form of $E\sigma^A$ or $E\sigma^A$ working in conjunction with an additional factor. If so, the requirement for *spo0H* could be understood as being indirect in controlling the expression of a gene whose product facilitates the utilization of the *spoIIE* and *spoIIG* promoters by the primary vegetative holoenzyme form. The *spoIIA* promoter, on the other hand, conforms to the consensus nucleotide sequence for σ^H-utilized promoters, and it is probable that it is under the direct control of σ^H-RNA polymerase, acting in conjunction with an additional regulatory protein, such as Spo0A (62).

Another intriguing aspect of the regulation of the *spoIIG* operon is the discovery that its transcription is strongly influenced by a mutation in *spoIIN*, whose predicted product is homologous to *ftsA*, a gene required in the cell division cycle of *Escherichia coli*

(P. Stragier and T. Leighton, personal communication; 1). If *spoIIN* (only a single, temperature-sensitive allele is known [65] and it is possible that it is an essential gene) is the *B. subtilis* equivalent of *ftsA*, then the activation of *spoIIG* operon transcription at the stage of sporulation septum formation could bring into play the action of a component of the vegetative cycle of binary fission.

spoIID

The *spoIID* gene is induced during h 2 of development (4, 47). The genetic requirements for this transcription are much more complex than those governing the expression of *spoVG*, *spoIIA*, *spoIIE*, and *spoIIG* in that the transcription of *spoIID* depends not only on the products of most stage 0 genes, but also on the products of many stage II genes. Despite this genetic complexity, the primary event in the induction of *spoIID* is probably simply the appearance of σ^E, the processed product of the promoter-distal gene (*sigE* or *spoIIGB*) of the *spoIIG* operon; *spoIID* is actively transcribed by $E\sigma^E$ in vitro (47), and the transcription of *spoIID* can be switched on during growth in cells that have been genetically engineered to produce σ^E in response to IPTG (52). A possible explanation for the complex genetic requirements for *spoIID* transcription is that σ^E must be proteolytically cleaved from its inactive precursor, pro-σ^E or P^{31} (34), the primary product of *spoIIGB* (*sigE*), and it is this processing step, as will be discussed below, that depends on the action of a large number of stage II genes (34, 52).

spoIIIG

The transcription of the *spoIIIG* gene, which encodes the forespore-specific sigma factor σ^G, is induced during h 3 of sporulation (28, 40). Karmazyn-Campelli and Bonamy (28) and Masuda et al. (40) have distinguished two modes of *spoIIIG* regulation. In the early mode, the transcription of *spoIIIG* originates from the *spoIIG* operon,

which is located immediately upstream from *spoIIIG*. The early mode is believed to "prime the pump" for transcription in the late mode, in which *spoIIIG* is autogenously regulated from its own promoter under the direction of its own gene product, σ^G (28). The significance of these early and late modes of regulation to the issue of the compartmentalization of the synthesis of σ^G and the compartmentalized expression of the genes whose transcription it directs is discussed below.

ssp, spoVA, gdh, gerA, and 0.3 kb

The *ssp* family of genes (*sspA* through *sspE*), the *spoVA* operon, the glucose dehydrogenase gene *gdh*, and the germination gene *gerA* are induced during h 3 of sporulation, and this transcription is confined to the forespore (12, 38, 46; A. Moir, personal communication). Several results support the view that the induction and compartmentalization of expression of these genes are a consequence of the appearance of the forespore-specific sigma factor σ^G (28, 55): (i) σ^G is present only in the forespore; (ii) it directs the transcription of *ssp* genes in vitro; (iii) σ^G is the product of *spoIIIG*, a null mutation which blocks the expression of *gdh*, *spoVA*, and *ssp* genes; and (iv) *ssp* and *gdh* transcription can be activated during growth in cells engineered to produce σ^G in response to IPTG. Interestingly, however, expression during sporulation is also completely dependent on the product(s) of the *spoIIIE* locus, even though *spoIIIE* is not required for the transcription of *spoIIIG* (12, 38; Stragier, personal communication).

Another gene that is switched on in the forespore compartment and depends on *spoIIIG* for its transcription is *0.3 kb*, a gene of unknown function that was identified on the basis of its 300-base transcript, which is present in late sporulating cells (41, 50). Like *ssp* genes, *0.3 kb* can be switched on during growth in cells engineered to produce σ^G in response to the *lac* inducer IPTG (42). In contrast to *ssp* and other known forespore-expressed genes, *0.3 kb* is switched on only after a delay upon the induction of σ^G synthesis in vegetative cells. Furthermore, *0.3 kb* is switched on significantly later (h 4) during sporulation than are other forespore-specific genes (h 3). These findings suggest that *0.3 kb* is representative of an additional temporal class of forespore gene expression. The regulatory protein that presumably governs the delayed induction of *0.3 kb* has not been identified.

spoIIIC and *spoIVCB*

spoIIIC and *spoIVCB* (one of two separately transcribed members of the *spoIVC* locus [15, 33; Y. Kobayashi, personal communication]) are mother-cell-specific genes required at stage IV of sporulation. Interestingly, the predicted product of *spoIIIC* is highly similar to the carboxyl terminus of sigma factors (10, 14), whereas the product of *spoIVCB* is highly similar to the NH$_2$ terminus of sigma factors (54). As discussed below, *spoIVCB* and *spoIIIC* become joined in frame by means of a chromosomal rearrangement at an intermediate stage of sporulation to create a composite gene (54), which has been identified as the structural gene (*sigK*) for the mother-cell-specific sigma factor σ^K (or σ^27; 32). Although the regulation of *spoIIIC* and *spoIVCB* has been the subject of separate investigations (33, 60; which revealed generally similar patterns of regulation for the two genes), the transcription of *spoIIIC* can now be understood as being a secondary consequence of its fusion to *spoIVCB*, whose promoter presumably controls the expression of the composite gene.

The transcription of *spoIVCB* (and hence the transcription of the *spoIVCB*-*spoIIIC* [*sigK*] composite gene) is restricted to the mother cell and is switched on at about 3.5 h after the start of sporulation (33). This RNA synthesis is abolished by stage 0 and II mutations and is influenced to various extents by mutations that cause blockage after stage II, but among late-blocking mutations, only a *spoIIID* mutation inhibits transcription very strongly. It was therefore inferred that the product of *spoIIID* is likely to be a direct regulator of *spoIVCB* transcription (see below). It was further inferred, on the basis of promoter deletion analysis, that this regulation is exerted at the level of transcription initiation (promoter activation).

Experiments (32) in which a cloned copy of *spoIVCB* was used as a template for in vitro RNA synthesis showed that specific transcription from the promoter for this gene could be achieved, albeit at a low level, by RNA polymerase containing the σ^K (σ^27) species of sigma factor. On this basis the transcription of the *spoIVCB*-*spoIIIC* composite gene is expected to be at least in part autoregulatory, and mutations in either *spoIVCB* or *spoIIIC* do reduce expression from the *spoIVCB* promoter (33). To prime the pump, however, the *spoIVCB* promoter must also be recognized by a holoenzyme form (e.g., Eσ^E) present earlier than Eσ^K.

In vitro transcription experiments also led to the identification and purification of a nonsigma protein of 14 kDa (called a "switch" protein for reasons considered below) that greatly stimulates transcription from the *spoIVCB* promoter by σ^K-RNA polymerase (32). The 14-kDa protein was assigned as the product of the *spoIIID* gene (B. Kunkel, L. Kroos, H. Poth, P. Youngman, and R. Losick, unpublished results) on the basis of a partial NH$_2$-terminal sequence of the protein (32). The finding that a *spoIIID* mutation inhibits *spoIVCB* expression more strongly than do mutations in *spoIVCB* or *spoIIIC* (33; B. Kunkel, unpublished results) suggests a role for the 14-kDa protein in the early, pump-priming mode of *spoIVCB* transcription. Thus, the *spoIIID*-encoded switch protein is a transcriptional activator of the *spoIVCB* promoter and should be considered a key component of the mother cell line of gene expression.

Spore Coat Genes *cotA*, *cotC*, *cotD*, and *cotE* and the Germination Gene *gerE*

Finally, we consider genes encoding structural proteins of the spore coat and the germination gene *gerE*, a putative transcriptional regulator of *cot* gene expression (see below). *cotA*, *cotC*, *cotD*, and *cotE* encode proteins of 65, 12, 11, and 24 kDa, respectively (8, 67). *cotA* (also known as *pig*) is responsible for the characteristic brown pigment of colonies of sporulating cells (8). *cotE* encodes a morphogenic protein required for the assembly of the electron-dense outer layer of the coat (67). These genes can be grouped into three temporal classes of gene expression, with *cotE* being turned on at about h 2, *cotA*, *cotD*, and *gerE* being turned on coordinately at about h 4.5, and *cotC* being activated at h 6, the latest of any known sporulation-associated gene (6, 48; S. Cutting, S. Panzer, and R. Losick, *J. Mol. Biol.*, in press; Zheng, unpublished results). It is inferred from their function in coat formation that all five genes are expressed selectively in the mother cell, but this has been established directly only for *cotA* and *gerE* (Cutting et al., in press).

Transcription experiments in vitro, based on the use of cloned copies of *cotA*, *cotD*, and *gerE*, indicate that all three genes are transcribed by RNA polymerase containing the mother-cell-specific sigma factor σ^K, which, as discussed above, acts in conjunction with the 14-kDa product of *spoIIID* to direct transcription from the *spoIVCB* promoter (32; L. Kroos, unpublished results). In contrast to *spoIVCB*, however, *cotA*, *cotD*, and *gerE* do not require the 14-kDa protein to be transcribed maximally; in fact, this small auxiliary transcription factor inhibits strongly transcription of the two *cot* genes but not *gerE* (32; Kroos, unpublished results). The 14-kDa protein has been called a switch protein because its inactivation or sequestering during the transition from stage IV to V could, in part, effect the temporal switch from *spoIVCB* transcription to transcription of the *cotA* and *cotD* genes (Fig. 2).

In vitro transcription studies have not been carried out on the *cotC* gene, but its induction is strongly impaired by mutations in *gerE* and *spoVP* (Zheng, unpublished results), a finding which suggests that the product of one or both genes is a direct or indirect activator of the *cotC* gene. A role for GerE in transcriptional regulation is also to be inferred from its amino acid sequence, which contains a region of significant homology to the DNA-binding domain of certain procaryotic repressors and activators (23) and which has high overall homology to the carboxyl-terminal region of the product of the *B. subtilis* regulatory gene *degU* (21). Conceivably, GerE or SpoVP or both act in conjunction with the $E\sigma^K$ form of RNA polymerase holoenzyme to direct *cotC* transcription. Another effect of a mutation in *gerE* is to cause a severalfold overexpression of the *cotA* gene (48). Thus, like *SpoIIID* (the "14-kDa" switch protein), GerE may be both an activator and a repressor of mother-cell-specific gene expression.

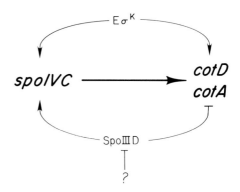

Figure 2. Proposed mechanism for temporal switching in the mother cell line of gene expression. SpoIIID is an activator of *spoIVC* transcription and a repressor of *cotA* and *cotD* transcription by σ^K-containing RNA polymerase. The inactivation or sequestering of SpoIIID by an unknown mechanism (the question mark) during the transition from stage IV to stage V turns off *spoIVC* expression and turns on *cot* gene expression.

COMPARTMENTALIZATION OF GENE EXPRESSION

After the formation of the sporulation septum, the sporangium is partitioned into two compartments, each of which has its own chromosome. As proposed by Lencastre and Piggot (35) on the basis of clever, but indirect, genetic experiments involving the correction by DNA-mediated transformation of mutations in genes required either in the forespore or in the mother cell, the two compartments selectively support (or, more precisely, selectively require) the expression of different sets of *spo* genes. This concept was confirmed and extended by Dancer and Mandelstam (7) in complementation experiments in which they fused protoplasts of various *spo* mutants blocked at stage III and later. Thus, certain genes (e.g., *spoVA* in the experiments of Lencastre and Piggot [35]) are deduced to act in the forespore compartment, whereas other genes (e.g., *spoIVC* in the experiments of Dancer and Mandelstam [7]) carry out their assigned function in the mother cell chamber.

Direct confirmation that postseptation gene expression is compartmentalized has come from the use of *lacZ* operon fusions and the application of methods for the subcellular localization of β-galactosidase. (Though direct, the use of *lacZ* fusions suffers in applications to weakly expressed genes because of the incompleteness and inefficiency of existing subcellular fraction methods and because of the instability of β-galactosidase in the mother cell [D. Foulger and J. Errington, personal communication].) Examples of genes induced after stage II whose expression, as judged by the use of *lacZ* fusions, is confined to the forespore are *gerA* (Moir, personal communication), *gdh* (46), *spoVA* (12), *sspA* through *sspE* (38), and *0.3 kb* (42). Examples of genes whose expression is believed to be confined to the mother cell are *gerE* (Cutting et al., in press), *spoIIIC* (60), *spoIVC* (33; to which *spoIIIC* becomes joined [54]), *spoVJ* (J. Errington, personal

communication), and *cotA* (Cutting et al., in press).

Compartment-Specific Sigma Factors

Differential gene expression in the two chambers of the sporangium is determined in part by the compartment-specific sigma factors σ^G and σ^K, which, as discussed above, direct the transcription of forespore- and mother-cell-expressed genes, respectively. The challenge of understanding compartmentalized gene expression reduces therefore to investigating the basis for the differential presence of the two sigma factors in the two compartments of the sporangium. As discussed above, the structural gene *spoIIIG* for the forespore-specific sigma factor σ^G is subject to early and late modes of regulation (28). In the early mode *spoIIIG* is transcribed by readthrough transcription from the adjacent *spoIIG* operon. This early mode serves to prime the pump for transcription in the late mode, which is directed by σ^G itself acting at a promoter at the beginning of the *spoIIIG* gene. Thus, the transcription of *spoIIIG* in the late mode is autoregulatory.

It is not known whether transcription in the early mode is compartmentalized, but transcription in the late mode is restricted to the forespore (28). Autoregulation in the late mode can be understood as being a means for amplifying the compartment-specific synthesis of σ^G in the late mode, but the important unsolved problem is the underlying mechanism for the late-mode, compartment-specific expression of *spoIIIG*, which autoregulation enhances. Karmazyn-Campelli and Bonamy (28) report that *spoIIIG* expression in the late mode is dependent on the *spoIIIA* operon, which is proposed to be a primary determinant of compartmentalized transcription of *spoIIIG*. However, studies by N. Illig and J. Errington (personal communication), on the other hand, indicate that at least one *spoIIIG*-dependent operon does not depend on *spoIIIA* for its transcription, and genetic experiments (35; J. Errington and

M. Young, personal communication) suggest that *spoIIIA* is a mother-cell-expressed operon and that its expression is not required in the forespore. Thus, the underlying basis for compartment-specific transcription of *spoIIIG* and the exact role of *spoIIIA* remain to be clarified.

Compartment-specific transcription of *spoIIIG* is not, however, the entire basis for forespore-specific gene expression, as the expression of forespore-specific genes also depends on the product(s) of the *spoIIIE* locus (12, 38), which is not required for *spoIIIG* transcription. Interestingly, *spoIIIE* is a vegetatively expressed locus, yet its product(s) is required specifically for gene expression in the forespore (Errington, personal communication).

The counterpart to σ^G in the mother cell is σ^K, which, as judged from the expression pattern of *spoIVCB* (33), is preferentially expressed in the mother cell. Interestingly, like *spoIIIG*, the σ^K structural gene is autoregulatory, which once again can be thought of as amplifying the compartment-specific synthesis of σ^K. Additionally, since *spoIVCB* transcription is stimulated by the *spoIIID* gene product and since the *spoIIID* product is not required for forespore gene expression, this stage III locus can be seen as being a critical determinant of the mother cell line of gene expression. Over and above autoregulation and activation by SpoIIID, however, recent evidence (54) indicates that the underlying basis for the selective synthesis of σ^K in the mother cell is a chromosomal rearrangement that serves to create the σ^K structural gene by the juxtaposition of its two truncated coding elements, *spoIVCB* and *spoIIIC*.

Split Genes and the Generation of the Structural Gene for the Mother Cell-Specific Sigma Factor σ^K by Chromosomal Rearrangement

Coding information for σ^K in the vegetative chromosome is split between two trun-cated genes, *spoIVCB*, which specifies the NH$_2$ terminus (32, 54), and *spoIIIC*, which specifies the COOH terminus (10, 54). The two incomplete coding elements are juxtaposed to create a composite gene by site-specific recombination between a 5-base-pair repeated sequence located near the end of *spoIVCB* and near the beginning of *spoIIIC* (54). A significant segment of the chromosome must be involved in the rearrangement because the two genes are at least 10 kilobases apart, as judged by the failure to detect linkage between them by DNA-mediated transformation (even though they are in the same region of the chromosome, as judged by phage PBS1-mediated transduction). The chromosomal rearrangement is developmental specific because it occurs at h 3 of sporulation and is prevented in certain sporulation mutants, such as *spoIIG* and *spoIIID* (54). Rearrangement does not require morphological events occurring after stage II, because the *spoIVCB*-*spoIIIC* composite gene is generated in a *spoIID* mutant (54). It is believed (but has not yet been directly demonstrated) that the rearrangement occurs only in the mother cell, because it is prevented by a mutation in *spoIIID*, a specific regulator of the mother cell line of gene expression, but not by a mutation in the forespore sigma factor gene *spoIIIG* (54). If the juxtaposition of *spoIVCB* and *spoIIIC* is restricted to the mother cell, then the rearrangement need not be reversible, because the mother cell chromosome is discarded after maturation of the spore is complete.

If the DNA rearrangement is limited to the mother cell, then it could be part of the underlying mechanism for compartment-specific transcription. Thus, transcription that is directed by Eσ^K may be restricted to the mother cell simply because the composite gene *sigK* is generated only in this compartment. Compartmentalization of σ^K could also be achieved, however, by selective utilization of the *spoIVCB* promoter in the mother cell or by compartment-specific processing of pro-σ^K, the hypothetical primary product of the composite gene (32). Insofar as transcription of the composite gene is autoregulatory (*spoIVCB* has a σ^K-utilized promoter), the previously observed (33) compartmentalization of *spoIVCB* transcription may be largely a secondary consequence of the compartmentalization of the rearrangement process. Thus, if the composite gene and hence σ^K is present only in the mother cell, then the transcription of *sigK* would necessarily be largely confined to the mother cell. It will be important to assess the relative contributions of DNA rearrangement, selective transcription, and pro-σ^K processing to the compartmentalization of σ^K and σ^K-directed gene expression.

As an alternative explanation for forespore-specific synthesis of σ^G, it would be interesting to determine whether the lack of *spoIIIG* transcription in the mother cell could be due to inactivation of the gene or the generation of a repressor of the gene by an additional chromosomal rearrangement in the mother cell.

COUPLING GENE EXPRESSION TO MORPHOGENESIS

Gene expression during sporulation is sometimes thought of as a free-running program in which sets of coordinately regulated genes are switched on in an ordered sequence without regard to the morphogenic events catalyzed by the products of developmental genes. Contrary to this view, recent work on the sporulation problem points to the existence of specific coupling mechanisms that link temporal and spatial aspects of the program of gene expression to landmark events in the morphogenic sequence. The emerging view is that the expression of sets of developmental genes is kept in register with the course of sporogenesis by feedback mechanisms in which the transcriptional apparatus somehow senses and responds to milestone morphological or physiological events in the metamorphosis of the vegetative cell into the endospore.

The most obvious example of morphological coupling is compartmentalized gene expression (considered above), in which the differential activation of mother-cell- and forespore-expressed genes depends on the partitioning of the sporangium into two chromosome-containing chambers by asymmetric septation. The mechanistic basis for this example of morphological coupling is not entirely understood, but it appears to be due in part (at least with respect to mother-cell-specific gene expression) to a compartment-specific chromosomal rearrangement, which we can presume is, in turn, coupled to some postseptation difference between the two compartments.

In the following sections we consider two additional examples of presumed morphological coupling.

Pro-σ^E Processing and Sporulation Septum Formation

As discussed above, the proteolytic cleavage of pro-σ^E (P^{31}) to its mature form, σ^E, occurs at the time of sporulation septum formation and depends on many genes whose products are required for normal septum formation. Labell et al. (34) have proposed that the processing enzyme sits in the septum and thereby serves to couple the attainment of a key morphological milestone to a regulatory change in RNA polymerase that governs the activation of postseptation gene expression. Stragier et al. (52) have independently advanced this idea and have elaborated on it by providing evidence that (i) the product of the upstream member (*spoIIGA*) of the σ^E-encoding *spoIIG* operon is the processing enzyme, in that the requirement for other stage II genes in σ^E activation can be bypassed by engineering the expression of *spoIIGA* and *spoIIGB* in vegetative cells, and (ii) SpoIIGA is a membrane protein, as judged from its deduced amino acid sequence.

Stragier et al. (52) speculate that the putative SpoIIGA processing enzyme must integrate into the properly assembled sporulation septum in order to catalyze the conversion of pro-σ^E to the active form of the sigma factor (Fig. 3). The requirement for the products of the *spoIIE* and *spoIIA* operons is understood as reflecting the role of these operons in the formation of the septum. Indeed, the nucleotide sequence of *spoIIE* suggests that at least one of its products is a membrane protein (Guzman, Wespheling, and Youngman, personal communication). The *spoIIA* operon encodes the RNA polymerase sigma factor σ^F, the predicted product of the promoter-distal gene *spoIIAC*. According to the model of Stragier et al. (52), σ^F directs the synthesis during sporulation of auxiliary proteins (possibly the products of certain vegetative septation genes) involved in the formation of the sporulation septum. If the appearance of the sporulation septum is the event that governs the activation of the SpoIIGA protease, it must be a subtle feature of the septum upon which the activity of SpoIIGA depends, as *spoIIA* and *spoIIE* mutants do produce septa, albeit septa that are slightly aberrant or improperly positioned. To explain how the production of SpoIIGA in vegetative cells bypasses the requirement for sporulation septum formation, Stragier et al. (52) propose that the division septum of growing cells

transiently plays the same role as the sporulation septum of differentiating cells, in that it resembles the stage II septum in the brief interval before it becomes filled with peptidoglycan.

Activation of *cotA*, *cotD*, and *gerE* Is Coupled to Completion of Morphological Stage IV

Spore coat genes *cotA* and *cotD* and the germination gene *gerE* are switched on at h 4 or 5 of sporulation in the mother cell chamber of the sporangium (6, 48; Cutting et al., in press; Zheng, unpublished results). Dependence studies have shown that the transcription of these genes is strongly impaired in almost all mutants blocked prior to stage V, the stage of coat synthesis and assembly. This is a striking result because the list of genes upon which *cotA*, *cotD*, and *gerE* transcription depends includes genes governing forespore-specific gene expression (*spoIIIA*, *spoIIIE*, and *spoIIIG*) as well as genes (e.g., *spoIIIC*, *spoIIID*, and *spoIVC*) of the mother cell pathway (48; Cutting et al., in press; Zheng, unpublished results). In light of the above discussion, it is attractive to imagine that the transcription of certain late-

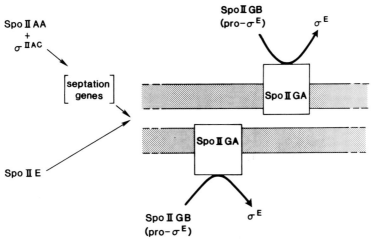

Figure 3. Proposed coupling of pro-σ^E processing to stage II septum formation. (Reprinted with permission from reference 52.)

activated, mother-cell-expressed genes is coupled in some way to a morphogenetic or
physiological event occurring at an intermediate stage of sporulation, such as a feature of the maturing forespore. In other
words, the developing sporangium may
have to proceed through stage IV in order
to be competent to activate cotA, cotD,
and gerE gene expression. This coupling
would ensure that the expression of genes
like cotA, cotD, and gerE is kept in register with the course of sporogenesis by a
feedback mechanism in which the transcriptional apparatus somehow senses a landmark event associated with the developing
sporangium.

The mechanistic basis for this coupling
is not known and it may involve more than
one coupling point. The discovery that the
NH$_2$ terminus of σ^K, the mother-cell-specific
sigma factor governing cotA, cotD, and gerE
transcription in vitro, is preceded by 20 residues in the predicted primary product of
spoIVCB-spoIIIC points to the possible existence of a pro-σ^K precursor (32, 54). By
analogy to the case of pro-σ^E, it is attractive
to imagine that the maturation of pro-σ^K depends on the completion of morphological
stage III (engulfment). Active σ^K-RNA polymerase may then act in conjunction with the
SpoIIID protein to transcribe genes involved
in cortex formation and stimulate further
pro-σ^K production from the spoIVCB-spo-
IIIC composite gene. A second possible coupling point, at least for the activation of cotA
and cotD, is inferred from the inhibition
by SpoIIID of σ^K-directed transcription in
vitro of these genes (32). Perhaps the mechanism that inactivates or sequesters the
SpoIIID protein (as suggested in Fig. 2) somehow senses the completion of the cortex and
mediates the transition from stage IV to stage
V. The discovery of mutants in which the
dependence of cotA induction on the forespore gene spoIIIG is bypassed may provide
a basis for testing the validity of these proposed coupling mechanisms (S. Cutting, unpublished results).

PATHWAYS OF DEVELOPMENTALLY REGULATED GENE EXPRESSION

To construct a network describing the
program of sporulation gene expression, we
have attempted to integrate the results of
studies from several laboratories on genes
whose time of induction and dependence on
other developmental genes have been studied
quantitatively. For simplicity, we have excluded spo0 genes, which are the subject of
a separate review (Smith, this volume), and
genes (out) that are turned on during spore
outgrowth. We have therefore only incorporated the results of studies on genes whose
transcription is induced during the course of
sporulation. In summarizing the results of dependence studies, we have arbitrarily restricted "dependence" to cases in which the
level of dependence of the expression of one
gene on another is substantial (that is, at least
5- to 10-fold lower in mutant cells than in
the wild type), although it is probable that
some (many) partial dependencies will turn
out to reveal important aspects of the regulatory network. Our proposed scheme, which
extends and updates an earlier effort (36),
should be regarded as provisional because it
is derived from results of several laboratories,
which have used partially overlapping sets of
mutants representing several different genetic
backgrounds and which have not in all cases
reported full quantitative data.

Figure 4 summarizes the dependence
patterns of 19 genes, which fall into at least
eight coordinately regulated gene sets. Regulation of the genes is depicted in a stepwise
hierarchy, a scheme devised by Kroos and
Kaiser (31) to describe patterns of developmental gene expression in Myxococcus
xanthus. The 19 genes whose regulation is
summarized in Fig. 4 are highlighted in bold-
face above the vertical arrows. The genes
upon which dependencies were determined
are indicated in italics in the boxes. The time
of induction of each gene is indicated by its
position along the abscissa. The position
along the ordinate represents the level of de-

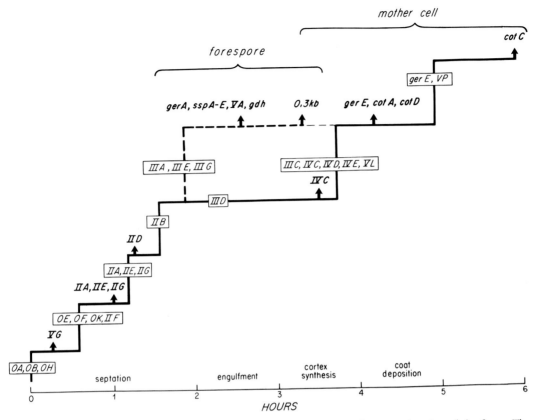

Figure 4. Proposed network for sporulation gene expression. See the text for an explanation of the figure. The indicated dependencies are based on work cited in the text or based on unpublished work from our laboratory or communicated to us by our colleagues in the field, including J. Errington, A. Moir, P. Setlow, and P. Stragier. The dependence of *spoIIG* expression on *spoOE*, *spoOF*, and *spoOK* and the dependence of *gerA* expression on *spoIIIG* are inferential and not based on direct measurement. Also inferential are some of the dependencies of genes high in the hierarchy on genes lower in the hierarchy. Recent results by Illig and Errington (personal communication) call into question the previously reported (12) dependence of *spoVA* expression on *spoIIIA*, and thus some uncertainty attaches to the role of *spoIIIA* in the forespore line of gene expression. IVC refers to the B gene (*spoIVCB*) and by inference to the composite gene *spoIVCB-spoIIIC*.

pendence on boxed-in genes beneath it in the step graph. Dependencies are cumulative with increasing numbers of vertical steps. Thus, a higher position along the ordinate indicates dependence on a greater number of genes.

The regulation of genes whose transcription is induced during the first 2 h or so of sporulation can be seen to fit into a simple linear pathway of cumulative dependencies. After h 2 the remaining genes are accommodated in the diagram by two principal branches, which reflect the separation of the mother cell and forespore genome and their differential expression. Thus, the thick broken line indicates two temporal classes of forespore-expressed genes (*gerA*, *gdh*, *spoVA*, and *sspA* through *sspE*; and *0.3 kb*), whereas the other branch indicates the existence of three temporal classes of mother-cell-expressed genes (*spoIVCB*; *gerE*, *cotA*, and *cotD*; and *cotC*). Note that in some cases candidate regulatory genes have not yet been identified to account for the different times of induction of certain genes (e.g., the delayed induction of *0.3 kb* in the forespore).

As discussed above, the activation of certain late mother-cell-specific genes (*gerE*, *cotA*, *cotD*, and *cotC*) depends on *spoIIIA*, *spoIIIE*, and *spoIIIG*, which are otherwise required only for the expression of genes on the forespore branch. This observation has been interpreted to indicate that late gene expression in the mother cell is somehow coupled to a morphological or physiological event associated with the maturation of the forespore. We have accommodated this link between the forespore and mother cell pathways of gene expression by the thin broken line, creating a loop that connects the forespore branch with the mother cell branch. Undoubtedly, other branches and other representatives of known classes of coordinately regulated gene sets will emerge as detailed information on the regulation of additional genes becomes available and as dependence relationships begin to be understood in terms of individual, *trans*-acting factors.

SUMMARY

The metamorphosis of a vegetative cell into the dormant cell type of the endospore involves the regulated activation of several scores of genes located at many scattered chromosomal positions. The expression of these genes is governed by an elaborate program of temporal, spatial, and morphological regulation. Sets of coordinately controlled genes are switched on in a highly ordered temporal cascade that lasts through the late stages of development. At intermediate to late stages of the cascade (after septation), gene expression becomes compartmentalized, with certain genes being expressed in the forespore chamber of the sporangium and others being expressed only in the mother cell. Emerging evidence points to the existence of regulatory mechanisms that couple the expression of certain genes to landmark events in sporogenesis, thereby keeping the program of sporulation gene expression in register with the morphological sequence of sporulation. On the basis of transcriptional studies of representative developmental genes and the identification of regulatory proteins of several kinds, it is now possible to describe the program of sporulation gene expression in terms of a rudimentary network of dependent pathways. The long-range challenge is to provide a complete description of the program of gene expression in a way that accounts for temporal, spatial, and morphological aspects of developmental regulation.

Acknowledgments. We thank A. L. Sonenshein, P. Stragier, P. Setlow, and P. Piggot for advice and suggestions.

This work was supported by Public Health Service grant GM18568 from the National Institutes of Health to R.L. and a postdoctoral fellowship from the Helen Hay Whitney Foundation to L.K.

LITERATURE CITED

1. **Beal, B., M. Lowe, and J. Lutkenhaus.** 1988. Cloning and characterization of *Bacillus subtilis* homologs of *Escherichia coli* cell division genes *ftsZ* and *ftsA. J. Bacteriol.* **170:**4855–4864.
2. **Carlson, H. C., and W. G. Haldenwang.** 1989. The σ^E subunit of *Bacillus subtilis* RNA polymerase is present in both forespore and mother cell compartments. *J. Bacteriol.* **171:**2216–2218.
3. **Carter, L., and C. P. Moran, Jr.** 1986. New RNA polymerase sigma factor under *spo0* control in *Bacillus subtilis. Proc. Natl. Acad. Sci. USA* **83:**9438–9442.
4. **Clarke, S., I. Lopez-Diaz, and J. Mandelstam.** 1986. Use of *lacZ* gene fusions to determine the dependence pattern of the sporulation gene *spoIID* in *spo* mutants of *Bacillus subtilis. J. Gen. Microbiol.* **132:**2987–2994.
5. **Connors, M. J., J. M. Mason, and P. Setlow.** 1986. Cloning and nucleotide sequencing of genes for three small, acid-soluble proteins from *Bacillus subtilis* spores. *J. Bacteriol.* **166:**417–425.
6. **Cutting, S., and J. Mandelstam.** 1986. The nucleotide sequence and the transcription during sporulation of the *gerE* gene of *Bacillus subtilis. J. Gen. Microbiol.* **132:**3013–3024.
7. **Dancer, B., and J. Mandelstam.** 1981. Complementation of sporulation mutations in fused protoplasts of *Bacillus subtilis. J. Gen. Microbiol.* **123:**17–26.
8. **Donovan, W., Z. Zheng, K. Sandman, and R. Los-**

ick. 1987. Genes encoding spore coat proteins from *Bacillus subtilis*. *J. Mol. Biol.* **196:**1–10.

9. Dubnau, E., J. Weir, G. Nair, L. Carter III, and C. Moran, Jr. 1988. *Bacillus* sporulation gene *spo0H* codes for σ³⁰ (σ^H). *J. Bacteriol.* **170:**1054–1062.

10. Errington, J. 1987. Two separable functional domains in the sigma subunit of RNA polymerase in *Bacillus subtilis? FEBS Lett.* **224:**257–260.

11. Errington, J., P. Fort, and J. Mandelstam. 1985. Duplicated sporulation genes in bacteria. *FEBS Lett.* **188:**184–188.

12. Errington, J., and J. Mandelstam. 1986. Use of a *lacZ* fusion to determine the dependence pattern and the spore compartment expression of sporulation operon *spoVA* in *spo* mutants of *Bacillus subtilis*. *J. Gen. Microbiol.* **132:**2977–2985.

13. Errington, J., and J. Mandelstam. 1986. Use of a *lacZ* fusion to study the dependence pattern of sporulation operon *spoIIA* in wild type *Bacillus subtilis* and in *spo* mutants. *J. Gen. Microbiol.* **132:**2987–2989.

14. Errington, J., S. Rong, M. S. Rosenkrantz, and A. L. Sonenshein. 1988. Transcriptional regulation and structure of the *Bacillus subtilis* sporulation locus *spoIIIC*. *J. Bacteriol.* **170:**1162–1167.

15. Farquhar, R., and M. D. Yudkin. 1988. Phenotypic and genetic characterization of mutations in the *spoIVC* locus of *Bacillus subtilis*. *J. Bacteriol.* **134:**9–17.

16. Ferrari, F. A., K. Trach, D. LeCoq, J. Spence, E. Ferrari, and J. R. Hoch. 1985. Characterization of the *spo0A* locus and its product. *Proc. Natl. Acad. Sci. USA* **82:**2647–2651.

17. Fliss, E. R., M. J. Connors, C. A. Loshon, E. Curiel-Quesada, B. Setlow, and P. Setlow. 1985. Small, acid-soluble spore proteins of *Bacillus*: products of a sporulation-specific, multigene family, p. 60–66. *In* J. A. Hoch and P. Setlow (ed.), *Molecular Biology of Microbial Differentiation*. American Society for Microbiology, Washington, D.C.

18. Fort, P., and P. Piggot. 1984. Nucleotide sequence of sporulation locus *spoIIA* in *Bacillus subtilis*. *J. Gen. Microbiol.* **130:**2147–2153.

19. Guzman, P., J. Westpheling, and P. Youngman. 1988. Characterization of the promoter region of the *Bacillus subtilis* spoIIE operon. *J. Bacteriol.* **170:**1598–1609.

20. Hackett, R. H., and P. Setlow. 1987. Cloning, nucleotide sequencing, and genetic mapping of the gene for small, acid-soluble spore protein γ of *Bacillus subtilis*. *J. Bacteriol.* **169:**1985–1992.

21. Henner, D. J., M. Yang, and E. Ferrari. 1988. Localization of *Bacillus subtilis sacU*(Hy) mutations to two linked genes with similarities to the conserved procaryotic family of two-component signalling systems. *J. Bacteriol.* **170:**5102–5109.

22. Hirschman, J., P.-K. Wong, K. S. Keener, and S.

Kustu. 1985. Products of nitrogen regulatory genes *ntrA* and *ntrC* of enteric bacteria activate *glnA* transcription *in vitro*: evidence that the *ntrA* product is a sigma factor. *Proc. Natl. Acad. Sci. USA* **82:**7525–7529.

23. Holland, S. K., S. Cutting, and J. Mandelstam. 1987. The possible DNA-binding nature of the regulatory proteins encoded by *spoIID* and *gerE* involved in the sporulation of *Bacillus subtilis*. *J. Gen. Microbiol.* **133:**2381–2391.

24. James, W., and J. Mandelstam. 1985. *spoVIC*, a new sporulation locus in *Bacillus subtilis* affecting spore coats, germination and the rates of sporulation. *J. Gen. Microbiol.* **131:**2409–2419.

25. Jenkinson, H. F. 1981. Germination and resistance defects in spores of a *Bacillus subtilis* mutant lacking a coat polypeptide. *J. Gen. Microbiol.* **127:**81–91.

26. Jenkinson, H. F. 1983. Altered arrangement of proteins in the spore coat of a germination mutant of *Bacillus subtilis*. *J. Gen. Microbiol.* **129:**1945–1958.

27. Jonas, R. M., E. A. Weaver, T. J. Kenney, C. P. Moran, Jr., and W. G. Haldenwang. 1988. The *Bacillus subtilis spoIIG* operon encodes both σ^E and a gene necessary for σ^E activation. *J. Bacteriol.* **170:**507–511.

28. Karmazyn-Campelli, C., and C. Bonamy. 1989. Tandem gene encoding sigma factors for consecutive steps of development in *Bacillus subtilis*. *Genes Dev.* **3:**150–157.

29. Kenney, T. J., P. A. Kirchman, and C. P. Moran, Jr. 1988. Gene encoding σ^E is transcribed from a σ^A-like promoter in *Bacillus subtilis*. *J. Bacteriol.* **170:**3058–3064.

30. Kenney, T. J., and C. P. Moran, Jr. 1987. Organization and regulation of an operon that encodes a sporulation-essential sigma factor in *Bacillus subtilis*. *J. Bacteriol.* **169:**3329–3339.

31. Kroos, L., and D. Kaiser. 1987. Expression of many developmentally regulated genes in *Myxococcus* depends on a sequence of cell interactions. *Genes Dev.* **1:**840–854.

32. Kroos, L., B. Kunkel, and R. Losick. 1989. Switch protein alters the specificity of RNA polymerase containing a compartment-specific sigma factor. *Science* **243:**526–528.

33. Kunkel, B., K. Sandman, S. Panzer, P. Youngman, and R. Losick. 1988. The promoter for a sporulation gene in the *spoIVC* locus of *Bacillus subtilis* and its use in studies of temporal and spatial control of gene expression. *J. Bacteriol.* **170:**3513–3522.

34. Labell, T. L., J. E. Trempy, and W. G. Haldenwang. 1987. Sporulation-specific σ factor σ²⁹ of *Bacillus subtilis* is synthesized from a precursor protein, P³¹. *Proc. Natl. Acad. Sci. USA* **84:**1784–1788.

35. Lencastre, H., and P. Piggot. 1979. Identification and different sites of expression for *spo* loci by trans-

formation of *Bacillus subtilis*. *J. Gen. Microbiol.* **114**:377–389.

36. **Losick, R., L. Kroos, J. Errington, and P. Youngman.** 1988. Pathways of developmentally regulated gene expression in *Bacillus subtilis*, p. 221–242. *In* D. Hopwood and K. Chater (ed.), *Genetics of Bacterial Diversity*. Academic Press, Inc., New York.

37. **Losick, R., P. Youngman, and P. J. Piggot.** 1986. Genetics of endospore formation in *Bacillus subtilis*. *Annu. Rev. Genet.* **20**:625–669.

38. **Mason, J. M., R. H. Hackett, and P. Setlow.** 1988. Regulation of expression of genes coding for small, acid-soluble proteins of *Bacillus subtilis* spores: studies using *lacZ* gene fusions. *J. Bacteriol.* **170**:239–244.

39. **Mason, J. M., and P. Setlow.** 1986. Essential role of small, acid-soluble spore proteins in resistance of *Bacillus subtilis* spores to UV light. *J. Bacteriol.* **167**:174–178.

40. **Masuda, E., H. Anaguchi, K. Yamada, and Y. Kobayashi.** 1988. Two developmental genes containing σ factor homologs are arranged in tandem in *Bacillus subtilis*. *Proc. Natl. Acad. Sci. USA* **85**:7637–7641.

41. **Ollington, J. F., and R. Losick.** 1981. A cloned gene that is turned on at an intermediate stage of spore formation in *Bacillus subtilis*. *J. Bacteriol.* **147**:443–451.

42. **Panzer, S., D. Sun, P. Setlow, and R. Losick.** 1989. Evidence for an additional temporal class of gene expression in the forespore compartment of sporulating *Bacillus subtilis*. *J. Bacteriol.* **171**:561–564.

43. **Perkins, J., and P. Youngman.** 1986. Construction and properties of Tn*917lacZ*, a transposon derivative that mediates transcriptional gene fusions in *Bacillus subtilis*. *Proc. Natl. Acad. Sci. USA* **83**:140–144.

44. **Piggot, P., and J. Coote.** 1976. Genetic aspects of bacterial endospore formation. *Bacteriol. Rev.* **40**:908–962.

45. **Piggot, P., A. Moir, and D. A. Smith.** 1981. Advances in genetics of *Bacillus subtilis* differentiation, p. 29–39. *In* H. S. Levinson, A. L. Sonenshein, and D. J. Tipper (ed.), *Sporulation and Germination*. American Society for Microbiology, Washington, D.C.

46. **Rather, P. W., and C. P. Moran, Jr.** 1988. Compartment-specific transcription in *Bacillus subtilis*: identification of the promoter for *gdh*. *J. Bacteriol.* **170**:5086–5092.

47. **Rong, S., M. Rosenkrantz, and A. L. Sonenshein.** 1986. Transcriptional control of the *spoIID* gene of *Bacillus subtilis*. *J. Bacteriol.* **165**:771–779.

48. **Sandman, K., L. Kroos, S. Cutting, P. Youngman, and R. Losick.** 1988. Identification of the promoter for a spore coat gene in *Bacillus subtilis* and studies

on the regulation of its induction at a late stage of sporulation. *J. Mol. Biol.* **200**:461–473.

49. **Sandman, K., R. Losick, and P. Youngman.** 1987. Genetic analysis of *Bacillus subtilis spo* mutations generated by Tn*917*-mediated insertional mutagenesis. *Genetics* **117**:603–617.

50. **Stephens, M. A., N. Lang, K. Sandman, and R. Losick.** 1984. A promoter whose utilization is temporally regulated during sporulation in *Bacillus subtilis*. *J. Mol. Biol.* **176**:333–348.

51. **Stock, A., D. E. Koshland, and J. Stock.** 1985. Homologies between the *Salmonella typhimurium* CheY protein and proteins involved in the regulation of chemotaxis, membrane protein synthesis and sporulation. *Proc. Natl. Acad. Sci. USA* **82**:7989–7993.

52. **Stragier, P., C. Bonamy, and C. Karmazyn-Campelli.** 1988. Processing of a sporulation sigma factor in *Bacillus subtilis*: how morphological structure could control gene expression. *Cell* **52**:697–704.

53. **Stragier, P., J. Bouvier, C. Bonamy, and J. Szulmajster.** 1984. A developmental gene product of *Bacillus subtilis* homologous to the sigma factor of *Escherichia coli*. *Nature* (London) **312**:376–378.

54. **Stragier, P., B. Kunkel, L. Kroos, and R. Losick.** 1989. Chromosomal rearrangement generating a composite gene for a developmental transcription factor. *Science* **243**:507–512.

55. **Sun, D., P. Stragier, and P. Setlow.** 1989. Identification of a new sigma factor involved in compartmentalized gene expression during sporulation of *Bacillus subtilis*. *Genes Dev.* **3**:141–149.

56. **Trach, K., J. W. Chapman, P. J. Piggot, and J. A. Hoch.** 1985. Deduced product of the stage 0 sporulation gene *spo0F* shares homology with the Spo0A, OmpR and SfrA proteins. *Proc. Natl. Acad. Sci. USA* **82**:7260–7264.

57. **Trempy, J. E., C. Bonamy, J. Szulmajster, and W. G. Haldenwang.** 1985. *Bacillus subtilis* sigma factor σ²⁹ is the product of sporulation gene *spoIIG*. *Proc. Natl. Acad. Sci. USA* **82**:4189–4192.

58. **Trempy, J. E., T. L. LaBell, G. L. Ray, and W. G. Haldenwang.** 1985. P³¹, a σ²⁹-like RNA polymerase binding protein of *Bacillus subtilis*, p. 162–169. *In* J. A. Hoch and P. Setlow (ed.), *Molecular Biology of Microbial Differentiation*. American Society for Microbiology, Washington, D.C.

59. **Trowsdale, J., S. M. H. Chen, and J. A. Hoch.** 1979. Genetic analysis of a class of polymyxin resistant partial revertants of stage 0 sporulation mutants of *Bacillus subtilis*: a map of the chromosomal region near the origin of replication. *Mol. Gen. Genet.* **173**:61–70.

60. **Turner, S. M., J. Errington, and J. Mandelstam.** 1986. Use of *lacZ* gene fusion to determine the dependence pattern of sporulation operon *spoIIIC* in *spo* mutants of *Bacillus subtilis*: a branched path-

way of expression of sporulation operons. *J. Gen. Microbiol.* **132**:2995–3003.

61. **Warburg, R. J.** 1981. Defective sporulation of a spore germination mutant of *Bacillus subtilis* 168, p. 98–100. *In* H. S. Levinson, A. L. Sonenshein, and D. J. Tipper (ed.), *Sporulation and Germination.* American Society for Microbiology, Washington, D.C.

62. **Wu, J.-J., M. G. Howard, and P. J. Piggot.** 1989. Regulation of transcription of the *Bacillus subtilis* spoIIA locus. *J. Bacteriol.* **171**:692–698.

63. **Yansura, D. G., and D. J. Henner.** 1984. Use of the *Escherichia coli lac* repressor and operator to control gene expression in *Bacillus subtilis. Proc. Natl. Acad. Sci. USA* **81**:439–443.

64. **Yoshikawa, H., J. Kazami, S. Yamashita, T. Chibazakura, H. Sone, et al.** 1986. Revised assignment for the *B. subtilis* spo0F gene and its homology with spo0A and two *Escherichia coli* genes. *Nucleic Acids Res.* **14**:1063–1072.

65. **Young, M.** 1976. Use of temperature-sensitive mutants to study gene expression during sporulation in *Bacillus subtilis. J. Bacteriol.* **126**:928–936.

66. **Youngman, P., P. Zuber, J. B. Perkins, K. Sandman, M. Igo, and R. Losick.** 1985. New ways to study developmental genes in spore forming bacteria. *Science* **228**:285–291.

67. **Zheng, L., W. Donovan, P. G. Fitz-James, and R. Losick.** 1988. Gene encoding a morphogenic protein required in the assembly of the outer coat of the *Bacillus subtilis* endospore. *Genes Dev.* **2**:1047–1054.

68. **Zuber, P., and R. Losick.** 1987. Role of AbrB in Spo0A- and Spo0B-dependent utilization of a sporulation promoter in *Bacillus subtilis. J. Bacteriol.* **169**:2223–2230.

Regulation of Procaryotic Development
Edited by Issar Smith, Ralph A. Slepecky, and Peter Setlow
© 1989 American Society for Microbiology, Washington, DC 20006

Chapter 12

Temporal and Spatial Control of Gene Expression during Sporulation: from Facts to Speculations

Patrick Stragier

Despite the abundant data recently accumulated about gene expression during sporulation of *Bacillus subtilis* (see R. Losick and L. Kroos, this volume), some of the most important links are still missing. For instance, five sigma factors are now identified as the products of *spo* genes (see Table 1 and C. P. Moran, Jr., this volume). However, it has not yet been demonstrated how they sequentially displace each other nor how they depend upon each other for their own synthesis (for instance, σ^H inducing *spoIIGB* and σ^E activating *spoIIIG*). Actually the existing data suggest a more complex and indirect pattern of dependence. Furthermore, the promoters recognized by some of these sporulation sigma factors are almost completely unknown, since only one bona fide *spo* gene has been unambiguously identified to be activated by σ^H or σ^E and none by σ^F (Table 1). Thus, the interweaving of sigma factors in the whole sporulation process is still far from being understood. Similarly, data are now available that can explain how compartmentalization of gene expression is maintained but not how it is established (11, 14, 31). I will review some of the unsolved aspects of gene expression during sporulation and will propose a few working hypotheses as stimuli for further experimentation.

ASYMMETRIC SEPTATION AND EARLY GENE EXPRESSION

The asymmetric septation that takes place during h 2 of sporulation leads to a morphological structure similar to the minicell phenotype induced in *Escherichia coli* by overproducing the FtsZ protein (see Fig. 2 in reference 18). This suggests that asymmetric septation in *B. subtilis* could be related to an increased level of *ftsZ* expression. The *B. subtilis* homologs of the *E. coli ftsA* and *ftsZ* genes have recently been cloned (1). The *ftsZ* gene of *B. subtilis* is located immediately downstream of *ftsA*, suggesting that both genes can at least partly be coordinately expressed. Transcription of the *B. subtilis ftsA* gene increases twofold around the end of exponential growth (C. Karmazyn-Campelli, unpublished data), an indication that synthesis of FtsZ could also increase at that time. In *E. coli* both gene products appear to act in a concerted way to trigger septation, although their exact roles are not known, and there is evidence that well-balanced concentrations of these two proteins are needed to insure correct septation (38). If some *ftsZ*-specific promoter is switched on at the onset of sporulation in *B. subtilis*, the FtsZ/FtsA ratio would be modified, which could induce asymmetric septation.

Patrick Stragier • Institut de Biologie Physico-Chimique, 13 rue Pierre et Marie Curie, 75005 Paris, France.

Table 1
Sporulation Sigma Factors

Sigma factor	Gene	Time of action	Target gene(s)	Physiological role
σ^H (σ^{30})	spo0H	Around t_0	spoVG	Entry in stationary phase
σ^F	spoIIAC	t_1–t_2 (?)	?	Asymmetric septation (?)
σ^E (σ^{29})	spoIIGB	$t_{1.5}$–$t_{3.5}$	spoIID	Engulfment; establishment of compartmentalization (?)
σ^G	spoIIIG	t_3–t_5	sspA–E, spoVA	Forespore gene expression
σ^K (σ^{27})	spoIVCB::spoIIIC	$t_{3.5}$–$t_{5.5}$	cotA, cotD	Mother cell gene expression

The *spoIIN279* mutation which arrests asymmetric septation is a missense mutation in *ftsA* (T. Leighton and B. Savelli, unpublished data). This thermosensitive sporulation mutant was isolated by M. Young, who showed that the *B. subtilis* cells have to be maintained at the permissive temperature during a short period of time, occurring about 1 h after the end of exponential growth, for sporulation to proceed normally (39). Identification of the *spoIIN* gene as *ftsA* fits with this phenotype since in *E. coli*, if cell division is to occur, synthesis of the FtsA protein is required during the 15-min period before the cells divide (8). These requirements suggest that the FtsA protein or one of its protein partners is very unstable and has to be used immediately after its synthesis. It might also reflect the fact that FtsA activity could somehow be needed at a critical time related to termination of DNA replication (33). Once sporulation started, such a synchrony could be achieved only during a short period and could be the molecular basis of the observations linking DNA replication and commitment to sporulation (20).

Interestingly, the same thermosensitive mutation in *B. subtilis ftsA* drastically reduces *spoIIG* transcription at the nonpermissive temperature (T. Leighton, personal communication). Both effects of the mutation are suppressed by a secondary mutation in σ^A, the major vegetative sigma factor (16). The simplest explanation for these results is to assume that the *ftsA* product may act as a transcriptional factor required for full expression of several genes, among which are *spoIIG* and maybe *ftsZ*; the absence of the *ftsA* product would then be compensated for by an alteration of σ^A. This interpretation is reinforced by the recent findings that transcription of some *fts* promoters in *E. coli* is modulated by the level of FtsA (W. Donachie, personal communication).

A common defect shown by many stage II mutants (namely, *spoIIA*, *spoIIE*, *spoIIF*, and *spoIIG*) is their "disporic" phenotype, a second asymmetric septum being laid down at the other pole of the cell, which might be interpreted as the result of FtsZ continuous activity. All these mutations interfere, one way or another, with the synthesis of σ^E in its active form (29; Leighton, personal communication). It seems plausible that σ^E activates the synthesis of a repressor of *ftsZ* or of a protein antagonizing FtsZ activity. Thus, σ^E would block any further septation, its own appearance in active form being the signal that the sporulation septum has been successfully completed (29).

PRO-σ^E PROCESSING AND ITS MECHANISMS

Coupling Gene Expression to a Morphological Stage

Six different gene products are known to be required for processing of pro-σ^E although pro-σ^E activation can be achieved, albeit inefficiently, by the SpoIIGA protein overproduced during vegetative growth (29). SpoIIGA and pro-σ^E are expected to be synthesized in a coordinate fashion since they are the products of the same operon. However, σ^E is accumulated during about 1 h before processing takes place (34), which indicates that some cofactor required for SpoIIGA proteolytic activity is provided only around $t_{1.5}$. The presence of this cofactor is dependent on the *spoIIAA*, *spoIIAC*, *spoIIEA*, *spoIIEB*, and *spoIIEC* products. Moreover, since *spoIIAC* codes for σ^F, it is likely to be involved by inducing the expression of some additional gene(s). These multiple requirements suggest that the stimulation of SpoIIGA processing activity is mediated through a complex signal.

A clue is given by the presence of five putative transmembrane segments in the SpoIIGA polypeptide (29; Fig. 1). All the existing data can be put together in a coherent picture by assuming that SpoIIGA becomes proteolytically active only when anchored in the septum, which induces a conformational change in the SpoIIGA cytoplasmic carboxy-terminal part, the presumptive peptidase domain (29). Many mechanisms can be thought of for discriminating between the normal cytoplasmic membrane and the septal double membrane structure, such as an inhibitory interaction of the periplasmic domain of SpoIIGA with the peptidoglycan that is absent from the sporulation septum, at least in its final state. Another possibility is implicit

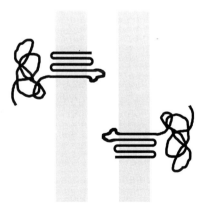

Figure 1. Topological model of SpoIIGA polypeptides inserted in the double membrane of the sporulation septum (adapted from reference 29).

in the scheme shown in Fig. 1: two monomers of SpoIIGA could interact with each other through their periplasmic domains, such a contact being made possible only in the septum.

The *spoIIA* and *spoIIE* mutants do septate but in an aberrant way, most of them showing a peptidoglycan layer deposited inside the septum (23, 37). Such a defect could interfere with the contacts between SpoIIGA polypeptides suggested in Fig. 1. Conversely, the weak processing activity of SpoIIGA obtained during growth could reflect a transient stimulation by the newly formed division septum before cell wall deposition. In that model, σ^F, the *spoIIAC* product, would control genes required for the elaboration of the asymmetric septum while *spoIIE* would encode proteins incorporated into the sporulation septum and playing a role in its specific properties.

Another Role for Pro-σ^E Processing?

Insertion of SpoIIGA in the membrane seems to be a convenient way for coupling septum formation to σ^E activation. Such a localization could have additional significance since, in procaryotes, amino-terminal processing of polypeptides is usually linked to their translocation across membranes. Could this be the case for σ^E? It is intriguing that SpoIIGA is predicted to contain five transmembrane domains, since only two are sufficient for transducing environmental signals in *E. coli* (13, 26). Moreover, these transmembrane regions are calculated to be arranged in an antiparallel β-pleated sheet structure (unpublished data), the secondary structure found in channel-forming proteins (12). This suggests that two SpoIIGA polypeptides, inserted in opposite membranes and contacting each other, could create a pore across the two membranes of the septum in a "Bayer-like" junction similar to the structure proposed to govern hemolysin secretion in *E. coli* (19). It is then striking that the first 20 residues of pro-σ^E have the potential to form an amphiphilic helix with basic amino acids clustered on one side and hydrophobic residues on the other (unpublished data). Such a feature is characteristic of mitochondrial targeting sequences that have been shown to be sufficient for translocating an attached polypeptide across the double membrane of mitochondria (36). Completion of the asymmetric septum could then be correlated with the formation of SpoIIGA pores and lead to the translocation of the pro-σ^E proteins in both directions with the release of active σ^E on the exit side of the pore. Such a mechanism would have the interesting consequence of switching tthe pro-σ^E molecules from one side of the septum to the other, transferring the higher amount of molecules initially enclosed in the large compartment to the small one and vice versa. As a result, σ^E would be present in the small compartment at a higher concentration than in the large one, which could be the first step to differentiate the two genomes from each other. However, this mechanism does not seem to be supported by experimental evidence (5).

ESTABLISHING COMPARTMENTALIZATION OF GENE EXPRESSION

A major issue is to understand how the two genomes of the sporulating cell are differentially transcribed as early as 60 to 90 min after asymmetric septation. A possible scheme can be proposed by taking into account once again the different sizes of the two compartments created at stage II. When the septum is closed the macromolecules present in the cell at that time are expected to be randomly segregated according to the volumes of the two compartments (one being three to four times larger than the other), and their initial concentrations should be identical on both sides of the septum. This is not the case for chromosomal DNA and DNA-bound proteins (including elongating RNA polymerase), which would partition equally in both compartments and reach a higher concentration in the small compartment than

in the large one. For a given gene, transcription will be initiated with the same frequency on both genomes and identical amounts of its product will be synthesized in each compartment. In the case of newly expressed genes, such as those turned on by σ^E, the concentration of their products will be at each instant three to four times higher in the small compartment. This gradient will even be enhanced if transcription is more active in the small compartment due to the higher concentration of RNA polymerase. Let's assume that some of the σ^E-activated genes encode DNA-binding proteins that will interact with their nucleotide targets only as a dimer or as a tetramer. Their concentration will progressively increase in both compartments, and the monomers will associate to form dimers or tetramers and then bind to DNA. Since the concentration of the DNA itself is higher in the small compartment, it follows that the concentration of bound dimers will be roughly 40 times (3^3 to 4^3) higher in the forespore chamber and that of bound tetramers will be roughly 500 times (3^5 to 4^5) higher. It is then easy to imagine that, after the time needed for accumulation of these DNA-binding proteins to their critical concentration, some promoters will be turned on or off specifically in the smaller compartment.

An obvious candidate for such a mechanism is the *spoIIIA* product, wich could activate the late *spoIIIG* promoter and induce a burst of σ^G synthesis only in the forespore (although there is some controversy about the requirement for the *spoIIIA* product in the forespore; see Losick and Kroos, this volume). In that case the postulated DNA-binding protein would switch on the synthesis of another regulatory protein, σ^G, which would in turn amplify the differentiation of the two genomes. Similarly, σ^K synthesis could be prevented in the forespore by the accumulation of a specific repressor. In this regard it is interesting that the *spoIIID* product, the 14-kilodalton (14K) protein which is both an activator and a repressor of mother cell gene expression (14), behaves as a tetramer on sizing columns (L. Kroos and R. Losick, personal communication). Combining positive and negative effectors would allow the specific activation of multiple gene sets in the forespore as well as in the mother cell.

LOOKING FOR ENGULFASE

The higher concentration of newly synthesized proteins in the forespore chamber will lead to an increasing difference of osmotic pressure between the two faces of the septum. This could be the basis of the bulging of the small compartment that accompanies the transition from stage II to stage III. It cannot, however, explain the specific migration of the mother cell around the forespore that culminates in its complete engulfment. Obviously a highly specific enzyme must be involved in that process. How could it work?

A putative scheme for engulfment is given in Fig. 2. Engulfase is shown as a protein synthesized in both compartments and pres-

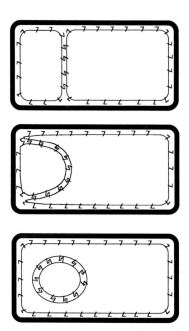

Figure 2. Schematic model for engulfment. Molecules of engulfase are symbolized by hooks stuck into the cytoplasmic membrane and pointing towards the outside.

ent on the whole external surface of the cytoplasmic membrane, including the septum, where individual polypeptides would be inserted in opposite orientations. In such a configuration, two facing engulfase molecules could contact and grasp each other. On the edges of the septum such an interaction would stick the two membranes together, bringing closer neighboring engulfase molecules that would in turn contact each other, and so on. With the help of the difference in osmotic pressure between the two sides of the septum, such interactions should lead to a progressive endocytosis of the small compartment by the large one until membranes collide and fuse. In this naive model the engulfase molecules would act as the teeth of a zipper or as the spines of Velcro.

What gene could code for engulfase? This gene should be activated by σ^E, since engulfment is correlated to the successful achievement of the asymmetric septum. According to the model proposed in the preceding section, compartmentalization of gene expression depends only on the volume difference between the two compartments and does not rely at all on engulfment. Thus, a mutation in the gene encoding engulfase is expected to create a morphological block at stage II but not to interfere with the immediately following cascade of gene activation. The *spoIID* gene fulfills these requirements: its transcription is controlled by σ^E (25, 29); it is expressed in both compartments, as determined by localization of β-galactosidase in a strain harboring a *spoIID-lacZ* fusion (A. Ryter, personal communication); and the *spoIID298* mutation does not block expression of *spoIIIG* (11), nor of *spoIVCB* (15), which indicates that both forespore and mother cell pathways are turned on. Moreover, in that mutant, engulfment does not proceed but the forespore chamber bulges in the mother cell (6), which might be a reflection of asymmetric macromolecular synthesis. Interestingly, the sequence of the SpoIID protein shows the presence of a highly hydrophobic amino-terminal domain (17). The comparison of the *B. subtilis* and *Bacillus amyloliquefaciens* sequences (35) indicates that this domain is not followed by a convincing cleavage site and that it could act as an anchor in the membrane rather than as a signal sequence, the main part of the SpoIID protein being located outside of the cytoplasm. However, since the *spoIID66* mutation blocks expression of the *spoIIIG* gene (11), it is possible that the *spoIID* locus might be more complex than usually thought, and its direct involvement in engulfment remains to be established.

spoIIIE, AN INTRUDER IN THE SPORULATION PROGRAM?

Mutations in *spoIIIE* block the expression of forespore-specific genes (9, 21) but not expression of *spoIIIG* itself, the gene encoding the forespore sigma factor (11). This suggests that *spoIIIE* codes for an auxiliary factor required for σ^G activity in vivo. When *spoIIIG* is placed under the control of the *spac* promoter, isopropyl-β-D-thiogalacto-pyranoside-induced synthesis of σ^G in vegetatively growing cells is followed by activation of σ^G-dependent promoters (22, 31). Identical results are obtained if these experiments are carried out in a *spoIIIE* mutant (P. Setlow and C. Karmazyn-Campelli, unpublished data), which indicates that the requirement for the *spoIIIE* product(s) can be bypassed, at least in certain conditions.

This problem has been made even more puzzling with the report that *spoIIIE* is expressed during growth while phenotypic consequences of mutations in that locus can only be observed at an intermediary stage of sporulation (D. Foulger and J. Errington, X Internat. Spores Conf., Woods Hole, Mass., 1988). Electron microscopy examination of the *spoIIIE36* strain has revealed a quite peculiar phenotype (Ryter, personal communication; see Fig. 3): engulfment proceeds normally, but as soon as a free forespore is released in the mother cell, its envelope appears to disintegrate and there are signs of

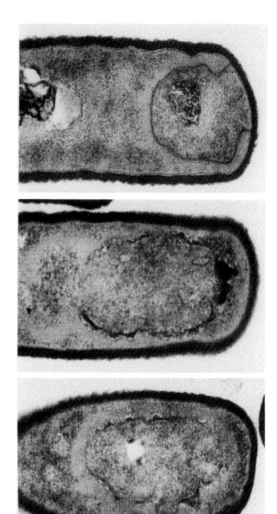

Figure 3. Electron micrograph of the *spoIIIE36* mutant during transition from stage II to stage III (top) and after the end of engulfment (middle and bottom) (provided by A. Ryter).

leakage between the two compartments, indicating that part of their contents could mix. It is reasonable to assume that the internal compositions of the two compartments are noticeably different when the cells reach stage III. Besides their protein content, it can be expected that they differ in other characteristics, such as pH or ionic strength and composition. Interestingly, σ^G activity seems to be very susceptible to environmental conditions: when σ^G synthesis is artificially induced after the start of sporulation its transcriptional activity is very weak (as measured by the expression of an *sspB-* or *sspE-lacZ* fusion), while it is quite high when induced during growth, as if the internal conditions in the starving cells were inhibitory (Setlow and Karmazyn-Campelli, unpublished data). It might be that the physicochemical environment required for σ^G activity is progressively created in the forespore during the hour following septation, a phenomenon that could by itself lead to the compartmentalized activation of the σ^G-dependent *spoIIIG* promoter. If this is actually the case, leaks between the two compartments would rapidly inactivate σ^G.

In this regard the requirements for the *spoIIIE* product(s) in forespore gene expression would reflect the need for the specific environment preserved by the forespore envelope. Nucleotide sequencing of the *spoIIIE* locus predicts the existence of two cistrons (4), but this is not supported by genetic evidence, which makes that result uncertain. The first part of the sequence suggests the presence of several membrane-spanning domains in (one of) the *spoIIIE* product(s). A mutation at the *spoIIIE* locus could create a defect in the membrane stability that would be cryptic during growth, owing to the presence of a rigid cell wall, but would have drastic consequences as soon as a free protoplast was released in the mother cell. According to this interpretation, and despite the apparent genetic evidence, *spoIIIE* would not be actually involved in the expression of forespore genes.

HOW THE TWO GENOMES COOPERATE

Building the Germ Cell Wall

Cortex synthesis is thought to result from enzymatic activity in both compartments. The cortex is composed of a thin layer of peptidoglycan called the germ cell wall, with the same composition as vegetative cell wall, and a thick outer layer of different chemical composition (24). Specific precursors for this outer layer and the corresponding enzymes have been found exclusively in the mother cell (32). From these data and structural considerations (the vegetativelike arrangement of germ cell wall around the inner membrane of the forespore), it is very likely that germ cell wall is synthesized specifically by forespore enzymes. In fact, among the *spoIII* genes belonging to the mother cell pathway, *spoIIIC* appears to arrest sporulation at stage IV (23), and no detailed examination of the *spoIIID* mutant has been published. Thus, a true stage III morphological blockage (forespore protoplast without any cell wall between the two membranes) could indicate a defect in the forespore genetic pathway preventing germ cell wall synthesis.

We may now anticipate that the enzymes involved in germ cell wall biosynthesis are the products of genes activated by σ^G. Paradoxically, among the numerous genes identified to be controlled by σ^G there is no *spoIII* gene and thus no candidate for germ cell wall enzymes (with the possible exception of the misclassified *spoIIIA59* mutation [S. Cutting, personal communication]). The simplest explanation is that these enzymes are the same that are used during vegetative growth. They would be encoded by vegetative genes (some of them being essential) whose expression would be shut off after the end of exponential growth and reactivated later through σ^G-controlled promoters. It seems critical that germ cell wall biosynthesis does not occur before the end of engulfment. So, in contrast with most of the σ^G-dependent genes, an additional control should exist related to the completion of engulfment (it may be significant that there is no germ cell wall synthesis in the *spoIID298* mutant, which contains active σ^G but is blocked at the engulfment stage). A simple chronological device with accumulation of an auxiliary protein (controlled by σ^G itself) could be imagined and seems at least adequate for the pattern of expression of the *0.3 kb* gene (22). Some more subtle

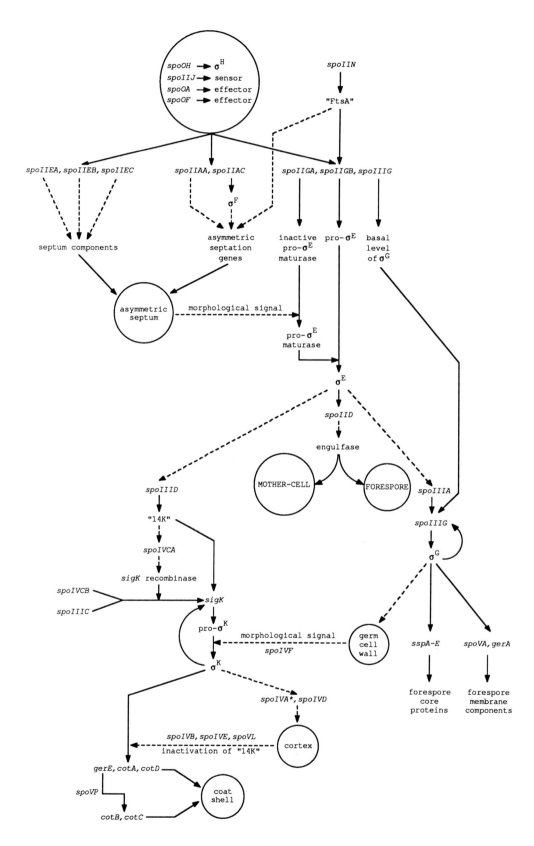

morphological coupling could also be created by the complete isolation of the forespore chamber from the outside medium.

From Germ Cell Wall to Cortex

Morphological studies suggest that genes *spoIIIC*, *spoIVA*, *spoIVCA*, *spoIVCB*, *spoIVD*, and *spoIVF* are involved, directly or indirectly, in cortex synthesis since the thick outer cortex layer is absent in strains mutated in one of these genes (23). *spoIVCB* and *spoIIIC* are spliced together in the mother cell by a chromosomal rearrangement that creates *sigK*, the gene for pro-σ^K, the putative precursor of the mother cell sigma factor (30). There is some evidence suggesting that *spoIVCA* could encode the *sigK* recombinase (30; B. Kunkel, R. Losick, and P. Stragier, unpublished results). Since σ^K enhances its own synthesis by activating the *spoIVCB* promoter (14), all mutations interfering somehow with synthesis of σ^K in its active form should similarly reduce *spoIVCB* transcription (about 20% of the wild-type level of β-galactosidase driven by the *spoIVCB* promoter is found in a *spoIVCB* strain [15]). A thorough analysis of the expression of the *spoIVCB* promoter in various backgrounds (15) points to *spoIVF* as a likely candidate for the gene encoding the pro-σ^K processing enzyme. Thus, *spoIIIC*, *spoIVCA*, *spoIVCB*, and *spoIVF* would all be involved in synthesis of σ^K. It may then be conjectured that *spoIVA* and *spoIVD* encode enzymes of the cortex biosynthetic pathway, their expression being controlled by σ^K.

The presumptive synthesis of σ^K as an inactive precursor offers an interesting additional level of control of mother cell gene expression. Synthesis of the thick outer cor-

tex layer might be strictly dependent on the prior synthesis of the inner germ cell wall if processing of pro-σ^K is coupled to that morphological step. The activity of the pro-σ^K processing enzyme could be triggered by the presence of the germ cell wall through a SpoIIGA-like mechanism. Such a control would ensure a logical sequence of enzyme synthesis leading to a multilayered morphological structure. It would also explain the low level of expression of the *spoIVCB* promoter in a *spoIIIG* mutant, which suggests the absence of active σ^K in that strain (15).

From Cortex to Spore Coats

gerE, *cotA*, and *cotD* define another temporal class of mother cell genes which are switched on around stage V (7, 27; L. Zheng and R. Losick, personal communication). In vitro transcription of *cotD* by σ^K-containing RNA polymerase is strongly inhibited by the 14K protein, the *spoIIID* product, suggesting that prior inactivation of this protein is required for expression of *cotD* in vivo (14). Expression of *cotA* and *gerE* is dependent on multiple genes that belong either to the forespore or to the mother cell pathway, among which *spoIVB*, *spoIVE*, and *spoVL* are not required for cortex synthesis (7, 23, 27, 28). If the latter gene products are involved in inactivation of the 14K protein it could be through a very indirect mechanism: transcription of *spoVL* starts at $t_{1.5}$ and culminates at t_3 (K. Sandman, Ph.D. Thesis, Harvard University, Cambridge, Massachusetts, 1986), which rules out a simple model of cascading transcriptional factors. Once again, some complex morphological feature, requiring multiple previous enzymatic steps in both compartments, could trigger inacti-

---

Figure 4. Hypothetical functional relationships between *spo* genes and their products. Arrows either indicate an activatory effect or a gene-product (or precursor-mature product) relationship. Dashed arrows are used for steps that are not supported by any data (and thus are completely speculative); solid arrows designate steps for which some genetic or biochemical evidence is available. Time increases from top to bottom; the various morphological stages are shown as circles. Due to their complex interdependence, no hierarchy was introduced among stage 0 genes (only those for which something of their functional role is known are indicated). The asterisk indicates the special case of the *spoIVA* product, which is required for cortex formation but not for coat synthesis.

vation of the 14K protein and switch on coat gene expression (27).

Two mutations are known that uncouple *cotA* expression from cortex completion. In the *spoIVA* mutant, which has a normal gene cell wall but no thick outer cortex layer, *cotA* is turned on while the coats are accumulated in the mother cell cytoplasm (6, 27). In the *spoVE* mutant, which apparently makes no cortex at all, *cotA* is expressed normally and the coats are synthesized and deposited around an otherwise stage III forespore (6, 27). Such mutations indicate that the requirement of *cotA* expression for cortex completion can be bypassed and that normal assembly of coats can take place in the absence of a cortex substratum. The *spoVE* gene encodes what appears to be an integral membrane protein that is synthesized at a very early stage of sporulation (2, 3). This protein, which should be present in both membranes of the forespore, is required neither for induction of forespore genes such as *spoVA* (9) nor for expression of mother cell genes such as *spoIVCB* (15). Sitting in the forespore envelope, the SpoVE protein could play a dual role, allowing the precursors of both cortex layers to be assembled between the forespore membranes and preventing coat deposition. Interaction with the completed cortex would inactivate the SpoVE protein and allow deposition of the coats on the outer membrane of the forespore.

CONCLUDING REMARKS

As emphasized in the preceding section, both genomes are actively involved in the elaboration of the mature spore. However, only the spore genome is conserved and transmitted to the next generation arising from the germinated spore. Thus, the mother cell genome, which is ultimately destroyed, can be subjected to irreversible rearrangements such as those described in specialized nitrogen-fixing cells of *Anabaena* sp. (10). A similar mechanism has now been found to create the *sigK* gene in the mother cell around t_3

(30). It is then quite possible that other chromosomal rearrangements take place in the mother cell, either to inactivate genes that have to be expressed only in the forespore, or to design new genes required at a given time in the mother cell.

A highly speculative overview of the functional relationships between some *spo* genes is shown in Fig. 4. Realization of this scheme was stimulated by the initial effort of J. Errington in his report of the Spores X conference (*Nature* [London] **333**:399, 1988) and has greatly benefited from unpublished results communicated by R. Losick and his co-workers. Although the gaps in our knowledge will have to be filled by experimental data, it is more and more obvious that bacterial sporulation is an excellent paradigm for developmental studies. Even if most of the hypotheses presented in this review are found to be incorrect, at least they show that unraveling the controls of gene expression during sporulation allows elaboration of exciting concepts and should still give us fun for the next (few?) years.

Acknowledgments. I tested most of the speculations presented in this review on my supportive colleagues, C. Karmazyn-Campelli and J. Bouvier. I enjoyed many discussions with R. Losick and his co-workers. It is a pleasure to acknowledge the friendly and stimulatory atmosphere they have created at the Biolabs during my visits. A first draft of the manuscript was reviewed by R. Losick and A. Grossman. The artwork was done and redone by B. Savelli, and electron micrographs were kindly provided by A. Ryter.

The work carried out in my laboratory was supported by funds from Centre National de la Recherche Scientifique (URA 136) and from Fondation pour la Recherche Médicale.

LITERATURE CITED

1. **Beal, B., M. Lowe, and J. Lutkenhaus.** 1988. Cloning and characterization of *Bacillus subtilis* homologs of *Escherichia coli* cell division genes *ftsZ* and *ftsA*. J. Bacteriol. **170**:4855–4864.

2. **Bugaichuk, U. D.** 1987. Studies of transcriptional regulation of the *Bacillus subtilis* developmental gene *spoVE*. *J. Gen. Microbiol.* **133**:2349–2357.

3. **Bugaichuk, U. D., and P. J. Piggot.** 1986. Nucleotide sequence of the *Bacillus subtilis* developmental gene *spoVE*. *J. Gen. Microbiol.* **132**:1883–1890.

4. **Butler, P. D., and J. Mandelstam.** 1987. Nucleotide sequence of the sporulation operon, *spoIIIE*, of *Bacillus subtilis*. *J. Gen. Microbiol.* **133**:2359–2370.

5. **Carlson, H. C., and W. G. Haldenwang.** 1989. The σ^E subunit of *Bacillus subtilis* RNA polymerase is present in both forespore and mother cell compartments. *J. Bacteriol.* **171**:2216–2218.

6. **Coote, J. G.** 1972. Sporulation in *Bacillus subtilis*. Characterization of oligosporogenous mutants and comparison of their phenotypes with those of asporogenous mutants. *J. Gen. Microbiol.* **71**:1–15.

7. Cutting, S., S. Panzer, and R. Losick. 1989. Regulatory studies on the promoter for a gene governing synthesis and assembly of the spore coat in *Bacillus subtilis*. *J. Mol. Biol.* **207**:393–404.

8. **Donachie, W. D., K. J. Begg, J. F. Lutkenhaus, G. P. C. Salmond, E. Martinez-Salas, and M. Vicente.** 1979. Role of the *ftsA* gene product in control of *Escherichia coli* cell division. *J. Bacteriol.* **140**:388–394.

9. **Errington, J., and J. Mandelstam.** 1986. Use of a *lacZ* gene fusion to determine the dependence pattern and the spore compartment expression of sporulation operon *spoVA* in *spo* mutants of *Bacillus subtilis*. *J. Gen. Microbiol.* **132**:2977–2985.

10. **Golden, J. W., S. J. Robinson, and R. Haselkorn.** 1985. Rearrangement of nitrogen fixation genes during heterocyst differentiation in the cyanobacterium *Anabaena. Nature* (London) **314**:419–423.

11. **Karmazyn-Campelli, C., C. Bonamy, B. Savelli, and P. Stragier.** 1989. Tandem genes encoding sigma factors for consecutive steps of development in *Bacillus subtilis. Genes Dev.* **3**:150–157.

12. **Kleffel, B., R. M. Garavito, W. Baumeister, and J. P. Rosenbusch.** 1985. Secondary structure of a channel-forming protein: porin from *E. coli* outer membranes. *EMBO J.* **4**:1589–1592.

13. **Krikos, A., N. Mutoh, A. Boyd, and M. I. Simon.** 1983. Sensory transducers of *E. coli* are composed of discrete structural and functional domains. *Cell* **33**:615–622.

14. **Kroos, L., B. Kunkel, and R. Losick.** 1989. Switch protein alters specificity of RNA polymerase containing a compartment-specific sigma factor. *Science* **243**:526–529.

15. **Kunkel, B., K. Sandman, S. Panzer, P. Youngman, and R. Losick.** 1988. The promoter for a sporulation gene in the *spoIVC* locus of *Bacillus subtilis* and its use in studies of temporal and spatial control of gene expression. *J. Bacteriol.* **170**:3513–3522.

16. **Leung, A., S. Rubinstein, C. Yang, L. Jing-Wei, and** T. Leighton. 1985. Suppression of defective-sporulation phenotypes by mutations in the major sigma factor gene (*rpoD*) of *Bacillus subtilis. Mol. Gen. Genet.* **201**:96–98.

17. **Lopez-Diaz, I., S. Clarke, and J. Mandelstam.** 1986. *spoIID* operon of *Bacillus subtilis*: cloning and sequence. *J. Gen. Microbiol.* **132**:341–354.

18. **Lutkenhaus, J.** 1988. Genetic analysis of bacterial cell division. *Microbiol. Sci.* **5**:88–91.

19. **Mackman, N., J.-M. Nicaud, L. Gray, and I. B. Holland.** 1986. Secretion of haemolysin by *Escherichia coli. Curr. Top. Microbiol. Immunol.* **125**:159–181.

20. **Mandelstam, J., J. M. Sterlini, and D. Kay.** 1971. Sporulation in *Bacillus subtilis*. Effect of medium on the form of chromosome replication and on initiation to sporulation in *Bacillus subtilis. Biochem. J.* **125**:635–641.

21. **Mason, J. M., R. H. Hackett, and P. Setlow.** 1988. Studies on the regulation of expression of genes coding for small acid-soluble proteins of *Bacillus subtilis* spores using *lacZ* gene fusions. *J. Bacteriol.* **170**:239–244.

22. **Panzer, S., R. Losick, D. Sun, and P. Setlow.** 1989. Evidence for an additional temporal class of gene expression in the forespore compartment of sporulating *Bacillus subtilis. J. Bacteriol.* **171**:561–564.

23. **Piggot, P. J., and J. G. Coote.** 1976. Genetic aspects of bacterial endospore formation. *Bacteriol. Rev.* **40**:908–962.

24. **Rogers, H. J.** 1977. Peptidoglycans (mucopeptides), structure, form and function, p. 33–54. *In* A. N. Barker, J. Wolf, D. J. Ellar, G. J. Dring, and G. W. Gould (ed.), *Spore Research 1976*. Academic Press, Inc. (London), Ltd., London.

25. **Rong, S., M. S. Rosenkrantz, and A. L. Sonenshein.** 1986. Transcriptional control of the *Bacillus subtilis spoIID* gene. *J. Bacteriol.* **165**:771–779.

26. **Ronson, C. W., B. T. Nixon, and F. M. Ausubel.** 1987. Conserved domains in bacterial regulatory proteins that respond to environmental stimuli. *Cell* **49**:579–581.

27. **Sandman, K., L. Kroos, S. Cutting, P. Youngman, and R. Losick.** 1988. Identification of the promoter for a spore coat protein gene in *Bacillus subtilis* and studies on the regulation of its induction at a late stage of sporulation. *J. Mol. Biol.* **200**:461–473.

28. **Sandman, K., R. Losick, and P. Youngman.** 1987. Genetic analysis of *Bacillus subtilis spo* mutations generated by Tn*917*-mediated insertional mutagenesis. *Genetics* **117**:603–617.

29. **Stragier, P., C. Bonamy, and C. Karmazyn-Campelli.** 1988. Processing of a sporulation sigma factor in *Bacillus subtilis*: how morphological structure could control gene expression. *Cell* **52**:697–704.

30. **Stragier, P., B. Kunkel, L. Kroos, and R. Losick.** 1989. Chromosomal rearrangement generating a

composite gene for a developmental transcription factor. *Science* **243**:507–512.

31. **Sun, D., P. Stragier, and P. Setlow.** 1989. Identification of a new sigma factor involved in compartmentalized gene expression during sporulation of *Bacillus subtilis*. *Genes Dev.* **3**:141–149.

32. **Tipper, D. J., and P. E. Linnett.** 1976. Distribution of peptidoglycan synthetase activities between sporangia and forespores in sporulating cells of *Bacillus sphaericus*. *J. Bacteriol.* **126**:213–221.

33. **Tormo, A., C. Fernandez-Cabrera, and M. Vicente.** 1985. The *ftsA* gene product: a possible connection between DNA replication and septation in *Escherichia coli*. *J. Gen. Microbiol.* **131**:239–244.

34. **Trempy, J. E., J. Morrison-Plummer, and W. G. Haldenwang.** 1985. Synthesis of σ^{29}, an RNA polymerase specificity determinant, is a developmentally regulated event in *Bacillus subtilis*. *J. Bacteriol.* **161**:340–346.

35. **Turner, S. M., and J. Mandelstam.** 1986. Cloning and sequencing of a gene from *Bacillus amyloliquefaciens* that complements mutations of the sporulation gene *spoIID* in *Bacillus subtilis*. *J. Gen. Microbiol.* **132**:3025–3035.

36. **von Heijne, G.** 1986. Mitochondrial targeting sequences may form amphiphilic helices. *EMBO J.* **5**:1335–1342.

37. **Waites, W. M., D. Kay, I. W. Dawes, D. A. Wood, S. C. Warren, and J. Mandelstam.** 1970. Sporulation in *Bacillus subtilis*. Correlation of biochemical events with morphological changes in asporogenous mutants. *Biochem. J.* **118**:667–676.

38. **Yi, Q.-M., S. Rockenbach, J. E. Ward, Jr., and J. Lutkenhaus.** 1985. Structure and expression of the cell division genes *ftsQ*, *ftsA* and *ftsZ*. *J. Mol. Biol.* **184**:399–412.

39. **Young, M.** 1976. Use of temperature-sensitive mutants to study gene expression during sporulation in *Bacillus subtilis*. *J. Bacteriol.* **126**:928–936.

Regulation of Procaryotic Development
Edited by Issar Smith, Ralph A. Slepecky, and Peter Setlow
© 1989 American Society for Microbiology, Washington, DC 20006

Chapter 13

Role, Structure, and Molecular Organization of the Genes Coding for the Parasporal δ-Endotoxins of *Bacillus thuringiensis*

Didier Lereclus, Catherine Bourgouin,
Marguerite M. Lecadet, André Klier,
and Georges Rapoport

The insecticidal crystal of *Bacillus thuringiensis* is accumulated in sporulating cells after stage II (5, 78). The production of the crystal has been demonstrated to be a sporulation-specific event by the timing of the appearance of crystal antigens and the phenotype of different asporogenous mutants, as well as its synthesis from a stable mRNA (26). In some subspecies the crystalline inclusion can account for 20 to 30% of the dry weight of the sporulated cells. These inclusions are composed of polypeptides, designated the δ-endotoxins, that are toxic when ingested by susceptible insect larvae. *B. thuringiensis* strains have been defined immunologically by the presence of specific flagellar antigens (17). Most of the *B. thuringiensis* serotypes are toxic to the larvae of many different lepidoptera. Some strains are active against mosquito and black fly larvae, while other strains are toxic for certain coleopteran larvae. Genetic studies have shown that the structural toxin genes are often located on one or several resident high-molecular-weight plasmids. The number of toxin genes in a strain varies from only one

to at least five. It appears likely that the insecticidal activity of a given strain is governed by several parameters, including the total number of toxin genes present, their relative levels of expression, and the structural differences between their products.

This review describes the structure of the different toxic polypeptides, deduced from the nucleotide sequences of the corresponding cloned genes. Comparison of the sequences demonstrates that they belong to one family of proteins and contain several conserved domains, whatever their host range specificity. The structural organization of the toxin genes and the regulation of their expression are also discussed.

THE INSECTICIDAL CRYSTAL PROTEINS AND THEIR MODE OF ACTION

The δ-Endotoxins

The crystal proteins of lepidopteran-active strains were the first to be well documented. Several different polypeptides, with

Didier Lereclus, Catherine Bourgouin, Marguerite M. Lecadet, André Klier, and Georges Rapoport • Unité de Biochimie Microbienne, Institut Pasteur, 25 rue du Docteur Roux, 75724 Paris Cedex 15, France.

molecular masses of 130 to 140 kilodaltons (kDa), were identified (10). The 130- to 140-kDa protoxins are converted by proteolytic enzymes in the insect midgut to small toxic components, the sizes of which are approximately 60 kDa. The identification of several classes of crystal protein genes, their coexistence in several *B. thuringiensis* subsp. *kurstaki* strains, and the characterization of their expression products (57, 104) have strengthened the notion of diverse δ-endotoxin components each of which may have different toxic host ranges (43, 46).

Yamamoto et al. (109) demonstrated that homologous genes encoding the 130-kDa polypeptides are expressed simultaneously at different levels in one strain. A similar situation was described for *B. thuringiensis* subsp. *aizawai* 7.29 and *B. thuringiensis* subsp. *entomocidus* 6.01 and 6.05, all of which are active against lepidopteran larvae belonging to the Noctuidae family (82, 94). Several 130- to 135-kDa-protein genes have been isolated from these strains, of which at least three were shown to be expressed in *B. thuringiensis* subsp. *aizawai* 7.29 (59). The toxicity to the Noctuidae lar-

vae was specifically associated with the presence of one particular crystal protein gene.

Among subspecies that are active against lepidoptera, there are particular strains that have a significant toxicity toward dipteran insect species (33). The activity against mosquitoes was correlated with the presence of cuboidal crystals which are easily distinguishable from the common bipyramidal crystals and were characterized as "mosquito factor" (110). The cuboidal crystals have been shown to contain a 66-kDa polypeptide, designated as the P2 protein, which differs markedly from the lepidopteran-active δ-endotoxins. However, the P2 protein also displays a significant activity against lepidoptera (see Table 1). Two other polypeptides of 29 and 67 kDa have been found to be associated with the cuboidal crystals (103).

Certain 130-kDa proteins produced by several strains were also shown to be toxic to both lepidopteran and dipteran larvae (97). Of particular interest in this respect is *B. thuringiensis* subsp. *aizawai* IC1 δ-endotoxin. A differential activation of the protoxin molecules with either lepidopteran or

Table 1
Composition and Activity Spectra of the Crystals of *B. thuringiensis*

Susceptible insects	*B. thuringiensis* subsp.:	Crystal protein composition (mol. mass,[a] kDa)
Lepidoptera	*kurstaki* *thuringiensis* *aizawai* *entomocidus*	130–140 (P1)
Diptera	*israelensis*	128, 134, 72, 27
	morrisoni PG14	*144, 125, 135, 65–68, 27*
	darmstadiensis } *kyushuensis* }	*130–140, 60–70, 23–28*
Diptera and Lepidoptera	*kurstaki* HD1	130–140 (P1), (66 [P2], 67, 29)[b]
	aizawai IC1, HD249	*130–140*
Coleoptera	*tenebrionis* } *san diego* }	66[c]

[a] The molecular size of the polypeptides was deduced from the nucleotide sequence of the corresponding genes, except for polypeptides indicated in italics.
[b] Cuboidal crystal.
[c] Coding capacity of the corresponding gene, 73 kDa.

dipteran midgut proteases determines the specificity to lepidopteran or to dipteran larvae, respectively (32).

The fate of the δ-endotoxin is still more complex in the case of the dipteran-active isolates, most of which belong to *B. thuringiensis* subsp. *israelensis*. The crystals, which are only toxic to mosquitoes and black flies, appear heterogenous in their shape (12) and their composition. Four main components with molecular masses of 134, 128, 72, and 27 kDa were characterized. The 72-kDa polypeptide, previously estimated as 65 to 68 kDa, was shown to correspond to distinct crystals (45). The carboxy-terminal halves of the 134-kDa and the 128-kDa polypeptides are homologous. They both belong to the 130-kDa protein family found in the lepidopteran-active strains (see following section) and are not antigenically related to any of the 72-kDa or 27-kDa polypeptides (77). Determining the role of each polypeptide in the overall toxicity of the crystals has proved complicated, mainly due to the difficulty of isolating each polypeptide and to the fact that these polypeptides may act in synergism (108). Cloning the genes encoding each of these polypeptides led to the finding that the 134-, the 128-, and the 72-kDa polypeptides are mosquitocidal whereas the 27-kDa protein is responsible for the nonspecific cytolytic activity associated with the *B. thuringiensis* subsp. *israelensis* crystals. The 134-kDa and the 128-kDa proteins possess different specificities against mosquito larvae (6, 19). Furthermore, it has been shown that the crystals from *B. thuringiensis* subsp. *israelensis* probably contain small amounts of another polypeptide with an apparent molecular mass of 58 kDa. This polypeptide, active against mosquito larvae, was identified as the open reading frame 1 (ORF1) product by Thorne et al. (92); its molecular mass, deduced from the nucleotide sequence of the gene, was estimated to be 72 kDa. There is evidence that this polypeptide acts in synergism with the 128-kDa protein against larvae of certain mosquito species (19).

Besides *B. thuringiensis* subsp. *israelensis*, other subspecies of *B. thuringiensis* are specifically active against mosquito and black fly larvae (see Table 1). Among them, strain PG14 (*B. thuringiensis* subsp. *morrisoni*) synthesizes crystals whose composition is similar to that of the *B. thuringiensis* subsp. *israelensis* crystals except for an additional polypeptide of 144 kDa. Three other dipteran-specific strains have been described: strains 72-10-2 and 72-10-16, belonging to *B. thuringiensis* subsp. *darmstadiensis*, and one strain from *B. thuringiensis* subsp. *kyushuensis*. The crystals of these strains contain polypeptides of sizes similar to those of *B. thuringiensis* subsp. *israelensis* crystal proteins (Table 1); however, they are not antigenically related to the latter proteins. Only the 68-kDa and the 26-kDa proteins from *B. thuringiensis* subsp. *darmstadiensis* have been characterized. The 68-kDa protein is larvicidal (47), whereas the 26-kDa protein was shown to be both larvicidal and hemolytic (23).

Two new strains have been isolated since 1983: *B. thuringiensis* subsp. *tenebrionis* and *B. thuringiensis* subsp. *san diego*, which are both active against coleopteran species such as the Colorado potato beetle, the corn root worm, and the cotton boll weevil (35, 54). These new strains extend the potential uses of *B. thuringiensis* as a pesticide. The larvicidal activity was associated with flat, parasporal inclusions which differ markedly from the lepidopteran-active crystals in structure and in their chemical properties (35, 71). They also differ in the sizes of the protein subunits, and the major component of 64 to 68 kDa has been characterized (4, 35). The gene encoding the active protein has been cloned and sequenced and encodes a polypeptide of 73 kDa (34, 42, 71, 88).

This brief survey of the different classes of crystal proteins highlights the extreme variability of these molecules. The narrow specificity of each molecular species, which is the main characteristic of the *B. thuringiensis* δ-endotoxins, has been a limitation to their use as pesticides. However, the multiplicity of crystal proteins now identified, and enhanced toxicities due to synergistic effects, promise to increase their potential uses.

Mode of Action

A feature common to the 130-kDa crystal protein family is the activation of the toxins, by the action of larval gut proteases of susceptible insects, into active 60- to 70-kDa polypeptides. These polypeptides have been identified as the N-terminal halves of the protoxin molecules (14, 72, 85). Similar activation of the 72-kDa proteins of *B. thuringiensis* subsp. *israelensis*, yielding active polypeptides of 30 to 35 kDa, has been reported (45, 77). The 27-kDa protein does not require activation to manifest activity in vitro (15). This proteolytic activation is the first step of the poisoning process that follows the ingestion of native crystals. Subsequently, a sequence of events takes place which has been detailed in several review articles (24, 58, 66). The most striking symptoms in lepidopteran larvae are paralysis of the gut and the mouth parts and changes in gut permeability. This latter phenomenon is also observed in dipteran larvae. In spite of differences in the manifestations of toxemia between the different groups of insects, the histological studies in vivo revealed that the larval midgut epithelial cells (particularly the brush border membrane) are the primary target of the toxins (11, 13, 66, 76). The important damage consists of swelling and destruction of the microvilli, swelling of the cells, progressive disintegration of the cell organelles, and a general intracellular vacuolization.

Most of the studies agree on the fact that after the protoxin has been activated, the toxin binds to specific receptors located on the cell surface. Knowles et al. (53), using *Choristoneura fumiferana* (CF1) cells, reported that the activated δ-endotoxin of *B. thuringiensis* subsp. *kurstaki* HD1 could bind in vitro to a specific glycoconjugate with a terminal N-acetylgalactosamine residue. Inhibition of the toxic effect after preincubation either of the toxin with N-acetylgalactosamine or of the cells with soybean lectin indicated that a lectinlike binding of the lepidopteran-active toxin to the gut epithelium is the initial step in the process. The hypothesis was supported by the identification of a putative receptor, a 146-kDa glycoprotein which is a membrane protein of CF1 cells (51). Several membrane proteins were also identified as putative receptors for the activated toxin from *B. thuringiensis* subsp. *aizawai* IC1 in dipteran and lepidopteran cell lines (30, 31).

Using a different approach, Hofmann and Lüthy (37) and Wolfersberger et al. (105) demonstrated the binding of labeled toxin from *B. thuringiensis* subsp. *thuringiensis* to midgut cells or to brush border membrane vesicles prepared from the midgut of *Pieris brassicae*. The binding was shown to be inhibited by a unique monoclonal antibody which acted as an inhibitor of the in vivo activity. It was suggested that the binding is rapidly followed by integration of the toxin into the membrane.

The binding sites on the membrane vesicles were further characterized (38), and it was shown that they are proteins or glycoproteins. However, the authors did not find that N-acetylgalactosamine was involved in the binding sites as suggested by Knowles et al. (53). The discrepancy could be due to differences between the experimental systems used. However, both groups did suggest that membrane proteins, probably glycosylated, are directly involved in the binding process. It was also demonstrated that for one insect species (*P. brassicae*) two distinct binding sites exist for two different δ-endotoxins (39).

The 27-kDa hemolytic and cytolytic toxin appears to act via a mechanism unlike that of the other crystal proteins. Thomas and Ellar (91) reported a possible interaction of the toxin with phospholipids of the membrane of dipteran cell lines, inducing a detergentlike rearrangement of the lipids followed by a disruption of the membrane integrity. This action is therefore a nonspecific process.

Changes of membrane permeability are believed to be involved in toxemia induced by the crystal proteins. As early as 1968, Angus reported that the antibiotic valinomycin induced in larvae symptoms similar to those caused by the crystal proteins in *Bombyx mori* (2). This suggested that the toxin, like valinomycin, acts as an ionophore, probably affecting K^+ permeability. A supporting observation came from Sacchi et al. (80), who showed, through the inhibition of the uptake of histidine by brush border membrane vesicles from midgut of *P. brassicae*, that the δ-endotoxin dissipated the transmembrane electric potential difference by increasing potassium permeability. Dow (22) suggested that such a drop in electric potential could result from an increase in the backflux of the ion, rather than from an inhibition of the active transport pump as previously proposed by Himeno et al. (36). More recently, Knowles and Ellar (52) proposed a two-step model of a colloid-osmotic lysis mechanism. After binding to specific membrane receptors of susceptible cells, the δ-endotoxin from lepidopteran- as well as from dipteran-active species would generate small pores in the plasma membrane, leading to colloid osmotic lysis of the cell.

It is clear that the identification of the primary site of action of the toxin at the molecular level will be greatly facilitated by a more precise knowledge of the domains of the proteins involved in the activity of the toxins.

THE δ-ENDOTOXIN GENES AND THEIR PRODUCTS

Since the first gene encoding a 130-kDa δ-endotoxin was cloned in 1981 by Schnepf and Whiteley (84), *B. thuringiensis* toxin genes have been cloned and sequenced by several laboratories. The analysis and the comparison of the different deduced amino

acid sequences led to the classification of the genes according to their relatedness and the toxic specificity of the proteins. The genes are grouped into five major classes, *cryI*, *cryII*, *cryIII*, *cryIV*, and *cyt*, as proposed by H. R. Whiteley and H. Höfte (personal communication). However, subsections (*A*, *B*, *C*, *D*) have been adopted for the classes *cryI* and *cryIV*, which regroup partially related genes. Moreover, a size criterion can be taken into account to distinguish the different toxins. The classification scheme is presented in Table 2.

The 130-kDa protein genes belonging to *cryI* and *cryIV* classes, the gene *cryIVC* (previously designated as ORF1-ORF2), and the 73-kDa coleopteran-specific protein gene (*cryIIIA*) are discussed first as these genes show evident molecular relatedness, irrespective of the host range specificity of their products.

130-kDa Protein Genes and Related Genes

From the biochemical studies discussed in the first section and from the genetic data reported here, it is evident that all the 130-kDa protein genes share a common feature. As described below, this common characteristic lies in the molecular structure of the genes which can be divided into two elements, the 5′ and the 3′ halves. The 5′ half corresponds to the toxin part of the molecule and constitutes a variable region which probably reflects the specificity of the gene product. The 3′ half is highly conserved among the 130-kDa genes and is presumably involved in the formation of the crystalline structure.

All the genes encoding 130- to 140-kDa polypeptides active against lepidopteran species were grouped into the *cryI* class by Whiteley and Höfte (see Table 2). Nevertheless, they do not display exactly the same toxicity spectra. Indeed, although a large majority of these δ-endotoxins are toxic to larvae of lepidoptera such as *Manduca sexta*, *P. brassicae*, or *B. mori* and poorly active

Table 2

Identification of the Toxin Genes of *B. thuringiensis*

Gene[a]		Size of gene products (kDa)	Toxicity	*B. thuringiensis* subsp.:	Other names	References
Class	Subclass					
cryI	A[b]	130–140	Lepidopteran larvae	*kurstaki*	"4.5," "5.3," "6.6"	57
				thuringiensis berliner	*Bt2*	41
	B	~130	Lepidopteran larvae	*thuringiensis* HD2	*cryA4*	81
	C	~135	Lepidopteran larvae	*aizawai* 7.29, *entomocidus* 6.05	*Bt VI, Bta*	82, 94
	D	~130	Lepidopteran larvae	*aizawai* 7.29, *entomocidus* 6.05	*Bt V, A4*	82, 94
cryII	A	66	Lepidopteran and dipteran larvae	*kurstaki* HD1	*cryB1*	21
cryIII	A	73	Coleopteran larvae	*tenebrionis, san diego*	*Bt.t* *BT SD* *Bt 13*	71 34 42
cryIV	A	134	Dipteran larvae	*israelensis*	"125"; IRSH4	6, 89
	B	128	Dipteran larvae	*israelensis*	"135"; IRSH3 *Bt8*	6, 89 16
	C	72	Dipteran larvae	*israelensis*	ORF1	92
	D	72	Dipteran larvae	*israelensis*	*cryD*	20
cyt		27	Cytolytic	*israelensis, morrisoni* PG14		

[a] Classification proposed by Whiteley and Höfte (personal communication).

[b] *cryIA* genes have been divided into three subclasses, (a), (b), and (c), previously described as "4.5," "5.3," and "6.6," respectively.

against larvae of other insects such as *Spodoptera littoralis* (Noctuidae), the product of the *cryIC* gene is preferentially toxic to the latter insect (82, 94).

Among the *cryIA* class, three distinct types of genes (a, b, and c), present in most of the *B. thuringiensis* strains, have been identified and sequenced (1, 25, 41, 73, 86, 90, 92, 96). The comparison of the deduced polypeptides revealed that they can be divided into constant and variable regions. The NH$_2$-terminal part of the molecule (amino acids 1 to 280) and the COOH-terminal domain are extensively conserved, whereas most of the substitutions occur in the second half of the NH$_2$-terminal domain (amino acids 281 to 618). These closely related genes, which could have evolved by homologous recombination, have previously been compared (3, 25, 41, 57, 60, 102).

The nucleotide sequence of the *Spodoptera*-specific gene (*cryIC*) originating from *B. thuringiensis* subsp. *entomocidus* 6.01 and 6.05 and *B. thuringiensis* subsp. *aizawai* 7.29 indicates that its product possesses a COOH-terminal domain almost identical to that of the *cryIA*-type proteins (Fig. 1; 44, 81). However, the NH$_2$ domain of the molecule (amino acids 1 to 618) is significantly different from that of the other δ-endotoxins belonging to this group. As described by Sanchis et al. (81), there is only 45% identity between these domains of the molecules, and the conserved positions are clustered in five boxes (Fig. 1 and 2).

Comparable results have been found upon comparison of the *cryIB* and *cryID* genes isolated from *B. thuringiensis* subsp. *thuringiensis* HD2 (8), *B. thuringiensis* subsp. *entomocidus* 6.01 and 6.05, and *B. thuringiensis* subsp. *aizawai* 7.29 (82, 94; Höfte, personal communication).

The determination of the nucleotide sequences has been valuable for the localization of the toxin part of these polypeptides. Experiments carried out with 5'- or 3'-trun-

cated genes indicated that the DNA region required for the synthesis of a toxic polypeptide stretches from codon 29 to codon 607 or 615, according to the gene (1, 41, 81, 85). Moreover, the truncated polypeptides retain the toxic specificity of the native δ-endotoxin (81). Thus, the role of the carboxy-terminal domain of these proteins is presumably structural and involved in the formation and the stability of the crystal. This hypothesis is supported by the fact that this part of the δ-endotoxin molecules contains the majority of the cysteine residues and therefore could be responsible for the crystalline structure. Sim-

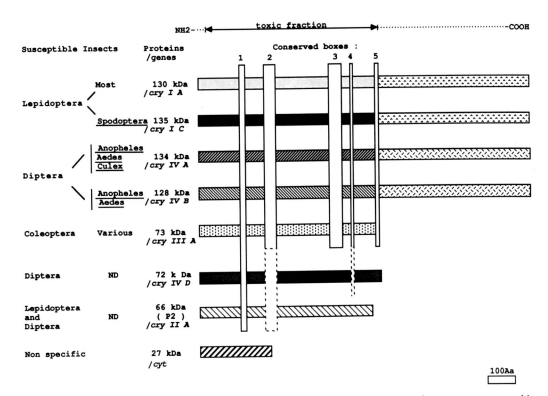

Figure 1. Amino acid sequence comparison of the different classes of δ-endotoxins. The five regions represented by vertical white boxes are highly conserved among the different proteins, as previously described by Sanchis et al. (81). The vertical dashed boxes indicate partial conservation of the amino acid sequences (see Fig. 2 for detailed description of the five boxes). In the N-terminal half of the sequences, the regions represented with different symbols are weakly or not at all related. However, although the COOH domains of the 128- to 135-kDa δ-endotoxins active against the lepidoptera and the diptera are not shown with identical symbols, they do show significant homology. The protein designated as the *cryIA* gene product represents the three different subclasses a, b, and c, although they vary between amino acid (Aa) 282 and amino acid 600. References corresponding to the different genes and proteins are cited in the text.

BOX 1

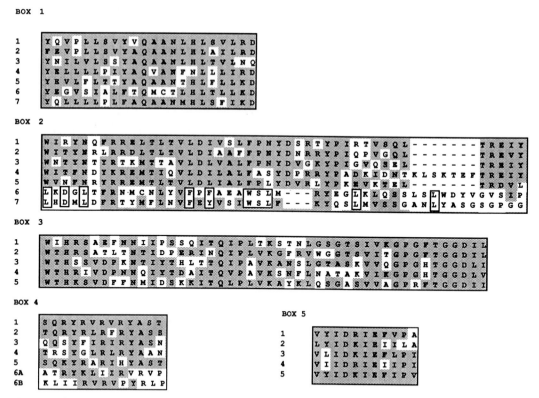

Figure 2. Alignment of the five amino acid sequences conserved in the lepidopteran-, dipteran-, or coleopteran-active δ-endotoxins. The relative positions of the five boxes, previously described by Sanchis et al. (81), are indicated in Fig. 1. Line numbers refer to the products of the following genes: 1, *cryIA*(b); 2, *cryIC*; 3, *cryIVA*; 4, *cryIVB*; 5, *cryIIIA*; 6, *cryIVD*; 7, *cryIIA*. 6A and 6B represent the two possible alignments of the putative box 4 of the *cryIVD* gene product. The genes and their products are described in the text and in Table 2. Identical and similar residues are boxed with gray (accepted conservative replacements used are I, L, V, and M; D and E; N and Q; A and G; R and K; S and T; and F and Y). In box 2 the similarities specifically observed between *cryIVD* and *cryIIA* gene products have been boxed in black.

ilar observations have been made concerning the 128- to 134-kDa δ-endotoxins which are specifically active against the diptera. The toxic fraction appears to be the NH_2-terminal domain of the molecules, from amino acids 42 to 634 (19, 40, 75).

The two genes (*cryIVA* and *cryIVB*) coding for the proteins active against *Aedes aegypti* larvae were cloned from *B. thuringiensis* subsp. *israelensis*, and their nucleotide sequences were determined (16, 89, 93, 99, 111). Comparison of the deduced amino acid sequences indicated that the two toxins are very similar in the COOH-termi-

nal regions (Fig. 1; 89, 100). In addition, significant identity (42%) was found between the carboxy-terminal domain of these proteins and the equivalent regions of the lepidopteran-specific δ-endotoxins (16, 100). However, the NH_2-terminal halves of these proteins are less similar. The similarities are largely located in five conserved regions, as established by Sanchis et al. (81) (Fig. 1 and 2). The extensive sequence differences between the toxic domains of the proteins encoded by the *cryIVA* and *cryIVB* genes might reflect their different activity spectra. Indeed, the *cryIVA* protein is toxic for *A. aegypti*,

Anopheles stephensi, and *Culex pipiens* whereas the product of the *cryIVB* gene is only active against *A. aegypti* and *A. stephensi* and not against *C. pipiens* (6, 19).

Another gene (*cryIVC*) encoding a mosquitocidal polypeptide with a deduced molecular mass of 72 kDa was cloned from *B. thuringiensis* subsp. *israelensis* (92). This gene, previously designated as ORF1, is adjacent to a second ORF, ORF2, which could be transcribed in the same orientation. The deduced amino acid sequences from these two contiguous genes present significant similarity to the 130-kDa δ-endotoxins. ORF1 and ORF2 taken together display striking similarity to the five conserved boxes of the NH$_2$-terminal domain and to a part of the carboxy-terminal moiety of the other gene products.

Another gene recently identified as closely related to the 130-kDa polypeptide family is the 73-kDa coleopteran-specific protein gene isolated from *B. thuringiensis* subsp. *tenebrionis* and *B. thuringiensis* subsp. *san diego*. The deduced amino acid sequence of the *cryIIIA* gene reveals limited but significant homology to the toxic fraction of the 130-kDa δ-endotoxins active against lepidopteran and dipteran larvae (34, 42, 71, 88). Amino acid sequences found in the 130-kDa polypeptides are also present in this protein (Fig. 1 and 2). Another common feature with the 130-kDa δ-endotoxins concerns the localization of the toxic domain, since the N terminus of the polypeptide isolated from the crystal of *B. thuringiensis* subsp. *tenebrionis* starts at position 58 of the deduced amino acid sequence (42). Similarly, the *cryIIIA* gene ends precisely at the codon corresponding to amino acid 607 or 615 in the CryIA or CryIC protein, respectively. Therefore, the coleopteran toxin does not possess the carboxy-terminal domain that is conserved in the 130-kDa lepidopteran and dipteran toxins but which does not contribute to the toxicity. The parasporal crystal composed of the 73-kDa protein is soluble in aqueous solutions and not internally linked by disulfide

bridges (4), and this strengthens the belief in the role of the COOH-terminal region of the 130-kDa proteins in the stability of the crystal.

The overall comparison of the sequences of the different toxins clearly shows that although the proteins display various activities, they belong to one family of proteins which is characterized structurally by five conserved domains. Beyond these molecular relationships, the hydropathic profiles of the members of this δ-endotoxin family display striking similarity. Such structural analogies suggest an identical ancestral origin for these toxins and an essential function of the conserved domains in their larvicidal activity.

Poorly Related Genes

Two genes are discussed in this section: the gene *cryIVD*, encoding a 72-kDa protein, and the gene *cryIIA*, which codes for a 66-kDa lepidopteran- and dipteran-active polypeptide previously designated as protein P2 (see Table 2). The *cryIVD* gene product, which is a component of the parasporal crystal from *B. thuringiensis* subsp. *israelensis*, is toxic to mosquito larvae. The *cryIIA* gene product forms, in association with at least two other polypeptides of 29 and 67 kDa, the cuboidal crystal of *B. thuringiensis* subsp. *kurstaki* HD1 (103). Solutions of this crystal are toxic to both lepidopteran and dipteran larvae (110).

Recently cloned and sequenced by Donovan et al. (20, 21), these two genes appear poorly related to the 130-kDa δ-endotoxin genes. However, Fig. 1 and 2 show that similarities are found in box 1 and in part in box 2. The *cryIVD* gene product also contains some similarity to box 4. This region is rich in arginine, and the repetitive nature of the sequence allows two possible alignments (Fig. 2). As a whole, these similarities are relatively weak but are probably significant since they occur at approximately the expected place along the sequence. It should be mentioned that greater similarity can be observed between *cryIIA* and *cryIVD* gene products in the box 2 region than between either of them and the 130-kDa polypeptide family. Although these two proteins appear to be only distantly related to the other δ-endotoxins, it is attractive to postulate that they retain some functional domains that are essential for the larvicidal toxicity.

Cytolysin Gene

The *cyt* gene encodes the 27-kDa polypeptide contained in the parasporal crystal from *B. thuringiensis* subsp. *israelensis*. The product of the cloned gene does not display a specific mosquitocidal activity but rather cytolytic properties which could act synergistically with other components of the crystal and enhance their toxicity to dipteran larvae (7, 101). The nucleotide sequence of the *cyt* gene was determined (95, 98), and the deduced amino acid sequence does not show any significant similarity to those of the other types of δ-endotoxins. Strikingly, the hydropathy plot of the protein reveals extensive hydrophobic regions, which is consistent with a cytolytic mechanism of action (98).

The analysis of the sequence data available for the crystal proteins has shown that they are all related, with the exception of the *cyt* gene product. All the members of the 66- to 73-kDa protein family and the 130- to 140-kDa protein family appear to be homologous. However, although conserved regions and variable regions have been identified, their functions and relevance to either the mode of action or specificity are largely speculative.

LOCALIZATION, ENVIRONMENT, AND STRUCTURAL ORGANIZATION OF THE CRYSTAL PROTEIN GENES

Early studies on the extrachromosomal DNA of various *B. thuringiensis* strains revealed the ubiquitous presence of large plasmids (>30 MDa) sharing homologous DNA sequences detectable by hybridization assays (29, 61). It was demonstrated, by means of curing experiments carried out by shifting the bacterial cultures to 42°C (29) and by means of a conjugationlike process (27), that specific large plasmids were involved in the crystal production. Subsequently, the cloning of the toxin structural genes and their utilization as radioactive probes confirmed the validity of these observations for most of the *B. thuringiensis* strains. Nevertheless, some crystal protein genes have only been identified in total DNA preparations and not in plasmid DNA preparations. This suggests that genes are located on the chromosome. However, no toxin gene has ever been found in association with a chromosomal auxotrophic marker, so it is possible that these genes are in fact carried on large plasmids (>150 MDa) which are indistinguishable from the chromosome by the usual techniques of DNA preparation. The variety of locations of the crystal protein genes naturally occurring in most of the *B. thuringiensis* isolates is summarized in Table 3. This presentation of the results emphasizes the plasmid location of the toxin genes, since it appears that only three *B. thuringiensis* subspecies, *dendrolimus*-type, *entomocidus*-type, and *wuhanensis* HD525, do not possess plasmids harboring a crystal gene. However, the presence of toxin genes both on a chromosome and on plasmids seems to be a predominant feature of *B. thuringiensis*. It is also remarkable that at least three strains (*B. thuringiensis* subsp. *kurstaki* HD1, *B. thuringiensis* subsp. *aizawai* 7.29, and *B. thuringiensis* subsp. *israelensis*-type) harbor more than four separate toxin genes.

This multiplicity of the δ-endotoxin genes within one strain is particularly well illustrated in *B. thuringiensis* subsp. *aizawai* 7.29 and *B. thuringiensis* subsp. *israelensis*-type, in which five toxin genes have been characterized (6, 7, 20, 82, 92). Figure 3 shows the simplified restriction maps of the five toxin genes present in *B. thuringiensis* subsp. *aizawai* 7.29. First, it appears that one gene, *cryIA*(b), is duplicated in this strain; one copy is located on a 45-MDa plasmid,

Table 3
Location of δ-Endotoxin Genes in Different *B. thuringiensis* Subspecies

H serotype[a]	*B. thuringiensis* subsp.:	Location of toxin genes		Minimum gene copy no.	References
		On plasmid (size, MDa)	On chromosome (or plasmid of >150 MDa)		
1	*thuringiensis* HD2	54–57; 75	ND[b]	3	27, 48, 57
1	*thuringiensis* berliner 1715	42	+	2	48
2	*finitimus*	98	+	2	18
3a	*alesti* HD4	100	ND	1	55
3a, 3b	*kurstaki* HD73	50	−	1	27, 55
3a, 3b	*kurstaki* HD1	44; 60; 110	−	5	21, 55, 103
4a, 4b	*sotto*-type	35	−	1	50, 55
4a, 4b	*dendrolimus*-type		+	1	48
4a, 4c	*kenyae*-type	40	ND	1	This laboratory, unpublished data
5a, 5b	*galleriae* HD8	150	ND	1	55
6	*entomocidus*-type		+	3	50, 82
6	*subtoxicus*-type	55	+	2	50
7	*aizawai* 7-29	45	+	5	50, 82
8a, 8b	*tanebrionis*	90	−	1	88
8a, 8b	*morrisoni*-type	145; 150	ND	2	55
9	*tolworthi*-type	44; 70; 100	ND	3	48, 55
10	*darmstadiensis*-type	60	ND	1	74
14	*israelensis*-type	72	−	5	6, 7, 20, 28, 100
ND	*wuhanensis* HD525		+	1	55

[a] Serotypes based on flagellar antigens as described by De Barjac (17).
[b] ND, Not determined.

whereas the other copy seems to be on the chromosomal DNA. In both cases, the *cryIA*(b) genes are located near a conserved genetic element flanked by insertion (IS) sequences (83); this transposonlike structure will be described below. Second, two other genes, *cryIC* and *cryID*, both of presumed chromosomal origin, are located close together and are transcribed in the same direction. The fifth gene, *cryIA*(a), has been identified on presumed chromosomal DNA and, according to its *Pvu*II sites, is related to the 4.5-kilobase (kb) class gene defined by Kronstad and Whiteley (57). Thus, *B. thuringiensis* subsp. *aizawai* 7.29 bears different types of genes encoding δ-endotoxins which display various specific activities against the lepidopteran insects (see Tables 1 and 2 and preceding paragraphs). It should also be noted that a similar structural organization of toxin genes has been found in

B. thuringiensis subsp. *entomocidus* 6.05 (94).

In contrast to the multiple locations of the crystal genes in these strains, it is now clear that all the dipteran-specific toxin genes of *B. thuringiensis* subsp. *israelensis*-type are located in one 72-MDa resident plasmid. Although the complete physical map has not been established, the compilation of results published by several laboratories allows us to map the different crystal genes of this strain to three separate DNA fragments (6, 7, 20, 92). Once again, it is evident (Fig. 3) that different genes are located close together; the gene *cryIVD* coding for the 72-kDa polypeptide and the gene for the cytolysin are transcribed divergently, whereas, in the second set, the gene *cryIVB* coding for the 128-kDa polypeptide and the gene *cryIVC* (ORF1-ORF2) are transcribed in the same direction. Such an accumulation of toxin

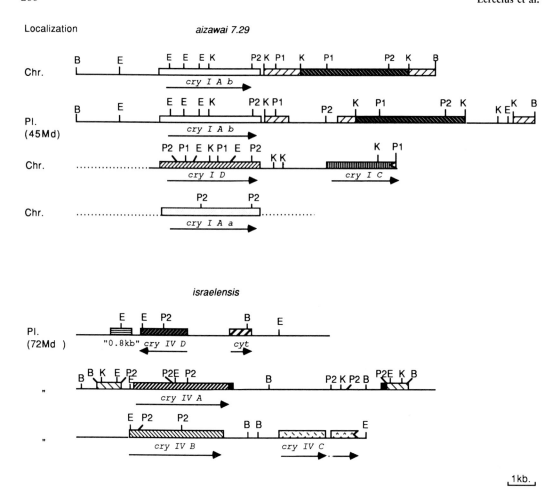

Figure 3. Location and structural organization of the δ-endotoxin genes in *B. thuringiensis* subsp. *aizawai* 7.29 and *B. thuringiensis* subsp. *israelensis*. The location and organization of the toxin genes identified in the two strains were described by Sanchis et al. (82, 83), Bourgouin et al. (6, 7), Thorne et al. (92), and Donovan et al. (20). Arrows show the direction of transcription of the genes. The boxes adjacent to the *cryIA*(b) and *cryIVA* genes indicate the transposable elements mentioned in the text and shown in Fig. 4 with the same symbols. The box "0.8 kb" adjacent to the *cryIVD* gene refers to the DNA regulator sequence mentioned in the text. Chr. and Pl. indicate whether the genes are chromosomally or plasmid encoded, respectively. Abbreviations used for the restriction endonuclease sites: B, *Bam*HI; E, *Eco*RI; K, *Kpn*I; P1, *Pst*I; P2, *Pvu*II. The dotted lines represent uncharacterized DNA segments.

genes on a conjugative plasmid (28) might confer an adaptive advantage to the species by ensuring the rapid spread of these genes to other *B. thuringiensis* strains.

Another remarkable characteristic concerning the location and the structural organization of the crystal genes is seen upon analysis of the DNA flanking these genes. Two previous studies carried out with *B. thuringiensis* subsp. *kurstaki* HD73 and *B. thu-*

ringiensis subsp. *thuringiensis* berliner 1715, using different approaches, showed that several plasmid genes encoding the lepidopteran-specific δ-endotoxins are flanked by two sets of inverted repeated (IR) sequences and a transposable element (Fig. 4A; 56, 64).

Nucleotide sequence analysis of three IR elements, belonging to the same family, revealed all the features of insertion sequences. The name IS*231* (a, b, or c) was therefore

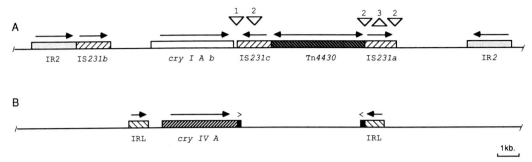

Figure 4. Structural organization of the transposable elements flanking δ-endotoxin genes in *B. thuringiensis*. (A) This model summarizes the different organizations of the five plasmids of 40 to 55 MDa from *B. thuringiensis* subsp. *thuringiensis* strains berliner 1715 (64) and HD2 (57), *B. thuringiensis* subsp. *kurstaki* strains HD1 and HD73 (57), and *B. thuringiensis* subsp. *aizawai* 7.29 (82, 83). As shown in Fig. 3, a similar situation is observed at a chromosomal location in *B. thuringiensis* subsp. *aizawai* 7.29. The triangles indicate the position of insertions (∇) or of deletions (△) of various sizes (1 to 5 kb) found in the plasmids from *B. thuringiensis* subsp. *kurstaki* HD73 (1), *B. thuringiensis* subsp. *aizawai* 7.29 (2), and *B. thuringiensis* subsp. *kurstaki* HD1 (3). IR, Inverted repeat; IS, insertion sequence. (B) DNA fragment of the 72-MDa resident plasmid from *B. thuringiensis* subsp. *israelensis* (6; Délécluse, personal communication). Arrows above the *cryIA*(b) and *cryIVA* genes indicate the direction of transcription. Arrows above the IR2, IS231, and IRL elements represent their relative orientation. The double arrow above Tn4430 signifies that the transposon can be found in opposite orientations in different strains. Arrowheads above the black squares represent IRS segments and indicate their relative orientation.

proposed for these iso-elements of 1,625 base pairs (bp) (68, 69). The second set of IR sequences, which are about 2,150 bp long and delimit the ends of the structure shown in Fig. 4A, were designated IR2 since they have not yet been characterized as IS.

The transposable element Tn4430, whose transposition was obtained in both *B. thuringiensis* (63) and *Escherichia coli* (62), is related to the Tn3 family by its transposition mechanism involving a transposase and its 38-bp terminal inverted repeats, whereas the protein which catalyzes the cointegrate resolution process belongs to the integrase family of site-specific recombinases (67). Thus, this transposon of 4,149 bp defines a novel type of class II mobile genetic elements.

The map presented in Fig. 4A is a summary of the different situations found within large plasmids of *B. thuringiensis* subsp. *kurstaki* strains HD73 and HD1, *B. thuringiensis* subsp. *thuringiensis* strains berliner 1715 and HD2, and *B. thuringiensis* subsp. *aizawai* 7.29 (57, 64, 82). Although some modifications (insertions or deletions) can be observed inside or at the ends of the IS231

elements, the genetic organization of the five plasmids is similar. These homologies are probably due to the conjugative properties of the plasmids which enable them to be transferred from one strain to another.

Recently, a comparable but significantly different transposonlike structure has been described for the *cryIVA* gene harbored by the 72-MDa plasmid of *B. thuringiensis* subsp. *israelensis* (6). The toxin gene is flanked by two sets of IR sequences designated IRL and IRS (Fig. 4B). IRL sequences, which are 865 bp long, code for a putative transposase and are delimited by two short terminal inverted repeats (A. Délécluse, personal communication). Thus, IRL appears to be a new insertion sequence unrelated to IS231 or IR2 sequences. IRS sequences, which are not longer than 200 bp, have not yet been characterized.

Although efficient transposition by IS231, IR2, and IRL sequences has not been demonstrated, the properties generally attributed to the insertion sequences strongly implicate their involvement in the mobility and the diversity of the crystal genes. In bacteria, the association of accessory genetic elements, such as plasmids, transposons, and genes determining pathogenicity or resistance to antibiotics, is frequently observed. Such associations are likely to enhance the spread of the adaptive genes and therefore contribute to the development of the bacterial species within a given ecological niche. In *B. thuringiensis*, the structural organization of the δ-endotoxin genes might confer an evolutionary advantage by enabling the bacteria to proliferate among several insect species.

REGULATION OF δ-ENDOTOXIN SYNTHESIS

Two striking characteristics concerning the synthesis of crystals in *B. thuringiensis* are, first, the coordination with sporulation and, second, its remarkable efficacy. Although the location of the toxin genes on plasmids and within a transposonlike structure has not been clearly demonstrated to be an important element of their regulation, it seems very likely. Indeed, the level of synthesis of the toxins is probably dependent on the plasmid copy number and on the multiplicity of the genes. However, no direct correlation has been established between these phenomena.

Several mechanisms of molecular regulation involved at the transcriptional or translational level have been described for the expression of the δ-endotoxin genes, as follows. First, studies of Spo⁻ mutants clearly indicated that the synthesis of the crystal is developmentally regulated and is dependent on sporulation (79). Spo⁻ mutants blocked at t_0 failed to make crystals, whereas those blocked after t_2 might produce crystals. Moreover, biochemical studies indicated that a sporulation-specific RNA polymerase is required for the synthesis of crystal protein mRNA (49).

By analogy with endospore formation in *B. subtilis* (65), the above results suggested that different sigma factors confer on RNA

polymerase the capacity to initiate transcription from sporulation-specific promoters of the crystal protein genes. This was recently demonstrated by Brown and Whiteley (9), who reported the isolation, from sporulating *B. thuringiensis* cells, of a RNA polymerase in which the core components are associated with a 35-kDa sigma factor which was not found in vegetative cells. In vitro, this polymerase ($E\sigma^{35}$) directed the transcription from the promoters of at least three different toxin genes, those encoding the 133-kDa δ-endotoxin of *B. thuringiensis* subsp. *kurstaki* HD1-Dipel, the 27-kDa cytolysin of *B. thuringiensis* subsp. *israelensis*, and the 29-kDa cuboidal crystal protein of *B. thuringiensis* subsp. *kurstaki* HD1 (9).

The transcriptional start sites for these genes were determined by S1 nuclease mapping or by primer extension with reverse transcriptase. Wong et al. (107) revealed that transcription of the 133-kDa δ-endotoxin gene started from two overlapping promoters used sequentially. Promoter *Bt*I was active during stages II and III of sporulation, and promoter *Bt*II was used from stage III onwards. Similar regions located about 100 nucleotides upstream from the coding sequence are found in front of all the lepidopteran-specific δ-endotoxin genes (1, 81, 90, 92, 96).

Two putative promoter regions were also identified for the gene encoding the 27-kDa cytolysin (95, 98). As shown in Table 4, which lists the −35 and −10 regions upstream of the different transcription initiation sites, the PBiI promoter identified by Ward and Ellar (98) is closely related to the *Bt*I promoter, whereas the PBiII region identified by Waalwijk et al. (95) resembles the PBc putative promoter of the gene encoding the 73-kDa coleopteran-specific δ-endotoxin (88). Thus, three types of promoters can be distinguished (Table 4). The *Bt*I and PBiI promoters and the putative promoter of the 29-kDa cuboidal crystal gene are recognized by $E\sigma^{35}$ (9). The sequence of *Bt*II is presented in Table 4, and although it differs from that of *Bt*I in the spacer between the −35 and

Table 4
Conserved Regions of Toxin Gene Promoters in *B. thuringiensis*[a]

Promoter	RNA polymerase	−35 region	Spacer length (bp)	−10 region	Distance (bp) to the transcription start
*Bt*I	Eσ[35]	GCATTT	15	CATATGTT	7
PBiI	Eσ[35]	GCATCT	14	CATAGAAT	4–6
P29	Eσ[35]	GCATAT	15	CATAGAAT	3–5
Consensus	Eσ[35]	GCAT–T	14 or 15	CATA–––T	3–7
*Bt*II	ND[b]	GCACTT	12	CATAAGAT	1–6
PBiII	ND	TGATTAA	18	CATAATTT	4–5
PBc	ND	TGATTAA	18	TATAAATT	4–5

[a] The promoter *Bt*I is used for the transcription of the 133-kDa δ-endotoxin gene of *B. thuringiensis* subsp. *kurstaki* HD1-Dipel during stages II and III of sporulation. After stage III the *Bt*II promoter is used (9, 107). The promoter P29 of the 29-kDa cuboidal crystal protein gene was described by Brown and Whiteley (9). PBiI and PBiII are the two promoters of the cytolysin gene of *B. thuringiensis* subsp. *israelensis*. PBiI was identified by Ward and Ellar (98), and PBiII by Waalwijk et al. (95). PBc is the promoter of the 73-kDa toxin gene from *B. thuringiensis* subsp. *tenebrionis* (88). The RNA polymerase Eσ[35] was identified by Brown and Whiteley (9).
[b] ND, Not determined.

−10 regions, it is not used as a promoter by Eσ[35] (9). Whether PBiII and PBc promoters require another RNA polymerase with a different σ subunit remains to be demonstrated.

It therefore appears that the regulation of crystal protein synthesis involves control at the level of transcription. Thus, the temporal control of the expression of the different toxin genes would be in part ensured by successive changes in the sigma factors. Nevertheless, recent work indicates that in *B. thuringiensis* subsp. *kurstaki* HD1 different toxin genes are not expressed in equal amounts (109). As these genes presumably harbor homologous promoter regions, this result indicates the involvement of other regulation systems in the synthesis of the crystal.

Such regulation mechanisms, apparently independent of the sporulation process, have been demonstrated for toxin genes cloned in *E. coli*. Schnepf et al. (87) showed that insertions or deletions located upstream of the promoter region enhance the transcription level of the *cryIA*(a) gene cloned from *B. thuringiensis* subsp. *kurstaki* HD1-Dipel. Similar results were also obtained with other cloned toxin genes (82). It is possible that this effect could result from the disruption or repositioning of a transcription-suppressing sequence (87). It should be noted that these experiments were performed in *E. coli*, so the relevance of the phenomenon to expression in *B. thuringiensis* is unclear. However, the promoter recognized in *E. coli* is located between the two promoters used in *B. thuringiensis*, and all three start sites are contained in a 20-bp region (107). It is conceivable, therefore, that a regulatory mechanism which probably negatively controls the promoter functional in *E. coli* is also operating in *B. thuringiensis*. It would be interesting to determine whether this negative regulation functions in *B. thuringiensis* during sporulation or only during the vegetative phase to prevent any residual expression of the toxin genes.

McLean and Whiteley (70) showed that a 0.8-kb DNA segment located 4 kb upstream of the gene encoding the 27-kDa cytolysin of *B. thuringiensis* subsp. *israelensis* enhanced the expression level of this protein in *E. coli*. Donovan et al. (20) cloned the 72-kDa toxin gene from the same strain, and it appears that the 0.8-kb DNA segment is located downstream from the *cryIVD* gene and upstream from the *cyt* gene (Fig. 3). This DNA region acts in *cis* or in *trans* on the *cyt* gene expression and might encode a 20-kDa polypeptide. The results suggested a post-transcriptional regulation, since the amount of cytolysin-specific mRNA is unchanged when the 0.8-kb sequence is present or is missing. However, the regulatory mechanism involved has not been determined, nor has it been shown to function in *B. thuringiensis* cells.

One of the hypotheses proposed to account for the high level of δ-endotoxin synthesis in *B. thuringiensis* is that the crystal protein mRNA is very stable; this has been shown by Glatron and Rapoport (26). Since this observation, it was shown that the transcriptional terminator of a toxin gene acts as a positive retroregulator and increases the mRNA half-life (106). Indeed, the fusion of this terminator sequence to the 3' end of heterologous genes enhances their expression level. It was postulated that the stem-and-loop secondary structure of the terminator protects the mRNA from exonuclease degradation. Consequently, the stability of the mRNA and the levels of protein synthesis are increased. It must be noted that the terminator sequences downstream of the different δ-endotoxin genes are frequently highly conserved.

These data as a whole reveal that a number of mechanisms are involved to control the expression of the toxin genes. The multiplicity of the systems reflects not only the essential role of the crystal for the *B. thuringiensis* species but also the fact that its synthesis is tightly linked to the sporulation process.

CONCLUSION

The main features of the results and observations described in this review can be summarized as follows.

(i) There may be a multiplicity of genes coding for the 130-kDa polypeptides within a strain of the different *B. thuringiensis* subspecies.

All the toxin gene products except the 27-kDa cytolysin of *B. thuringiensis* subsp. *israelensis* are homologous. Regions of identity are found in several boxes in the N-terminal half of the polypeptide, which corresponds to the toxic fraction of about 65 kDa. The percentages of identity are much higher in the C-terminal domain of each toxin class (*cryI* and *cryIV*, for example).

The diversity of genes which is found in *B. thuringiensis* represents a striking example of "divergent evolution" which confers to this bacterium a strong adaptive potential in terms of its target insects.

The crystal protein genes are associated with transposable elements. The result of this association is the multiplicity of genes found on various different plasmids, and this may also account for their divergence.

Recombination events between homologous genes could be postulated for the 130-kDa crystal protein genes of *B. thuringiensis*. Such homologous recombinations may also have been involved in the evolution of this diversity.

(ii) There are many insects susceptible to the *B. thuringiensis* toxins, including not only lepidoptera but also diptera and coleoptera. It is likely that the mode of action of the different types of toxins is basically the same in the different hosts. This view is reinforced by the fact that one toxin may possess two different pathotypes on lepidopteran and dipteran larvae; an example is the P2 protein.

For a better understanding of crystal toxins it is essential to obtain information about the mechanism of action of the toxin at the molecular level. Areas that require future study include the determination of the nature of the various cellular receptors involved in toxemia and analysis of the relationship between the conserved domains and toxicity, and perhaps also between the variable domains and host specificity. Site-directed mutagenesis of the above-mentioned domains should allow the identification of the polypeptide sequences required both for specificity and toxicity.

These studies will be facilitated by the

genetic tools which are now available in *B. thuringiensis*, allowing the reintroduction of cloned genes into the host. Moreover, by using new transformation techniques such as electroporation and the use of resident *B. thuringiensis* plasmids as cloning vectors, it is now possible to study the expression of the δ-endotoxin genes and their complex mechanism of regulation in *B. thuringiensis*.

Acknowledgments. We are deeply indebted to Armelle Delécluse, Herman Höfte, and Vincent Sanchis for providing information before publication. We thank Alex Edelman for correcting the manuscript. We are grateful to Elisabeth Cracy and Danièlle Lefèvre for typing the manuscript.

The research from our laboratory (directed by Raymond Dedonder) reported in this review was supported by Institut Pasteur, Centre National de la Recherche Scientifique (UA 1300), Institut National de la Recherche Agronomique, Université Paris VII, Fondation pour la Recherche Médicale, and the United Nations Development–World Health Organization Special Programme for Research and Training in Tropical Diseases.

LITERATURE CITED

1. **Adang, M. J., M. J. Staver, T. A. Rocheleau, T. Leighton, R. F. Barker, and D. V. Thompson.** 1985. Characterized full-length and truncated plasmid clones of the crystal protein of *Bacillus thuringiensis* subsp. *kurstaki* HD-73 and their toxicity to *Manduca sexta*. *Gene* **36**:289–300.
2. **Angus, T. A.** 1968. Similarity of effect of valinomycin and *Bacillus thuringiensis* parasporal protein in larvae of *Bombyx mori*. *J. Invertebr. Pathol.* **11**:145–146.
3. **Aronson, A. I., W. Beckman, and P. Dunn.** 1986. *Bacillus thuringiensis* and related insect pathogens. *Microbiol. Rev.* **50**:1–24.
4. **Bernhard, K.** 1986. Studies on the delta-endotoxin of *Bacillus thuringiensis* var. *tenebrionis*. *FEMS Microbiol. Lett.* **33**:261–265.
5. **Betchell, D. B., and L. A. Bulla, Jr.** 1976. Electron microscopic study of sporulation and parasporal crystal formation in *Bacillus thuringiensis*. *J. Bacteriol.* **127**:1472–1481.
6. **Bourgouin, C., A. Delécluse, J. Ribier, A. Klier, and G. Rapoport.** 1988. A *Bacillus thuringiensis* subsp. *israelensis* gene encoding a 125-kilodalton larvicidal polypeptide is associated with inverted repeat sequences. *J. Bacteriol.* **170**:3575–3583.
7. **Bourgouin, C., A. Klier, and G. Rapoport.** 1986. Characterization of the genes encoding the haemolytic toxin and the mosquitocidal delta-endotoxin of *B. thuringiensis israelensis*. *Mol. Gen. Genet.* **205**:390–397.
8. **Brizzard, B. L., and H. R. Whiteley.** 1988. Nucleotide sequence of an additional crystal protein gene cloned from *Bacillus thuringiensis* subsp. *thuringiensis*. *Nucleic Acids Res.* **16**:2723–2724.
9. **Brown, K. L., and H. R. Whiteley.** 1988. Isolation of a *Bacillus thuringiensis* RNA polymerase capable of transcribing crystal protein genes. *Proc. Natl. Acad. Sci. USA* **85**:4166–4170.
10. **Calabrese, D. M., K. W. Nickerson, and L. C. Lane.** 1980. A comparison of protein crystal subunit sizes in *Bacillus thuringiensis*. *Can. J. Microbiol.* **26**:1006–1010.
11. **Charles, J. F., and H. de Barjac.** 1981. Histopathologie de l'action de la δ-endotoxine de *Bacillus thuringiensis* var. *israelensis* sur les larves d'*Aedes aegypti* (Dip.: Culicidae). *Entomophaga* **26**:203–212.
12. **Charles, J. F., and H. de Barjac.** 1982. Sporulation et cristallogénèse de *Bacillus thuringiensis* var. *israelensis* en microscopie électronique. *Ann. Inst. Pasteur Microbiol.* **133A**:425–442.
13. **Charles, J. F., and H. de Barjac.** 1983. Action des cristaux de *Bacillus thuringiensis* var. *israelensis* sur l'intestin moyen des larves d'*Aedes aegypti* L, en microscopie électronique. *Ann. Inst. Pasteur Microbiol.* **134A**:197–218.
14. **Chestukhina, G. G., L. I. Kostina, A. L. Mikhailova, S. A. Tyurin, F. S. Klepikova, and V. M. Stepanov.** 1982. The main features of *Bacillus thuringiensis* δ-endotoxin molecular structure. *Arch. Microbiol.* **132**:159–162.
15. **Chilcott, C. N., and D. J. Ellar.** 1988. Comparative toxicity of *Bacillus thuringiensis* var. *israelensis* crystal proteins *in vivo* and *in vitro*. *J. Gen. Microbiol.* **134**:2551–2558.
16. **Chungjatupornchai, W., H. Höfte, J. Seurinck, C. Angsuthanasombat, and M. Vaeck.** 1988. Common features of *Bacillus thuringiensis* toxins specific for Diptera and Lepidoptera. *Eur. J. Biochem.* **173**:9–16.
17. **de Barjac, H.** 1981. Identification of H-serotypes of *Bacillus thuringiensis*, p. 35–43. *In* H. D. Burges (ed.), *Microbial Control of Pests and Plant Diseases 1970–1980.* Academic Press, Inc. (London), Ltd., London.
18. **Debro, L., P. C. Fitz-James, and A. Aronson.** 1986. Two different parasporal inclusions are produced by *Bacillus thuringiensis* subsp. *finitimus*. *J. Bacteriol.* **165**:258–268.
19. **Delécluse, A., C. Bourgouin, A. Klier, and G. Ra-**

poport. 1988. Specificity of action on mosquito larvae of *Bacillus thuringiensis israelensis* toxins encoded by two different genes. *Mol. Gen. Genet.* **214**:42–47.

20. **Donovan, W. P., C. Dankocsik, and M. P. Gilbert.** 1988. Molecular characterization of a gene encoding a 72-kilodalton mosquito-toxic crystal protein from *Bacillus thuringiensis* subsp. *israelensis. J. Bacteriol.* **170**:4732–4738.

21. **Donovan, W. P., C. C. Dankocsik, M. P. Gilbert, M. C. Gawron-Burke, R. G. Groat, and B. C. Carlton.** 1988. Aminoacid sequence and entomocidal activity of the P2 crystal protein. An insect toxin from *Bacillus thuringiensis* var. *kurstaki. J. Biol. Chem.* **263**:561–567.

22. **Dow, J. A. T.** 1986. Insect midgut function, p. 107–328. *In* P. D. Evans and V. B. Wigglesworth (ed.), *Advances in Insect Physiology.* Academic Press, Inc. (London), Ltd., London.

23. **Drobniewski, F. A., T. J. Sawyer, and D. J. Ellar.** 1986. The mode of action of the mosquitocidal δ-endotoxin of *Bacillus thuringiensis* var. *darmstadiensis,* p. 55–56. *In* P. Falmagne, J. E. Alouf, F. J. Fehrenbach, J. Jeljaszewicz, and M. Thelestam (ed.), *Bacterial Protein Toxins,* Suppl. 15. Gustav Fischer Verlag, Stuttgart.

24. **Fast, P. G.** 1981. The crystal toxin of *Bacillus thuringiensis,* p. 223–248. *In* H. D. Burges (ed.), *Microbial Control of Pests and Plant Diseases 1970–1980.* Academic Press, Inc. (London), Ltd., London.

25. **Geiser, M., S. Schweitzer, and C. Grimm.** 1986. The hypervariable region in the genes coding for entomopathogenic crystal proteins of *Bacillus thuringiensis:* nucleotide sequence of the *kurhd1* gene of subsp. *kurstaki* HD1. *Gene* **48**:109–118.

26. **Glatron, M. F., and G. Rapoport.** 1972. Biosynthesis of the parasporal inclusion of *Bacillus thuringiensis:* half-life of its corresponding messenger RNA. *Biochimie* **54**:1291–1301.

27. **Gonzalez, J. M., Jr., B. J. Brown, and B. C. Carlton.** 1982. Transfer of *Bacillus thuringiensis* plasmids coding for delta-endotoxin among strains of *B. thuringiensis* and *B. cereus. Proc. Natl. Acad. Sci. USA* **79**:6951–6955.

28. **Gonzalez, J. M., Jr., and B. C. Carlton.** 1984. A large transmissible plasmid is required for crystal toxin production in *Bacillus thuringiensis* var. *israelensis. Plasmid* **11**:28–38.

29. **Gonzalez, J. M., Jr., H. T. Dulmage, and B. C. Carlton.** 1981. Correlation between specific plasmids and δ-endotoxin production in *Bacillus thuringiensis. Plasmid* **5**:351–365.

30. **Haider, M. Z., and D. J. Ellar.** 1987. Characterization of the toxicity and cytopathic specificity of a cloned *Bacillus thuringiensis* crystal protein using insect cell culture. *Mol. Microbiol.* **1**:59–66.

31. **Haider, M. Z., and D. J. Ellar.** 1988. An investigation of relationships between structure and insecticidal specificity for a *Bacillus thuringiensis aizawai* delta-endotoxin, p. 53–54. *In* F. J. Fehrenbach, J. E. Alouf, F. Falmagne, W. Goebel, J. Jeljaszewicz, D. Jürgens, and R. Rappuoli (ed.), *Bacterial Protein Toxins,* Suppl. 17. Gustav Fischer Verlag, Stuttgart.

32. **Haider, M. Z., E. S. Ward, and D. J. Ellar.** 1987. Cloning and heterologous expression of an insecticidal delta-endotoxin gene from *Bacillus thuringiensis* var. *aizawai* IC1 toxic to both lepidoptera and diptera. *Gene* **52**:285–290.

33. **Hall, I. M., K. Y. Arakawa, H. T. Dulmage, and J. A. Correa.** 1977. The pathogenicity of strains of *Bacillus thuringiensis* to larvae of *Aedes* and to *Culex* mosquitoes. *Mosq. News* **37**:246–251.

34. **Herrnstadt, C., T. E. Gilroy, D. A. Sobieski, B. D. Bennett, and F. H. Gaertner.** 1987. Nucleotide sequence and deduced aminoacid sequence of a coleopteran-active delta-endotoxin gene from *Bacillus thuringiensis* subsp. *san diego. Gene* **57**:37–46.

35. **Herrnstadt, C., G. G. Soares, E. R. Wilcox, and D. L. Edwards.** 1986. A new strain of *Bacillus thuringiensis* with activity against coleopteran insects. *Bio/Technology* **4**:305–308.

36. **Himeno, M., N. Koyama, T. Funato, and T. Komano.** 1985. Mechanism of action of *Bacillus thuringiensis* insecticidal delta-endotoxin on insect cells *in vitro. Agric. Biol. Chem.* **49**:1461–1468.

37. **Hofmann, C., and P. Lüthy.** 1986. Binding and activity of *Bacillus thuringiensis* delta-endotoxin to invertebrate cells. *Arch. Microbiol.* **146**:7–11.

38. **Hofmann, C., P. Lüthy, R. Hütter, and V. Pliska.** 1988. Binding of the delta-endotoxin from *Bacillus thuringiensis* to brush-border membrane vesicles of the cabbage butterfly (*Pieris brassicae*). *Eur. J. Biochem.* **173**:85–91.

39. **Hofmann, C., H. Vanderbruggen, H. Höfte, J. Van Rie, S. Jansens, and H. Van Mellaert.** 1988. Specificity of *Bacillus thuringiensis* δ-endotoxins is correlated with the presence of high-affinity binding sites in the brush border membrane of target insect midguts. *Proc. Natl. Acad. Sci. USA* **85**:7844–7848.

40. **Höfte, H., W. Chungjatupornchai, J. Van Rie, S. Jansens, and M. Vaeck.** 1988. Molecular organization and applications of *Bacillus thuringiensis* delta-endotoxins, p. 435–440. *In* F. J. Fehrenbach, J. E. Alouf, P. Falmagne, W. Goebel, J. Jeljaszewicz, D. Jürgens, and R. Rappuoli (ed.), *Bacterial Protein Toxins,* Suppl. 17. Gustav Fischer Verlag, Stuttgart.

41. **Höfte, H., H. de Greve, J. Seurinck, S. Jansens, J. Mahillon, C. Ampe, J. Vandekerckhove, H. Vanderbruggen, M. Van Montagu, M. Zabeau, and**

M. Vaeck. 1986. Structural and functional analysis of a cloned delta-endotoxin of *Bacillus thuringiensis berliner* 1715. *Eur. J. Biochem.* **161**:273–280.

42. Höfte, H., J. Seurinck, A. Van Houtven, and M. Vaeck. 1987. Nucleotide sequence of a gene encoding an insecticidal protein of *Bacillus thuringiensis* var. *tenebrionis* toxic against Coleoptera. *Nucleic Acids Res.* **15**:7183.

43. Höfte, H., J. Van Rie, S. Jansens, A. Van Houtven, H. Vanderbruggen, and M. Vaeck. 1988. Monoclonal antibody analysis and insecticidal spectrum of three types of Lepidopteran-specific insecticidal crystal proteins of *Bacillus thuringiensis. Appl. Environ. Microbiol.* **54**:2010–2017.

44. Honée, G., T. van der Salm, and B. Visser. 1988. Nucleotide sequence of crystal protein gene isolated from *B. thuringiensis* subspecies *entomocidus* 60.5 coding for a toxin highly active against *Spodoptera* species. *Nucleic Acids Res.* **16**:6240.

45. Ibarra, J. E., and B. A. Federici. 1986. Isolation of a relatively nontoxic 65-kilodalton protein inclusion from the parasporal body of *Bacillus thuringiensis* subsp. *israelensis. J. Bacteriol.* **165**:527–533.

46. Jarrett, P. 1985. Potency factors in the δ-endotoxin of *Bacillus thuringiensis* var. *aizawai* and the significance of plasmids in their control. *J. Appl. Bacteriol.* **58**:437–448.

47. Kim, K. H., M. Ohba, and K. Aizawa. 1984. Purification of the toxic protein from *B. thuringiensis* serotype 10 isolate demonstrating a preferential larvicidal activity to the mosquito. *J. Invertebr. Pathol.* **44**:214–219.

48. Klier, A., F. Fargette, J. Ribier, and G. Rapoport. 1982. Cloning and expression of the crystal protein genes from *Bacillus thuringiensis* strain *berliner* 1715. *EMBO J.* **1**:791–799.

49. Klier, A., M.-M. Lecadet, and G. Rapoport. 1978. Transcription *in vitro* of sporulation-specific mRNA's by RNA polymerase from *Bacillus thuringiensis*, p. 205–212. *In* G. Chambliss and J. C. Vary (ed.), *Spores VII.* American Society for Microbiology, Washington, D.C.

50. Klier, A., D. Lereclus, J. Ribier, C. Bourgouin, G. Menou, M.-M. Lecadet, and G. Rapoport. 1985. Cloning and expression in *Escherichia coli* of the crystal protein gene from *Bacillus thuringiensis* strain *aizawai* 7-29 and comparison of the structural organization of genes from different serotypes, p. 217–224. *In* J. A. Hoch and P. Setlow (ed.), *Molecular Biology of Microbial Differentiation.* American Society for Microbiology, Washington, D.C.

51. Knowles, B. H., and D. J. Ellar. 1986. Characterization and partial purification of a plasma membrane receptor for *Bacillus thuringiensis* var. *kurstaki* lepidopteran-specific δ-endotoxin. *J. Cell Sci.* **83**:89–101.

52. Knowles, B. H., and D. J. Ellar. 1987. Colloid-osmotic lysis is a general feature of the mechanism of action of *Bacillus thuringiensis* δ-endotoxins with different insect specificity. *Biochim. Biophys. Acta* **924**:509–518.

53. Knowles, B. H., W. E. Thomas, and D. J. Ellar. 1984. Lectin-like binding of *Bacillus thuringiensis* var. *kurstaki* lepidopteran-specific toxin is an initial step in insecticidal action. *FEBS Lett.* **168**:197–202.

54. Krieg, A., A. M. Huger, G. A. Langenbruch, and W. Schnetter. 1983. *Bacillus thuringiensis* var. *tenebrionis*: ein neuer gegenuber Larven von Coleopteren wirksamer Pathotyp. *Z. Angew. Entomol.* **96**:500–508.

55. Kronstad, J. W., H. E. Schnepf, and H. R. Whiteley. 1983. Diversity of locations for *Bacillus thuringiensis* crystal protein genes. *J. Bacteriol.* **154**:419–428.

56. Kronstad, J. W., and H. R. Whiteley. 1984. Inverted repeat sequences flank a *Bacillus thuringiensis* crystal protein gene. *J. Bacteriol.* **160**:95–102.

57. Kronstad, J. W., and H. R. Whiteley. 1986. Three classes of homologous *Bacillus thuringiensis* crystal protein genes. *Gene* **43**:29–40.

58. Lecadet, M.-M. 1970. *B. thuringiensis* toxins. The proteinaceous crystal, p. 437–471. *In* T. C. Montie, S. Kades, and S. J. Ajl (ed.), *Microbial Toxins,* vol. 2. Academic Press, Inc., New York.

59. Lecadet, M.-M., V. Sanchis, G. Menou, P. Rabot, D. Lereclus, J. Chaufaux, and D. Martouret. 1988. Identification of a δ-endotoxin gene product specifically active against *Spodoptera littoralis* Bdv. among proteolysed fractions of the insecticidal crystals of *Bacillus thuringiensis* subsp. *aizawai* 7–29. *Appl. Environ. Microbiol.* **54**:2689–2698.

60. Lereclus, D. 1988. Génétique et biologie moléculaire de *Bacillus thuringiensis. Bull. Inst. Pasteur* **86**:337–371.

61. Lereclus, D., M.-M. Lecadet, J. Ribier, and R. Dedonder. 1982. Molecular relationships among plasmids of *Bacillus thuringiensis*: conserved sequences through 11 crystalliferous strains. *Mol. Gen. Genet.* **186**:391–398.

62. Lereclus, D., J. Mahillon, G. Menou, and M.-M. Lecadet. 1986. Identification of Tn*4430*, a transposon of *Bacillus thuringiensis* functional in *Escherichia coli. Mol. Gen. Genet.* **204**:52–57.

63. Lereclus, D., G. Menou, and M.-M. Lecadet. 1983. Isolation of a DNA sequence related to several plasmids from *Bacillus thuringiensis* after a mating involving the *Streptococcus faecalis* plasmid pAMβ1. *Mol. Gen. Genet.* **191**:307–313.

64. Lereclus, D., J. Ribier, A. Klier, G. Menou, and

M.-M. Lecadet. 1984. A transposon-like structure related to the δ-endotoxin gene of *Bacillus thuringiensis*. *EMBO J.* **3**:2561–2567.

65. Losick, R., P. Youngman, and P. J. Piggot. 1986. Genetics of endospore formation in *Bacillus subtilis*. *Annu. Rev. Genet.* **20**:625–669.

66. Lüthy, P., and H. R. Ebersold. 1981. *Bacillus thuringiensis* delta-endotoxin: histopathology and molecular mode of action, p. 235–267. *In* E. W. Davidson (ed.), *Pathogenesis of Invertebrate Microbial Diseases*. Allanheld, Osmun & Co. Publishers, Inc., Totowa, N.J.

67. Mahillon, J., and D. Lereclus. 1988. Structural and functional analysis of Tn*4430*: identification of an integrase-like protein involved in the cointegrate-resolution process. *EMBO J.* **7**:1515–1526.

68. Mahillon, J., J. Seurinck, J. Delcour, and M. Zabeau. 1987. Cloning and nucleotide sequence of different iso-IS*231* elements and their structural association with the Tn*4430* transposon in *Bacillus thuringiensis*. *Gene* **51**:187–189.

69. Mahillon, J., J. Seurinck, L. Van Rompuy, J. Delcour, and M. Zabeau. 1985. Nucleotide sequence and structural organization of an insertion sequence element (IS*231*) from *B. thuringiensis* strain berliner 1715. *EMBO J.* **4**:3895–3899.

70. McLean, K. M., and H. R. Whiteley. 1987. Expression in *Escherichia coli* of a cloned crystal protein gene of *Bacillus thuringiensis* subsp. *israelensis*. *J. Bacteriol.* **169**:1017–1023.

71. McPherson, S. A., F. J. Perlak, R. L. Fuchs, P. G. Marroe, P. B. Lavrik, and D. A. Fischhoff. 1988. Characterization of the coleopteran-specific protein gene of *Bacillus thuringiensis* var. *tenebrionis*. *Bio/Technology* **6**:61–66.

72. Nagamatsu, Y., Y. Itai, C. Hatanaka, G. Funatsu, and K. Hayashi. 1984. A toxic fragment from the entomocidal crystal protein of *Bacillus thuringiensis*. *Agric. Biol. Chem.* **48**:611–619.

73. Oeda, K., K. Oshie, M. Shimizu, K. Nakamura, H. Yamamoto, I. Nakamaya, and H. Okhawa. 1987. Nucleotide sequence of the insecticidal gene of *Bacillus thuringiensis* strain *aizawai* IPL7 and its high-level expression in *Escherichia coli*. *Gene* **53**:113–119.

74. Ozawa, K., and H. Iwahana. 1986. Involvement of a transmissible plasmid in heat-stable exotoxin and delta-endotoxin production in *Bacillus thuringiensis* subspecies *darmstadiensis*. *Curr. Microbiol.* **13**:337–340.

75. Pao-Intara, M., C. Angsuthanasombat, and S. Panyim. 1988. The mosquito larvicidal activity of 130 kDa delta-endotoxin of *Bacillus thuringiensis* var. *israelensis* resides in the 72 kDa amino-terminal fragment. *Biochem. Biophys. Res. Commun.* **153**:294–300.

76. Percy, J., and P. G. Fast. 1983. *Bacillus thuringiensis* crystal toxin: ultrastructural studies of its effect on silkworm cells. *J. Invertebr. Pathol.* **41**:86–98.

77. Pfannenstiel, M. A., G. A. Couche, E. J. Ross, and K. W. Nickerson. 1986. Immunological relationships among proteins making up the *Bacillus thuringiensis* subsp. *israelensis* crystalline toxin. *Appl. Environ. Microbiol.* **52**:644–649.

78. Ribier, J., and M.-M. Lecadet. 1973. Etude ultrastructurale et cinétique de la sporulation de *B. thuringiensis* Berliner 1715. Remarques sur la formation de l'inclusion parasporale. *Ann. Inst. Pasteur Microbiol.* **124A**:311–344.

79. Ribier, J., and M.-M. Lecadet. 1981. *Bacillus thuringiensis* var. Berliner 1715. Isolement et caractérisation de mutants de sporulation. *C. R. Acad. Sci. Ser. III Sci. Vie* **292**:803–808.

80. Sacchi, V. F., P. Parenti, G. M. Hanozet, B. Giordana, P. Lüthy, and M. Wolfersberger. 1986. *Bacillus thuringiensis* toxin inhibits K⁺-gradient-dependent amino acid transport across the brush border membrane of *Pieris brassicae* midgut cells. *FEBS Lett.* **204**:213–218.

81. Sanchis, V., D. Lereclus, G. Menou, J. Chaufaux, S. Guo, and M.-M. Lecadet. 1989. Nucleotide sequence and analysis of the N-terminal coding region of the *Spodoptera* active delta-endotoxin gene of *Bacillus thuringiensis aizawai* 7.29. *Mol. Microbiol.* **3**:229–238.

82. Sanchis, V., D. Lereclus, G. Menou, J. Chaufaux, and M.-M. Lecadet. 1988. Multiplicity of δ-endotoxin genes with different insecticidal specificities in *B. thuringiensis aizawai* 7.29. *Mol. Microbiol.* **2**:393–404.

83. Sanchis, V., D. Lereclus, J. Ribier, G. Menou, D. Martouret, and M.-M. Lecadet. 1988. Cloning of several δ-endotoxin genes from *B. thuringiensis* strains active against *Spodoptera littoralis*, p. 253–254. *In* F. J. Fehrenbach, J. E. Alouf, P. Falmagne, W. Goebel, J. Jeljaszewicz, D. Jürgens, and R. Rappuoli (ed.), *Bacterial Protein Toxins*, Suppl. 17. Gustav Fischer Verlag, Stuttgart.

84. Schnepf, H. E., and H. R. Whiteley. 1981. Cloning and expression of the *Bacillus thuringiensis* crystal protein gene in *Escherichia coli*. *Proc. Natl. Acad. Sci. USA* **78**:2893–2897.

85. Schnepf, H. E., and H. R. Whiteley. 1985. Delineation of a toxin-encoding segment of a *Bacillus thuringiensis* crystal protein gene. *J. Biol. Chem.* **260**:6273–6280.

86. Schnepf, H. E., H. C. Wong, and H. R. Whiteley. 1985. The amino acid sequence of a crystal protein from *Bacillus thuringiensis* deduced from the DNA base sequence. *J. Biol. Chem.* **260**:6264–6272.

87. Schnepf, H. E., H. C. Wong, and H. R. Whiteley. 1987. Expression of a cloned *Bacillus thuringiensis*

crystal protein gene in *Escherichia coli*. *J. Bacteriol.* **169**:4110–4118.

88. Sekar, V., D. V. Thompson, M. J. Maroney, R. G. Bookland, and M. J. Adang. 1987. Molecular cloning and characterization of the insecticidal crystal protein gene of *Bacillus thuringiensis* var. *tenebrionis*. *Proc. Natl. Acad. Sci. USA* **84**:7036–7040.

89. Sen, K., G. Honda, N. Koyama, M. Nishida, A. Neki, H. Sakai, M. Himeno, and T. Komano. 1988. Cloning and nucleotide sequences of the two 130 kDa insecticidal protein genes of *Bacillus thuringiensis* var. *israelensis*. *Agric. Biol. Chem.* **52**:873–878.

90. Shibano, Y., A. Yamagata, N. Nakamura, T. Iizuka, H. Sugisaki, and M. Takanami. 1985. Nucleotide sequence coding for the insecticidal fragment of the *Bacillus thuringiensis* crystal protein. *Gene* **34**:243–251.

91. Thomas, W. E., and D. J. Ellar. 1983. Mechanism of action of *Bacillus thuringiensis* var. *israelensis* insecticidal δ-endotoxin. *FEBS Lett.* **154**:362–368.

92. Thorne, L., F. Garduno, T. Thompson, D. Decker, M. Zounes, M. Wild, A. M. Walfield, and T. J. Pollock. 1986. Structural similarity between the Lepidoptera- and Diptera-specific insecticidal endotoxin genes of *B. thuringiensis* subsp. "*kurstaki*" and "*israelensis*." *J. Bacteriol.* **166**:801–811.

93. Tungpradubkhul, S., C. Settasatien, and S. Panyim. 1988. The complete nucleotide sequence of a 130 kDa mosquito-larvicidal delta-endotoxin gene of *Bacillus thuringiensis* var. *israelensis*. *Nucleic Acids Res.* **16**:1637–1638.

94. Visser, B., T. van der Salm, W. van den Brink, and G. Folkers. 1988. Genes from *Bacillus thuringiensis entomocidus* 60.5 coding for insect-specific crystal proteins. *Mol. Gen. Genet.* **212**:219–224.

95. Waalwijk, C., A. M. Dullemans, M. E. S. Van Workum, and B. Visser. 1985. Molecular cloning and the nucleotide sequence of the Mr 28000 crystal protein gene of *Bacillus thuringiensis* subsp. *israelensis*. *Nucleic Acids Res.* **13**:8207–8217.

96. Wabiko, H., K. C. Raymond, and L. A. Bulla, Jr. 1986. *Bacillus thuringiensis* entomocidal protoxin gene sequence and gene product analysis. *DNA* **5**:305–314.

97. Ward, E. S., D. J. Earp, B. H. Knowles, S. M. Pinnavaia, C. N. Nicholls, T. J. Sawyer, P. E. Granum, and D. J. Ellar. 1988. Diversity of structure and specificity of mosquitocidal δ-endotoxins, p. 67–68. *In* F. J. Fehrenbach, J. E. Alouf, P. Falmagne, W. Goebel, J. Jeljaszewicz, D. Jürgens, and R. Rappuoli (ed.), *Bacterial Protein Toxins*, Suppl. 17. Gustav Fischer Verlag, Stuttgart.

98. Ward, E. S., and D. J. Ellar. 1986. *Bacillus thu-*

ringiensis var. *israelensis* δ-endotoxin. Nucleotide sequence and characterization of the transcripts in *Bacillus thuringiensis* and *Escherichia coli*. *J. Mol. Biol.* **191**:1–11.

99. Ward, E. S., and D. J. Ellar. 1987. Nucleotide sequence of a *Bacillus thuringiensis* var. *israelensis* gene encoding a 130 kDa delta-endotoxin. *Nucleic Acids Res.* **15**:7195.

100. Ward, E. S., and D. J. Ellar. 1988. Cloning and expression of two homologous genes of *Bacillus thuringiensis* subsp. *israelensis* which encode 130-kilodalton mosquitocidal proteins. *J. Bacteriol.* **170**:727–735.

101. Ward, E. S., D. J. Ellar, and J. A. Todd. 1984. Cloning and expression in *Escherichia coli* of the insecticidal δ-endotoxin gene of *Bacillus thuringiensis* var. *israelensis*. *FEBS Lett.* **175**:377–382.

102. Whiteley, H. R., and H. E. Schnepf. 1986. The molecular biology of parasporal crystal body formation in *Bacillus thuringiensis*. *Annu. Rev. Microbiol.* **40**:549–576.

103. Whiteley, H. R., W. R. Widner, and H. E. Schnepf. 1988. Genes encoding peptides of the small cuboidal crystals of *Bacillus thuringiensis*, p. 239–244. *In* A. T. Ganesan and J. A. Hoch (ed.), *Genetics and Biotechnology of Bacilli*, vol. 2. Academic Press, Inc., San Diego, Calif.

104. Wilcox, D. R., A. G. Shivakumar, B. E. Melin, F. Mahlon, M.-F. Miller, T. A. Benson, C. W. Schopp, D. Casuto, G. J. Gundling, T. J. Bolling, B. B. Spear, and J. L. Fox. 1986. Genetic engineering of bioinsecticides, p. 395–413. *In* M. Inouye and R. Sarma (ed.), *Protein Engineering: Applications in Science, Medicine and Industry*. Academic Press, Inc., New York.

105. Wolfersberger, M. G., C. Hofmann, and P. Lüthy. 1986. Interaction of *Bacillus thuringiensis* delta-endotoxin with membrane vesicles isolated from lepidopteran larval midgut, p. 237–238. *In* P. Falmagne, J. E. Alouf, F. J. Fehrenbach, J. Jeljaszewicz, and M. Thelestam (ed.), *Bacterial Protein Toxins*, Suppl. 15. Gustav Fischer Verlag, Stuttgart.

106. Wong, H. C., and S. Chang. 1986. Identification of a positive retroregulator that stabilizes mRNAs in bacteria. *Proc. Natl. Acad. Sci. USA* **83**:3233–3237.

107. Wong, H. C., H. E. Schnepf, and H. R. Whiteley. 1983. Transcriptional and translational start sites for the *Bacillus thuringiensis* crystal protein gene. *J. Biol. Chem.* **258**:1960–1967.

108. Wu, D., and F. N. Chang. 1985. Synergism in mosquitocidal activity of 26 and 65 kDa proteins from *Bacillus thuringiensis* subsp. *israelensis* crystal. *FEBS Lett.* **190**:232–236.

109. Yamamoto, T., A. Ehmann, J. M. Gonzalez, Jr., and B. C. Carlton. 1988. Expression of three genes

coding for 135 kilodalton entomocidal proteins in *Bacillus thuringiensis kurstaki. Curr. Microbiol.* **17:**5–12.

110. **Yamamoto, T., and R. E. McLaughlin.** 1981. Isolation of a protein from the parasporal crystal of *Bacillus thuringiensis* var. *kurstaki* toxic to the mosquito larvae *Aedes taeniorhynchus. Biochem.* *Biophys. Res. Commun.* **103:**414–421.

111. **Yamamoto, T., I. A. Watkinson, L. Kim, M. V. Sage, R. Stratton, N. Akande, Y. Li, D. P. Ma, and B. A. Roe.** 1988. Nucleotide sequence of the gene coding for a 130-kDa mosquitocidal protein of *Bacillus thuringiensis israelensis. Gene* **66:**107–120.

Regulation of Procaryotic Development
Edited by Issar Smith, Ralph A. Slepecky, and Peter Setlow
© 1989 American Society for Microbiology, Washington, DC 20006

Chapter 14

Sporulation in *Streptomyces*

Keith F. Chater

INTRODUCTION

The last progenitor shared by *Escherichia coli* and *Streptomyces* species probably died several billion years ago, when the lines leading to gram-negative and gram-positive organisms diverged, and streptomycetes, which represent the division of gram-positive organisms with high G + C in their DNA (ca. 73% for streptomycetes), have probably evolved separately from *Bacillus subtilis* (at 43%, typifying the division with low G + C in their DNA) for more than 700 million years (64, 84). The evolutionary separation of these gram-positive soil bacteria is presumably ultimately bound up with their ecology and morphology. *B. subtilis*, a motile rod growing by binary fission, is probably adapted to allow opportunistic colonization of relatively wet niches. Its sporulation is perhaps more an aid to, than an obligatory part of, dispersal. On the other hand, *Streptomyces* species are able to colonize relatively dry insoluble organic debris in soil and have adapted to this niche by evolving a large repertoire of extracellular hydrolytic enzymes, combined with a mycelial growth habit. Mycelial growth obligatorily necessitates a dispersive phase, which is manifested in *Streptomyces* species, as in filamentous eucaryotic molds, by the growth of specialized spore-bearing aerial branches (Fig. 1). Such multicellular differentiation is unusual among procaryotes

and therefore worthy of study. It has recently become amenable to molecular analysis owing to the development of systems for DNA cloning in *Streptomyces* species (31), with the result that important advances have been made since the last extensive review of *Streptomyces* differentiation (11). For the purposes of this article, the term sporulation is taken to encompass all of the events leading up to the formation of mature spores and therefore includes a discussion of the initiation and growth of aerial hyphae.

General Account of Mycelial Growth and Differentiation in *Streptomyces*

Recent studies have largely confirmed results reviewed previously (11) that cell wall synthesis is most active at hyphal tips (57, 67). However, this reflects only about 30% of the total wall synthetic activity, the remainder being dispersed along the length of the hyphae (57). In the absence of significant amounts of branching, hyphae growing from recently germinated spores of *Streptomyces antibioticus* show a periodic doubling of the hyphal extension rate and of rates of incorporation of cell wall and DNA precursors, at ca. 40-min intervals (57), even though the number of tips is almost constant. In older hyphae (of *Streptomyces coelicolor*), recently incorporated [^{3}H]thymidine, although show-

Keith F. Chater • John Innes Institute and AFRC Institute of Plant Science Research, Colney Lane, Norwich NR4 7UH, United Kingdom.

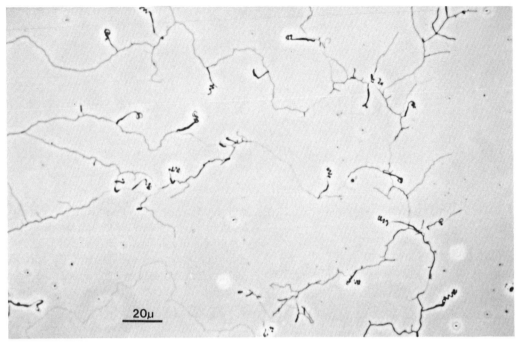

Figure 1. Mycelial growth and development of *S. coelicolor*. This phase-contrast photomicrograph shows branching substrate mycelium at the edge of a nutrient-limited culture, supporting the growth of individual phase-dark aerial hyphae. The tightly coiled tips of the aerial hyphae would normally form spore chains. In this micrograph, a *whiI* mutation (Table 2) has interrupted this process, preserving the aerial hyphae in the coiled stage. (Reproduced from reference 9 with permission.)

ing a transient partial association with hyphal tips, is rapidly dispersed throughout the hyphae (67).

Putting together these results (which were obtained with hyphae of different ages and species and under different growth conditions), it seems possible that wall growth zones are associated with replicating chromosomes and that these zones may be relatively close together near hyphal tips, where each chromosome might possess multiple replication forks (especially in germlings). On the other hand, there is some suggestion in older hyphae of a dislocation of DNA synthesis and cell wall growth. This raises questions about the localization of biosynthesis of the membrane that intervenes between the cell wall and the DNA. In view of the probable coupling of the DNA replication apparatus to the membrane, at least in *E. coli* (86), it may be that membrane extension is not obligatorily coupled spatially to cell wall extension.

Septa are not abundant in vegetative hyphae and never occur near growth tips. Thus, many tens of chromosomes occupy most hyphal compartments. The abundance and location of cross walls are closely, but not absolutely, related to the position of branches (which are also seldom found closer than 60 μm behind growing tips). The branch nearest a hyphal tip is often not associated with a cross wall (67), implying that the formation of cross walls is not a prerequisite for branching, but probably often occurs after a branch has formed.

Clearly, the questions raised by these observations demand that a fuller, genetically based investigation of the normal cell cycle in all its aspects should be made. Without knowledge of normal DNA replication, cell wall and membrane synthesis, septation, and branching, one cannot expect to understand fully the developmental biology of the aerial mycelium. A similar dearth of information about the normal cell cycle of *B. subtilis* is rapidly becoming a major limitation in studies of endospore formation, where understanding the differences between normal and asymmetrical septation has become crucial in the analysis of developing switching.

In response (it is usually supposed) to nutrient limitation, ageing colonies (perhaps 48 h old) typically show three kinds of change: electron-transparent granules, some of which in at least one species appear to contain glycogen, are produced in the cytoplasm (5); aerial hyphae grow; and antibiotic and pigment production becomes apparent. The aerial hyphae grow away from the substrate to a considerable distance (sometimes for more than 100 times their diameter), posing questions about the supply of nutrients to the tips (assuming that most growth is at the tip) and the maintenance of turgor in cells removed from an aqueous, solute-rich medium. The growing aerial hyphae do not usually contain observable "storage granules," but at the next stage of development—metamorphosis of hyphae into chains of spores—granules reappear, finally to disappear during spore maturation (4, 5, 82).

Sporulation of an aerial hypha involves the synchronous and regularly spaced ingrowth of specialized sporulation septa, coupled with orderly segregation of the population of chromosomes present in the hyphal tip into compartments each containing a single genome. These compartments then undergo changes in shape and thickening of their walls to give spore chains (26, 53, 83). Aerial hyphae and spores have hydrophobic surfaces, which may function both to avoid desiccation and to allow the ready dispersal of spores on water droplets (see reference 11 for references).

This developmental sequence raises numerous questions. What metabolic signal(s) activates genetic events that lead to differentiation? Is there communication among hyphae within a mycelium, or is developmental regulation autonomous within a hyphal compartment? Do morphological differentiation and the various facets of physiological differentiation (such as storage granule formation and antibiotic synthesis) depend on the same trigger(s)? Why do aerial hyphae grow away from the substrate? What controls the timing and localization of sporulation septation? Some of these questions are addressed in the remainder of this article.

Physiological and Genetic Approaches to the Study of *Streptomyces* Sporulation Are Finding Common Ground

Sporulation can take place in submerged culture in some *Streptomyces* species and under some conditions, providing greater possibilities for physiological analysis. The most well-known examples are in various *Streptomyces griseus* strains, in some of which suitable nutritional downshifts can give rise to moderately well-synchronized sporulation (42). Recently, work in Leon (J. A. Gil, personal communication) and at Harvard University (R. Santamaria, personal communication) has led to the development of conditions that give rise to submerged sporulation in many species, including *S. coelicolor* A3(2), the main strain for genetics. Thus, it is possible that the long-standing gap between physiologists and geneticists working on *Streptomyces* species, caused by the different technical possibilities of different organisms, may be bridged. This gap is also closing as a result of recombinant DNA technology: effective cloning is now possible in many species, including *S. griseus* (2), and its application to comparative studies means that—at long last—it is becoming possible to make wide generalizations about the genetics and physiology of sporulation.

Despite these advances in submerged sporulation, it has to be said that one of the most interesting aspects of *Streptomyces* development is the phenomenon of growth into the air, a condition that is not amenable to study in submerged culture and is therefore very difficult to analyze physiologically. The use of genetics to facilitate the localization of gene products in situ should be valuable in this respect. The principal tools for this purpose will presumably be immunolocalization (often using antisera obtained with recourse to information or materials arising from molecular genetics), hybridization of labeled probes to specific nucleic acid sequences, and fusions of developmentally regulated promoters to readily detected reporter genes. As yet, only the last of these has been employed, in the exploitation of the *luxA* and *luxB* genes that normally determine light emission by the marine organism *Vibrio harveyi*. These genes, which specify luciferase, have been incorporated into *Streptomyces* plasmid vectors in such a way that their transcription depends on the insertion of a promoter into an upstream restriction site polylinker (68–70). Transcription and translation lead to a potential for light emission that is made real by exposure to tiny amounts of vapor from a long-chain aldehyde, *n*-decanal. Initial results have shown that different promoters can direct light emission from different parts of colonies and in some cases at very specific times (68). Most encouraging is the discovery of promoters which cause only the aerial mycelium to luminesce, proving that the externally added undecanal substrate and the internally generated reduced flavin mononucleotide cofactor can both gain access to this isolated tissue in quantities sufficient to permit detectable light emission. Exposure to *n*-decanal has no apparent effect on growth or differentiation, so that colonies can subsequently be reexamined or subcultured. Moreover, the luciferase enzyme is not highly stable in hyphae, at least over 24-h periods, reducing the problem of distinguishing between recent and less recent synthesis in accounting for luminescence at any given time point. For many laboratories the main difficulty in applying *lux* technology may be access to suitable apparatus: sensitive light detection systems coupled to very low envi-

ronmental background levels of light are necessary for many applications. (A cautionary note about the use of *lux* genes and other reporter genes in *Streptomyces* species is given below [The *bldA* Gene of *S. coelicolor* Specifies a tRNA for a Rare Codon].)

GROWTH OF THE AERIAL MYCELIUM

What Metabolic Signal Initiates Differentiation in *Streptomyces* Species?

In *B. subtilis* there is compelling evidence that a fall in the guanine nucleotide pool, caused by starvation for suitable nutrients, is a crucial trigger for sporulation (23). It is not clear how this signal is perceived by the cell. In several *Streptomyces* species a role of the guanine nucleotide pool levels in triggering sporulation has also been suggested from four kinds of experiments.

(i) In surface cultures of *Streptomyces griseoflavus*, GTP levels fall much more sharply than do ATP, CTP, or UTP levels at the onset of aerial mycelium formation, while ppGpp levels increase (63).

(ii) In an *S. griseus* strain (13189) which is capable of extensive sporulation in submerged culture in response to nitrogen limitation (see above), downshift is rapidly followed by a specific sharp fall in the GTP pool and a rise in ppGpp (60).

(iii) Submerged sporulation of *S. griseus* 13189 (60, 62) and aerial mycelium formation by *Streptomyces* sp. strain MA406-A-1 (58) and *S. griseus* 13189 (62) are stimulated by decoyinine, an inhibitor of GMP synthetase.

(iv) Thiopeptin-resistant mutants of *Streptomyces* sp. strain MA406-A-1 (59), *S. griseus* 13189 (60, 62), *S. antibioticus* (61), and *S. griseoflavus* (63), which display a phenotype like that of *rel* mutants of *E. coli* in failing properly to reduce RNA synthesis or to accumulate ppGpp in response to amino acid limitation, exhibit marked delays in the onset of aerial mycelium formation. Decoyinine induces efficient sporulation even of a putative *rel* mutant of *S. griseus* which does not accumulate very much ppGpp after decoyinine treatment. This indicates that a reduced GTP pool may be more directly involved in initiating sporulation than an increased pool of ppGpp. Indirect effects of ppGpp levels on the GTP pool may be expected, since ppGpp inhibits the enzyme IMP dehydrogenase, which generates GMP.

Recent experiments (20) on submerged culture in *S. griseus* NRRL 2682, which suggest that a clock mechanism, rather than nutrient limitation, triggers sporulation, seem to complicate the picture. However, it would be of some value to make more direct comparisons of the strains and conditions that were used to obtain the apparently contrasting results of the Ensign and Ochi laboratories.

There have been numerous reports of the frequent spontaneous occurrence, in different strains, of arginine auxotrophs which show aerial mycelium deficiencies variously suppressible by supplementation with arginine or citrulline (see reference 11 for a review). Ochi (63) presented some physiological data suggesting that GTP pool levels are correlated with the suppressive effect. However, it is important to note the difficulty of properly attributing causality in these physiological experiments. For example, the convincing identification of the crucial importance of the GTP pool size for *B. subtilis* sporulation involved the study of large numbers of mutants deficient in different aspects of primary and intermediary metabolism (23).

In Some *Streptomyces* Species Endogenously Generated Extracellular Factors May Stimulate Sporulation of Hyphal Populations

It is likely that some of the taxonomic and genetic diversity of streptomycetes arises from, or may be reflected in, different responses to environmental circumstances. In sporulation, for example, the relative sur-

vival advantages of rapid sporulation of a whole population of hyphae as an early response to nutritional stress, compared with piecemeal sporulation responses by individual hyphae to starvation, would be very dependent on detailed microecological factors. Thus, it should not be surprising that some streptomycetes—notably, *S. griseus* spp.—have evolved mechanisms for population responses, whereas in many others such mechanisms have not been detected. To give a flavor of this phenomenology, I here briefly review three different examples that have received sustained analysis.

A-factor

In most *S. griseus* strains, sporulation (and indeed production of streptomycin) depends on the presence of a diffusible extracellular molecule, A-factor (2S-isocapryloyl-3S-hydroxymethyl-γ-butyrolactone) (43). This discovery was initially made possible by the occurrence of mutants that could sporulate only when stimulated by a product of a sporulating culture (44). It has now become clear that there is a large family of A-factor-like compounds, different members of which are produced by different streptomycetes (19). The experiments of Hara, Horinouchi, and Beppu and their co-workers have led to the following overall picture of A-factor genetics.

(i) A-factor synthesis from intermediary metabolites (possibly from a glycerol derivative and a 10-carbon β-keto acid) probably involves a single gene, *afsA*, which has been cloned from *Streptomyces bikiniensis* (35, 36) and sequenced, revealing an open reading frame (ORF) for a 32.6-kilodalton protein with no obvious similarity to proteins present in major data bases (37a). There is no information about how the chemical heterogeneity of the A-factor family is biochemically and genetically determined.

(ii) A-factor-negative mutants of *S. griseus* and *S. bikiniensis* (which occur at a rather high frequency at least in some strains) are usually deleted for *afsA*, perhaps because of the loss of a plasmid carrying the gene (25, 33).

(iii) In *S. coelicolor*, which produces an A-factor-like compound (25, 43) but does not seem to require it for sporulation, the *afsA* gene for A-factor biosynthesis has been identified genetically but not cloned. However, another gene, *afsB*, is required for A-factor synthesis (25). On the basis of the phenotype of *afsB* mutants and DNA sequence analysis (37), it is proposed that *afsB* specifies a transcriptional activator needed for the synthesis of several secondary metabolites, including A-factor. In the simplest case the *afsB* dependence of A-factor synthesis could reflect the activation of *afsA* transcription. It is interesting that *afsB* itself may well be regulated by an adjacent gene, *afsC* (33, 34), suggesting a complex regulatory cascade for secondary metabolism (see also references 13, 29, and 30). It is not clear whether *afsB* and *afsC* homologs are present in *S. griseus* nor, therefore, whether they also influence sporulation in that species.

Factor C

In another *S. griseus* strain (52-1), experiments analogous to those leading to the discovery of A-factor led to the identification of an extracellular 34-kilodalton protein, factor C, capable of restoring sporulation to a nonsporulating mutant (3). Factor C is widespread among streptomycetes. There are as yet no genetic studies of factor C, but there are some physiological and immunological results concerning its effects and localization. The availability of a factor C-specific monoclonal antibody (76) should now make it possible to clone the factor C gene.

Pamamycin

Spontaneous aerial mycelium-deficient mutants of *Streptomyces ambofaciens* which are phenotypically suppressed by a diffusible substance, pamamycin, produced by the par-

ent strain have been described (51). Pamamycin is the only clear example of a compound that plays a developmental role in the producing organism and also has antibiotic activity. There are several related pamamycins (51), but the structure has been determined for only one—pamamycin-607—which is an anion-transfer agent (47).

A General Survey of Mutants of *S. coelicolor* A3(2) and *S. griseus* NRRL 2682 Deficient in Aerial Mycelium Formation Suggests that Some Functional Redundancy Operates During the Formation of Aerial Hyphae

Unfortunately, many of the mutants described in the previous section have been isolated in strains known and studied for little other than their properties as antibiotic producers. In two cases, however [*S. coelicolor* A3(2) and *S. griseus* NRRL 2682], mutants defective in aerial mycelium formation (*bld*, for bald, mutants) have been isolated and studied in a broader background of physiological and genetic analysis. Table 1 summarizes information currently available about these various *bld* mutants.

For *S. coelicolor* at least seven classes of *bld* genes have been described. For three of these (*bldC*, *bldD*, and *bldF*), there is only a single mutation known, suggesting that other *bld* genes remain to be discovered. Of all of the mutations, only that in *bldC* is not overtly pleiotropic (the remainder all cause defects also in antibiotic synthesis). This paucity of antibiotic production-proficient *bld* mutants suggests either that very few gene products are involved as essential regulatory or structural components specific to the morphological development of aerial mycelium or that some of these functions can be supplied by more than one gene product, so that more than one mutation would be needed to give a Bld phenotype. A precedent for this situation is found in *B. subtilis*: although the main regulatory cascade that determines the se-

Table 1
Genes Involved in Aerial Mycelium Formation in *S. coelicolor* A3(2) and *S. griseus* NRRL B-2682

Gene	Phenotype of mutant
S. coelicolor	
bldA,[a] *bldD,*[a] *bldG*[b]	Wettable, soft (fragmented) colony surface on minimal medium not containing glucose (but normal hydrophobic aerial mycelium on mannitol); no synthesis of antibiotics under any conditions; *bldA* cloned,[c] specifies rare tRNA[d]
bldB[a]	Hard nonfragmenting colonies; weak aerial mycelium and antibiotic production develop only after prolonged incubation; no effect of carbon source; *bldB* cloned,[e] not all *bldB* mutants complemented by cloned fragment
bldH[b]	Hard, fragmenting, antibiotic-nonproducing bald colonies on glucose; both morphological and antibiotic defects corrected on mannitol
bldC[a]	Smooth bald colonies, no fragmentation; no deficiencies in antibiotic production; no effect of carbon source; sporulation restored by diffusible factor from sporulating cultures[a,f]
bldE[f]	Mutants isolated by selection for ability to use agar as sole carbon source in the presence of homoserine; bald and red on glucose; sporulation on mannitol; map location not determined
bldF[g]	Smooth bald colonies; no production of actinorhodin, methylenomycin, or calcium-dependent antibiotic; undecylprodigiosin produced abundantly; no carbon source effect; map location not determined
sapA,[b] *sapC,*[i] *sapD*[i]	Mutant phenotypes not studied; genes cloned by reverse genetics; all encode spore (aerial mycelium?)-associated proteins
S. griseus	
bldI[j]	No sporulation under any growth condition (3 mutations, not tested for possible allelic relationships)
bldII[j]	No sporulation induction by changes in growth conditions, but stimulated to sporulate by nearby sporulating colonies; resemble classical A-factor-deficient mutants but retain streptomycin resistance (5 mutations, not tested for allelic relationships); where tested, sporulation restored by plasmid vector pIJ702 without additional inserts
bldIII[j]	No sporulation on sporulation medium, but sporulation induced on glucose-ammonia minimal agar or in phosphate-limited submerged culture; red pigment deficiency (of 13 mutations, 4 were tested for complementation with cloned fragment, and all 4 gave positive result; therefore, *bldIII* mutations are probably alleles of the same locus)
bldIV[j]	One representative; differs from *bldIII* mutants only in making red pigment; complemented by same cloned fragment

[a] Reference 56.
[b] Reference 8.
[c] Reference 66.
[d] Reference 48.
[e] Reference 65.
[f] D. A. Hodgson, Ph.D. thesis, University of East Anglia, Norwich, United Kingdom, 1980.
[g] A.-M. Puglia, personal communication.
[h] Reference 24.
[i] Santamaria et al., personal communication.
[j] Reference 2.

quential stage- and compartment-specific activation of sporulation genes involves many nonredundant gene functions, partial functional redundancy has been observed for many spore-associated structural proteins (49). A route to the isolation of further genes (potentially including redundant ones) involved in aerial mycelium development is provided by the use of *lux* fusion vectors: for example, Schauer et al. (68) described a promoter (cloned in plasmid pS355) that activated luminescence specifically at the colony surface when aerial hyphae were about to emerge.

A survey of *bld* mutants in *S. griseus*

NRRL B-2682 (2) led to the recognition of four phenotypic classes: class I mutants (3 representatives) at least superficially resemble *bldB* mutants of *S. coelicolor*; class II mutants (5 representatives) are capable of sporulating when grown near the wild type or class III or IV mutants and thus superficially resemble A-factor-nonproducing (*afsA*) mutants of other *S. griseus* strains (but note that the *afsA* mutants, unlike the class II *bld* mutants, have lost streptomycin resistance); class III mutants (12 representatives) are bald only on complex medium, and at least 4 of them are complemented by a particular cloned DNA fragment; and a single class IV mutant resembles class III mutants (and is complemented by the same DNA fragment) except that it retains the wild-type pigmentation. It is not clear from published results whether the *S. griseus bld* mutants have deficiencies in secondary metabolism.

It is notable that most (although not all) of these *bld* mutants in both species can be induced to sporulate normally by growth on certain carbon sources. Typically, growth on glucose gives the Bld phenotype, and growth on mannitol gives sporulation (with glucose repressing sporulation even in the presence of mannitol). Again one sees suggestions that alternative genes, but with different regulation, can apparently fulfill functionally homologous roles. Whether these functions are enzymatic, structural, or regulatory is largely unknown.

The *bldA* Gene of *S. coelicolor* Specifies a tRNA for a Rare Codon

The *bldA* gene was the first *Streptomyces* morphological gene to be cloned (66). A library of *S. coelicolor* DNA was constructed in a bacteriophage φC31-based vector, and the resulting phage plaques were introduced by replica plating into a *bldA* mutant, giving several phages that complemented the *bldA* mutation. Subcloning analysis narrowed the region carrying *bldA* down to an 850-base-pair segment which could comple-

ment all *bldA* mutants tested and which was then sequenced from the wild type and from various mutant alleles. Surprisingly, it turned out that the *bldA* gene specifies a tRNA. This tRNA would translate the leucine codon UUA, which is very rare in *Streptomyces* genes (see below). The evidence for these statements is as follows.

(i) Each of five independent *bldA* mutant alleles contained a single base change within a DNA segment whose transcription product can be folded into a tRNA-like structure (Fig. 2a) (48).

(ii) The putative tRNA would have the anticodon 5'-UAA-3' (assuming no modifications), hence corresponding to the UUA codon, and would share nearly all of the conserved features of leucyl-tRNAs from other organisms (Fig. 2b) (13, 48).

(iii) Transcripts corresponding to the tRNA were observed in the salt-soluble fraction of samples of RNA from *S. coelicolor* cultures, as expected of tRNA-sized molecules, but no complementary transcripts could be detected (48).

(iv) Genes containing TTA codons are specifically impaired in expression in *bldA* mutant hosts (13; B. Leskiw, unpublished data).

The TTA Codon Is Probably Confined to a Very Small Number of Genes

Among the known *bldA* mutations, one—*bldA39*—would change the anticodon of tRNA$_{TTA}^{Leu}$ from 5'-UAA-3' to 5'-UGA-3', which would recognize the UCA codon for serine and would clearly prevent the reading of UUA codons by the mutant tRNA (48). Cloned genes containing TTA codons are not expressed in the *bldA39* mutant (with a partial exception discussed in reference 13), irrespective of the carbon source, so there appears to be no alternative way in which UUA codons can normally be read. Nevertheless, many (uncharacterized) genes, for all of the normal growth functions, are expressed in this mutant, since it is fully viable. It follows

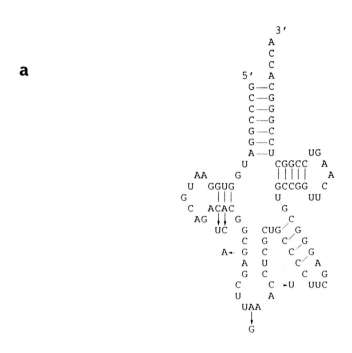

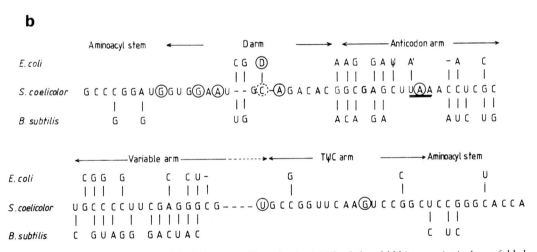

Figure 2. tRNA-like gene product of the *bldA* gene of *S. coelicolor*. (a) The deduced *bldA* transcript is shown folded into a tRNA-like "cloverleaf" form. Five point mutations that give a *bldA* mutant phenotype are shown (from reference 48). (b) Comparison of the deduced *bldA* gene product with equivalent tRNAs from *E. coli* (directly determined [85]) and *B. subtilis* (deduced from DNA sequencing [80]). Spacing follows the numbering of residues used by McClain and Nicholas (52). A dash indicates the absence of a residue from this position. For the *E. coli* and *B. subtilis* tRNAs, only those positions different from the *bldA* gene product are shown. Rings indicate bases particularly characteristic of leucyl-tRNAs of coliform bacteria (52), and the dotted ring indicates a departure from this pattern. Residue A′ is not completely characterized. (From reference 13.)

that no genes essential for the normal growth of *S. coelicolor* contain TTA codons. Similar arguments allow one to conclude that genes in plasmids SCP1, SCP2, and pIJ101 and in phage φC31 which are essential for their replication and maintenance do not contain TTA (a conclusion borne out by sequence analysis for pIJ101, in which the only TTA codon in 11 ORFs is in a dispensable region of the plasmid [41], and part of φC31 [72]). Since *bldA* mutants can sporulate with mannitol as the carbon source, the essential sporulation genes are also free from TTA codons. Finally, there seem to be no TTA codons in the sets of genes for the biosynthesis of at least two of the antibiotics (actinorhodin and undecylprodigiosin) whose production is eliminated by *bldA* mutations, since the production of each of these pigmented antibiotics can be activated in *bldA* mutants by cloned, putative regulatory segments of the relevant gene cluster (A. M. Puglia and E. P. Guthrie, personal communications) or, for undecylprodigiosin, by mutation (E. P. Guthrie, personal communication). The TTA codon has not been found among the sequences of 19 *S. coelicolor* or *S. lividans* genes available at the time of this writing. It has, however, been found in a few genes from other *Streptomyces* species, but with one exception these genes are connected with antibiotic resistance (*hyg* [87], *sph* [78], and *carB* [21]) or with regulation of antibiotic production (*strR* [17]). (The location of the TTA codon in both *hyg* and *sph* is upstream from a potential second translation start site, giving rise to speculation that the use of this start site could give rise to a functional resistance protein, lacking the "normal" NH$_2$ terminus, before *bldA* is activated; the subsequent availability of tRNA$_{TTA}^{Leu}$ would permit the synthesis of the additional NH$_2$-terminal domain, which might play a role in switching on the synthesis of the cognate antibiotic [12, 13].) The exception is an *S. griseus* ORF that complements the class III and IV *bld* mutants of *S. griseus* (2; M. J. Babcock and K. E. Kendrick, personal communication); since all of the avail-

able evidence points to the idea that TTA codons are confined to very few genes, which probably mostly have important regulatory roles in developmental processes, it will be of great interest to identify the *S. coelicolor* homolog of this *S. griseus bld* gene.

A Technical Problem: TTA Codons Are Present in Several Reporter Genes Potentially Useful for Studying *Streptomyces* Development

As already discussed, the use of fusion to reporter genes is an important tool in analyzing pathways of dependence for developmental gene expression. The *lacZ* gene has been elegantly exploited for this purpose in many organisms (71). Previous studies (46) showed that *lacZ* is very poorly expressed in *S. coelicolor* and *Streptomyces lividans*, and one can now speculate that the presence of many TTA codons in *lacZ* probably accounts for this problem. At least two other reporter genes potentially useful for developmental studies also contain TTA codons (*ampC* [39] and *luxAB* [40]). Although this does not seem to prevent the detectable expression of either gene in colonies approaching or undergoing aerial mycelium development, it does prevent their expression in a *bldA* mutant (B. Leskiw, personal communication; J. Willey, M. Ranes, and R. Losick, personal communication), thereby interfering with the determination of *bldA* dependence of expression of promoters fused to these reporter genes. However, translational dependence on *bldA* can also be turned to an advantage, since the activation of *luxAB*-dependent luminescence or *ampC*-dependent β-lactamase can be used to detect the presence of an active *bldA* gene product; thus, the discovery of *luxAB* fusions that give luminescence in the aerial mycelium (68) and of others that give luminescence in the substrate mycelium shows that tRNA$_{TTA}^{Leu}$ can be present in both "tissues."

Recently, the *xylE* gene from the *Pseudomonas* TOL plasmid has been used as a reporter gene in *Streptomyces* species (6). This gene does not contain TTA codons, and its use to detect the expression of developmentally regulated promoters is being investigated in several laboratories.

How Is the Expression of *bldA* Controlled?

Dot-blot studies of *bldA* RNA abundance (48) and time course studies of light emission by colonies containing the *luxAB* genes transcriptionally coupled to the *bldA* promoter region (68) both indicated that *bldA* was detectably transcribed only relatively late in colony growth, when aerial mycelium was about to appear. One would therefore expect that transcriptional fusion of the *bldA* promoter to a resistance gene (such as Tn5 *neo* in pIJ486, a high-copy-number *Streptomyces* promoter-probe plasmid [79]) should not confer significant kanamycin resistance on young cultures. Surprisingly, however, such a construction does confer fairly high resistance (E. J. Lawlor, Ph.D. thesis, University of East Anglia, Norwich, United Kingdom, 1987). To reconcile these apparently conflicting observations, it may be proposed that conditions that reduce either protein synthesis (perhaps activating a stringent response) or growth rate could signal *bldA* expression. In either case, note the implication that this tRNA gene would be subject to regulation in a direction precisely opposite that generally seen with tRNA genes in, for example, *E. coli*, which would normally be poorly expressed under these conditions.

When *bldA* was cloned on a high-copy-number (30 to 100) plasmid, the amount of gene product detected was scarcely greater than that detected from a single copy of the gene (48). This suggests that some kind of autoregulation occurs, or that *bldA* expression is limited by the availability of some transcriptional activator, or that the cloned fragment also contains a negative regulatory element. In vitro transcription of a *bldA*-containing template with purified RNA polymerase gave clear evidence of transcription from a specific start (Lawlor, Ph.D. thesis), tending to argue against the transcriptional activator hypothesis.

bldB, a Gene Necessary for Normal Development in *S. coelicolor* but Absent from Other Species

The strategy used to clone *bldA* also allowed the identification of clones that complemented a *bldB* mutant (65). The recombinational pickup of *bldB* mutations and the use of these mutant clones to insert the mutations into another genetic background (as previously described for *bldA* [66] and *whiG* [55]) proved that the cloned DNA corresponded to *bldB*. While several "*bldB*" mutations fell in the same cistron in classical complementation tests, some did not (W. Champness, unpublished data [quoted in reference 65]). Thus, *bldB* appears to be a complex locus. The cloned *bldB* gene gives rise to a monocistronic transcript which, so far as the analysis has progressed, is apparently constitutive. The gene is unstable on certain high-copy-number vectors in *S. coelicolor*, but not in a closely related *S. lividans* strain. Remarkably, *bldB* is absent from *S. lividans* and from each of several other strains tested (65).

The implied absence of *bldB* from most streptomycetes raises the legitimate question of how far developmental regulatory mechanisms are shared among different species: after all, A-factor, factor C, and pamamycin are important only in a limited range of species, too. There probably are at least some widely conserved mechanisms, since a *bldA*-like DNA sequence is widespread among *Streptomyces* strains (2; Lawlor, Ph.D. thesis), and defined *bldA* mutants of *S. lividans* which have a phenotype like that of *S. coelicolor bldA* mutants have been constructed (E. J. Lawlor, unpublished data). Furthermore, *S. griseus* DNA that complements class III or IV *bld* mutants hybridizes to DNA from other species, including *S. coelicolor*. The *whiG* gene of *S. coelicolor*, which is discussed in later sections, also hybridizes to DNA from most *Streptomyces* strains tested (K. F. Chater, C. J. Bruton, K. A. Plaskitt, M. J. Buttner, C. Méndez, and J. Helmann, submitted for publication). At this stage it seems plausible that the differences among strains are mainly in the signals needed to activate differentiation, whereas the core processes of differentiation may involve genes of wide occurrence among streptomycetes, perhaps including *bldA* and *S. griseus* class III or IV *bld* genes. If this generalization holds for *bldB*, then its gene product may be involved in generating or sensing such signals rather than as a central regulatory or structural component of a developmental pathway.

What Physiological Attributes Are Likely To Be Uniquely Important in the Growth of Aerial Hyphae?

For hyphae to grow into the air, there are several minimal physiological requirements: sensing of growth direction, provision of nutrients to growth zones, and maintenance of turgor. The failure of any of these three systems would give rise to a Bld phenotype. There is almost no information about any of them. The aerial hyphae certainly can be, and perhaps always are, nutritionally parasitic on the substrate mycelium (54), which has usually undergone extensive lysis in mature colonies. Since storage granules, some of which in at least some species contain glycogen (5), are produced in the ageing substrate mycelium, it is likely that they are a significant source of nutrients for the aerial mycelium. The observed absence of such granules from growing aerial hyphae (5, 82) would be consistent with a supply of nutrients from the granules in the form of soluble breakdown products. This could provide a means by which turgor is maintained, since the breakdown of storage polymers would generate increased osmotic pressure. Thus, the control of polymer metabolism could be important in morphological differentiation for both nutritional and physicochemical reasons. It is interesting to speculate that the *bldA* gene could exert its effects on

morphology at this level, since mature *bldA* mutant colonies are covered with what appear to be aberrant aerial hyphae lying prostrate on the surface and undergoing poorly regulated sporulation processes (14). Such a phenotype could be explained by a failure to generate sufficient turgor for growth into the air, perhaps because of a failure of *bldA*-dependent polymer breakdown systems. Reversal of this phenotype by growth on mannitol might then be readily accounted for by derepression of a hypothetical second system for, say, glycogen breakdown, which would normally be repressed by glucose. Comparable explanations might account for the phenotypic suppression of other *bld* mutants.

The protection of aerial hyphae against dehydration may be an important function of the extra surface layer (or layers) found on aerial hyphae, which imparts a characteristic hydrophobicity to differentiated colonies. Contributions to the surface layer may perhaps be made by some of the spore-associated proteins isolated by Guijarro et al. (24) and R. Santamaria (personal communication) by nonlethal detergent washing of spore preparations. Several of these proteins (SapA, SapB, SapC, etc.) have been studied, and at least two appear to be glycosylated (R. Santamaria, J. Guijarro, and R. Losick, personal communication). One could readily imagine a role for such surface proteins in aerial mycelium formation: the Sap proteins would be secreted by ageing substrate mycelium, but form a structured surface layer only if they are not in contact with water. This might provide both a direction and a scaffolding for the erection of aerial hyphae.

Genes for Proteins Associated with the Outer Surface of Aerial Hyphae and Spores

The genes for Sap proteins described in the previous section are potentially readily accessible to the "reverse genetics" approach to gene isolation: very few proteins are washed from spores in significant amounts, therefore greatly simplifying the purification of each (principally by electrophoresis). Microsequencing techniques have allowed the N-terminal sequences of several Saps to be determined, leading to the preparation of oligonucleotide probes and the cloning of the relevant genes (24; Santamaria et al., personal communication).

The most detailed study so far of what promises to be a rich vein of material has been of *sapA*, which specifies the 13-kilodalton SapA protein (24). The accumulation of *SapA* mRNA is delayed during colony growth until aerial mycelium begins to appear, as judged both by quantitative nuclease protection experiments with in vivo RNA and by the results of *sapA-luxAB* fusion experiments. In the latter experiments, light emission was localized to those parts of the surface of colonies on which aerial mycelium could be seen. (In relation to earlier caveats about the use of TTA-containing reporter genes, this pattern is not a consequence merely of *bldA*-dependent activation of *luxAB* translation, since other *luxAB* fusions showed significantly earlier, and different spatial patterns of, luminescence.) Accumulation of *sapA* mRNA, as measured by nuclease protection with in vivo mRNA, was not entirely prevented by any known morphological mutations, although somewhat reduced amounts were noted in representative *bldC*, *bldD*, and *whiH* mutants (24). This suggests that there may be more than one regulatory cascade operating at this stage of colony development.

The SapB protein, of only 3 kilodaltons in size, has not proved to be readily amenable to sequencing, and the reverse genetics approach has yet to succeed for it. However, the use of polyclonal antiserum to SapB in Western blot (immunoblot) analysis has shown that the SapB antigen is absent from all of the *bld* mutants tested. It remains to be determined whether this control is exerted at the transcription level, a question whose answer will probably depend on the isolation of the *sapB* gene. SapB is produced by several mutants that make aerial hyphae but not spores; SapB is thus associated both with aerial hyphae and with spores (Santamaria et al., personal communication). Interestingly, only very small amounts of SapB are produced by certain *S. coelicolor* mutants that produce spores which, while otherwise morphologically apparently normal, stay white instead of turning grey as they mature (*whiE* mutants [9, 53]) (R. Santamaria and N. K. Davis, personal communication). The *whiE* gene has recently been cloned on a low-copy-number plasmid, and the cloned DNA dramatically increases SapB levels (N. K. Davis, personal communication). However, it is still quite uncertain that *whiE* directly encodes SapB.

METAMORPHOSIS OF AERIAL HYPHAE INTO SPORES

Genetic Analysis of Sporulation in Aerial Hyphae

In *S. coelicolor* A3(2), the young aerial mycelium imparts a fuzzy white surface to colonies. As sporulation proceeds, the aerial mycelium turns grey, a color associated with mature spores. The study of colonies that remain white upon prolonged incubation (*whi* mutants [32]) has led to the recognition of at least eight *whi* genes that are scattered around the *S. coelicolor* chromosome (9; Table 2). Most *whi* mutants fail to make spores, and representative mutations of five *whi* genes were used to construct double mutants for epistasis tests (10). These tests indicated that, in terms of final morphology, a *whiG* mutation—giving long, straight, undifferentiated aerial hyphae— was epistatic to a *whiH*, *whiA*, *whiB*, or *whiI* mutation; a *whiH* mutation (giving loosely coiled aerial hyphae) was epistatic to a *whiA*, *whiB*, or *whiI* mutation; and *whiA* and *whiB* mutants and their double mutant shared the same phenotype of long, unfragmented, tightly coiled aerial hyphae and were epistatic to a *whiI* mutation. Mutants defective in *whiG*, *whiH*, *whiA*, or *whiB* generally make no sporulation septa,

Table 2
Genes Involved in Differentiation of *S. coelicolor* Aerial Hyphae into Spores

Gene	Phenotype of mutant
whiA and *whiB*[a,b,c,d]	Aerial hyphae coil,[a] but few or no sporulation septa detectable[c]; *whiB* cloned, specifies small protein with many charged residues
whiC[a]	Aerial hyphae mostly uncoiled; very low level of sporulation[a]
whiD[a,c]	Chains of spores formed, but spore wall thin[a,c]
whiE[a,c,d]	Chains of morphologically normal, but unpigmented, spores[a,c]; very low levels of SapB spore-associated protein[d,e]; cloned *whiE* gene confers high SapB levels[d]
whiF[a,c]	Rod-shaped, unpigmented spores, reduced in abundance; possibly synonymous with *whiG*[a,c]
whiG[a,b,c,f,g]	Straight aerial hyphae,[a] no sporulation septation[c]; mutations all probably in a single gene[f] that specifies a sigma factor[g]
whiH[a,b,c]	Loosely coiled, partially fragmented aerial hyphae[a,c]
whiI[a,b]	Tightly coiled,[b] partially fragmented aerial hyphae[a]

[a] Reference 9.
[b] Reference 10.
[c] Reference 53.
[d] Davis, personal communication.
[e] Santamaria et al., personal communication.
[f] Reference 55.
[g] Chater et al., submitted.
[h] See Fig. 1.

but in *whiI* mutants these septa are made at a reduced frequency, giving rise to short, curled fragments. Two other *whi* genes, *D* and *E*, are involved in later stages of spore maturation; the single known *whiD* mutant makes thin-walled spores, and *whiE* mutants make white, but otherwise apparently normal, spores. (The *whiE* gene was discussed in the previous section.) Several other *whi* mutations (*whi-53*, *whi-77*, and *whiC193*) give rise to oligosporogeny and have not been studied in detail.

Cloning *whi* genes has not proved to be easy, at least partly because the mutants themselves are inconvenient to handle, but—in addition to *whiE* (see above)—*whiG* and *whiB* have been cloned and analyzed, and they will be discussed in the following sections.

The Deduced *whiG* Gene Product Is a Sigma Factor Crucial in Determining the Developmental Fate of Hyphae

The *whiG* gene was cloned by using a library constructed in a cloning vector derived from phage ϕC31 to restore a grey color to a *whiG* mutant. Subsequently, some *whiG*

mutations were picked up by recombination of the chromosomes of various *whiG* mutants with the cloned gene, allowing classical complementation testing to be done. Those *whiG* mutations tested all fell into a single group (55). The deduced protein encoded by an ORF corresponding to *whiG* was found to resemble known RNA polymerase sigma factors of other bacteria sufficiently closely, in all of the regions generally conserved among sigma factors, to make it virtually certain that the *whiG* gene product is itself a sigma factor, termed sigma-*whiG* (Chater et al., submitted). Sigma-*whiG* particularly closely resembles σ^D ($= \sigma^{28}$) of *B. subtilis* (which has no role in sporulation), but it is comparatively unrelated to σ^H, the sigma factor that initiates the regulatory cascade leading to sporulation in *B. subtilis* (18) (Fig. 3). In an end-to-end alignment, 38% of sigma-*whiG* and σ^D residues were identical, a closer resemblance than is shown in any pairwise comparison of characterized sigma factors of *B. subtilis*. Regions of sigma factors believed to make sequence-specific contacts with the −10 and −35 regions of cognate promoters have recently been pinpointed (23a, 70a, 88), and in these regions sigma-*whiG* and σ^D are noticeably more

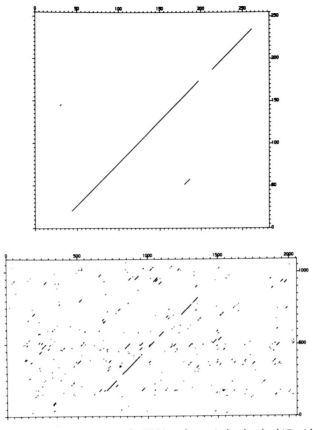

Figure 3. Comparisons at the DNA and protein levels of *whiG* with the *B. subtilis sigD* gene (specifying σ^D or σ^{28} [28]). The panels show DOTPLOT analyses (window, 40 residues; stringency, 19.0). Horizontal axes, *whiG*; vertical axes, *sigD*. Top, Protein comparison; bottom, DNA comparison. The *whiG* DNA sequence used in the lower panel includes flanking sequences; the *whiG* ORF itself extends from positions 551 to 1390. Note that a strong resemblance is detected even at the DNA level, even though the genes differ greatly in base composition (*whiG*, 70% G+C; *sigD*, 42% G+C). (From Chater et al., submitted.)

similar to each other than to other sigma factors (which themselves also fall into recognizable groups) (Fig. 4). It therefore seems likely that the two sigma factors may recognize similar features in their cognate promoters. Disruption of *whiG* caused a typical *whiG* mutant phenotype, but had no perceptible effect on vegetative growth (Chater et al., submitted). Hence, *whiG* appears to encode a sigma factor essential for the transcription of at least one gene specifically needed at an early stage in sporulation.

Further evidence suggests that sigma-*whiG* sensitively determines the amount and the location of sporulation. When cloned in a high-copy-number vector, the *whiG* gene causes sporulation to be earlier and more abundant (55) and even to occur in substrate mycelium or during submerged culture (Fig. 5; Chater et al., submitted). On the other hand, a reduction in the level of sporulation results from the introduction of many copies of a fragment of *B. subtilis* DNA carrying a promoter recognized by the σ^D RNA poly-

−10 recognition

```
          •  •        •       ↑↑      •      •
Sig32    F  A  V  H  W  I  K  A  E  I  H  E  Y  V  L  R  N  W  R  I  V  K  V  A

Sig70    Y  A  T  W  W  I  R  Q  A  I  T  R  S  I  A  D  Q  A  R  T  I  R  I  P
SigA     Y  A  T  W  W  I  R  Q  A  I  T  R  A  I  A  D  Q  A  R  T  I  R  I  P

SigB     F  A  I  P  T  I  I  G  E  I  K  R  F  L  R  D  K  T  W  S  V  H  V  P
SigF     Y  A  V  P  M  I  I  G  E  I  Q  R  F  I  R  D  D  G  -  T  V  K  V  S
SigG     F  A  V  P  M  I  I  G  E  I  R  R  Y  L  R  D  N  N  -  P  I  R  V  S
SigE     Y  A  S  R  C  I  E  N  E  I  L  M  Y  L  R  R  N  N  K  -  I  R  S  E
SigK     Y  A  A  R  C  I  E  N  E  I  L  M  H  L  R  A  L  K  K  -  T  K  K  D

SigH     F  A  E  L  C  I  T  R  Q  I  I  T  A  I  K  T  A  T  R  Q  K  H  I  P

SigD     Y  A  S  F  R  I  R  G  A  I  I  D  G  L  R  K  E  D  W  L  P  R  T  S
WhiG     Y  A  I  T  R  I  R  G  A  M  I  D  E  L  R  A  L  D  W  I  P  R  S  V
```

−35 recognition

```
                        ↑  ↑        ↑  ↑
          •  •  •  •     •           •        •
Sig32    G  V  S  A  E  R  V  R  Q  L  E  K  N

Sig70    D  V  T  R  E  R  I  R  Q  I  E  A  K
SigA     G  V  T  R  E  R  I  R  Q  I  E  A  K

SigB     G  I  S  Q  M  H  V  S  R  L  Q  R  K
SigF     G  I  S  Q  V  Q  V  S  R  L  E  K  K
SigG     G  I  S  Q  A  Q  V  S  R  L  E  K  K
SigE     G  I  S  Q  S  Y  I  S  R  L  E  K  R
SigK     G  I  S  R  S  Y  V  S  R  I  E  K  R

SigH     N  R  H  V  K  S  I  D  N  A  L  Q  R

SigD     N  L  S  T  S  R  I  S  Q  I  H  S  K
WhiG     G  V  T  E  S  R  V  S  Q  I  H  T  K
```

Figure 4. Comparison of *whiG* and other sigma factors for putative recognition regions for −10 and −35 sequences of cognate promoters. The sequences are arranged so that apparently similar sigma factors are grouped. Residues conserved within groups but not among all sigma factors are boxed. The groupings for −10 and −35 recognition regions are very similar. Residues in positions marked with daggers are believed to make direct contact with bases at particular positions in −10 and −35 regions of promoters (see the text for references). Bullets mark residues previously noted to be relatively highly conserved among sigma factors (27, 74; these are also the sources on which these alignments are based). Apart from σ^{70} and σ^{32} (both from *E. coli*) and *whiG* (the product of the *whiG* gene of *S. coelicolor*), the other sigma factors (A through K) are from *B. subtilis*.

merase holoenzyme, as if the sigma-*whiG* holoenzyme were recognizing this promoter and being partially sequestered from its proper targets, the promoters for certain sporulation genes. Thus, the quantity of sigma-*whiG* appears to limit sporulation, and the normal absence of sporulation in substrate hyphae seems to be entirely due to an insufficient amount of sigma-*whiG* holoenzyme (Chater et al., submitted).

Application of Transposon Mutagenesis to the Analysis of the Cloned *whiB* and *whiE* Genes, and the Sequencing of the *whiB* Gene

DNA complementing *whiB* and *whiE* mutants was cloned by Davis (personal communication), using self-transmissible low-copy-number SCP2*-derived vectors (pIJ922 [50] and pIJ698 [45]). In both cases, analysis

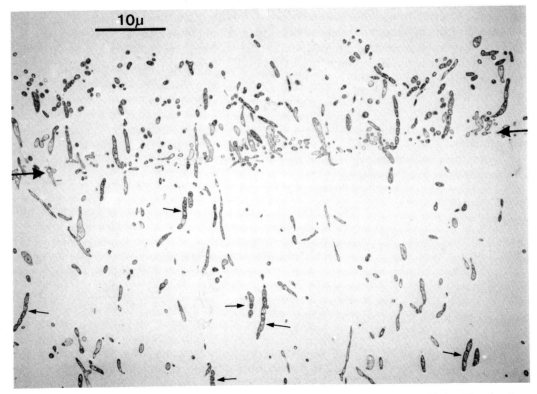

Figure 5. "Homeotic" effect of introducing additional copies of the *whiG* gene into *S. coelicolor*. The plate is an electron micrograph of a thin section through a 3-day-old colony of *S. coelicolor* containing pIJ4412 (55), a high-copy-number plasmid containing the cloned *whiG* gene. Specimen preparation was as described by Chater et al. (submitted). Small arrows indicate spore chains below the agar surface, and large arrows indicate the position of the agar surface. (Micrograph courtesy of K. A. Plaskitt.)

by subcloning was supplemented and accelerated by the use of a system of transposon mutagenesis suggested by S. T. Chung, using the Tn3-like *Streptomyces* transposon Tn*4560*, which encodes viomycin resistance (15, 16). The transposon was introduced on a non-transmissible, nonmobilizable plasmid into strains carrying the target replicons. Mating between the resulting strains and a suitable recipient strain, followed by selection for transfer of viomycin resistance into the recipient, gave rise to an apparently fairly random array of transposon insertions into the target replicon. The sequence around a mutagenic insertion (coupled with information from subcloning) revealed that *whiB* would encode a small protein rich in charged residues which is not recognizably similar to any protein in available data bases (Davis, personal communication). Thus, although it has been surmised that *whiB* may specify a structural component of sporulation septa (14), its product is unlikely to be intimately associated with any membranes. The *whiB* protein also shows no obvious sequence features reminiscent of sigma factors or DNA-binding proteins, making it unlikely to be directly involved in gene regulation.

CODA

The Underlying Regulatory Elements for Sporulation in *Streptomyces* and *Bacillus* Species Are Probably Not Homologous

Bacilli and actinomycetes are both groups of gram-positive sporulating bacteria which arose from a common progenitor which had already separated from the evolutionary branch leading to gram-negative bacteria (84). Did this ancestor undergo a kind of ur-sporulation from which the processes of endospore formation and aerial sporulation have both developed? If so, clearly much use could be made by *Streptomyces* researchers of the detailed knowledge of endospore formation now available.

Sporulation in *Streptomyces* and in *Bacillus* species is probably initiated as a response to nutritional deprivation. In the case of *B. subtilis*, the whole subsequent process takes place in a preformed sacculus in an aqueous environment. The cell contents are redistributed into the more concentrated, tightly organized form of an endospore through a sequence of events principally characterized by asymmetrical septation and further membrane reorganization to form the prespore, and sequential activation of sets of genes by a regulatory cascade that includes sigma factors specific to the mother cell and to the prespore. Toward the end of the process, almost all of the remaining usable protein of the mother cell is converted into coat proteins that are targeted to the prespore outer membrane.

In *Streptomyces* species the first stage in sporulation is a new round of growth in which aerial hyphae are formed, at least partially, by the reuse of material from the substrate mycelium. Subsequently, in response to the occurrence of a critical concentration of sigma-*whiG*, growth of the aerial hyphae ceases and the materials of the hyphae are reorganized into spores through the action of what appear to be relatively few specific genes. The thick surface layers that develop around spores must be formed entirely from materials present in the compartments formed during sporulation septation. There can be no supply of materials such as coat proteins from outside sources, in contrast to the situation for endospores. The final spore product, although desiccation resistant, can tolerate scarcely more heat than can the vegetative hyphae.

Thus, there are fairly extensive morphological and physiological differences between the two processes and their end products, but both processes critically involve the activation of specific gene sets by sigma factors. Is there an underlying genetic homology? Perhaps the best evidence for this would be to find that sporulation-specific sigma factors are homologous between the groups. How do their sigma factor repertoires compare? Vegetatively growing *B. subtilis* cells contain several different sigma factors: σ^A, the principal sigma factor, homologous with *E. coli* σ^{70}; σ^B, which is not needed for growth or sporulation; σ^H, the product of the *spo0H* gene, which is necessary for early events in sporulation but is not needed for growth; σ^C, which is of unknown function; and σ^D, which transcribes genes for motility and chemotaxis and is also necessary for an autolytic activity that allows cell separation (18, 27). Four further sigma factors are specifically induced during sporulation (49, 74; R. Losick and L. Kroos, this volume; P. Stragier, this volume; C. P. Moran, Jr., this volume). Vegetatively growing *S. coelicolor* contains at least three, and possibly up to seven or more, different sigma factors, one of which resembles σ^B in its promoter specificity (81) and at least one and probably four of which are homologs of σ^A (7, 77, 81) (Table 3). No σ^D-like transcribing activity could be detected in purified RNA polymerase preparations from vegetative cultures of *S. coelicolor* or its close relative *S. lividans* (Chater et al., submitted). This is no surprise in view of the absence of motility, chemotaxis, or cell separation during vegetative growth.

It is difficult to look for sporulation-specific sigma factors in *Streptomyces* species because of the tiny amounts of material that can be obtained as aerial mycelium compared with the almost unlimited amounts obtainable from large-scale liquid cultures of sporulating bacilli. For this reason no direct attempts have been made (although sporulation of *S. griseus* during submerged culture appears to offer opportunities), and efforts have

Table 3
Streptomyces Sigma Factors

Gene (protein)	Comments	Reference(s)
hrdA (?), *hrdB* (?), *hrdC* (?), *hrdD* (?)	*S. coelicolor* genes identified by hybridization as being similar to principal sigma factor genes; probably all recognize similar -10 and -35 regions of promoters; one of the *hrd* genes probably specifies σ^{35}	7, 77, 81; M. J. Buttner, personal communication
? (σ^{49})	Identified biochemically; recognizes promoters like those recognized by σ^B of *B. subtilis*	7, 81
? (σ^{28})	Identified biochemically by recognition of the *dagAp2* promoter of the *S. coelicolor* agarase gene	7
whiG (?)	Required for sporulation in *S. coelicolor*; resembles σ^D of *B. subtilis*	Chater et al., submitted

focused on a genetic approach. The first sporulation-specific sigma factor to emerge, sigma-*whiG*, has turned out to be more closely related to σ^D than to any sporulation sigma factor of *B. subtilis*, and both sigma factors appear to have descended from a σ^A ancestor, independently from the evolution of known sporulation sigma factors (Chater et al., submitted). Thus, the first clear-cut information about the underlying regulation of *Streptomyces* sporulation suggests nonhomology with endospore formation. The possible interconnection of *Streptomyces* sporulation and *Bacillus* motility can be rationalized. In both cases multinucleate cells that have ceased to grow are subdivided into uninucleate dispersive cells, in one case adapted to swim toward favorable new environments and in the other furnished with a hydrophobic surface that should allow a very rapid spread on the surface of water droplets (dispersal by air currents is probably also important and may be the reason for the development of "hairy" spores in some species). It is interesting that some relatives of *Streptomyces* species, such as *Actinoplanes* species, produce motile spores.

Outlook

The availability of ever-improving systems for the cloning and functional analysis of *Streptomyces* genes, which was still merely offering the potential for breakthroughs in

the last extensive review of developmental biology of these organisms (11), has now led to the discovery of the transcriptional and translational regulatory roles of *afsB*, *afsC*, *whiG*, and *bldA* and the development of sensitive probes for developmentally regulated promoters. Significant progress is now being made on the molecular genetics of the major regulatory circuits of primary metabolism that may be important in sensing nutrient limitation (e.g., reference 22 for nitrogen regulation and references 6, 38, and 73 for carbon catabolite regulation). In another 5 years a review of this area will no doubt describe the regulatory phenomena that govern the expression of the regulatory genes so far described, as well as the elucidation of some of the targets of their action. Information gained from fusions to reporter genes such as *luxAB* will begin to clarify the dependence of developmental genes on both intracellular regulatory elements and extracellular factors such as A-factor. More speculatively, the analysis of the regulation and influence of synthesis and degradation of storage materials may have made progress, and the potential developmental role of such pervasive bioregulators as Ca^{2+} ions may have been addressed (a Ca^{2+}-binding protein has been discovered in *Streptomyces* species-like *Saccharopolyspora erythraea* [75], and roles for Ca^{2+} in differentiation have recently been hinted at [47]). It will be of considerable interest to compare progress in *Streptomyces* developmental biology with that in the mold *Aspergillus nidulans*. The ecological and morphological convergence of these two evolutionarily unrelated genera may well provide illuminating physiological resemblances, if current results are any guide: the critical role played by sigma-*whiG* concentration in determining the switch from growth to sporulation in *S. coelicolor* is paralleled by closely similar findings for a putative DNA-binding protein, the *brlA* gene product, in *A. nidulans* (1).

Finally, this review has been written with hardly a mention of secondary metabolites, a topic recently reviewed elsewhere (12). However, it is obvious that morphological differentiation and physiological differentiation are interdependent (30), and their interplay and common regulation will form a significant part of any serious program of research on *Streptomyces* developmental biology.

Acknowledgments. This article has profited from discussions with and unpublished information from numerous colleagues. They include Marty Babcock, Mervyn Bibb, Alfredo Braña, Mary Brawner, Celia Bruton, Mark Buttner, Wendy Champness, Shiau-Ta Chung, Brendan Daly, Nigel Davis, Olga Efremenkova, Jose Gil, David Gray, Pepe Guijarro, Ellen Guthrie, John Helmann, David Hopwood, Sueharu Horinouchi, Bengtåke Jaurin, Kathy Kendrick, Liz Lawlor, Brenda Leskiw, Rich Losick, Benjamin Manzanal, Carmen Méndez, Jacqueline Piret, Kitty Plaskitt, Jim Prosser, Anna-Maria Puglia, Monica Ranes, Ramon Santamaria, Al Schauer, Patrick Stragier, Jan Westpheling, and Jo Willey. I also thank David Hopwood and Mervyn Bibb for their comments on the manuscript and Meredyth Limberg for typing it.

LITERATURE CITED

1. **Adams, T. H., M. T. Boylan, and W. E. Timberlake.** 1988. *brlA* is necessary and sufficient to direct conidiophore development in *Aspergillus nidulans. Cell* 54:353–362.

2. **Babcock, M. J., and K. E. Kendrick.** 1988. Cloning of DNA involved in sporulation of *Streptomyces griseus. J. Bacteriol.* 170:2802–2808.

3. **Biró, S., I. Bekesi, S. Vitális, and G. Szabó.** 1980. A substance effecting differentiation in *Streptomyces griseus.* Purification and properties. *Eur. J. Biochem.* 103:359–363.

4. **Braña, A. F., M. B. Manzanal, and C. Hardisson.** 1982. Characterization of intracellular polysaccharides of *Streptomyces. Can. J. Microbiol.* 28:1320–1323.

5. **Braña, A. F., C. Méndez, L. A. Diaz, M. B. Manzanal, and C. Hardisson.** 1986. Glycogen and trehalose accumulation during colony development in *Streptomyces antibioticus. J. Gen. Microbiol.* 132:1319–1326.

6. **Brawner, M., C. Ingram, J. Fornwald, H. Lichenstein, and J. Westpheling.** 1988. Structure of the *Streptomyces* galactose operon's catabolite controlled promoter, p. 35–40. *In* Y. Okami, T. Beppu, and H. Ogawara (ed.), *Biology of Actinomycetes '88.* Japan Scientific Societies Press, Tokyo.

7. **Buttner, M. J., A. M. Smith, and M. J. Bibb.** 1988. At least three RNA polymerase holoenzymes direct transcription of the agarase gene (*dagA*) of *Streptomyces coelicolor* A3(2). *Cell* 52:599–607.

8. **Champness, W. C.** 1988. New loci required for *Streptomyces coelicolor* morphological and physiological differentiation. *J. Bacteriol.* 170:1168–1174.

9. **Chater, K. F.** 1972. A morphological and genetic mapping study of white colony mutants of *Streptomyces coelicolor. J. Gen. Microbiol.* 72:9–28.

10. **Chater, K. F.** 1975. Construction and phenotypes of double sporulation deficient mutants in *Streptomyces coelicolor* A3(2). *J. Gen. Microbiol.* 87:312–325.

11. **Chater, K. F.** 1984. Morphological and physiological differentiation in *Streptomyces*, p. 89–115. *In* R. Losick and L. Shapiro (ed.), *Microbial Development.* Cold Spring Harbor Laboratory, Cold Spring Harbor, N.Y.

12. **Chater, K. F., and D. A. Hopwood.** 1989. Antibiotic biosynthesis in *Streptomyces*, p. 129–150. *In* D. A. Hopwood and K. F. Chater (ed.), *Genetics of Bacterial Diversity.* Academic Press, Inc. (London), Ltd., London.

13. **Chater, K. F., E. J. Lawlor, C. Méndez, C. J. Bruton, N. K. Davis, K. Plaskitt, E. P. Guthrie, B. L. Daly, H. A. Baylis, and K. Vu Trong.** 1988. Gene expression during *Streptomyces* development, p. 64–70. *In* Y. Okami, T. Beppu, and H. Ogawara (ed.), *Biology of Actinomycetes '88.* Japan Scientific Societies Press, Tokyo.

14. **Chater, K. F., and M. J. Merrick.** 1979. *Streptomyces*, p. 93–114. *In* J. H. Parish (ed.), *Develop-*

mental Biology of Prokaryotes. Blackwell Scientific Publications, Ltd., Oxford.

15. **Chung, S.-T.** 1987. Tn*4556*, a 6.8-kilobase-pair transposable element of *Streptomyces fradiae. J. Bacteriol.* **169**:4436–4441.

16. **Chung, S.-T.** 1988. Transposition of the *Streptomyces fradiae* transposon Tn*4556*, p. 127–132. *In* Y. Okami, T. Beppu, and H. Ogawara (ed.), *Biology of Actinomycetes '88.* Japan Scientific Societies Press, Tokyo.

17. **Distler, J., A. Ebert, K. Mansouri, K. Pissowotzki, M. Stockmann, and W. Piepersberg.** 1987. Gene cluster for streptomycin biosynthesis in *Streptomyces griseus:* nucleotide sequence of three genes and analysis of transcriptional activity. *Nucleic Acids Res.* **15**:8041–8056.

18. **Dubnau, E., J. Weir, G. Nair, L. Carter III, C. Moran, Jr., and I. Smith.** 1988. *Bacillus* sporulation gene *spo0H* codes for $\sigma^{30}(\sigma^H)$. *J. Bacteriol.* **170**:1054–1062.

19. **Efremenkova, O. V., L. N. Anisova, and Y. E. Bartoshevich.** 1985. Regulators of differentiation in Actinomycetes. *Antibiot. Med. Biotekhnol.* **9**:687–707. (In Russian.)

20. **Ensign, J. C.** 1988. Physiological regulation of sporulation of *Streptomyces griseus,* p. 309–315. *In* Y. Okami, T. Beppu, and H. Ogawara (ed.), *Biology of Actinomycetes '88.* Japan Scientific Societies Press, Tokyo.

21. **Epp, J., S. G. Burgett, and B. E. Schoner.** 1987. Cloning and nucleotide sequence of a carbomycin resistance gene from *Streptomyces thermotolerans. Gene* **51**:73–83.

22. **Fisher, S. H.** 1988. Nitrogen assimilation in *Streptomyces,* p. 47–51. *In* Y. Okami, T. Beppu, and H. Ogawara (ed.), *Biology of Actinomycetes '88.* Japan Scientific Societies Press, Tokyo.

23. **Freese, E.** 1982. Initiation of bacterial sporulation, p. 1–12. *In* H. S. Levinson, A. L. Sonenshein, and D. J. Tipper (ed.), *Sporulation and Germination.* American Society for Microbiology, Washington, D.C.

23a. **Gardella, T., H. Moyle, and M. M. Susskind.** 1989. A mutant *E. coli* σ^{70} subunit of RNA polymerase with altered promoter specificity. *J. Mol. Biol.* **206**:579–590.

24. **Guijarro, J., R. Santamaria, A. Schauer, and R. Losick.** 1988. Promoter determining the timing and spatial localization of transcription of a cloned *Streptomyces coelicolor* gene encoding a spore-associated polypeptide. *J. Bacteriol.* **170**:1895–1901.

25. **Hara, O., S. Horinouchi, T. Uozumi, and T. Beppu.** 1983. Genetic analysis of A-factor synthesis in *Streptomyces coelicolor* A3(2) and *Streptomyces griseus. J. Gen. Microbiol.* **129**:2939–2944.

26. **Hardisson, C., and M. B. Manzanal.** 1976. Ultra-

structural studies of sporulation in *Streptomyces. J. Bacteriol.* **127**:1443–1454.

27. **Helmann, J. D., and M. J. Chamberlin.** 1988. Structure and function of bacterial sigma factors. *Annu. Rev. Biochem.* **42**:839–872.

28. **Helmann, J. D., L. M. Márquez, and M. J. Chamberlin.** 1988. Cloning, sequencing, and disruption of the *Bacillus subtilis* σ^{28} gene. *J. Bacteriol.* **170**:1568–1574.

29. **Hopwood, D. A.** 1988. Understanding the genetic control of antibiotic biosynthesis and sporulation in *Streptomyces,* p. 3–10. *In* Y. Okami, T. Beppu, and H. Ogwara (ed.), *Biology of Actinomycetes '88.* Japan Scientific Societies Press, Tokyo.

30. **Hopwood, D. A.** 1988. Towards an understanding of gene switching in *Streptomyces,* the basis of sporulation and antibiotic production. *Proc. R. Soc. London B* **235**:121–138.

31. **Hopwood, D. A., M. J. Bibb, K. F. Chater, T. Kieser, C. J. Bruton, H. M. Kieser, D. J. Lydiate, C. P. Smith, J. M. Ward, and H. Schrempf.** 1985. *Genetic Manipulation of Streptomyces. A Laboratory Manual.* The John Innes Foundation, Norwich, United Kingdom.

32. **Hopwood, D. A., H. Wildermuth, and H. M. Palmer.** 1970. Mutants of *Streptomyces coelicolor* defective in sporulation. *J. Gen. Microbiol.* **61**:397–408.

33. **Horinouchi, S., and T. Beppu.** 1987. A-factor and regulatory network that links secondary metabolism with cell differentiation in *Streptomyces,* p. 41–48. *In* M. Alačević, D. Hranueli, and Z. Toman (ed.), *Proceedings of Fifth International Symposium on the Genetics of Industrial Microorganisms,* part B. Pliva, Zagreb, Yugoslavia.

34. **Horinouchi, S., and T. Beppu.** 1988. Regulation of secondary metabolism in *Streptomyces,* p. 71–75. *In* Y. Okami, T. Beppu, and H. Ogwara (ed.), *Biology of Actinomycetes '88.* Japan Scientific Societies Press, Tokyo.

35. **Horinouchi, S., Y. Kumada, and T. Beppu.** 1984. Unstable genetic determinant of A-factor biosynthesis in streptomycin-producing organisms: cloning and characterization. *J. Bacteriol.* **158**:481–487.

36. **Horinouchi, S., M. Nishiyama, H. Suzuki, Y. Kumada, and T. Beppu.** 1985. The cloned *Streptomyces bikiniensis* A-factor determinant. *J. Antibiot.* **38**:636–641.

37. **Horinouchi, S., H. Suzuki, and T. Beppu.** 1986. Nucleotide sequence of *afsB,* a pleiotropic gene involved in secondary metabolism in *Streptomyces coelicolor* A3(2) and *Streptomyces lividans. J. Bacteriol.* **168**:257–269.

37a. **Horinouchi, S., H. Suzuki, M. Nishiyama, and T. Beppu.** 1989. Nucleotide sequence and transcriptional analysis of the *Streptomyces griseus* gene (*afsA*)

responsible for A-factor biosynthesis. *J. Bacteriol.*
171:1206–1210.

38. **Ikeda, H., E. T. Seno, C. J. Bruton, and K. F. Chater.**
1984. Genetic mapping, cloning and physiological
aspects of the glucose kinase gene of *Streptomyces
coelicolor. Mol. Gen. Genet.* **196**:501–507.

39. **Jaurin, B., and T. Grundström.** 1981. *ampC* ceph-
alosporinase of *Escherichia coli* K-12 has a different
evolutionary origin from that of β-lactamases of the
penicillinase type. *Proc. Natl. Acad. Sci. USA*
78:4897–4901.

40. **Johnston, T. C., R. B. Thompson, and T. O. Bald-
win.** 1986. Nucleotide sequence of the *luxB* gene
of *Vibrio harveyi* and the complete amino acid sequence of the β subunit of bacterial luciferase. *J.
Biol. Chem.* **261**:4805–4811.

41. **Kendall, K. J., and S. N. Cohen.** 1988. Complete
nucleotide sequence of the *Streptomyces lividans*
plasmid pIJ101 and correlation of the sequence with
genetic properties. *J. Bacteriol.* **170**:4634–4651.

42. **Kendrick, K. E., and J. C. Ensign.** 1983. Sporulation
of *Streptomyces griseus* in submerged culture. *J.
Bacteriol.* **155**:357–366.

43. **Khokhlov, A. S.** 1985. Actinomycete autoregula-
tors, p. 791–798. *In* G. Szabó, S. Bíró, and M.
Goodfellow (ed.), *Proceedings of the Sixth International Symposium on Actinomycete Biology.*
Akadémiai Kiadó, Budapest.

44. **Khokhlov, A. S., L. N. Anisova, I. I. Tovarova, F.
M. Kleiner, I. V. Kovalenko, O. I. Krasilnikova,
E. Y. Kornitskaya, and S. A. Pliner.** 1973. Effects
of A-factor on the growth of asporogenous mutants
of *Streptomyces griseus*, not producing this factor.
Z. Allg. Mikrobiol. **13**:647–655.

45. **Kieser, T., and R. Melton.** 1988. Plasmid pIJ699,
a multi-copy positive selection vector for *Streptomyces. Gene* **65**:83–91.

46. **King, A. A., and K. F. Chater.** 1986. The expression
of the *Escherichia coli lacZ* gene in *Streptomyces.
J. Gen. Microbiol.* **132**:1739–1752.

47. **Kondo, S., K. Yasni, M. Natsume, M. Katayama,
and S. Marumo.** 1988. Isolation, physico-chemical
properties and biological activity of pamamycin-
607, an aerial mycelium-inducing substance from
Streptomyces alboniger. J. Antibiot. **41**:1196–1204.

48. **Lawlor, E. J., H. A. Baylis, and K. F. Chater.** 1987.
Pleiotropic morphological and antibiotic deficiencies result from mutations in a gene encoding a
tRNA-like product in *Streptomyces coelicolor* A3(2).
Genes Dev. **1**:1305–1310.

49. **Losick, R., L. Kroos, J. Errington, and P. Youngman.** 1988. Pathways of developmentally regulated
gene expression in the spore-forming bacterium *Bacillus subtilis*, p. 221–242. *In* D. A. Hopwood and
K. F. Chater (ed.), *Genetics of Bacterial Diversity.*
Academic Press Inc. (London), Ltd., London.

50. **Lydiate, D. J., F. Malpartida, and D. A. Hopwood.**

1985. The *Streptomyces* plasmid SCP2*: its functional analysis and development into useful cloning
vectors. *Gene* **35**:223–235.

51. **McCann, P. A., and B. M. Pogell.** 1979. Pamamycin: a new antibiotic and stimulator of aerial my-
celia formation. *J. Antibiot.* **32**:673–678.

52. **McClain, W. H., and H. B. Nicholas, Jr.** 1987. Differences between transfer RNA molecules. *J. Mol.
Biol.* **194**:635–642.

53. **McVittie, A.** 1974. Ultrastructural studies on sporulation in wild-type and white colony mutants of
Streptomyces coelicolor. J. Gen. Microbiol. **81**:291–
302.

54. **Méndez, C., A. F. Braña, M. B. Manzanal, and C.
Hardisson.** 1985. Role of substrate mycelium in colony development in *Streptomyces. Can. J. Microbiol.* **31**:446–450.

55. **Méndez, C., and K. F. Chater.** 1987. Cloning of
whiG, a gene critical for sporulation of *Streptomyces coelicolor* A3(2). *J. Bacteriol.* **169**:5715–5720.

56. **Merrick, M. J.** 1976. A morphological and genetic
mapping study of bald colony mutants of *Streptomyces coelicolor. J. Gen. Microbiol.* **96**:299–315.

57. **Miguélez, E. M., M. C. Martin, M. B. Manzanal,
and C. Hardisson.** 1988. Hyphal growth in *Streptomyces*, p. 490–495. *In* Y. Okami, T. Beppu, and
H. Ogawara (ed.), *Biology of Actinomycetes '88.*
Japan Scientific Societies Press, Tokyo.

58. **Ochi, K.** 1986. A decrease in GTP content is associated with aerial mycelium formation in *Streptomyces* MA406-A-1. *J. Gen. Microbiol.* **132**:299–
305.

59. **Ochi, K.** 1986. Occurrence of the stringent response
in *Streptomyces* sp. and its significance for the initiation of morphological and physiological differ-
entation. *J. Gen. Microbiol.* **132**:2621–2631.

60. **Ochi, K.** 1987. Changes in nucleotide pools during
sporulation of *Streptomyces griseus* in submerged
culture. *J. Gen. Microbiol.* **133**:2787–2795.

61. **Ochi, K.** 1987. A *rel* mutation abolishes the enzyme
induction needed for actinomycin synthesis by
Streptomyces antibioticus. Agric. Biol. Chem.
51:829–835.

62. **Ochi, K.** 1987. Metabolic initiation of differentiation and secondary metabolism by *Streptomyces
griseus*: significance of the stringent response (ppGpp)
and GTP content in relation to A-factor. *J. Bacteriol.* **169**:3608–3616.

63. **Ochi, K.** 1988. Nucleotide pools and stringent response in regulation of *Streptomyces* differentiation,
p. 330–337. *In* Y. Okami, T. Beppu, and H. Ogawara (ed.), *Biology of Actinomycetes '88.* Japan
Scientific Societies Press, Tokyo.

64. **Ochman, H., and A. C. Wilson.** 1987. Evolutionary
history of enteric bacteria, p. 1649–1654. *In* F. C.
Neidhardt, J. L. Ingraham, K. B. Low, B. Magasanik, M. Schaechter, and H. E. Umbarger (ed.),

Escherichia coli and Salmonella typhimurium: Cellular and Molecular Biology. American Society for Microbiology, Washington, D.C.

65. Piret, J., V. Bernan, M. Harasym, and M. Brandt. 1988. Characterization of cloned bald genes of *Streptomyces coelicolor* A3(2), p. 321–323. *In* Y. Okami, T. Beppu, and H. Ogawara (ed.), *Biology of Actinomycetes '88*. Japan Scientific Societies Press, Tokyo.

66. Piret, J. M., and K. F. Chater. 1985. Phage-mediated cloning of *bldA*, a region involved in *Streptomyces coelicolor* morphological development, and its analysis by genetic complementation. *J. Bacteriol.* 163:965–972.

67. Prosser, J. I., D. I. Gray, and G. W. Gooday. 1988. Cellular mechanisms for growth and branch formation in streptomycetes, p. 316–323. *In* Y. Okami, T. Beppu, and H. Ogawara (ed.), *Biology of Actinomycetes '88*. Japan Scientific Societies Press, Tokyo.

68. Schauer, A., M. Ranes, R. Santamaria, J. Guijarro, E. Lawlor, C. Méndez, K. Chater, and R. Losick. 1988. Visualizing gene expression in time and space in the morphologically complex, filamentous bacterium *Streptomyces coelicolor*. *Science* 240:768–772.

69. Schauer, A. T. 1988. Visualizing gene expression with luciferase fusions. *Trends Biotechnol.* 6:23–27.

70. Schauer, A. T., and H. Im. 1988. Visualizing gene expression in *Streptomyces coelicolor*, p. 346–350. *In* Y. Okami, T. Beppu, and H. Ogawara (ed.), *Biology of Actinomycetes '88*. Japan Scientific Societies Press, Tokyo.

70a. Siegele, D. A., J. C. Hu, W. A. Walter, and C. A. Gross. 1989. Altered promoter recognition by mutant forms of the σ^{70} subunit of *Escherichia coli* RNA polymerase. *J. Mol. Biol.* 206:591–603.

71. Silhavy, T. J., and J. R. Beckwith. 1985. Uses of *lac* fusions for the study of biological problems. *Microbiol. Rev.* 49:398–418.

72. Sinclair, R. B., and M. J. Bibb. 1988. The repressor gene (c) of the *Streptomyces* temperate phage φC31: nucleotide sequence, analysis and functional cloning. *Mol. Gen. Genet.* 213:269–277.

73. Smith, C. P., and K. F. Chater. 1988. Structure and regulation of controlling sequences for the *Streptomyces coelicolor* glycerol operon. *J. Mol. Biol.* 204:569–580.

74. Stragier, P., B. Kunkel, L. Kroos, and R. Losick. 1989. Chromosomal rearrangement generating a composite gene for a developmental sigma factor. *Science* 243:507–512.

75. Swan, D. G., R. S. Hale, N. Dhillon, and P. F. Leadlay. 1987. A bacterial calcium-binding protein ho-

mologous to calmodulin. *Nature* (London) 329:84–85.

76. Szabó, G., F. Szeszak, S. Vitális, and F. Toth. 1988. New data on the formation and mode of action of factor C, p. 324–329. *In* Y. Okami, T. Beppu, and H. Ogawara (ed.), *Biology of Actinomycetes '88*. Japan Scientific Societies Press, Tokyo.

77. Tanaka, K., T. Shiina, and H. Takahashi. 1988. Multiple principal sigma factor homologs in eubacteria: identification of the "rpoD box." *Science* 242:1040–1042.

78. Vögtli, M., and R. Hütter. 1987. Characterisation of the hydroxystreptomycin phosphotransferase gene (*sph*) of *Streptomyces glaucescens*: nucleotide sequence and promoter analysis. *Mol. Gen. Genet.* 208:195–203.

79. Ward, J. M., G. R. Janssen, T. Kieser, M. J. Bibb, M. J. Buttner, and M. J. Bibb. 1986. Construction and characterisation of a series of multi-copy promoter-probe plasmid vectors for *Streptomyces* using the aminoglycoside phosphotransferase gene from Tn5 as indicator. *Mol. Gen. Genet.* 203:468–478.

80. Wawrousek, E. F., N. Narashima, and J. N. Hansen. 1984. Two large clusters with thirty-seven transfer RNA genes adjacent to ribosomal RNA sets in *Bacillus subtilis*. *J. Biol. Chem.* 259:3694–3702.

81. Westpheling, J., M. Ranes, and R. Losick. 1985. RNA polymerase heterogeneity in *Streptomyces coelicolor*. *Nature* (London) 313:22–27.

82. Wildermuth, H. 1970. Development and organisation of the aerial mycelium in *Streptomyces coelicolor*. *J. Gen. Microbiol.* 60:43–50.

83. Wildermuth, H., and Hopwood, D. A. 1970. Septation during sporulation in *Streptomyces coelicolor*. *J. Gen. Microbiol.* 60:57–59.

84. Woese, C. R. 1987. Bacterial evolution. *Microbiol. Rev.* 51:221–271.

85. Yamaizumi, Z., Y. Kuchino, F. Harada, S. Nishimura, and J. A. McCloskey. 1980. Primary structure of *Escherichia coli* tRNA$^{Leu}_{UUR}$. *J. Biol. Chem.* 255:2220–2225.

86. Yung, B. Y., and A. Kornberg. 1988. Membrane attachment activates dnaA protein, the initiation protein of chromosome replication in *Escherichia coli*. *Proc. Natl. Acad. Sci USA* 85:7202–7205.

87. Zalacain, M., G. González, M. C. Guerrero, R. J. Mattaliano, F. Malpartida, and A. Jiménez. 1986. Nucleotide sequence of the hygromycin B phosphotransferase gene from *Streptomyces hygroscopicus*. *Nucleic Acids Res.* 14:1565–1581.

88. Zuber, P., J. Healy, H. L. Carter III, S. Cutting, C. P. Moran, and R. Losick. 1989. Mutation changing the specificity of an RNA polymerase sigma factor. *J. Mol. Biol.* 206:605–614.

Index

aprE (subtilisin) gene, 131–146, 200–201

Bacillus cereus T
  conductivity, 50
  cortex, 50
  requirement for dipicolinic acid, 58
  sporulation proteins, 219
Bacillus megaterium
  cortex, 50–51
  demineralization in, 55
  effect of sublethal heating, 58
  heat resistance, 45–46
  sporulation proteins, 219
  triggering of germination, 89–108
Bacillus stearothermophilus
  demineralization in, 55
  effect of sublethal heating, 57–58
  requirement for dipicolinic acid, 59
  thermal adaptation, 56
  water activity, 48–49
Bacillus subtilis
  cloning
    DNA, 74
    ger genes, 100
    φ105 cloning system, 75–77
    SPβ cloning system, 77–80
    use of Tn917 derivatives, 74–75
  competence regulon, 147–166
  control of gene expression during sporulation, 243–254
  dependence pathways for gene expression, 223–241, *see also* Gene expression, dependence pathways
  endospore formation, 223–241
  forespore-specific genes, 211–221
    gdh, 215–216, 229
    gerA, 217–219
    0.3 kb, 218–219, 229–230
    spoIIG, 213, 229
    spoVA, 215–216, 229
    ssp, 215, 217, 219, 229
    techniques for analyzing, 211–213
  functional analysis of cloned genes, 80–84
    copy number manipulation, 82–83
    creation of permanent chromosomal deletions, 82
    high-resolution gene disruption, 80–81
    integrative gene disruption, 80
    reintroduction of altered DNA, 81–82
    use of phleomycin resistance vectors, 83
    use of SPAC cassettes, 83

  genetic manipulation
    generation of *spo* mutations, 65–66
    identification of regulated genes, 65–73
    mutagenesis, 65–69
    shuttle mutagenesis, 68–69
    use of insertional vectors, 68, 70–71
    use of *lac-cat* fusions, 71–72
    use of *lac* fusions, 69–71
    use of the SPAC expression cassette, 72–73
    use of Tn917 derivatives, 66–68, 70
  initiation of sporulation, 171–174, 185–210
  metabolic regulation of sporulation, 109–130
  regulation of sporulation, 294–295
  regulation of transcription, 167–184
  sigma factors, 167–184
  subtilisin gene, 131–146
  triggering of germination, 89–108
Bacillus subtilis 168
  requirement for dipicolinic acid, 58
  revised genetic map, 1–41
    genetic markers, 3–21
    map, 2
  small, acid-soluble proteins, 59
Bacillus subtilis subsp. *niger*
  demineralization, 55–56
  requirement for dipicolinic acid, 59
Bacillus thuringiensis δ-endotoxins, 255–276
  genes and gene products, 259–264
    cytolysin gene, 264
    130-kilodalton protein gene and related genes, 259–263
    poorly related genes, 263–264
  insecticidal crystal proteins, 255–259
    description, 255–258
    genes encoding, 259–268
    mode of action, 258–259
  regulation of synthesis, 268–270

Clostridium perfringens
  cortex, 51, 94–95
  localization of minerals, 53–54
Competence, 147–166
  com and *com* genes
    mutants, 150–151, 156–159
    organization and sequence, 152–154
    promoters, 154
    triggering of competence by, 154–155
  components of the machinery for, 155–156
  definition, 147
  din genes, 160–162

DNA transport, 159–160
"early" mutants, 150–151
early studies on, 149–150
epistatic interactions, 151–152
evolution, 162–163
genetic nomenclature, 149
"late" mutants, 151
mutants impaired in binding and uptake, 156–159
physiology, 148–149
relationship to sporulation, 161–162
transformation pathway, 147–148
triggering, 154–155

δ-Endotoxin genes, 255–276
Escherichia coli
　phosphate-regulated genes, 123–124
　use in cloning of *Bacillus subtilis* DNA, 65, 68–74

Gene expression, control during sporulation, 243–254
　asymmetric septation and early gene expression, 243–244
　cooperation of genomes, 249–252
　　building the germ cell wall, 249–251
　　cortex synthesis, 251
　　formation of spore coats, 251–252
　engulfment, 247–248
　establishing compartmentalization of, 246–247
　pro-σ^H processing, 244–246
　　coupling gene expression to a morphological stage, 244–245
　　insertion of SpoIIA in membrane, 245–246
　role of engulfase, 247, 248
　role of *spoIIIE*, 248–249
Gene expression, dependence pathways, 223–241
　compartmentalization, 232–234
　　identification, 232
　　sigma factors involved in, 232–234
　　split genes, 233–234
　　structural gene for σ^H, 233–234
　coupling gene expression to morphogenesis, 234–236
　　activation of *cotA*, *cotD*, and *gerA*, 235–236
　　pro-σ^E processing, 234
　developmentally regulated genes, 227–230
　　gdh, 215–216, 219, 229
　　ger, 217–219, 229
　　0.3 kb, 218–219, 229–230
　　spoIIA, 228
　　spoIID, 229
　　spoIIE, 228
　　spoIIG, 213, 229
　　spoIIIC, 230
　　spoIIIG, 229
　　spoIVCB, 230
　　spoVA, 215–216, 219, 229

spoVG, 227–228
ssp, 215, 217, 229
germination gene *gerA*, 231
identification of genes involved in sporulation and germination, 224–225
network, 236–238
regulatory genes and products, 225–227
spore coat genes *cotA*, *-C*, *-D*, and *-E*, 231
stages of endospore formation, 223–224

Sigma factors
　σ^A, 114, 194–195
　σ^B, 114, 174
　σ^C, 114, 175
　σ^D, 114, 175, 201–202, 290–292
　σ^E, 175–179, 247
　σ^F, 179–180
　σ^G, 180–181, 213–215, 218–219, 229, 232–233, 248–251
　σ^H, 114, 121–122, 124, 195–198, 228, 169–175
　σ^K, 181, 230–231, 233–234, 236, 251
　$E\sigma^{35}$, 268–269
　in compartmentalization of gene expression, 214–215, 232–234
　in expression of endospore formation genes, 228–236
　in initiation of sporulation, 171–174, 195–198
　in regulation of competence, 154
　in regulation of δ-endotoxin synthesis, 268
　in regulation of gene expression, 244–251
　in regulation of sporulation, 114, 121–122, 124, 194–198, 200–202, 213–215, 219, 243–251, 290–292, 295–296
　in regulation of transcription
　　as bridge between promoters and RNA polymerase, 168–171
　　role of σ^B, 174
　　role of σ^C, 175
　　role of σ^E, 175–179
　　role of σ^G, 180–181
　　role of σ^F, 179–180
　　role of σ^H, 171–175
　　role of σ^K, 181
　pro-σ^E, 228–229, 245–246
　pro-σ^H, 244–245
Sporulation
　control of gene expression during, 243–254, *see also* Gene expression, control during sporulation
　dependence pathways for gene expression during, 223–241, *see also* Gene expression, dependence pathways
　description, 109–111
　genes
　　cloning, 73–80
　　direct isolation, 124–125
　　functional analysis, 80–84

genetic manipulation, 65–73
in *Streptomyces*, 277–299, *see also Streptomyces*
    spp., sporulation
initiation, 171–174, 185–210
    alternative pathways, 200–202
    effector mechanisms, 193–197
    environmental signals, 186–193
    fail-safe mechanisms and conditional loops, 198–
        202
    in organisms other than *B. subtilis*, 193
    related sensory transduction systems, 193
    repressors, 198
    role of AbrP, 199–200
    role of Eσ^A, 194, 200
    role of Eσ^H, 195–197, 200–203
    role of Hpr, 189–199
    role of Sin, 199
    role of *spo0A*, 189–191, 200–203
    role of *spo0B*, 189–192
    role of *spo0E*, 189–191, 193
    role of *spo0F*, 189–191, 194
    signal transduction, 186–193
metabolic regulation, 109–130
paradigms for, 115
regulation of genes during, 116–124
    genes regulated by carbon, 116
    genes regulated by nitrogen, 122–123
    genes regulated by phosphate, 123–124
    inducible, glucose-stimulated genes, 122
    Krebs cycle genes, 118–122
    noninducible genes, 117–118
relationship to competence, 161–162
role of GTP concentration, 111–113
role of sigma factors, 114, 121–122,
    124
role of the *spo0A* product, 113
unanswered questions, 125–126
Staphylococcus aureus agr gene, 193
Stationary-phase phenomena, 109–130
Streptomyces ambofaciens, production of pamamycin,
    282
Streptomyces antibioticus
    aerial mycelium formation, 280
    mycelial growth and differentiation, 277
Streptomyces bikiniensis, A-factor, 281–282
Streptomyces coelicolor
    A-factor, 282
    bldA gene, 284
    bldB gene, 287–288
    metamorphosis of aerial spores into hyphae, 289–
        290
    mutants deficient in aerial mycelium formation,
        282–284
    mycelial growth and differentiation, 278
    regulation of sporulation, 294–295
Streptomyces griseoflavus, aerial mycelium formation,
    280, 281

Streptomyces griseus
    aerial mycelium formation, 280
    A-factor, 281–282
    factor, 282
    mutants deficient in aerial mycelium formation,
        282–284
Streptomyces spp., sporulation
    approaches to study, 279–280
    growth of the aerial mycelium, 280–289
        genes involved in, 284–289
        metabolic signals for differentiation, 280–281
        mutants deficient in, 282–284
        physiological attributes important to, 288–289
        regulatory elements, 294–295
        role of *bldA* gene, 284–287
        role of *bldB* gene, 287–288
        role of *sapA* gene, 288–289
        role of TTA codon, 284–287
    metamorphosis of aerial spores into hyphae, 289–
        294
        analysis of *whiB* and *whiE* genes, 293–294
        genetic analysis, 289–294
        role of deduced *whiG* gene product, 290–292
        use of transposon mutagenesis for analysis, 293–
            294
    mycelial growth and differentiation, 277–279
    stimulation by extracellular factors, 281–282
        A-factor, 281–282
        factor C, 282
        pamamycin, 282
Subtilisin gene (*aprE*), 131–146
    description, 131–132
    models for expression, 142–143
    negative control of expression, 140–142
        by *hpr*, 141–142
        by *sin*, 140–141
    regulation
        by *aprB*, 139–140
        by *degU-degS*, 134–136
        by *spo0A*, 139–140
        by *spo0H*, 139–140
        description, 132–133
    regulatory region of, 133–134
    transcriptional activators
        degQ, 137, 138–139
        degR, 137–139
        degU-degS, 134–136
        requirement for, 142–143
        senN, 138–139

Thermoresistance mechanisms, 43–63
    dehydration, 44–52
        background, 44
        correlation with heat resistance, 45–46
        measurement in protoplasts, 44–45
        molecular mechanisms, 52
        refractility, 46–47

resistance after further hydration, 49
  role of cortex, 49–51
  role of membranes, 51
  water activity, 47–49
mineralization, 52–56
  background, 52
  correlation with heat resistance, 52–53
  exchange and effectiveness of cations, 55–56
  localization, 53–54
  physical states, 55
  uptake during sporulation, 54–55
other possible determinants, 58–59
  dipicolinic acid, 58–59
  smal, acid-soluble proteins, 59
thermal adaptation, 56–58
  correlation with heat resistance, 56–57
  resistance after sublethal heating, 57–58
Tn917
  for insertional mutagenesis, 66–68
  to identify sporulation-specific genes, 70, 71–75
  to isolate com mutants, 149–150, 157–159

Transcription, sigma factors in regulation, 167–184
Transposon mutagenesis, 293–294
Trigger mechanism of germination, 89–108
  activation of germination, 90
  functional approaches to understanding, 91–100
    autolysin activation, 96–97
    biochemical analysis, 92
    cortex lytic enzyme model, 94–96
    degradation of spore components, 97–98
    determination of sequencing events, 93–94
    general spore metabolism, 93
    germinant metabolism, 92–92
  future prospects, 101–102
  genetical analysis, 98
    cloning of ger genes, 100
    essential genes, 98
    mutants blocked after triggering, 99–100
    mutants blocked in trigger response, 98–99
  germinants, 90
  model for, 100–101
  physiological and biochemical events of germination, 91